Introduction to

Biochemical Engineering

Second Edition

About the Author

Dubasi Govardhana Rao is currently working as scientist and Head of CFTRI (Central Food Technological Research Institute) Resource centre at Hyderabad for the last 18 years. He obtained a PhD in Chemical Engineering from IIT Bombay after completing BTech and MTech in Chemical Engineering from Andhra University, Waltair. Subsequently, he did an MBA (Strategic Management) from the Institute of Public Enterprise (OU), Hyderabad, in the year 2002.

After a brief stint at Banaras Hindu University, Varanasi, he started his professional career in the year 1982 as a Scientist at the erstwhile Regional Research Laboratory (presently known as NEIST) at Jorhat, Assam. He then shifted to CFTRI, Mysore, and later to Hyderabad.

Dr Rao has put in 28 years of professional experience specializing in homogeneous catalysis, food processing, food engineering and bioprocessing. He has published 38 research papers in national and international journals in the fields of Chemical Engineering, Food Technology and Biotechnology, and presented papers in a number of national conferences. He also has 7 patents to his credit.

Dr Rao is a visiting faculty in a number of Indian universities teaching biochemical engineering, food engineering and chemical engineering. He is also a member of the Board of Studies (BOS) in a number of universities in the faculties of biotechnology and food technology. He was also an expert member for syllabus-formulation committees of various universities for the biotechnology and food technology curricula.

Introduction to Biochemical Engineering

Second Edition

D G Rao

Scientist and Head
CFTRI (Central Food Technological Research Institute) Resource Centre
Hyderabad

Tata McGraw Hill Education Private Limited
NEW DELHI

McGraw-Hill Offices
New Delhi New York St Louis San Francisco Auckland Bogotá Caracas
Kuala Lumpur Lisbon London Madrid Mexico City Milan Montreal
San Juan Santiago Singapore Sydney Tokyo Toronto

Tata McGraw Hill

Published by the Tata McGraw Hill Education Private Limited,
7 West Patel Nagar, New Delhi 110 008.

ISBN (13): 978-0-07-068109-5
ISBN (10): 0-07-068109-0

Managing Director: *Ajay Shukla*

Head—Higher Education Publishing: *Vibha Mahajan*
Manager—Sponsoring: *Shalini Jha*
Editorial Executive: *Surabhi Shukla*
Asst Development Editor: *Devshree Lohchab*
Executive—Editorial Services: *Sohini Mukherjee*
Sr Manager—Production: *P L Pandita*

General Manager: Marketing—Higher Education: *Michael J Cruz*
Senior Product Manager: SEM & Tech Ed: *Biju Ganesan*
Asst Product Manager: *Amit Paranjpe*

General Manager—Production: *Rajender P Ghansela*
Asst General Manager—Production: *B L Dogra*

Typeset at The Composers, 260, C.A. Apt., Paschim Vihar, New Delhi 110 063 and printed at
Lalit Offset Printer 219, FIE, Patparganj Industrial Area, Delhi 110 092

Cover: Rashtriya Printers

RZZLCRQFDXLLD

The McGraw·Hill Companies

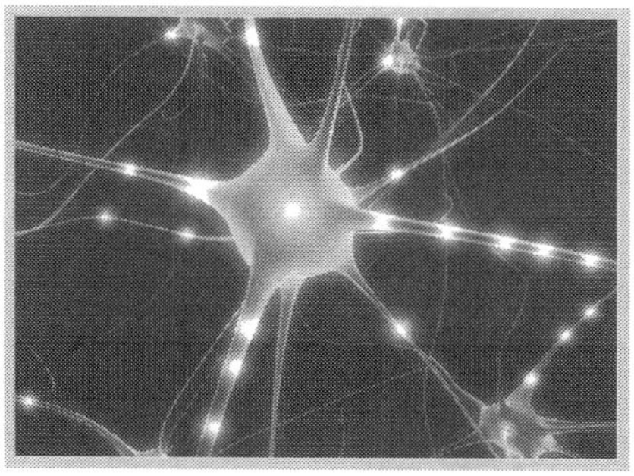

Contents

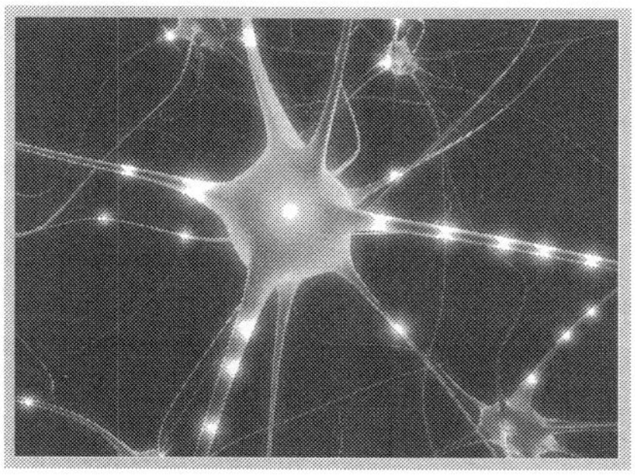

Preface

I am thankful to all the readers for the encouragement they have given to the first edition of the book.

Based on the interaction with various students and faculty in the last five years, the present edition is revised bringing in a few changes and additions. Two new chapters are included on **Heat Transfer,** one dealing with heat transfer in bioprocessing and the other on heat transfer equipment. This was felt necessary as the earlier edition did not deal with heat transfer. Also, a third chapter on **Bioprocess Economics** was added at the end. A biochemical-engineering textbook is not complete without a mention of bioprocess economics which is so vital for any project on engineering operations.

The topic of Enzyme Kinetics in Chapter 9 was merged with Chapter 4 which exclusively deals with Enzymes. Similarly, Chapter 6 on Media Design for Fermentation was merged with Chapter 5 dealing with Fermentation. Kinetics of Homogeneous Reactions was shifted to Chapter 7 on **Microbial Growth and Biochemical Reactors**.

In addition to the above, based on the response from students, a number of **worked examples** are included in various chapters. It is hoped that students who go through the worked examples will definitely appreciate the applicability of various engineering principles and calculations in the design and operation of fermentation processes. As has been done in the first edition, all efforts have been made to judiciously blend the chemical-reaction engineering principles for successful design and operation of fermenters.

The book is ideally suited for the UG students of biotechnology pursuing a BTech or masters degree (MSc) in biotechnology. Even though all efforts have been made to present the engineering principles without an overdose of mathematical derivations, knowledge of mathematics at the 10 + 2 level would definitely be an added advantage. The book is also ideally suited for students of BTech (Chemical Engineering) and MSc (Food Technology) for their first course in Biochemical Engineering. The book adequately covers bioprocessing principles and biochemical engineering. Various chapters in the book can be subdivided as follows:

Bioprocessing	(Ch 1 → 4 → 5 → 6 → 7→ 16)
Biochemical Engineering	(Ch 8 → 9 → 10 → 11 → 12 → 13 → 14 →15 → 19 → 20)
Downstream Processing	(Ch 17 → 18)
Process Engineering Principles	(Ch 1 → 5 → 8 → 10 → 11 → 12 → 13 → 14)
Enzyme Engineering	(Ch 4 → 5 → 8 → 15)

Chapter 1 introduces the reader to bioprocessing, viz., the historical developments in bioprocessing, overview of traditional and modern applications of biotechnology and various steps outlined in bioprocess-

ing. The understanding of biochemical engineering is strengthened with a brief discussion on the fundamentals of microbiology and biochemistry in **Chapters 2** and **3**. **Chapter 4** on Enzymes outlines the enzyme classification, and how enzymes specifically catalyse certain bioprocesses. The concept of biocatalysts is introduced here. Some common immobilization methods and industrial applications of enzymes are also described. **Chapter 5** introduces Fermentation with its requirements, range of processes, and different types of fermentation, viz., aerobic, anaerobic, solid state and submerged. Design and construction aspects are also highlighted. A brief mention about different types of fermenters is made at the end of this chapter.

Media design, their requirements, and various sources of carbon, nitrogen and the energy required for the medium are also discussed. Sterilization, which is an important component in any biochemical/biotechnological process, is discussed in **Chapter 6**. Various methods of sterilization of media, viz., batch and continuous, and of calculation of batch sterilization time are also explained. **Chapter 7** deals with the kinetics of microbial growth and biochemical reactions. The general principles of biochemical kinetics, and the methods to evaluate the rate and order of reaction are stressed in this chapter.

The biochemical engineering principles are highlighted in **Chapters 8-15** and **Chapter 19**. **Chapter 8** starts with a description of ideal reactors, viz., batch reactor, plug flow reactor and continuous flow reactor. Design features of these reactors are also highlighted in this chapter. **Chapter 9** classifies the biochemical reaction systems into simple, parallel, series reactions and a combination of these. This classification is akin to that done by Deindoerfer (1960). Efforts were made to cite examples of biochemical reactions that fall into each of these categories. Non-isothermal reaction systems and the criterion for achieving thermal stability are also discussed. After having discussed the ideal reactions and ideal reactor systems, non-ideal reactors (i.e., reactors with non-ideal flow) are highlighted in **Chapter 10**. This chapter is very important because of the fact that most of the fermenters behave non-ideally.

The concept of non-ideality was explained with RTD (residence time distribution) studies. **Chapter 11** discusses the rheology and mixing in fermentation broths. **Chapter 12** discusses Heat Transfer in Bioprocessing through the methods of conduction, convection and radiation. **Chapter 13** deals with Heat Transfer Equipment like double-pipe, shell and tube heat exchangers. **Chapter 14** highlights the oxygen mass transfer phenomena, which is an important aspect in the aerobic fermentation processes.

In a way, **Chapter 14** is a treatment of homogeneous systems with the dissolution of oxygen in the fermentation broths, while heterogenous systems involving immobilized bioreactor systems are highlighted in **Chapter 15**. The diffusional limitations and how the mass transfer affects the immobilized cell reactions are also highlighted. Thus, Chapter 15 discusses the engineering applications of enzymes dealt in Chapter 4, and may be considered as an extension to the same. Bioreactors and fermentation processes are discussed in **Chapter 16**. This chapter blends the bioreactor operations with fermentation processes. Various bioseparation steps (downstream processing steps) for separation and purification of the metabolic products are discussed in **Chapter 17**. General principles of effluent treatment with the concept of Central Effluent Treatment Plants (CETP) are discussed in **Chapter 18**. In the biochemical engineering aspect, the design and analysis of bioreactors are discussed in **Chapter 19**. The scale-up criterion for bioreactors is also discussed in this chapter. The choice of different bioreactors is highlighted based on the physico-chemical properties of the fermentation broths. Finally, Bioprocess Economics is discussed in **Chapter 20**.

I am greatly indebted to the publishers, Tata McGraw Hill Education Pvt. Ltd., New Delhi, for their constant encouragement during the revision, and their efforts in bringing out the book in its present form by providing excellent editorial support. I am grateful to the various authors and publishers who have accorded permissions to utilise their materials in this book.

The numerous reviewers who took out time to review the book also deserve a special mention. Their names are mentioned below.

T R Sreekrishnan	*IIT Delhi, Delhi*
Ashish M Gujrathi	*BITS Pilani, Rajasthan*
Susmita Dutta	*NIT Durgapur, West Bengal*
Amalesh Sirkar	*HIT Haldia, West Bengal*
Sameer Jadhav	*IIT Bombay, Mumbai, Maharashtra*
T Panda	*IIT Madras, Chennai, Tamil Nadu*
P Ravichandran	*St Joseph's College of Engineering, Chennai, Tamil Nadu*
A V N Swamy	*JNTU College of Engineering, Anantpur, Andhra Pradesh*

A heartfelt thanks to my young daughters, Asha Latha and Priyanka, who have cheerfully typed the additional chapters and worked examples from my illegibly handwritten manuscripts. I would also like to acknowledge the help I have received from the staff of JNTU, particularly Prof. M Lakshmi Narasu, Head, Centre for Biotechnology and Mr D Kiran Kumar for helping me at different times during the course of preparation of this book. A special thanks to Mrs V Rohini for helping me in the writing of Chapter 2. My thanks are also due to a few of my relatives and friends who constantly encouraged me in this endeavour. Last but not the least, I wish to appreciate the role of my wife, Sai Leela, for her ungrudging endurance during the period of writing this book.

It has always been my endeavour to keep the contents of the book crisp and easily comprehensible. I feel that the contents in the book are by no means exhaustive, and am afraid that certain mistakes and omissions may have inadvertently crept in. Readers may kindly indicate them to me for future improvement. Any suggestions for improving this book are most welcome. Users of the book can contact me at dgrao1950@rediffmail.com.

Dated 3rd July, 2009 **D G Rao**

Visual Walkthrough

New Chapters

Two new chapters are included on *Heat Transfer*, Chapter 12 on *Heat Transfer in Bioprocessing* and Chapter 13 on *Heat Transfer Equipment*.

A new chapter on *Bioprocess Economics* included in the second edition at the end would help engineers comprehend the role of economics and finance in process design and project engineering.

Project economics for citric acid manufacture given in Chapter 20 would help biochemical engineers understand the basic of cost estimation, and analyse the project from the investment point of view and profits.

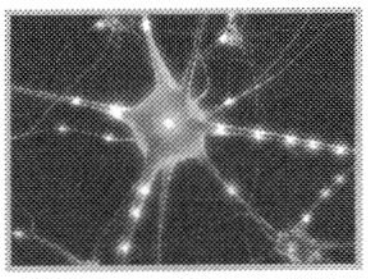

chapter 12

Heat Transfer in Bioprocessing

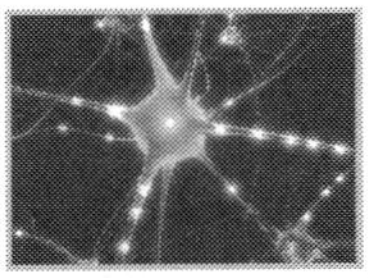

chapter 13

Heat Transfer Equipment

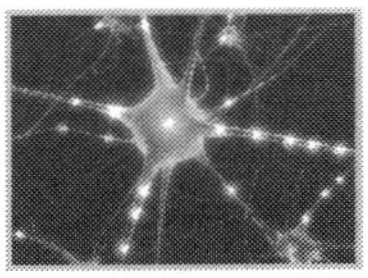

chapter 20

Bioprocess Economics

464 *Biochemical Engineering*

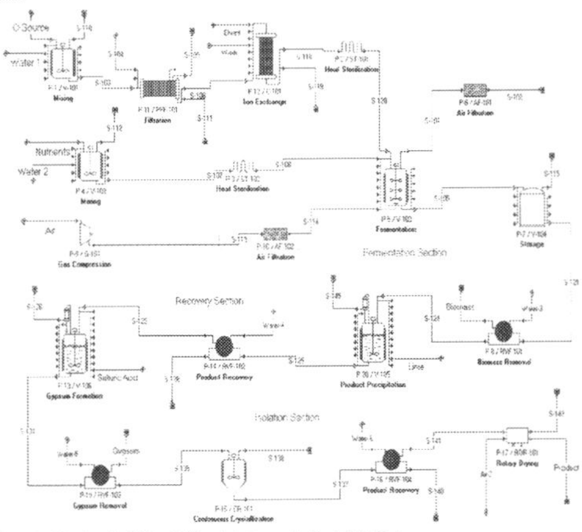

Fig. 20.8 *Flowsheet for Citric acid (Copyright © 2009 by Demitri Petrides)*

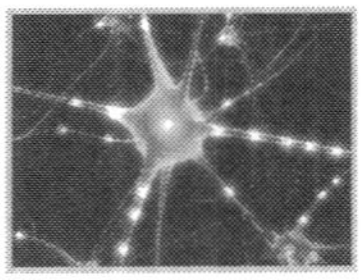

chapter 15

Heterogeneous Reaction Systems

As we have seen in Sec. 4.5, reaction systems may be of two types:

(i) homogeneous
(ii) heterogeneous.

In heterogeneous reaction systems, we encounter two phases; mostly the biocatalyst is in one phase and the substrate in another phase. Invariably, the biocatalyst is an immobilized enzyme system or immobilized cells in the solid phase, and the substrate is usually in the liquid phase. In a way, we can say that the heterogeneous reaction systems are mostly encountered in immobilized enzyme systems or in immobilized bioreactors. Hence, the discussion in this chapter is mostly restricted to immobilized enzyme systems in the solid phase (Section 4.6), and may be considered as a continuation to Chapter 4.

The reaction systems involving solid biocatalysts are important in bioprocessing operations, and they are complex for analysis in view of the fact that they require high level of mathematical treatment. The solid biocatalysts could be

- immobilized cells
- immobilized enzymes in the form of pellets
- flocks of cells
- clumps formed naturally by bacteria or fungi.

Introduction

Most of the chapters begin with an introduction to indicate what has been covered in the earlier chapters and what is going to be dealt in the present chapter.

Detailed Topics

The decision tree while starting a project is a helpful tool for entrepreneurs for making investments in a project.

In oxygen mass transfer which is an important aspect of biotechnology, the effect of air bubble size on mass transfer is shown in tabular forms.

The classification of various types of bioreactors in Chapter 5 gives a clear understanding about the concept of flow behavior in ideal bioreactor systems.

Non–ideal flow which normally prevails in most of the bioreactors and is an important concept of biochemical reaction engineering is exclusively covered in the form of a full chapter.

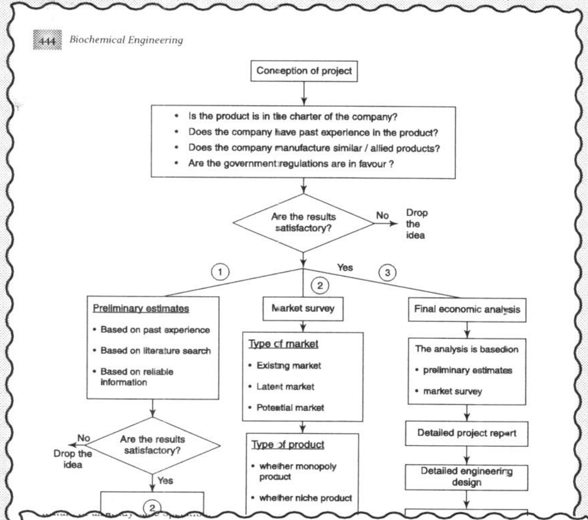

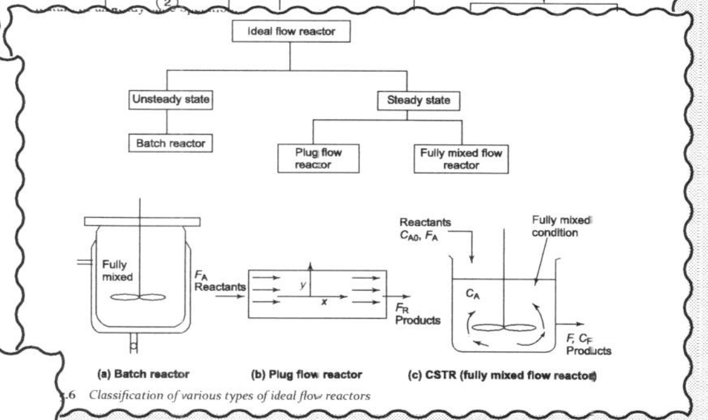

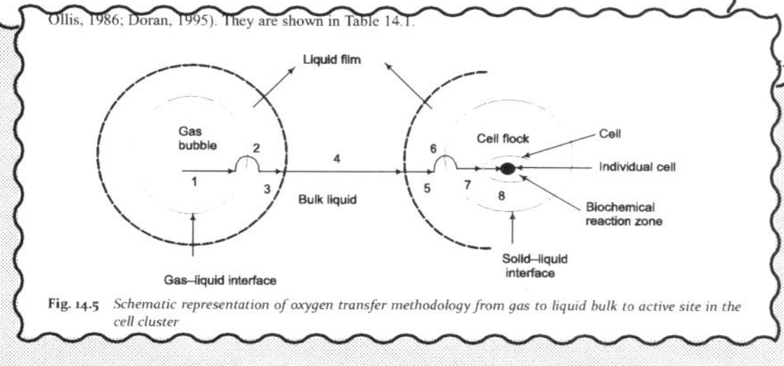

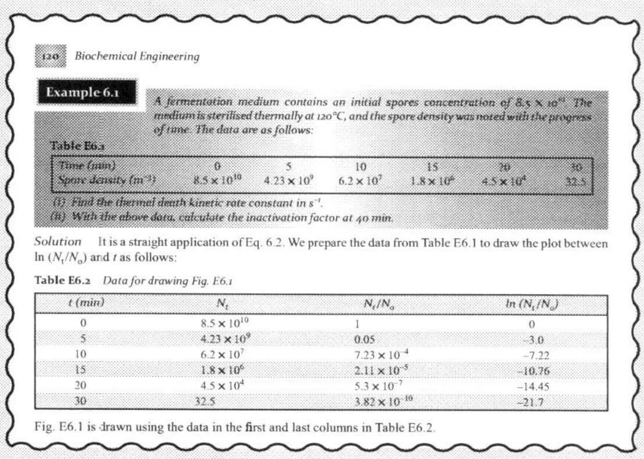

120 Biochemical Engineering

Example 6.1 *A fermentation medium contains an initial spores concentration of 8.5×10^{10}. The medium is sterilised thermally at 120°C, and the spore density was noted with the progress of time. The data are as follows:*

Table E6.1

Time (min)	0	5	10	15	20	30
Spore density (m^{-3})	8.5×10^{10}	4.23×10^{9}	6.2×10^{7}	1.8×10^{6}	4.5×10^{4}	32.5

(i) Find the thermal death kinetic rate constant in s^{-1}.
(ii) With the above data, calculate the inactivation factor at 40 min.

Solution It is a straight application of Eq. 6.2. We prepare the data from Table E6.1 to draw the plot between ln (N_t/N_o) and t as follows:

Table E6.2 *Data for drawing Fig. E6.1*

t (min)	N_t	N_t/N_o	ln (N_t/N_o)
0	8.5×10^{10}	1	0
5	4.23×10^{9}	0.05	–3.0
10	6.2×10^{7}	7.23×10^{-4}	–7.22
15	1.8×10^{6}	2.11×10^{-5}	–10.76
20	4.5×10^{4}	5.3×10^{-7}	–14.45
30	32.5	3.82×10^{-10}	–21.7

Fig. E6.1 is drawn using the data in the first and last columns in Table E6.2.

Worked Examples

Worked examples in various chapters at appropriate locations help the students understand the applicability of engineering calculations for design and operation of fermenters.

Explanatory Figure

The whole gamut of bioprocessing is depicted by one single figure in Chapter 1, based on which the whole text is developed.

Introduction to Bioprocessing Fundamentals 5

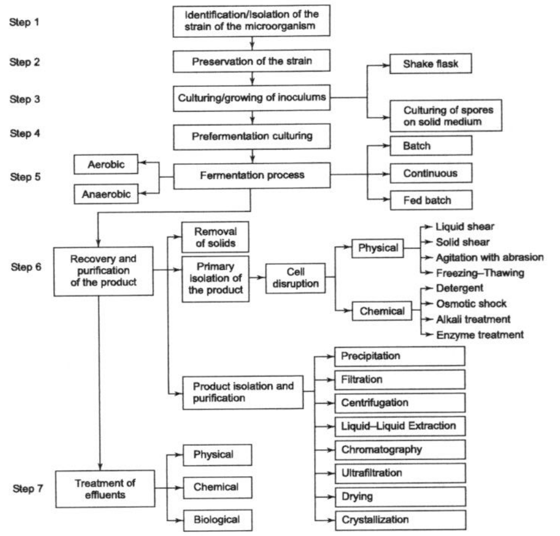

Fig. 1.1 *Integrated bioprocessing steps*

370 *Biochemical Engineering*

Fig. 17.5 *Plate and Frame Filter Press used to separate calcium phytate cake from corn steep liquid (CSL) which is the starting raw material for manufacture of Inositol (a Vitamin B). The filter press size is 36″ ×36″ ×2″, No. of plates is 36, and capcity is 1000 kg of CSL cake (Courtesy: M/s Siddartha Enterprises, Nizamabad, AP)*

Photographs

Photographs of the equipments given in some chapters help students have a visual grasp of the equipment.

Explanatory Table

The criteria for selection of a bioreactor based on physicochemical and thermal properties of the fermentation broth are given comprehensively in one table which is useful for both biologists and engineers.

438 *Biochemical Engineering*

Table 19.4 *Some tips for selection of bioreactors*

Physico-chemical property/cellular type	*Reactor configuration*	*Remarks*
Low-viscosity broths	Stirred tank, air-lift, bubble column	The selection does not matter much
Large fermenters (50 –500 m^3) with low-viscosity broths	Bubble columns	
Volume of the fermenter >500 m^3		Impossible to operate stirred tanks
High-viscosity broths	Stirred tanks	CSTR alone is suitable
Fermentation broths with particle attrition	Packed beds	Stirred tanks are not suitable
High heat of reaction systems (such as SCP from methanol)	Fluidised beds, bubble columns	Stirred tanks are not suitable
Fermentations with product inhibition/substrate inhibition	Fed-batch reactors	
Immobilized cells/enzymatic reactions	Packed towers, fluidised beds, CSTR	The system is akin to fluid–solid systems
Immobilized systems	Packed towers, fluidised beds, air-lift	Low hydrodynamic shear to be ensured
Liquid recycling systems	Packed beds, fluidised beds, Trickle beds	
Plant cells of low cell density	Air-lift reactors	
Plant cells of high cell density	Stirred tank with paddle/ helical ribbon impeller	If the plant cells twist around impeller, proper choice of impeller is very important
Anchorage-dependent animal cells	Fluidized beds, packed beds, CSTF	Similar to immobilised systems, high S/V ratios are desired
Non anchorage-dependent animal cells	Stirred tank, air-lift, bubble column	Viscosities are moderate
Animal cells used for production of monoclonal antibodies	Fed-batch reactors	
Dense cultures/filamentous organisms	Bioreactors with anchor-type stirrer/helical screw stirrer	The fermentation broth is non-Newtonian in nature
Organisms producing extracellular products	Bioreactors with anchor-type stirrer, helical screw stirrer	The fermentation broth is non-Newtonian in nature
Pelletised filamentous organisms	Packed beds, Fluidised beds	
Shear-sensitive mammalian/ insect/ plant cells	Stirred tanks are not suitable	Reactors with low shear but high mass transfer rates are preferred

Fermentation 113

References

5.1 Aiba. S., Humphrey, A.E. and Millis. N.F. (1965), *Biochemical Engineering*, Academic Press. New York. pp. 4–15.
5.2 Atkinson, B. (1974), *Biochemical Reactors*, Pion Ltd., London, p. 4.
5.3 Bailey, J.E., and Ollis, D.F. (1986), *Biochemical Engineering Fundamentals*, 2nd Edn., McGraw-Hill, Inc., New York, pp. 595–606.
5.4 Bartholomew, W.H., Karow, E.O., and Sfat, M.R. (1950). "Design and operation of a laboratory fermenter", *I&EC*, 42, 1827–1830.
5.5 Beaman, R.G. (1967), "Vinegar fermentation", *Microbial Technology*, Papper, H.J. (Ed.), Reinhold, New York, pp. 344–359.
5.6 Bisio, A., and Kabel, R.L. (1985), *Scale-up of Chemical Processes*, John Wiley & Sons, New York.
5.7 Borders, D. (1992), "Antibiotics survey", *Kirk–Othmer ECT.*, Kroschwitz, J.I.(Executive Ed.), 4th Edn., Vol. 2, pp. 893–904.
5.8 Casida Jr., L.E. (1996), *Industrial Microbiology*, New Age International (P) Ltd., New Delhi, Ch. 7, pp. 117–135.
5.9 Charles, M. (1985), "Fermenter design and scale-up", *Comprehensive Biotechnology*, Moo-Young, M. (Chief Ed.) Vol. 2, Pergamon Press, pp. 57–75.
5.10 Chisti, M.Y. (1989), *Air-lift Bioreactors*, Elsevier Applied science, Barking.
5.11 Coughanowr, D.R (1991), *Process Systems Analysis and Control*, McGraw-Hill, Inc., New York, pp. 111–163.
5.12 Crueger, W., and Crueger, A. (1989), Biotechnology: A *Textbook of Industrial Microbiology*, 2nd Edn, Sinauer Association, Inc., Sunderland, pp. 59–63.
5.13 Doran, P.M. (1995), *Bioprocess Engineering Principles*; Academic, Press, London, pp. 4–5, 143
5.14 Erickson, L.E., and Stephanopoulos, G. (1987), *Chemical Reaction and Reactor Engineering*, Carberry, J.J., and Varma, A.(Eds), Marcel Dekker, Inc., New York, ch. 12, p. 799–838.

References

A comprehensive list of references in the form of textbooks and journals with the title of the article is given at the end of each chapter.

Review Questions

Review questions are provided at the end of each chapter. They cover the contents of the entire chapter. These review questions are very helpful to teachers in setting up examination papers and class work assignments.

Review Questions

7.1 How are biochemical reactions different from chemical reactions?
7.2 What are the various complexities one comes across in the case of biochemical reactions/processes?
7.3 Describe the various phases of cell growth in a batch culture, with a neat diagram.
7.4 What is Malthus' law?
7.5 Describe the kinetic equations for cell growth.
7.6 How do you calculate the specific growth rate (μ)?
7.7 Define the yield factor.
7.8 What is meant by 'limiting substrate concentration'?
7.9 Describe the Monod model for specific growth rate.
7.10 Explain how do you find out the μ_m and K_s in Monod model by linearizing it.
7.11 Describe the growth-associated and non-growth-associated product formation in fermentation process.
7.12 Describe a substrate-inhibited cell growth.
7.13 Describe a product inhibited cell growth.
7.14 Describe the concept of structured models.
7.15 Define and explain the term "dilution rate".
7.16 Write the material balance equation for CSTR and explain it with a neat schematic diagram.
7.17 In a CSTR (chemostat), show how recycling improves the production rate.
7.18 Explain how do you determine the rate of a reaction.
7.19 Explain how do you calculate the first-order reaction rate constant.
7.20 Describe how do you find out the order of a reaction.

Problems

12.1 A furnace wall is composed of a metallic wall and a brick wall. The temperature inside the furnace is 500°C and that outside is 40°C. The thermal conductivity of the brick wall is 0.21 W/ m K. If the tolerable heat loss from the furnace wall is 1.1 kW/m^2, what should be the thickness of the brick wall?
(Ans : 8.8 cm)
12.2 A double-walled jacketed fermenter with agitation is used to heat a dilute medium solution. The wall of the fermenter is maintained at 100°C. The tank diameter is 2 m, and the agitator diameter is 1/3 of the tank diameter. The agitator rotates at 90 rpm. Find the heat-transfer coefficient assuming that the physico-chemical properties of the dilute medium solution may be taken as those of water.
(Ans : h_j = 1.76 kJ/m^2 °C)
12.3 In the above fermenter, 1400 litres of the feed solution at 30°C is heated to a final temperature of 85°C. Find the time taken by the feed solution.
(Ans : t = 19.5 min)
12.4 The outside surface of a copper vessel is at a temperature of 80°C. Calculate the radiative heat emitted per unit area of the vessel if the emissivity of copper is 0.03 W/m^2 K^4.
(Ans : 26.36 W)
12.5 A block of baker's yeast is kept in a deep freezer at –10°C. The initial temperature of the block is 30°C. The average exposed area of block is 0.08 m^2 and its emissivity is 0.55. Calculate the radiative heat-transfer rate.
(Ans : 9W/m^2)

Problems

A set of problems (with answers) is given in most of the chapters. These help the students test their comprehension if they attempt to solve them.

List of Symbols

A list of symbols with units used is given at the end of each chapter. These help students quickly comprehend the mathematical expression.

Symbols

a	: stoichiometric coefficient for A
A, B, C, D	: reactants [Eqs. (8.4) and (8.5)]
b	: stoichiometric coefficient for B
c	: stoichiometric coefficient for C
C	: concentration g/l or mol/m^3)
d	: stoichiometric coefficient for D
F	: molar flow rate (mol/s or kg mol/h)
k_2	: second-order reaction rate constant
N	: number of moles of component, number of CSTRs in series in Eq. (8.55)
r	: rate of reaction (mol/(m^3 s) or mol/(m^3 h))
s	: space velocity (1/s)
t	: time (s or h)
t_m	: average residence time in mixed flow reactor (s)
v	: volumetric flow rate (m^3/s)
V, V_1, V_2	: reactor volume (m^3)
V_i	: volume of ith reactor [Eq. (8.54)]
X	: conversion
$Y_{C/A}$	: yield factor of C based on A

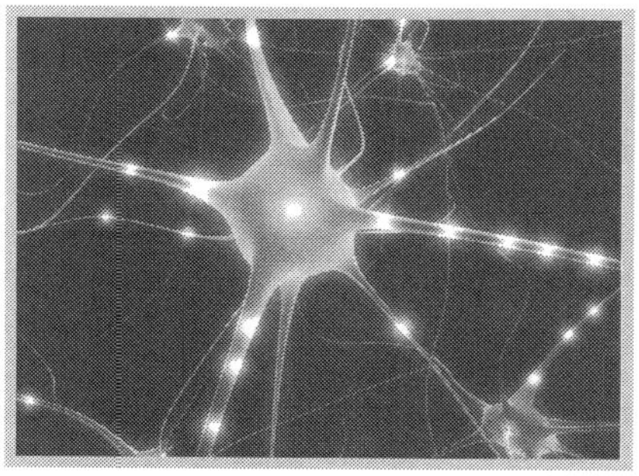

chapter 1

Introduction to Bioprocessing Fundamentals

1.1 HISTORICAL DEVELOPMENTS OF BIOPROCESSING TECHNOLOGY

Biotechnology is the art and science of converting reactants (substrate) into useful products by the action of microorganisms or enzymes. Whether we are aware of it or otherwise, microorganisms have been honestly serving the mankind. Thus, any process in which microbes or living organisms play a vital role in getting transformation of the feed into useful products is termed as **bioprocessing.** It could be by way of converting milk into curds, or fruit juices into wines, or sugars into alcohol or the application of yeast in bread-making etc.

Greeks were known to have produced wine as early as 7000 BC, and Egyptians were baking bread in 4000 BC. However, an understanding of the role played by the living cells was not emanated until the **'cell theory'** was proposed in 1838 by Schleiden and Schwann. But, the foundations of modern biotechnology as a distinct field were laid as the outcome of microbiology in the late 19th century.

The emergence of **Biochemical Engineering** is of more recent origin, since the biological industries did not recognize the importance of engineering inputs until the experience of penicillin manufacture. It is the manufacture of antibiotics which gave a boost to the biochemical industries. Even though the world's first successful commercial production of citric acid was commenced by Pfizer in 1923, nothing worth mentioning happened subsequently for some years. The origin of antibiotics can be traced to the first classical isolation of *penicillin* strain as a contaminant by Alexander Fleming in 1928 while he was working with *Staphylococcus aureus*. A mould of the *penicillium* family occurred as a contaminant in a petridish around which the *Styphylococcus* organism could not grow at all. This tremendous invention, however, did not receive much attention till the Second World War when there was a great demand for penicillin to heal the wounds of the war soldiers. It is at this stage again that a chance discovery of a new strain of penicillin, viz. *Penicillium chrysogenum*, gave 200 times more yield of penicillin than the Fleming's mould which was cultured in deep sterile aerated tanks.

Thus, the need arose to operate the submerged fermentation tanks with aeration under sterile conditions, and it became the job of the biochemical engineers to provide such sterile atmosphere and sterile air to meet the requirements of the submerged fermentation process. Aiba et al. (1965) observed *"perhaps the most notable contribution the engineer made was in the advancement of sterile techniques, or the "contamina-*

tion-free" philosophy, in the design and operation of the fermentation vessel and its associated maze of piping". Thus, the penicillin production ushered in a new era of biotechnology, even though the first successful fermentation plant for production of citric acid from sugar by using *Aspargillus niger* was established in 1923 by Pfizer.

The chronological developments in biotechnology was classified by Atkinson (1974) into three stages:

- the period of ignorance (pre-1800)
- the period of discovery (1800–1900)
- the period of industrial development (post-1900)

However, Stanbury and Whitaker (1993) gave a more exhaustive division into five stages which are shown in Table 1.1. This is similar to the one proposed by Perlman (1977) as *"A century of growth of industrial fermenters"* and summarised by Sengha (1993).

1.2 OVERVIEW OF TRADITIONAL AND MODERN APPLICATIONS OF BIOTECHNOLOGY

The traditonal application of biotechnology/bio-processing are mostly in the areas of foods, bakery products and alcoholic beverages. The earliest known application was in the production of wine from fruit juices. Fermentation of milk to make curd was also being practised in India from time immemorial. However, baking of bread by leavening the dough of wheat flour using yeast has brought biotechnology into day-to-day use. Most of the traditional processes in which the microorganisms were playing a vital role in bringing about the bioconversion/biotransformation was not well-recognised until the cell theory was proposed in 1838, and subsequently until Louis Pasteur proposed his classical theory in the middle of the 19th century that some diseases in animals and plants were caused by microorganisms. This discovery was a turning point in the medical history, and laid the foundation of **Microbiology.**

The modern fermentation industry has its roots in modern microbiology and also in the ancient technology. Particularly, when it comes to brewing, the breweries still use the old recipes. The improvement is only in the technological practices. Hoogerheide (1977) felt that the brewing industry had accepted the technological progress, but the scientific progress only with hesitation. The era of modern applications of biotechnology was ushered in with the commercial production of **antibiotics**, particularly penicillin, during the Second World War. Subsequently, the tentacles were spread over to diverse areas of health and hygiene, for the production of vaccines, enzymes, various fermented foods, organic acids, etc. The world market of various fermented foods is of the order of $30 billion out of which one-third is the share of US alone during 1990s.

The commercial products of fermentation are expanding day by day. Even during 1960–1970, about two dozens of antibiotics, and a large number of enzymes, polysaccharides, flavour-enhancing compounds were included in the list of fermented products. The area of antibiotics continues to dominate the scene. More than 10,000 antibiotics and similar bioactive natural substances are known which are obtained from microorganisms. Out of them about 100 products have found applications.

Perlman (1977) gave a list of 145 fermentation industries of which most are multinationals having their operations in more than one country and manufacturing a large number of products.

1.3 INTERDISCIPLINARY APPROACH TO BIOPROCESSING

Bioprocessing or biotechnological operations are truly interrelated with a large number of faculties in science. Particularly the wedding of engineering sciences with life sciences is an interesting feature of biotechnology. Until the time of Louis Pasteur in 1857, alcoholic fermentation was considered to be a chemical process in

Table 1.1 *The stages in the chronological development of fermentation industry*

Stage	*Main products*	*Vessels*	*Process control*	*Culture method*	*Quality control*	*Pilot plant facilities*	*Strain selection*
1 (Pre-1900)	Alcohol	Wooden upto 1500 barrels capacity. Copper used in later breweries	Use of thermometer, hydrometer and heat exchanger	Batch	Virtually nil	Nil	Pure yeast cultures used at the Carlsberg brewery (1896)
	Vinegar	Barrels, shallow trays, trickle filters		Batch	Virtually nil	Nil	Fermentations inoculated with 'good' vinegar
2 (1900-1940)	Baker's yeast, glycerol, citric acid and acetone/ butanol	Steel vessels upto 2000 hectalitres for acetone, butanol. Air spargers used for bakers' yeast. Mechanical stirring used in small vessels	pH electrodes with off-line control. Temperature control	Batch and fed-batch systems	Virtually nil	Virtually nil	Pure cultures used
3 (1940-till date)	Penicillin, streptomycin, other antibiotics, gibberelin, amino acids, nucleotides, transformations, enzymes	Mechanically aerated vessels, operated aseptically-true fermenters	Sterilizable pH and oxygen electrodes. Use of control loops which were later computerized	Batch and fed-batch common. Continuous culture introduced in brewing and for some primary metabolites	Very important	Becomes common	Mutation and selection programmes essential
4 (1960-till date)	Single-cell protein using hydrocarbon and other feed stocks	Pressure cycle and pressure jet vessels developed to overcome gas and heat-exchange problems	Use of computer linked control loops	Continuous culture + medium recycle	Very important	Very important	Genetic engineering of production strains
5 (1979-till date)	'Foreign' compounds, not normally produced by microbial cells, e.g. insulin, interferon	Fermenters developed in stages 3 and 4.	Control and sensors developed in Stages 3 and 4	May be batch, fed-batch or continuous	Very important	Very important	Introduction of foreign genes into microbial hosts using genetic engineering techniques

(Reprinted from "Principles of Fermentation Technology", Stanbury, P.F., and Whitaker, A., p. 6 , Copyright © (1993), with permission from Elsevier.)

which the sugars were getting converted to alcohol. Subsequently, it was realized that the conversion was affected by microorganisms which were the silent workers. This has opened up a wide range of interrelationships between various disciplines in science.

The commercial success of production of the penicillin during the Second World War period showed the path of advantages of engineering miracles. Subsequently, the biochemical engineers, who were responsible for providing an ideal and contamination-free environment, were considered as a part of bioprocessing.

The recent advances in **genetic engineering**, which help genetically engineer a cell to produce

- large amounts of recombinant proteins
- metabolites other than the proteins,

have opened up new vistas of mathematical modeling of the cell as a micro batch reactor. The cell can be forced, by applying genetic engineering principles, to produce high amounts of the limiting enzymes. The over-production of a particular enzyme or protein does not help the organism in any way. On the contrary, it may disrupt the cell metabolic process, which may ultimately lead to the cell death (Georgiou, 1988). Hence, to optimize the production, the culture has to be maintained and forced to produce the desired enzymes without being perished in the process.

An advanced knowledge of cell growth and cell morphology, which are the subject matters of microbiology, biochemistry and cell physiology, is a useful tool for precisely controlling the biotechnological process. Modern biotechnological techniques such as recombinant DNA techniques, gene manipulation, cell fusion and tissue culture offer useful tools to improve the existing processes. Sophisticated medicines, kits, cultured human tissues and organs, biochips for new-age computers, environmentally compatible pesticides, and powerful pollutant-degrading microbes herald a revolution in the role of biology in the industry (Doran, 1995).

A successful scale-up technique can alone translate a scientific invention into a commercial reality. Since the behaviour of biological systems is always complex, their scale-up is not that easy. Hence a rigorous evaluation and monitoring of various process parameters in sterile environmental conditions can alone result in realizing the commercial reality of the bioprocess. Application of mathematical modelling principles, backed by adequate pilot plant information will help in commercialization of the process.

Thus, the application of engineering principles to the biological processes to achieve commercial success is the subject matter of biochemical engineering, and constitutes the core substance matter for this textbook.

1.4 OUTLINES OF INTEGRATED BIOPROCESS

Integrated bio-processing consists of various steps, some of them sequential and others concurrent. They are shown in Fig.1.1, and will be discussed here briefly in general. The specific details will be discussed separately in respective chapters.

Step 1. Isolation of the strain

The very first step is the identification and isolation of the strain of the microorganism which brings out the desired bioconversion. This is generally the task of a microbiologist. This basic work which involves some shake flask experimentation, etc., will be the basis for taking up any process. The techno-economic viability of the process will be taken up subsequently if the results of Step 1 are encouraging.

Step 2. Preservation of the strain

Preservation of the strain for future use is the basic requirement for the success of commercial ventures. The high-yielding strains, identified and isolated in Step 1, will be preserved for future use. It is essential that the strain is not only preserved, but its ability to perform and yield the products should also be preserved, without which the whole exercise is set at naught. The strains can be stored:

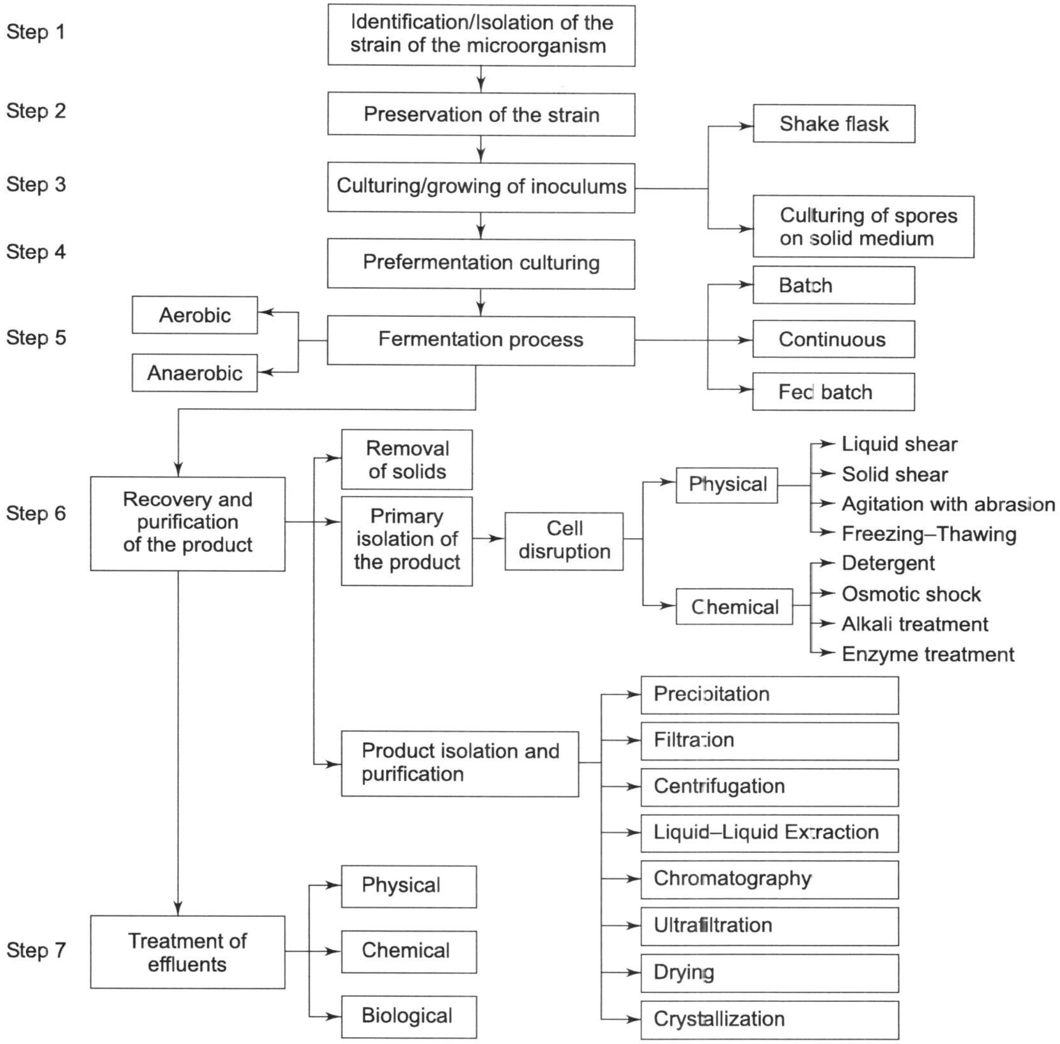

Fig. 1.1 *Integrated bioprocessing steps*

(i) at refrigerated temperatures (2 – 6 °C)
(ii) at frozen storage temperatures (–18 to – 80°C)
(iii) by freeze drying (lyophilization)

Step 3. Growth of inoculums

Before using the strain in the fermenter, the inoculum is cultured for growth. The preserved strain/culture is revived by growth in shake flasks or by solid-state fermentation on the solid-surfaces (Section 5.6.1).

The standard growth time for lyophilized cultures is approximately 4–10 days, whereas the frozen cultures and refrigerated cultures take different times for different cultures, viz. bacteria, actinomycetes and fungi (Crueger and Crueger, 1989), as shown in Table 1.2

Table 1.2 *Growth time for frozen cultures and refrigerated cultures*

Culutres	*Growth time*	
	Frozen culture	*Refrigerated culture*
Bacteria	4–48 h	4–24 h
Actinomycetes	1–5 days	1–3 days
Fungi	1–7 days	1–5 days

Step 4. Prefermentation culturing

Before the cells are admitted into the fermenter for performing bioconversions, they are initially grown separately with nutrients and the substrate so that the cells can multiply and proliferate. This will help increase the cell density. The optimal cell concentrations in a fermenter are as follows (Crueger and Crueger, 1989)

Bacteria	:	0.1 – 3.0%
Actinomycetes	:	5 – 10%
Fungi	:	5 – 10%

Step 5. Fermentation:

The major activity in industrial operation in the whole process is fermentation, in which biochemical engineers play an active role. Fermentation is the heart of the bio-processing operation. The fermenter sizes vary from 1 to – 450m^3, depending upon the type of fermentation process. Based on the capacity, tonnage and nature of the fermentation product, it could be taken up in:

(i) batch fermenter
(ii) fed-batch fermenter
(iii) continuous fermenter.

The conditions of fermentation could be aerobic (with bubbling of air) or anaerobic (in the absence of air). Various fermentation operations will be dealt with in detail in the subsequent chapters.

Step 6. Recovery and purification of the product

The recovery and purification of the fermented product from the fermenter are crucial steps, and sometimes even decide the economic viability of the process. The cost of recovery can vary anywhere between 20 and 60% of the total manufacturing costs (Stanbury and Whitaker, 1993). Various stages of operations involved in this step are shown in Fig.1.1. The methodology of primary isolation of the product depends upon the nature of the cells and how they hold the product, whether it is intracellular (the product is held within the cells) or extra cellular (the product is excreted from the cells).

Any solids, etc. are removed by simple filtration or centrifugation techniques. The intracellular material is obtained by cell disruption techniques. Some of these techniques which seem to be simple and feasible at laboratory scale, may not be so at the commercial scale, particularly for intracellular enzyme extractions (Wang, et al., 1979).

For purification and isolation of the final product, different unit operations are employed which will be dealt in more detail in the subsequent chapters.

Step 7. Treatment of effluents

Fermentation processes release large amounts of effluents which may be rich in organic matter, and may be a potential hazard for the environment if let out as they are. They need to be treated before being released into the outside world. Hence, our ability to economically treat the effluents before they are let out into the environment makes or mars the viability of the product manufacture. In 1996, Supreme Court of India has ordered the closure of 400 tanneries in Tamilnadu which were highly polluting. Central Leather Research Institute (CLRI, Chennai) has taken up the cause of tanneries and successfully demonstrated the cleaner leather processing technologies. The efforts of CLRI has resulted in reducing the BOD (Biological Oxygen Demand) and COD (Chemical Oxygen Demand) load by 35%, neutral salt discharges by 20–25%, sulphide by 60% and chromium by 98–99%[1]. In this direction, the R&D inputs of CLRI are commendable, which have saved the livelihood of thousands of cobblers living in that area.

The various effluent-treatment techniques can be classified as:

(i) physical
(ii) chemical
(iii) biological.

The final choice depends upon the individual cases and local circumstances (Stanbury and Whitaker, 1993).

1.5 UNIT OPERATIONS IN BIOPROCESS

The whole bioprocessing operations can be broadly classified into two classes/types of operations. They are:

(i) fermentation
(ii) product recovery and effluent treatment.

Fermentation is a biochemical operation and can be classified as unit process. All the rest of the physical operations carried out on the fermentation broth for product recovery and downstream operations including the effluent treatment may be classified as unit operations. Even some of the operations carried out on the fermenter such as fluid dynamics, mixing, air sparging, heat transfer to the fluids, etc., may also be categorized as unit operations.

Since it involves mixing and separation of various phases, streams and fluids, we can broadly comprehend the systems encountered in the product recovery and downstream processing lines as:

- solid-liquid
- liquid-liquid
- liquid-liquid-solid
- liquid-gas
- solid-liquid-gas
- solid-solid

The separation may be carried out with or without phase change.

Mixing is also one of the most important unit operations in bio-processing. It could be for

- gas-liquid
- liquid-liquid
- liquid-solid
- gas-liquid-solid systems

Design of various types of impellers for mixing (see Fig.13.13) is a challenging task for biochemical engineers. Particularly the efficient mixing of gas (oxygen/air) with fermentation broth decides the progress of reaction as the dissolved oxygen gas in the broth alone is available for the utilization by the cells to respire, grow, proliferate and perform.

Oldshue's paper (1966) on fermentation mixing and scale-up techniques highlights the scale-up criteria after the classical work of Rushton et al. (1950).

A comprehensive treatise on various unit operations can be found in standard text books (Mc-Cabe et al., 1993; Doran, 1995, pp. 218-259).

---

[1]Source: *Science Tech. Entrepreneur*, February 2001.

References

1.1 Aiba, S., Humphrey, A.E., and Millis, N.F. (1965), *Biochemical Engineering*, Academic Press, New York, p.7.

1.2 Atkinson, B. (1974), *Biochemical Reactors*, Pion Ltd., London, p.1.

1.3 Crueger, W. and Crueger, A. (1989). *Biotechnology: A Textbook of Industrial Microbiology*, Brock, T.D.(Ed), 2nd Edn (English), Sinauer Associates, Inc., Sunderland, pp. 104-105.

1.4 Doran, P.M. (1995), *Bioprocess Engineering Principles*, Academic Press, London, p. 3.

1.5 Georgiou, G. (1988), "Optimizing the production of recombinant proteins in microorganisms", *AIChEJ*, 34(8), 1233-1248.

1.6 Hoogerheide, J.C. (1977), "Traditional fermentation in modern dress", *Chemtech*, 17, 94-97.

1.7 Mc-Cabe, W.L., Smith, J.C, and Harriott, P(1993), "*Unit Operations of Chemical Engineering*, 5th Edn., McGraw-Hill, Inc., New York.

1.8 Oldshue, J.Y. (1966), "Fermentation mixing scale-up techniques", *Biotechnol. Bioeng*., 8, 3-14.

1.9 Perlman, D(1977), "Fermentation industries…..quo vadis?", *Chemtech*, 17, 434-443.

1.10 Rushton, J.H., Costich, E.W. and Everett, H.J. (1950). "Power characteristics of mixing impellers", *Chem. Eng. Prog.*, 46, 395-404, 467-476.

1.11 Sengha, S.S. (1994). "Fermentation" in Kirk-Othermer ECT, Kroschwitz, J.I.(Executive Ed), 4th Edn, Vol.10, John Wiley, New York, pp. 361-381.

1.12 Stanbury, P.F, and Whitaker, A. (1993), *Principles of Fermentation Technology*, Pergamon Press, Oxford, pp.193-230.

1.13 Wang, D.I.C., Cooney, C.L., Demain, A.L., Dunnill, P., Humphrey, A.E., and Lilly, M.D. (1979). "Enzyme isolation",. *Fermentation and Enzyme Technology*, Wiley, New York, pp. 238-310.

Review Questions

1.1 What are the traditional and modern applications of biotechnology?

1.2 "Biotechnology is truly an interdisciplinary subject." Justify the statement.

1.3 What are the various steps outlined in integrated bioprocessing?

1.4 Distinguish between "Unit operation" and "Unit process".

1.5 What are various unit operations involved in downs stream processing?

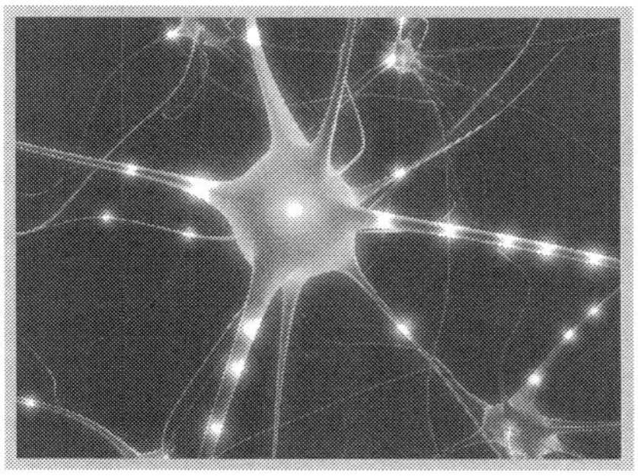

chapter 2

Overview of Microbiology

Microbiology is the study of microorganisms, which are not only microscopic and exist as single cells, but also those ultramicroscopic organisms which are not cellular and hence cannot exist independently, e.g. viruses. Microbiology deals with the study of functioning of cells, their diversity and evolution, their interaction with environment, other living organisms and man. Thus, it forms the basis for the complimentary development in cell biology, biochemistry, molecular biology and genetics.

Microbiology has evolved as an important branch of science, and is studied with respect to two major aspects:

(i) as a basic biological science
(ii) as an applied biological science.

As a basic biological science, it provides a system to understand the nature of life processes, the principles behind it, and the genetics which is involved in the inheritance of traits to the next generation.

Basic microbiology also consists of various sub-sections as shown in Table 2.1.

Table 2.1 *Various branches of basic microbiology*

Branch	*Areas of coverage*
Medical microbiology	Study of pathogenic microorganisms that cause diseases, and ways to eliminate them.
Agricultural microbiology	Study of plant diseases, understanding various beneficial interactions with plant systems, like soil fertility, crop-protection and increasing yield.
Environmental microbiology	Study of relationship of microorganisms with its habitat; pollution effect and its impact on environment from the stand points of ecological balance and health.
Food & Dairy microbiology	Study of microorganisms that produce various food and dairy products; study of control of microorganisms in foods and transmission of food-borne diseases.

As an applied biological science, it deals with the study of useful microorganisms as well as that of pathogenic organisms. The orientation of applied microbiology is in several fields, and it uses the basic information and techniques of different sub-divisions. Industrial microbiology is the study of exploitation of the biochemical potential of microbes for the production of various products like antibiotics, vaccines, steroids, solvents, vitamins, leaching of minerals, etc. It is also serving as a base for **Genetic engineering** which is an *upcoming branch of Biotechnology* and deals with *improving the performance of microorganisms.*

2.1 HISTORIC BACKGROUND

One of the most important discoveries in the history of biology occurred in the year 1665 with the help of studies using a relatively crude microscope. An English scientist, Robert Hooek, reported small structural units called cell, by fruiting structure of mould, and published a major work called *Micrographie*. But the first person to see the microorganisms in some detail was Antoni van Leeuwenhoek (1674) of Holland. He gave a reasonably detailed description of yeast cells in brewing of beer; but did not realize that they were living organisms. It has taken almost one and half centuries to realize the existence of the yeast cells which were responsible for the production of beer. Hoogerheide (1977) noted that the derailing of progress by at least a quarter century was mainly due to the then powerful and scholarly chemists like Gay-Lussac, Berzelius, Wohler von Liebig, Berthelot and Mitscherlich. Justis Liebig, in the year 1839, reported that all the living activities, which were being reported till that time, were mainly due to chemical and physical reactions in the medium.

Robert Hooke's writings awakened the learned Europe, and many other scientists were quick to follow his lead. The discoveries by Hooke, Swammerdom and other scientists showed that microscope was an important tool for unlocking the secrets of the world of small creatures. Antoni van Leeuwenhoek, in the year 1674, was credited to be the first person ever to make a crude microscope. He used to grind pieces of glass during his spare times into fine lenses by placing them between two silver or brass plates riveted together and added an adjustable device for holding tiny specimens, and could magnify an object over 200 times. Initially he used the lenses to inspect the quality of cloth and observed tiny organisms which he called 'animal cules'. These studies opened up the doors to a completely new concept dealing with living things.

However, microbiology as a subject did not develop until the 19th century, despite the development of microscope, due to inefficient techniques for studying microbes. It achieved significant importance in the second half of the 19th century with the theory of **spontaneous generation**. With the significant work of Louis Pasteur, there was an explosion of discoveries in microbiology. Hence, the period from 1857 to 1914 has been named as the 'golden age of microbiology' (Table 2.2). During this period, there were many discoveries, which included the study of pathogens and the role of immunity in the prevention and cure of diseases, improved techniques for performing microscopy and culturing of microorganisms. The need for a good microscope to have a vision of the microorganisms was felt all the way.

2.2 MICROSCOPY

Microorganisms and their structural components are measured in smaller units such as micrometers or microns (10^{-6} m), nanometers or millimicrons (10^{-9} m) and angstroms (10^{-10} m). The earliest report on the use of magnifying lens (telescope) was that of Conrad Gesber in the year 1558. A Dutch spectacle maker, Zaccharias Jonssen is credited with making the first compound microscope, but in 1830, better microscopes were developed by Joseph Jackson Lister. Various improvements were made on Lister's microscope that resulted in the development of modern compound microscope. Table 2.3 shows some of the limits of microscopic resolutions. Subsequent microscopic studies of live specimens have revealed dramatic interactions between microbes.

Table 2.2 *Important discoveries in the golden age* of microbiology

Year of discovery	*Name of the Scientist*	*Discovery*
1857	Louis Pasteur	Fermentation
1861	Louis Pasteur	Disproved spontaneous generation
1864	Louis Pasteur	Pasteurization
1867	Joseph Lister	Aseptic surgery
1870	Ernst Karl Abbe	Condenser and oil immersion objective
1876	Robert Koch	Germ theory of diseases
1879	Neisser	*Neisseria gonorrhoeae* (organism)
1880	Louis Pasteur	Immunization technique
1881	Robert Koch	Pure cultures
1882	Robert Koch	Microbacterium tuberculae
1882	Walter Hess	Agar (solid) media
1883	Robert Koch	*Vibro cholera*
1884	Elic Metchnikoff	Phagocytosis
1884	Christian Gram	Gram staining procedure
1884	Escherich	*Escherichia coli*
1887	Petri	Petri dish
1890	Von Bering	Diphtheria antitoxin
1890	Paul Ehrlich	Theory of immunity
1892	Winogradsky	Sulphur cycle
1898	Shiga	*Shigella dysenteriae*
1910	Paul Ehrlich	Syphilis

Table 2.3 *Limits of microscopic resolutions*

Size range of specimen	*Type of specimen*	*Operation range of microscope*
1 cm		Unaided eye
1 mm		Unaided eye
0.1 mm	Most unicellular organisms, algae, fungi, protozoa	Light microscope
100 –10 μm	Most eucaryotic cells	Light microscope
10–1 μm	Most bacteria, mitochondria	Light microscope/Electron microscope
1–100 nm	Mycoplasma	Light microscope/Electron microscope
100–10 nm	Viruses, ribosomes	Light microscope/Electron microscope
10–1 nm	Proteins, lipids	Electron microscope
1–0.1 nm	Small molecule	Electron microscope
0.1 nm	Atoms	Electron microscope

2.3 MICROBIAL TAXONOMY

Taxonomy is the subject of classification of biological entities. Earlier, all living organisms were classified under two main kingdoms, namely, *Plantae* and *Animalia*—by Swedish botanist Carolus Linnaeus, in the 18th century, in his *Systema Nature*. Later in 1886, German naturalist Earnst H. Haeckel separated microorganisms from plant and animal kingdoms, and placed them under *protista* kingdom which included microorganisms, fungi, protozoa and microscopic algae (Bailey and Ollis, 1986). Many taxonomists opposed this, and there have been constant endeavours to revise this system of classification.

Finally, in 1969, Robert H. Whittaker proposed a five-kingdom classification which received wide acceptance. According to this system, the five kingdoms are:

(i) *Monera,* which include true bacteria and blue green algae.
(ii) *Protista,* which include eucaryotes such as protozoa which are unicellular, contain chlorophyll (plant like) and possess movement (animal like).
(iii) *Fungi,* which include non-green eucaryotic organisms like slime moulds, bread moulds, sac fungi. The composition of cell wall differed from that of plants, and had adjacent cells without separation (coenocytic condition), and hence not considered as true multicellular organisms.
(iv) *Plantae,* which include algae, all mosses, ferns, conifers, and flowering plants.
(v) *Animalia,* sponges,[1] worms, insects, and vertebrates.

The taxonomical classification of all the microorganisms is shown in Fig. 2.1. The entire taxonomical classification of Fig. 2.1 can be simplified to a tabular form shown in Table 2.4.

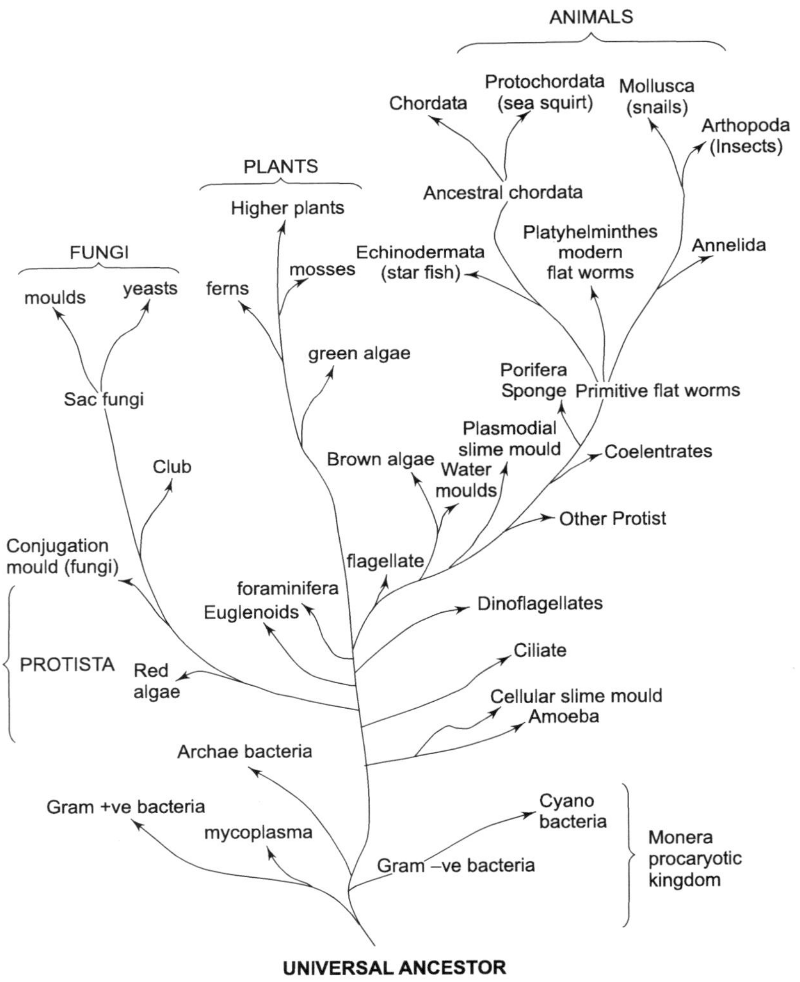

Fig. 2.1 *Taxonomical classification of microorganisms*

[1] Sponges are primitive multicellular animals, incapable of moving. They belong to the phylum porifera, which is the next higher level to protozoa.

Table 2.4 *Simplified classification of universal Ancestor*

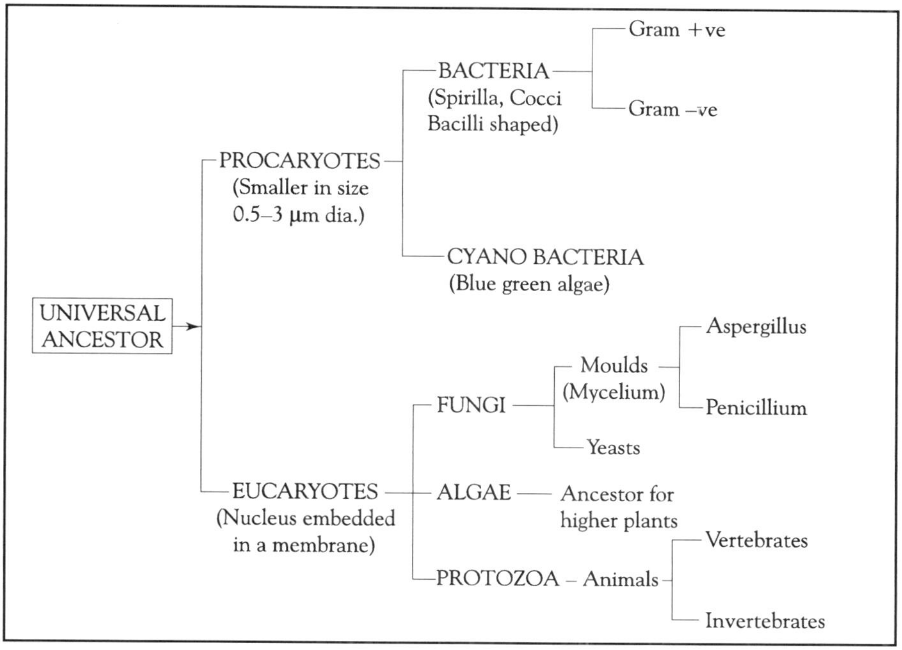

2.3.1 Microbial Diversity

Evolutionary roots of cells form a basis for understanding the microbial diversity (which requires an appreciation of the evolutionary roots of cells). The evolution has shaped all life on earth which is responsible for structural and functional diversity in cells. Microbial diversity can be seen in terms of variations in cell size, morphology, metabolism, adaptation, growth, etc. Several evolutionary branches occur in bacteria, as some of them are pathogenic, while many others are found in soil, water, and animal digestive tract; some of them use light as energy source, while others use organic and inorganic chemicals as fuel for cellular metabolism. They are observed in oxic and anoxic environments. They also thrive under extreme environmental conditions like hot springs, salty water bodies, highly acidic or alkaline soils.

2.3.2 Microbial Nomenclature

In microbiology, the basic taxonomic unit is species which can be defined as a collection of similar strains. The species concept gives a formal taxonomic identity to the selected strains.

Groups of species are collected into *genera*. They share major properties, but differ by the presence or absence of a few characteristics. Groups of genera are classified into families, families into orders, and orders into division.

Following the binomial system of nomenclature, microorganisms are given genus and species name, which are either Latin or Greek derivations of some descriptive property appropriate for that species. For example, *Bacillus subtilus* or *Bacillus* meaning 'rod shape', which indicates its slender nature; *B. cereus* meaning 'waxen'; *B. stearothermophilus* meaning 'heat loving'; and *B. acidoceldarius* meaning 'acid-thermal'. They are written in italics with the genus name starting with capital letter and the species name with a small letter. The rules for bacterial nomenclature are described in a publication called "*The International Code of Nomenclature of Bacteria*" which spells the rules for naming newly isolated organisms as new genera or new species (Lapage, et al, 1975).

There are many ways of grouping (classification) of procaryotes. The characteristics of taxonomic approach include morphology, Gram reaction[2], nutrition, cell wall chemistry, pigments, storage products, ability to use various substrates, fermentation product, gaseous needs, temperature, pH requirement, tolerances, sensitivity, pathogenicity, symbiotic relationships, immunological characteristics and habitat.

Molecular taxonomy includes guanine and cytosine ratio (GC ratio) in organisms, the extent of DNA: DNA hybridization, amino acid sequencing and protein analysis. *Bergey's Manual of Systematic Bacteriology* is a compendium of classical and molecular information of all recognized species of procaryotes and contains a number of dichotomous keys for identification.

2.4 CHEMICAL COMPOSITION

Cells are composed of small molecules as well as macromolecules, and are made up of four basic elements, viz. carbon, oxygen, hydrogen and nitrogen. Apart from these elements, phosphorous, iron, sulphur, zinc, manganese, copper, molybdenum and cobalt are also present. Water accounts for about 90% of the weight of the cells. Of the macromolecules, proteins are most abundant by weight (55%). The chemical compositions of a procaryotic cell are given in Table 2.5. The dry weight of an actively growing *E. coli* cell $\approx 2.8 \times 10^{-13}$ g.

Table 2.5 *Chemical composition of a procaryotic cell*

Molecule	*Percent of dry weight*	*Molecules per cell*	*Different kinds*
Total macromolecules	96	24,610,000	~2500
Protein	55	2,350,000	~1850
Polysaccharide	5	4,300	2*
Lipid	9.1	22,000,000	4**
DNA	3.1	2.1	1
RNA	20.5	255,500	~660
Total monomers	3.5		~350
Amino acids and precursors	0.5		~100
Sugars and precursors	2		~50
Nucleotides and precursors	0.5		~200
Inorganic ions	1		18
Total (%)	100		

* Peptidoglycan and glycogen are assumed to be major polysaccharides.

** There are several classes of phospholipids, each of which exists in many kinds because of variability in fatty acid composition between species and due to different growth conditions.

(Reprinted from "Brock Biology of Microgranisms", Madigan, M.M. Martinko, J., and Parker, J., 8th Edn, p . 112, copyright © (1997), with permission from Pearson Education, Inc., Upper Saddle River, NJ.)

[2]Gram reaction is a classical technique to identify the basic difference in the cell wall structure of the bacteria. The cells are stained with crystal violet dye, and then treated with iodine solution. The cells are washed with alcohol. Those which retain the blue crystal violet colour are classified as gram positive, and those which do not retain the colour are called as gram negative species. Named after the scientist Christion Gram.

Most procaryotes require an organic compound of some sort as their source of carbon. After carbon, the most abundant element in cells is nitrogen (12% by dry weight), which is a major constituent of proteins and nucleic acids. Nitrogen is found in organic and inorganic forms, and is mostly assimilated as ammonia (NH_3), nitrates (NO_3) or N_2.

2.5 NUTRITIONAL REQUIREMENTS

Nutritional studies revealed that organisms can assimilate various organic compounds, and use them to make new cell wall material, which may be amino acids, fatty acids, organic acids, sugars, nitrogen bases and aromatic compounds. Other macronutrients that are required are:

(i) **phosphorous**, which is required by the cell during nucleic acid synthesis and phospholipids
(ii) **sulphur**, which has a structural role in amino acids like methionine, cysteine, and vitamins like thiamine, biotin, lipoic acid and co-enzyme A
(iii) **potassium**, which is involved in the protein synthesis and for proper functioning of enzymes
(iv) **magnesium**, which stabilizes ribosomes, cell membrane, nucleic acid and is required for the activity of enzymes
(v) **calcium**, which is essential for growth, stabilizes cell walls and is responsible for the stability of endospores.

Table 2.6 highlights the various macronutrients in culture media.

Table 2.6 *Macronutrients in nature and culture media*

Element	*Usual form of nutrient found in the environment*	*Chemical form supplied in culture media*
Carbon (C)	CO_2 organic compounds	Glucose, malate, acetate, pyruvate, etc. or in the form of yeast extract, peptone, etc.
Hydrogen (H)	H_2O organic compounds	H_2O organic compounds
Oxygen (O)	H_2O, O_2 organic compounds	H_2O, O_2 organic compounds
Nitrogen (N)	NH_3, NO_3, N_2 organic nitrogen compounds	*Inorganic:* NH_4Cl, $(NH_4)_2SO_4$, KNO_3, N_2 *Organic:* Amino acids, nitrogen bases of nucleotides, many other N-containing organic compounds
Phosphorus (P)	PO_4^{3-}	KH_2PO_4, Na_2HPO_4
Sulphur (S)	H_2S, SO_4^{2-}, organic sulphur compounds, metal sulphides (FeS, CuS, ZnS, NiS, and so on)	Na_2SO_4, $Na_2S_2O_3$, Na_2S, cysteine, or other organic sulphur compounds
Potassium (K)	K+ in solution or as various K salts	KCl, KH_2PO_4
Magnesiun (Mg)	Mg^{2+} in solution or as various Mg salts	$MgCl_2$, $MgSO_4$
Sodium (Na)	Na^+ in solution or as NaCl, or other Na salts	NaCl
Calcium (Ca)	Ca^{2+} in solution or as $CaSO_4$, or other Ca salts	$CaCl_2$
Iron (Fe)	Fe^{2+} or Fe^{3+} in solution or as FeS, $Fe(OH)_3$, or many other Fe salts	$FeCl_3$, $FeSO_4$, various chelated iron solutions (Fe^{3+} EDTA, Fe^{3+} citrate, and so on)

Various micronutrients are also needed by living organisms and are tabulated in Table 2.7.

Table 2.7 *Micronutrients required by cells*

Element	*Cellular function*
Chromium (Cr)	Required by mammals for glucose metabolism; no known microbial requirement
Cobalt (Co)	Vitamin B_{12}; transcarboxylase (propionic acid bacteria)
Copper (Cu)	Certain proteins, notably those involved in respiration (for example, cytochrome *c* oxidase), or in photosynthesis (for example, plastocyanin) some superoxide dismutases
Manganese (Mn)	Activator of many co-enzymes; present in certain superoxide dismutases and in water-splitting enzyme of photosystem II in oxygenic phototrophus.
Molybdenum (Mo)	Present in various flavin-containing enzymes; also in molybdenum nitrogenase, nitrate reductase, sulphite oxidase, DMSO-TMAO reductases, some formate dehydrogenase, oxotransferases
Nickel (Ni)	Most hydrogenases; co-enzyme F_{430} of methanogens; carbon monoxide dehydrogenase; urease
Selenium (Se)	Formate dehydrogenase; some hydrogenases; the amino acid selenocysteine
Tungsten (W)	Some formate dehydrogenases; oxotransferases of hyperthermophiles (for example, aldehyde: ferridoxin oxidoreductase of *Pyrococcus furiosus*)
Vanadium (V)	Vanadium nitrogenase; bromoperoxidase
Zinc (Zn)	Present in the enzymes carbonic anhydrase, alcohol dehydrogenase, RNA and DNA polymerases, and many DNA-binding proteins
Iron (Fe)[b]	Cytochromes,catalases, peroxidases, iron–sulphur proteins (for example, ferredoxin), oxygenases, all nitrogenases

[b] Needed in greater amounts than other metals, not generally considered as a trace element.
(Reprinted from "Brock Biology of Microorganisms", Madigan, M.M., Martinko, J., and Parker, J., 8th Edn., p. 114, Copyright © (1997), with permission from Pearson Education, Inc., Upper Saddle River, NJ.)

Growth factors They are the organic compounds required in small amounts for growth and metabolism, and these include vitamins, amino acids, purines and pyramidines. Vitamins are not only needed for growth, but also play an important role as co-enzymes. Vitamins and their functions are tabulated in Table 2.8. Most of the vitamins are synthesized by organisms themselves, but some of them depend on external sources. (See also Section 3.5.)

Table 2.8 *Vitamins and their functions*

Vitamin	*Function*
P-Aminobenzoic acid	Precursor of folic acid
Folic acid	One-carbon metabolism; methyl group transfer
Biotin	Fatty acid biosynthesis; β-decarboxylations; CO_2 fixation
Cobalamin (B_{12})	Reduction of and transfer of single-carbon fragments; synthesis of deoxyribose
Lipoic acid	Transfer of acyl groups in decarboxylation of pyruvate and α-ketoglutarate
Nicotinic acid (niacin)	Precursor of NAD^+ ; electron transfer in oxidation-reduction reactions
Pantothenic acid	Precursor of co-enzyme A; activation of acetyl and other acyl derivatives
Riboflavin	Precursor of FMN, FAD in flavoproteins involved in electron transport
Thiamine (B_1)	α -Decarboxylations ; transketolase
Vitamins (B_6) (pyridoxal-pyridoxamine group)	Amino acid and keto acid transformations
Vitamin K group; quinones	Electron transport; synthesis of sphingolipids
Hydroxamates	Iron-binding compounds; solubilization of iron and transport into cell
Co-enzyme M (CoM)	Required by certain methanogens; plays a role in methanogenesis

2.6 METABOLISM

The life support activity of simple organisms involves complex biochemical reactions. Metabolism is a sum of such biochemical reactions in living organisms, which include release of energy by breakdown of complex organic compounds into simpler forms (known as **catabolism**)and release energy to form new molecules like synthesis of proteins from amino acids, sugars to polysaccharides (known as **anabolism**).

Catabolism reactions furnish the energy needed to drive anabolic reactions; this coupling of energy-requiring reactions and electron-releasing reactions is made possible through ATP (Adenosine tri-phosphate) which stores the energy released by catabolic reactions and makes it (energy) available for anabolic reactions and other cellular works. That is why ATP is known as the currency of chemical energy in cells.

2.6.1 Catabolism

Nutrient molecules have some energy associated with electrons that form bonds between atoms. Organisms release and store energy from organic molecules by a series of controlled reactions which are mostly oxidation–reduction reactions. In fact, the whole life-cycle is known as one huge redox reaction (Webb, 1964).

Most organisms oxidize carbohydrates, of which glucose is the most common carbohydrate used as an energy source. Organisms take up "respiration and fermentation" path to produce energy from glucose. Respiration and fermentation both start with the **glycolic pathway** where glucose is broken down to pyruvic acid in a series of steps (Fig. 2.2). The pyruvic acid thus formed can undergo Kreb's cycle (Fig. 2.3) or fermentation. Pentose phosphate pathway is an alternative pathway taken up by the cell for oxidation of glucose to pyruvic acid with the production of ATP and energy containing NADH (Fig. 2.4 and Table 2.9). During Kreb's cycle, pyruvic acid is oxidized to carbon dioxide along with the production of ATP, NADH and another reduced electron- carrier $FADH_2$. NADH and $FADH_2$ are oxidized in electron transport chain and e^- are carried from substrate to a cascade of oxidation and reduction reactions involving generation of ATP.

Lipids are catabolized in organisms with lipases, and they are broken down to fatty acids and glycerol. Glycerol undergoes a series of reactions to form dihydroxy acetone phosphate which is metabolized by glycolysis through Kreb's cycle (Fig. 2.2 & 2.3). Fatty acid undergoes *β*-oxidation resulting in the formation of acetyl Co-A which takes part in Kreb's cycle. Proteins are usually catabolized in the presence of proteases and peptidases to form amino acids which undergo deamination where the amino group is removed and converted to NH_4^+ which is excreted. The remaining organic portion enters Kreb's cycle.

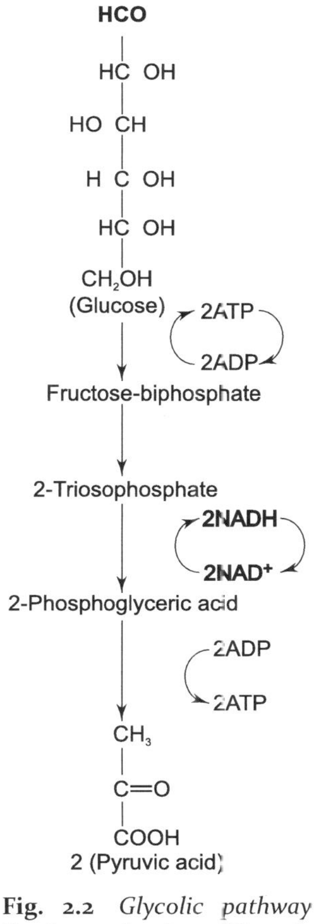

Fig. 2.2 *Glycolic pathway for the production of Pyruvic acid from glucose*

2.6.2 Anabolism

The metabolic oxidation of glucose to CO_2 and H_2O is considered effective, but 45% of the energy so produced is lost as heat, and rest is used for anabolic process. Some amount of energy is also used for the active transport of substances and for movement (by flagella or celia) in motile organisms. Autotrophs build their organic compounds by fixing carbon dioxide in Calvin–Benson cycle, but heterotrophs must have external source of organic compounds.

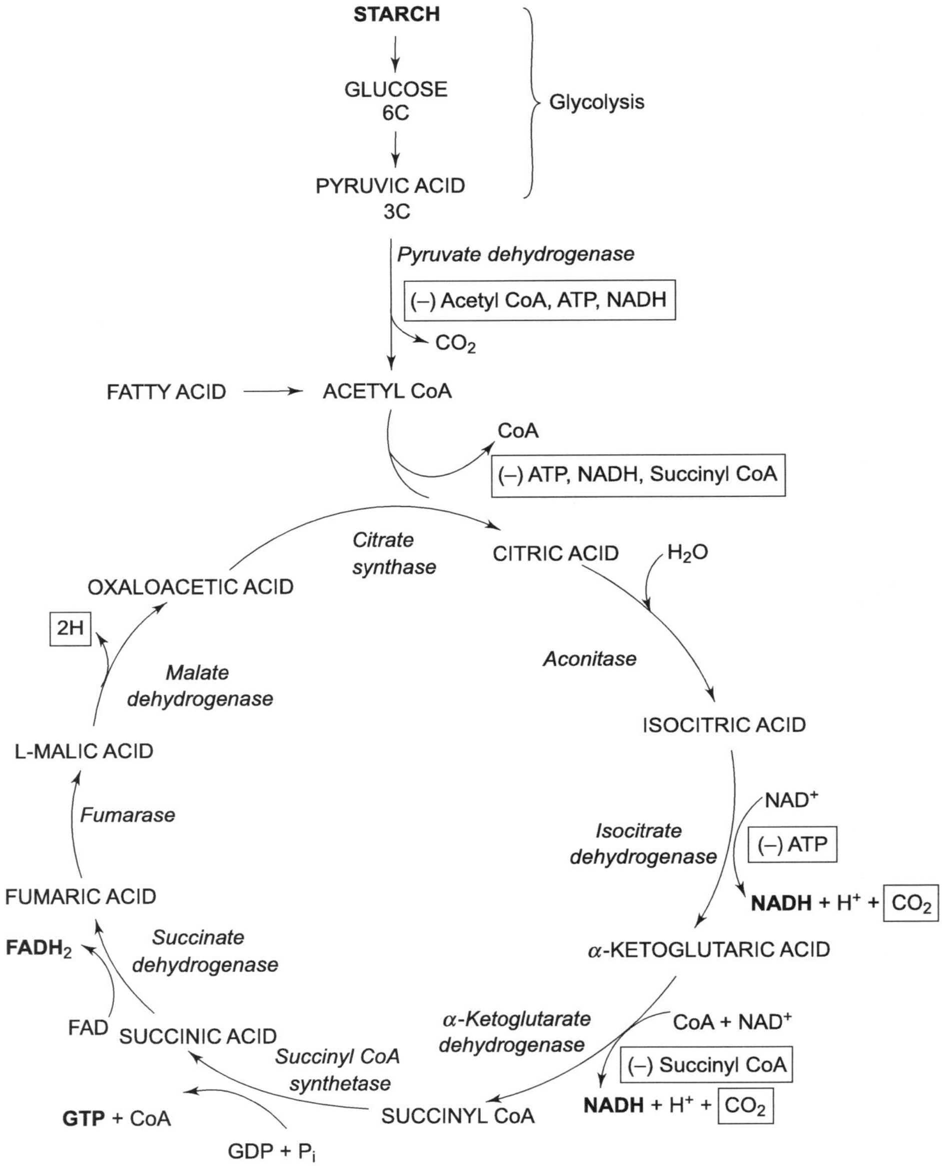

Fig. 2.3 *Kreb's cycle (also known as Tricarboxylic acid (TCA) cycle)*

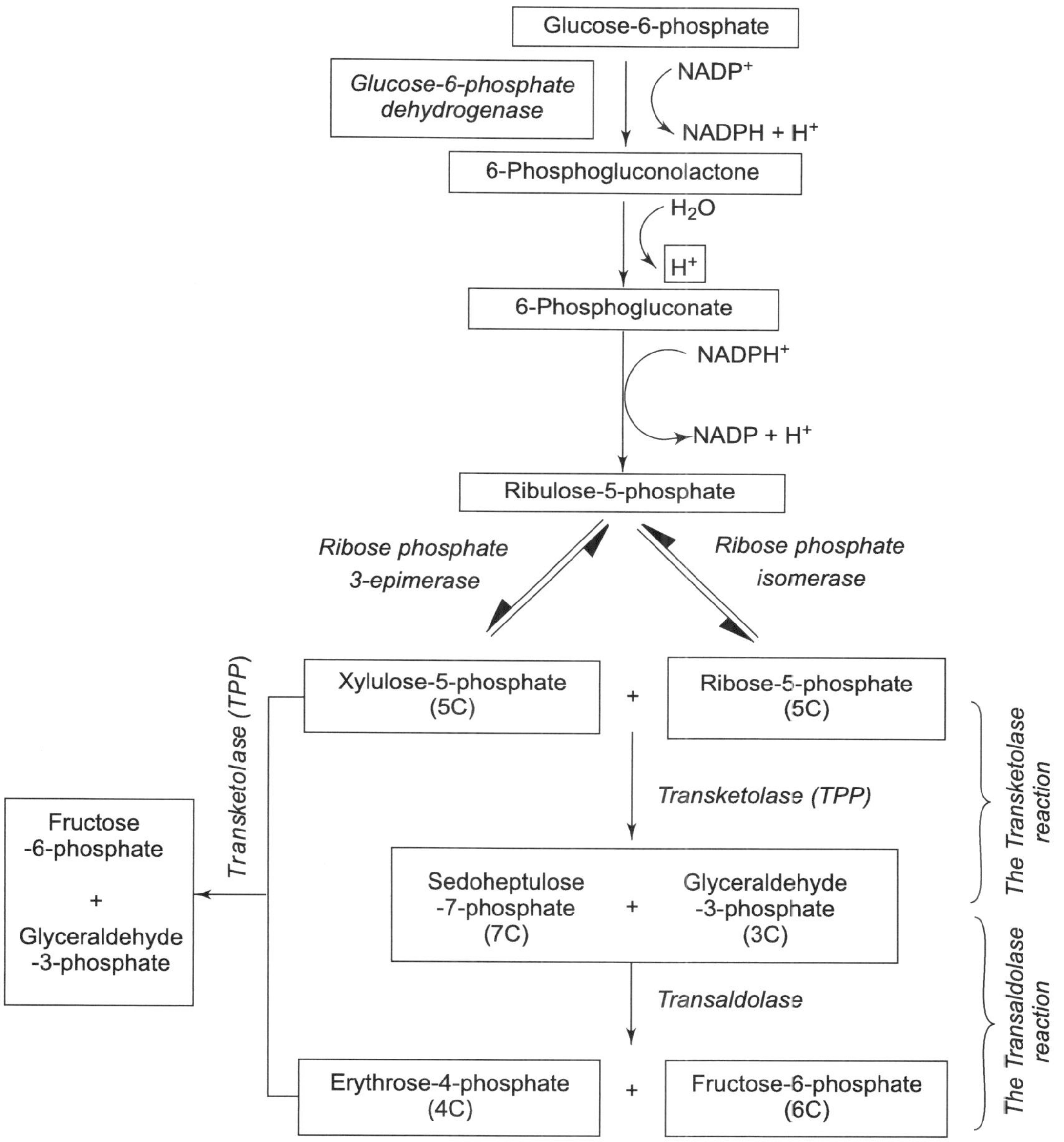

Fig. 2.4 *The Pentose phosphate pathway*

Table 2.9 *Names and chemical structures of various compounds participating in Pentose phosphate pathway*

S. No.	*Name*	*Chemical structure*
1	Glucose-6-phosphate	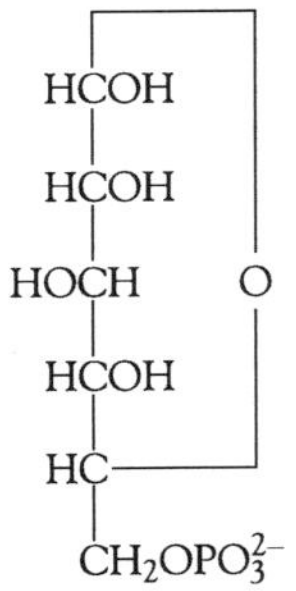
2	6-Phosphogluconolactone	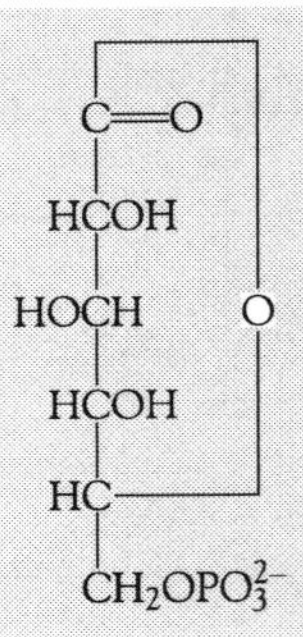
3	6-Phosphogluconate	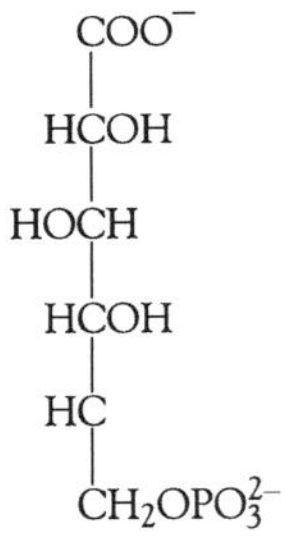
4	Ribulose-5-phosphate	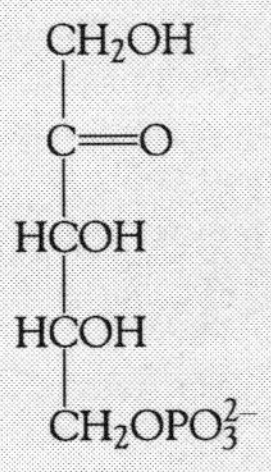

(*Contd.*)

(*Contd.*)

5	Xylulose-5-phosphate	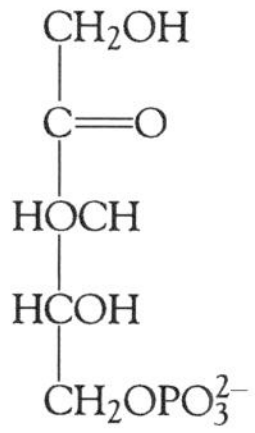
6	Ribose-5-phosphate	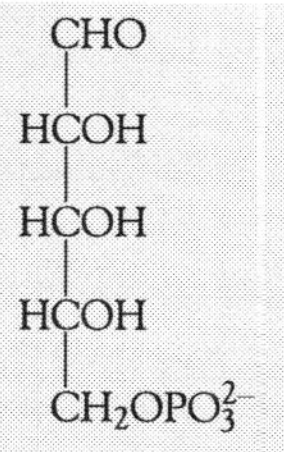
7	Sedoheptulose-7-phosphate	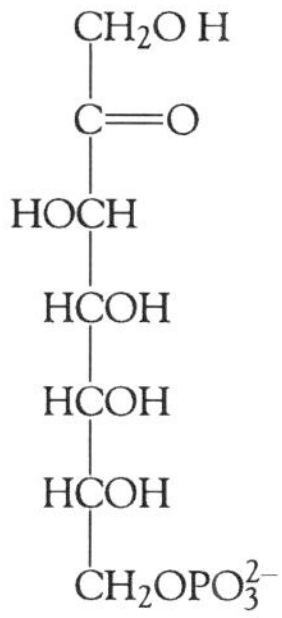
8	Glyceraldehyde-3-phosphate	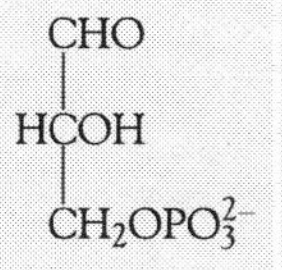
9	Erythrose-4-phosphate	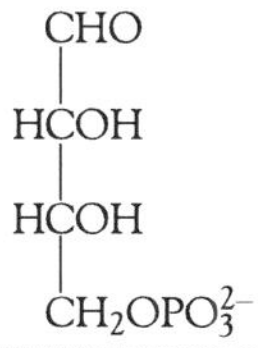
10	Fructose-6-phosphate	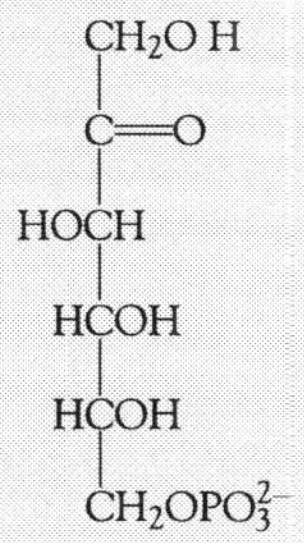

Thus, based on the mode of the nutrient, organisms can be broadly classified as follows:

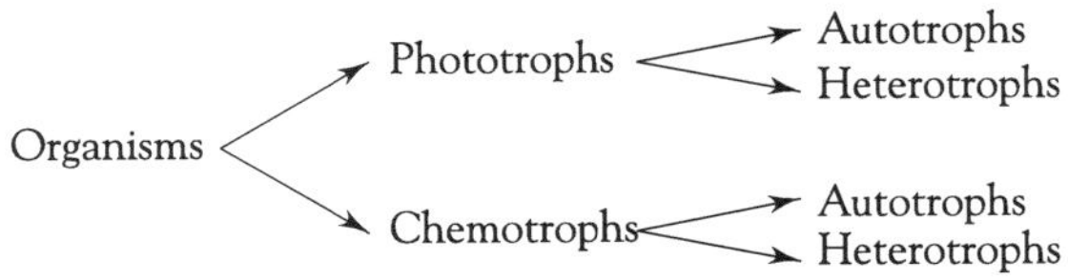

Based on the type of chemicals used as energy source, they are classified as:

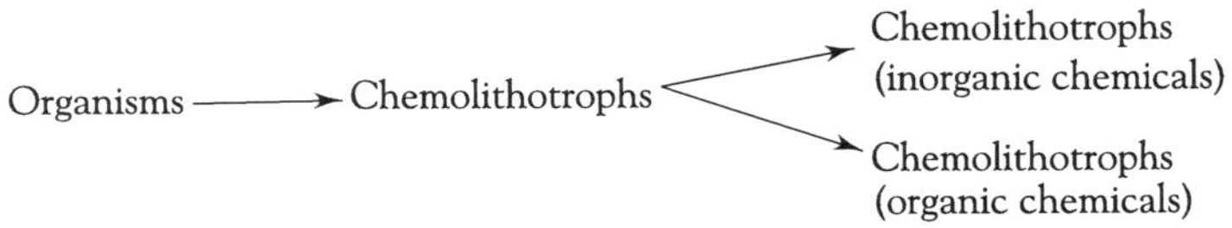

Photoautotrophs are organisms which use light as the primary energy source, and chemotrophs are organisms which undergo chemical reactions to derive their energy requirements. Thus, phototrophs can biosynthesize all cell material from CO_2 as their sole carbon source. They usually contain chlorophyll which is responsible for photosynthesis, e.g. green sulphur bacteria, purple sulphur bacteria, cyanobacteria. Photoheterotrophs cannot convert CO_2 to sugars, but can utilize alcohols, fatty acids, carbohydrates and organic acids as their carbon source, e.g., green non-sulphur bacteria, purple non-sulphur bacteria.

Chemoautotrophs use electrons from reduced inorganic compounds as a source of energy, and use CO_2 as the principal carbon source. Inorganic sources of energy include:

(i) H_2S – *Beggiatoa*
(ii) S – *Thiobacillus thioxidons*
(iii) NH_3 – *Nitrosomonas*
(iv) NO_2 – *Nitrobacter*

Chemotrophs mostly include saprophytes and parasites. They use electrons from hydrogen atoms in organic compounds.

Anabolic reactions include biosynthesis of different macromolecules, e.g.

(i) synthesis of polysaccharides (glycogen → uridine diphosphoglucose) which are the starting material for peptidoglycan
(ii) biosynthesis of lipids
(iii) biosynthesis of amino acids and proteins.

2.7 PROCARYOTIC CELL

The word **procaryote** means primitive nucleus in Greek, and has the following features:

(i) the genetic material is not enclosed within membrane
(ii) they lack membrane-bound organelles
(iii) cell walls contain complex polysaccharide peptidoglycon
(iv) they have a simple method of reproduction.

The procaryotes include a vast heterogeneous group of very small unicellular organisms which comprise eubacteria and archaeobacteria.

Size Most procaryotes range from 0.5 to 3μm in diameter and 2–8μm in length. They have a volume of 10^{-12} ml per cell and may weigh approximately 10^{-12} g per cell.

Shape They are variously shaped like cocci (meaning berry), spherical to oval, which may be attached in a chain pattern, streptococci, or may be found as tetrads, or as a bunch of grapes (groups of eight sarcinae), or may be rod-shaped bacillus (meaning little staffs) or may be spiral. Curved rods are called vibrios and the helical-shaped ones are called spiralla. Usually, the shapes are determined by genetic factors, and most of them maintain single shape (monomorphic), but a few undergo alteration, e.g. corny bacterium, rhizobium.

2.7.1 External Appendages

These appendages are present external to the cell wall, and include glycocalyx, flagella, axial filaments, fimbriae and pili. They are described in the following paragraphs.

Glycocalyx is a viscous gelatinous polymer composed of polysaccharide which may be loosely attached (slime layer) or firmly attached (capsule).

Flagella (meaning whip) is a long filamentous appendage which helps in movement and is variously arranged, viz. monotrichous (single polar flagellum), amphitrichous (two or more at one end of cell), or peritrichous (around the cell). It comprises of three basic parts:

(i) Long outermost filament made of spherical protein flagellin arranged as helix around a hollow core.
(ii) A hook which connects the external filament and basal body.
(iii) The basal body which anchors flagellum to cell wall and plasma membrane.

Procaryotic flagellum is simple compared to eucaryotes. It is semi-rigid, and helps in movement of cell similar to the shaft of an electric motor. The mobility enables the bacterium to move towards favourable environment.

Axial filaments are composed of fibril bundles, which arise at the end of the cell and beneath the outer sheath. They are mainly present in spirochetes, and the rotation of these filaments produces spiral motion similar to cork-screw moving through a cork.

Fimbriae are usually present in gram –ve bacteria. These are short and thin appendages, mainly useful for attachment rather than for motility. Fimbriae occur at poles of bacteria or may be present around the cell. These appendages may be few to several hundreds, and are useful for colonization.

Pili are usually longer than fimbriae (1–2 per cell), and are made up of special proteins called pilin. Pili help bacterial cells for transfer of DNA from one cell to another, and hence they are occasionally referred to as sex pili.

2.7.2 Cell Wall

The bacterial cell wall is complex, semi-rigid and is responsible for characteristic shape. It protects cell organelles from external environment and helps in anchorage. It is credited for its taxonomic value for classification of major types of bacteria.

The cell wall is composed of a macromolecular network called peptidoglycan (murein) which consists of a repeated disaccharide (N-acetyl glucosamine (NAG) and *N*-acetyl muramic acid (NAM)) attached by polypeptides to form a lattice (Fig. 2.5). Alternating NAM and NAG molecules are linked in rows of 10–65 sugars to form a carbohydrate backbone (glycon portion of peptidoglycan). The adjacent rows are linked by polypeptides (peptide portion of peptidoglycan). Although the structure of polypeptide link varies, it always includes tetrapeptide side chains which consist of four amino acids attached to NAMs in the backbone. The amino acids occur in alternating patterns of D and L forms.

Certain primitive group of procaryotes does not possess at all or possesses a very little cell wall material, e.g. mycoplasma which are smallest bacteria. They possess unique plasma membrane having lipids known as sterols, which help in prevention of osmotic lysis. Archaebacteria lack or have unusual cell walls of poly-

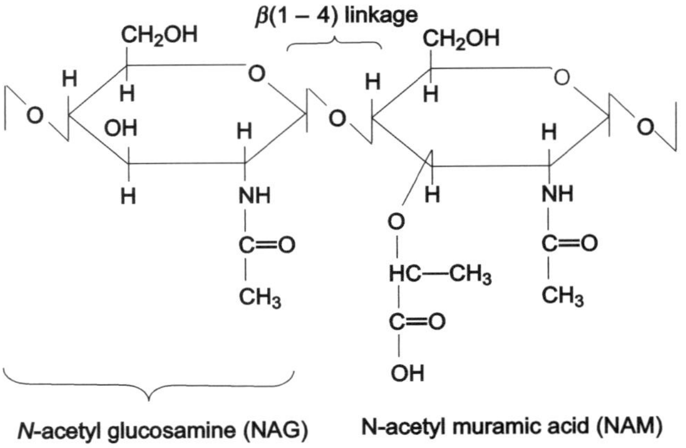

Fig. 2.5 *Macromolecular network of peptidoglycan.*

saccharides and proteins without peptidoglycan or contain pseudomurein which is similar to peptidoglycan. Pseudomurein is made up of *N*-acetyl alosaminuronic acid instead of NAM and lacks D-amino acids.

Gram +ve bacterial cell wall consists of thick, rigid structure due to several peptidoglycan layers, and contains teichoic acids (such as glycerol or ribitol and phosphate). Teichoic acid is mainly helpful in the movement of cations across the membrane, prevents cell lysis, and provides antigenic specificity. There are two classes of teichoic acids:

(i) lipoteichoic acid, which extends over the peptidoglycan layer and links to the plasma membrane.
(ii) wall teichoic acid, which links to the peptidoglycan layer.

Gram -ve cell wall contains a few layers of peptidoglycan and outer membrane (which is made up of lipoproteins, lipo- saccharide and phospholipids). The peptidoglycan is bonded to lipoproteins in outer membrane and is in the periplasmic space (in between outer membrane and plasma membrane). The periplasmic space contains a high concentration of degradative enzymes and transport proteins. Absence of teichoic acid and fewer layers of peptidoglycan make the cell more susceptible to mechanical breakage. The negative charge on the outer membrane acts as a defence mechanism from the evading phagocytes.

2.7.3 Cytoplasmic Membrane

It is a thin phospholipid bilayer surrounding the cell (about 8nm thick) acting as a barrier from external environment. It is a highly selective membrane enabling the cell to concentrate specific metabolites and excrete waste. It primarily consists of phospholipids and proteins and lacks sterols; and hence it is less rigid than eucaryotic organisms. The membrane has phospholipid bilayer and proteins embedded in it with hydrophobic region towards inside. Hydrophilic region is exposed to the external environment. The overall structure of the cytoplasmic membrane is stabilized by hydrogen bonds and hydrophobic interactions, cations like Ca^{2+} and Mg^{2+} ionic bonds with negatively charged phospholipids.

The outer surface of the membrane faces the environment. In certain bacteria, it makes contact with a variety of proteins that serve to bind substrates, or process large molecules for transport into cell. The inner side of the membrane faces the cytoplasm and interacts with proteins involved in energy yielding reactions. Although the cytoplasmic membrane appears rigid, it is actually fluid in nature, and allows phospholipid and protein molecules to have freedom to move about the membrane. The membrane has selective permeability, and prevents destruction by antimicrobial agents. It permits active and passive transport of nutrients.

2.7.4 Cytoplasm

It is the polyphasic, colloidal, living system of cell, which is thick, semitransparent and elastic in nature, and has higher water content. It contains hydrophilic components (protein particles, carbohydrates and salts) and hydrophobic components (globules of lipid or fats). The physical properties of cytoplasm are:

(i) sol–gel phase reversal during metabolic activities
(ii) molecules form large aggregates due to cohesive forces
(iii) absorption and excretion
(iv) cytoplasm shows resistance towards flow, which makes the particles to be in a state of suspension rather than settling down due to gravitational pull
(v) it exhibits Brownian movement.

2.7.5 Nuclear Area

Procaryotes do not have nuclear envelope and do not show cytoplasmic organelles like endoplasmic reticulum, Golgi complex, mitochondria and lysosomes. The bacterial cell shows single chromosome in cytoplasm which consists of single, circular, supercoiled naked DNA (i.e. not associated with histones); this closed loop of DNA contains hereditary information of cell, and this region is referred to as nucleoid.

Smaller molecules of DNA exist apart from chromosome loops called as plasmids, contain few genes which are not essential for growth and metabolism, but confer special adaptive methods for better survival, e.g. resistance factor (R-factor) for drugs or extreme environmental conditions. Plasmids replicate independent of cell DNA, and transfer from one cell to another by the recombination process.

Ribosomes are usually present in cytoplasm and are more active in protein-synthesizing cells. They contain RNA and protein, and are essential machinery for protein synthesis. Procaryotic ribosome (70S) comprises two subunits—50s and 30s. 50s subunit is spherical to dome-shaped in structure, and 30s subunit is ellipsoid to prolate in structure.

The procaryotic cytoplasm has many inclusions, commonly called as inclusion bodies, which store nutrients for starvation periods. Some of these are as follows:

(i) ***Metachromatic granules:*** These are large inclusions, stained with a dye like methylene blue. They are also known as volutin granules. They are reservoirs of inorganic phosphate, and used in the synthesis of ATP.
(ii) ***Polysaccharide granules:*** They are reservoirs of glycogen and starch.
(iii) ***Lipid inclusions/granules:*** They are reservoirs of mainly polymer poly β–hydroxy butyric acid.
(iv) ***Sulphur granules:*** They are reservoirs of elemental sulphur.
(v) ***Carboxysomes:*** They are polyhedral, hexagonal inclusions, and mostly contain enzyme–ribulose 1, 5– diphosphate carboxylase.
(vi) ***Gas vacuoles:*** These are hollow cavities in the cytoplasm, useful in maintenance of buoyancy at depth in water, and allow cells to float up and down in water.
(vii) ***Magnetosomes:*** They are intracellular crystal particles of iron mineral magnetite Fe_3O_4 responsible for magnetotaxis in cell.

2.7.6 Endospores

Mostly Grams +ve bacteria produce highly resistant special structures called endospores (mostly by members of genera *Bacillus* and *Clostridium*). These organisms grow, mature and reproduce for several hours as vegetative cells. Endospores contain a little amount of water and exhibit very few reactions. They have a large amount of dipicolinic acid which stabilizes the proteins.

The spore formation begins under unfavourable conditions. The bacterial chromosome replicates, a small amount of cytoplasm gathers with it, and the membrane grows to seal off the developing spore; then thick layers of peptidoglycan coat it and vegetative cell degenerates. On return of favourable conditions, protective layers break down, and spores germinate to vegetative cells. Thick layers of peptidoglycan form and coat it, and the vegetative cell disintegrates. As a result, the spore is freed.

2.8 EUCARYOTIC CELL

The word eucaryote means "true nucleus". Eucaryotes consist of three main components:

(i) cell membrane
(ii) cytoplasm
(iii) nucleus.

Eucaryotic cells are larger and more complex than the procaryotic cells. They have the genetic material enclosed in a specialized membrane. They contain distinct structures called cell organelles within cytoplasm (Fig. 2.6). Typical eucaryotic cells range from 2 to 200 μm in diameter.

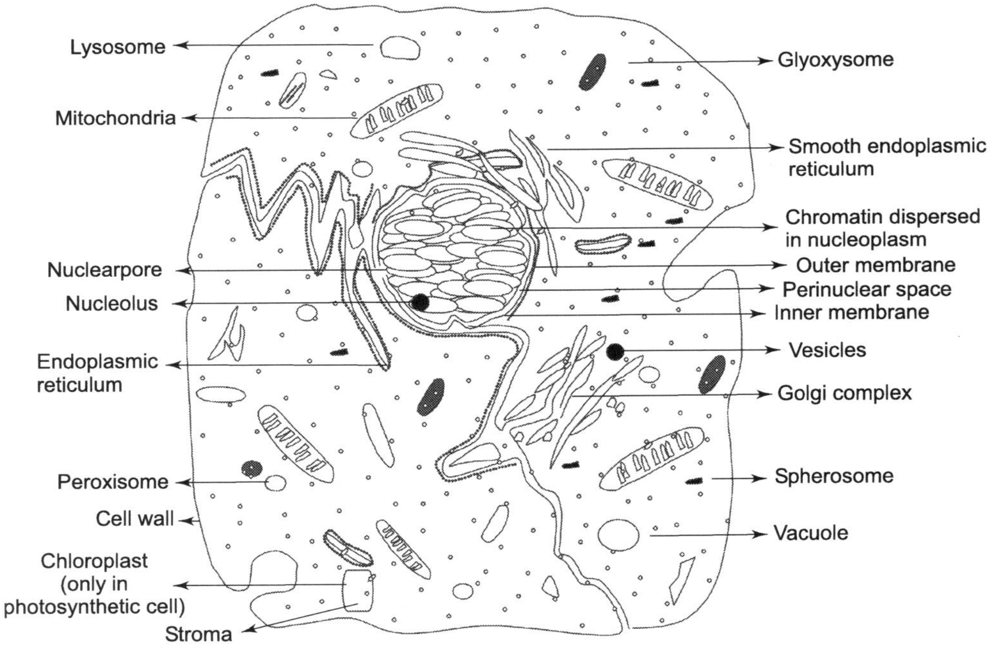

Fig. 2.6 *A typical Eucaryotic cell*

2.8.1 Cell Membrane

The general structure of most biological membranes consists of a phospholipid bilayer; but the major difference in the chemical composition of membranes of eucaryotes as compared to that of procaryotes is that the eucaryotic membranes have sterols. Five to twenty-five percent of the sterols constitute the total lipids of eucaryotes. Sterols are rigid and planar molecules which stabilize membrane structure. They are responsible for providing rigidity to the membrane, which is essential to endure against physical stress in the absence of cell wall. However, sterols can be easily destabilized by certain groups of antibiotics like polyenes (which include filipin, nystatin, candicidin). Hence, absence of sterols in the membrane of procaryotes protects them.

The membrane thickness is generally of the order of 70–100 Å and the thickness of the lipid layer is 30Å, while that of the protein layer is 25 Å. The protein part is responsible for wettability and flexibility. The important lipids in the phospholipid membrane are phospholipids, sphingolipids, and cholesterol glyophosphaticles.

2.8.2 Cytoplasm

The cytoplasm is a complex, colloidal, viscous, semitransparent, jelly-like substance and it is pervaded by a complex endomembrane system. The endomembrane system consists of endoplasmic reticulum, Golgi complex and nuclear envelope. This system is involved in secretion, processing, package, and transport of substances within the cell.

The cytoplasm present outside the endomembrane system is called ground cytoplasm or cytoplasmic matrix proper. The cytoplasm of eucaryotes includes a number of organelles which are absent in procaryotes, such as endoplasmic reticulum, Golgi complex, mitochondria, lysosomes, peroxysomes, glyoxysomes, microtubules and true nucleus.

(i) Endoplasmic Reticulum Endoplasmic reticulum gives mechanical support to the cytoplasm and acts as a circulatory system. It is an extensive complex of vesicles and tubules of vacuolar nature, and it connects plasma and nuclear membrane. It is of two types:

(a) rough endoplasmic reticulum (associated with ribosomes)
(b) smooth endoplasmic reticulum (without ribosomes).

Rough endoplasmic reticulum, present in cells which actively synthesize proteins, and represents an intricate arrangement of membrane-bound spaces in the form of channels with distinct inside and outside paths. Thus, the division of cytoplasm into compartments allows no interference of different metabolic reactions. Numerous enzymatic proteins are associated with endoplasmic reticulum membrane which provides increased surface area for biochemical reactions. Ionic gradients and electrical potentials help in conduction of impulses within the cell.

(ii) Golgi Complex The main function of Golgi complex is storage and absorption; it stores mainly proteins, lipids, hormones, enzymes, vitamins, gums, mucilages, and hemicellulose. Iron, silver and copper accumulate in Golgi membrane. Golgi complex participates in recycling of plasma membrane, helps to concentrate secretory materials and packs them into secretory residues, and mainly secretes proteins, lipids, pectin, cellulose, steroids, etc.

Several units of Golgi complex are scattered in cytoplasm and each unit is called dictyosome. Each dictyosome consists of cisternae, clusters of tubules, vesicles and large vacuoles. Cisternae or lamellae are flattened fluid–filled sacs, curved long (20–30 nm wide), arranged parallelly, bound to a single membrane and are connected to smooth endoplasmic reticulum. Vesicles are small drop-like structures of about 60 nm diameter, associated with convex forming face of cisternae. Golgi vesicles form cell plate during cytokinesis, and play an important role in metabolic activities of cell. High concentration of enzymes like acid phosphatase are transferred from Golgi complex to lysosomes.

Vacuoles are rounded sacs associated with maturing face of cisternae.

(iii) Lysosomes They are single membrane bound, spherical structures (0.4μm) containing hydrolytic enzymes. They enclose fine granular fluid, rich in digestive enzymes and originate from tubules of endoplasmic reticulum or from cisternae of Golgi complex. They release, due to external forces like freezing, thawing, blending or UV radiations, acid phosphates, ribonuclease, cathepsin, collagenase, α-glucosidase, β-glucosidase, α-mannosidase, sulphatase, etc.

Lysosomes digest larger food molecules which enter the cell (heterophagy) or digest stored food materials within the cell (autophagy); they also digest damaged or ageing cells by releasing hydrolytic enzymes (autolysis). They even digest the extracellular materials by releasing hydrolytic enzymes outside the cell.

(iv) Ribosomes They are ribonucleoprotein particles of cytoplasm associated with protein synthesis. They are basophilic granules (mostly present on the endoplasmic reticulum) with a spheroidal structure of about 23 nm and consist of two unequal subunits. Their size is measured by the speed with which they sediment in centrifuge and is given by **Swedberg** unit.

Ribosomes occur in all organisms including PPLO (Pleuro Pneumonia like organisms—mycoplasma). The eucaryotic 80s ribosome consists of 60s and 40s subunits, and they assemble into ribosome only during protein synthesis. Ribosome provides the surface area for the synthesis of protein, and guide multiple interactions involved in protein synthesis.

(v) Mitochondria They are granular or filamentous cytoplasm organelles and are found in the form of minute globules, vesicles, threads and rods. Their number is more in actively metabolizing cell and is about 7 μm length and 0.5 μm in diameter. It is surrounded by two membranes: (a) outer membrane (6 nm thick, smooth and lipoproteinaceous) and (b) inner membrane (6 nm thick, lipoproteinaceous and projects into mitochondrial matrix in the form of fine finger-like foldings called 'mitochondrial crests' or 'cristac'). This increases the surface area for biochemical reactions. The inner surface of cristac contains oxysomes or F_1 particles which play an important role in electron transport and hydrogen transfer systems.

Mitochondria is made up of proteins (65–75%), lipids (25–30%), RNA, DNA, etc.; the proteins constitute 70% isoenzymes and 30% structural proteins, whereas lipids constitute 90% phospholipids and 10% cholesterol.

Due to the presence of DNA and RNA, mitochondria are regarded as self-duplicating organelles. The F_1 particle is about 8.5 nm long and has basal piece, stalk piece, and head piece. The head piece consists of a special enzyme called ATP synthetase which catalyzes synthesis of ATP from ADP and Pi.

Mitochondria are involved in the oxidation of food materials, and the so-obtained potential energy is converted into kinetic energy. They are the centers for production of ATP, and hence they are called as 'power houses' of the cell.

(vi) Plastids Plastids are special cytoplasmic organelles present mostly in eucaryotic plant cells, and are classified into the following

(a) Leucoplasts
(b) Chloroplasts
(c) Chromoplasts

(a) **Leucoplasts:** They are colourless plastids, which lack ribosomes and thylakoids, and their main function is storage. Depending on the type of food stored, they are further classified into:
  - AMYLOPLASTS, which store starch
  - ELAIOPLASTS, which store oils and fats (also called as oleosomes)
  - PROTEOPLASTS, which store protein.

(b) **Chloroplast:** They are green plastids and are mainly present in photosynthetic organisms. A chloroplast mainly shows three main components:
  - Outer envelope containing two membranes, each of 4–6 nm thickness, lipoproteinaceous materials and a small amount of xanthophylls.
  - 'Stroma' is a heterogeneous proteinaceous matrix, colourless, gel-like with 70s ribosomes, RNA, highly twisted circular DNA, starch grains, granules and enzymes of Calvin cycle.

  Chloroplast is a self-duplicating organelle (like mitochondria) due to the presence of DNA and RNA. It is the seat of photosynthesis which is two phased; light phase takes place in grana and dark phase in stroma.

(c) **Chromoplast:** It is a coloured plastid (other than green), usually containing yellow (xanthophyll) or orange (carotene) pigments. It is variously shaped as needle of rhomboid. Chromoplast of red algae contains phycocyanin, phycoerythrin (phycobilins), while those of brown algae contain fucoxanthin (a type of xanthophyll).

(vii) Peroxisomes They are small, oval-shaped, membrane-bound structures of cytoplasm containing oxidizing enzymes (like peroxidases, catalases or oxidases) in fine granular matrix. They carry out photorespiration and protect the cell from toxic effects of hydrogen peroxide, which is produced as a byproduct within the cell. They are involved in β-oxidation of fatty acids and initial steps of phospholipid synthesis.

(viii) Glyoxysomes They are small, single membrane-bound and oval-shaped (0.6 μm), enclosing an amorphous protein matrix containing enzymes of glyoxylate cycle. They are involved in degradation of fats to carbohydrates via glyoxylate cycle.

(ix) Spherosomes They are small (2.5 μm), single membrane-bound organelle associated with synthesis and storage of lipids.

(x) Cytoskeleton The cytoplasmic matrix proper shows cytoskeleton formed by the following :

(a) **Microtubules:** They are 25 μm in diameter and several microns long. They occur freely in most eucaryotic cells, made up of special protein called tubulin. They provide mechanical support to cytoplasm, play an important role in cell division and are associated with intracellular transport. During cell division, they form spindle fibres and help in contraction and movement of chromosomes.

(b) **Microfilament:** It is thin (5–7nm wide), just beneath the cell membrane, made up of contractile proteins (like actin and myocin), and helps in cytoplasmic streaming.

(c) **Intermediate filaments:** They are intermediate in size between microtubules and microfilaments (10 nm thick), form a cage around the nucleus, and connect the nucleus with the cell surface. They play an important role in shaping a cell compartmentation within the cell.

(xi) Vacuole It is a cavity of cytoplasm filled with a watery fluid called cell sap and surrounded by a membrane called tonoplast. The cell sap contains 98% water and 2% secretory products, stored products, excretory products, which include organic materials like carbohydrates, proteins, alkaloids, glucosides, tannins, etc. inorganic substances such as sulphates, phosphates, nitrates, carbonates, oxalates, etc., and sometimes pigments also. Vacuoles regulate the process of osmosis, maintain turbidity of cells, and store water, food and byproducts (toxic sometimes) of metabolism.

2.8.3 Nucleus

It is a highly specialized structure of protoplasm. It is the dynamic center of the cell, and the site of storage and replication of hereditary materials. The position and number of nuclei vary from organism to organism. For example:

(i) Acetabularia is located at the basal region. Meristematic cells are centrally located. Many cells are peripherally located.

(ii) Most of the cells have single nucleus called as uninucleate. Ascogenous **hyphal** cells and secondary mycellium of *Basidiomyceles* are binucleate. *Rhizopus* is multinucleate.

Nucleus shows four main parts:

(i) nuclear envelope
(ii) nucleoplasm
(iii) chromatin
(iv) nucleolus

(i) **Nuclear envelope:** It consists of two concentric membranes, called outer and inner membranes, with perinuclear space in between. The outer surface is rough and stuffed with ribosomes, and the inner membrane is smooth and a layer of fibrous protein called nuclear lamina is attached to it. The nuclear lamina helps in disassemblage and reassemblage of nuclear envelope during cell division. Nuclear envelope has a number of pores where the two membranes fuse together; the pore is not a wide open channel but plugged with protein to form annulus which allows transport between nucleoplasm and cytoplasm.

(ii) **Nuleoplasm:** It is a semi-fluid substance in which glycoproteins, ribonucleoproteins, and hydrolytic enzymes are suspended, and has dispersed chromatin.

(iii) **Chromatin:** It is a viscous gelatinous substance, made out of DNA, RNA, histone proteins and non-histone proteins. It is formed by repeating units called nucleosome, which is nothing but tightly coiled DNA along with histone proteins.

(iv) **Nucleolus:** It has deeply stained, prominent body consisting of central fibrillar (made up of 5–10 nm fibrals) and peripheral granular zone (granules of 10–20 nm in diameter which are precursors of ribosomes). It is composed of RNA, DNA and proteins. It disappears during prophase and reappears at telophase stage of cell division.

Centrosome They are present in lower plants, algae and fungi. They are minute inclusions near nucleus and have deeply stained granules called centriole, which is embedded in a mass of hyaline substance (called centrosphere). Centrosome is involved in formation of bipolar spindle during cell division, and also in the formation of cilia.

2.9 VIRUSES

Viruses are not classified as a part of any five kingdoms. According to International Committee of Taxonomy of Viruses, there are no orders, divisions and kingdoms established for viruses.

Viruses lack formal names, and most of them acquired their names from source. For example, measles virus derived its name after the disease; Coxsackie virus after the place Coxsackie, New York, where it was originally isolated; Epstein–Barr virus after the researchers who studied it, etc. But the viruses are categorized into genera, and each genus name ends with a suffix. Virus and genera have been organized into 73 families each ending with-*viridae*.

Viruses are not composed of cells, and use anabolic machinery within living host cells to multiply. A viral genome can direct biosynthesis inside a host cell, and can get incorporated into host genome. The niche of a virus is its specific host cell, so they may be more closely related to their hosts rather than to other viruses.

2.10 FUNGI

Fungi are a diverse group of eucaryotes lacking chlorophyll and are filamentous, with the cell wall made up of chitin and coenocytic (i.e. cytoplasm mingles between adjacent walls through pores). They are typically aerobic, found in damp and dark places. They are active producers of various hydrolytic enzymes, and hence they can decompose cellulose and lignin (like paper and wood products). They can also grow on the surfaces of electrical insulators and make them transmit electricity.

Fungi are mostly saprophytic with complex life cycles involving spore formation. They are mainly classified into:

(i) moulds

(ii) yeasts.

Fungi consist of masses of intertwined filaments called as 'hypha'(plural of hyphae) which have vegetative structures. Thick mass of hyphae is known as mycelium. The cell walls of the fungi are composed of cellulose and a large amount of chitin (a polymer of acetylglucosamine unit). Fungi play an important role in ecological cyles, as they rapidly decompose the dead and release carbon and minerals back to the environment.

Yeasts are capable of anaerobic growth and survive in different environments. If access is given to oxygen, yeasts perform aerobic respiration to metabolize carbohydrates to CO_2 and H_2O. In the absence of oxygen, they ferment carbohydrates to ethanol and CO_2; hence they are used in brewing, wine making and baking industries.

2.11 ALGAE

Algae are eucaryotic, photosynthetic, mostly multicellular, branching organisms and are non-motile. They are more similar to plants, i.e. most algae have branching structures (leaf), attached to substrate by root-like hold fasts, and some species also contain pigments (photosynthetic). The cell organization, process of photosynthesis and storage materials link them strongly to higher plants, whereas their mode of nutrition (in some cases, it is non-photosynthetic) and reproduction have resemblance close to animals.

Algae are considered as primary food producers as they convert chemically simple nutrients to complex organic matter. Their dead cells contribute to the formation of a complex organic material called 'humus' which is provided as food to other microbes.

2.12 PROTOZOA

Protozoa mean "first animals". Except the flagellate algae, the phylum protozoa are the only animal group in the entire kingdom of Protista. Protozoans seem to have evolved in the direction of complexity of physiological function without differentiation of specialized organs such as brain, liver, heart, etc. They are heterogeneous group with highly specialized cell structure, mode of life and reproduction. They greatly vary in size; cell wall is absent in them. Their outermost covering is pellicle or periplast and is composed of lipoproteins.

They reproduce asexually by cell fission or by genetic recombination and multiplication, which result in the rejuvenation of an asexual process called "Conjugation". During dormant conditions, cells lose water, produce thick cell wall to form cysts which is known as "encystment". All protozoa are chemoorganotrophic, capable of taking solid particles into cell. They respond to stimuli such as heat, chemicals, gravity and electricity. They accept or reject food particles with discrimination. Certain organisms even have eye spot which can distinguish light of different wavelengths.

2.13 IMPORTANCE OF MICROBIOLOGY

As has been mentioned earlier, microbiology is studied mainly for two major purposes:

(i) as a basic biological science to understand life process.

(ii) as an applied biological science which deals with exploitation of biochemical potential of microbes.

2.13.1 Microbiology as a Basic Science

To understand the principles behind the life process, microbes serve as an excellent model, with the following major advantages:

(i) simplicity

(ii) easily propagatable nature on simple media in the laboratory
(iii) smaller size and less complex genome
(iv) rapid rate of multiplication
(v) ease of handling and analysis of mutants.

Of course, these organisms may not be useful in studying the metabolism and development of higher organisms, but the studies on these organisms had laid the foundation for understanding the fundamentals of biochemistry and molecular biology, of which most could be applied to higher organisms like plants and animals. The organisms so far used as models are as follows:

(i) Escherichia Coli *E.coli* is an ideal model for understanding the fundamental aspects of biochemistry and molecular biology. It was studied widely because of its ease of propagation and fast multiplication. *E.coli* grows on simple media, mutant selection and analysis are easy, and can be easily isolated. The size of the genome is small, i.e. four million base pairs will approximately encode to 4000 different proteins. Nearly half the total numbers of *E.coli* gene functions have been determined. (Imagine the magnitude of the complex problem with the human genome size being three billion base pairs, and encoding to one lakh different proteins!)

(ii) Yeast Although bacteria (*E.coli*) have been an invaluable model for studies of many conserved properties of cells, yeasts are the simplest eucaryotes having similar advantages to those of *E.coli* for understanding the unique properties of eucaryotic cell. The size of the genome is in the million base pairs (three times that of *E.coli*). It also readily grows, but the division is not as fast as that of bacteria, by dividing for every 2 hours.

(iii) Dictystelium discoideum It is a simple unicellular slime mould. Its genome is 10 times larger than *E.coli,* simple to grow and amenable to a variety of genetic manipulations. It is a highly mobile cell and hence is an important model for studying molecular mechanisms responsible for movement of animals. An interesting feature of it is the ability of single cells to aggregate into multicellular structure. Under starvation conditions, it forms into slugs (i.e. about one lakh cells function as a single unit) and this is very important for understanding cell signaling and cell–cell interactions.

Thus, microorganisms play a critical role in understanding the principles of life.

2.13.2 Microbiology as an Applied Science

Microbiology scores also as an applied science for the study of useful and pathogenic organisms. It mainly deals with the production of valuable products from cheaper or disposable substrates using microbes. The pathogenic strains are studied for the prevention of destruction of valuable products by contamination.

Industrial microbiology is the exploitation of the biochemical potential of microbes for the production of valuable products either by bioconversion or by production of biomass, which have economic benefits. Apart from industrial activity, microbes play an important role in the field of biotechnology which uses living organisms to carry out a defined chemical process for industrial applications. Some such industrially important fermentation products are listed in Tables 5.1 and 5.2.

The biotechnological exploitation of microbes to produce useful metabolic products from otherwise not much useful substrates is achieved by the application of genetic engineering principles. This is done by deliberate expression of a foreign gene into the organism for production of specific proteins.

Molecular cloning is a base for most genetic engineering procedures. The purpose of molecular cloning is to isolate large quantities of specific or desired genes in pure form and their expression into simpler organisms. Various steps involved in the development of new products using genetic engineering techniques are shown in Fig. 2.7.

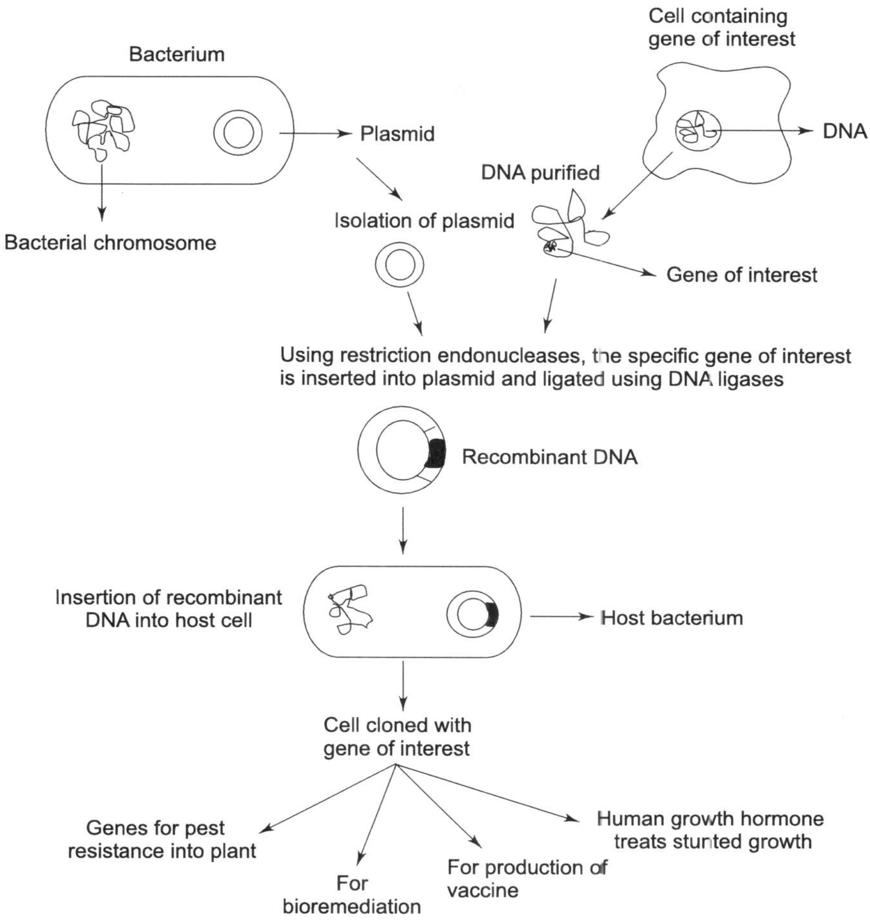

Fig. 2.7 *Development of new products using Genetic Engineering*

The advent of genetic engineering owes to the microbes because it all started with the discovery and isolation of a group of bacterial enzymes called **endonucleases** which cleave phosphate—sugar bonds of nucleic acid to insert a foreign gene. These are also called as restriction enzymes.

The development of diverse DNA products resulted consequent to the discovery of plasmids and DNA ligases (an enzyme that combines two DNA fragments).

Plasmids are natural vectors (carriers) which replicate independently of host chromosomes and carry other genes that confer important properties on their hosts. Apart from plasmids, bacteriophage lambda and yeasts serve as important vectors.

r-DNA Technology Products There is a number of products so far produced by r-DNA technology. Some of these are given as follows:

- Humulin is produced by inserting insulin genes into plasmids.
- Interferon is an antiviral drug produced by genetically altering *E.coli*.
- Human growth hormones (used for treating pituitary dwarfism) can be produced by DNA technology by using new biochemical methods.

- Gene transfer experiments in plants can be successfully performed by using transfer of plasmid (t-DNA) from *Agrobacterium tumefaciens*.
- Bacteria coding for bovine growth hormones injected into cows increased milk production.
- Several types of vaccines can be produced.

2.14 CONCLUDING REMARKS

The cell structure, metabolism and function of the cells constitute the core of microbiology, and its application to make economically useful metabolic products constitute the subject matter for study in **Biochemical Engineering**. An understanding of the subject and our ability to interweave it with process engineering skills would ultimately lead us to design bioreactors as successful biochemical engineers to steer the cells to produce useful metabolic products from organic substrates. Thus, for successful production and economization of a fermentation process, it requires both industrial microbiologist as well as a biochemical engineer, along with constant experimental work and patience.

References

2.1 Aleamo, E.I. (1997), *Fundamentals of Microbiology*, 5th Edn. The Benjamin/Cummings Publishing Company, Inc. Bridge Parkway.

2.2 Bailey, J.E, and Ollis, D.F. (1986), *Biochemical Engineering Fundamentals*, 2nd Edn; McGraw-Hill, Inc., New York, p.12.

2.3 Hoogerheide, J.C. (1977), "Traditional fermentation in modern dress", *Chemtech*, February, 94–97.

2.4 Kala, V.V., and Bhusari, K.P. (2001), *Applied Microbiology*, Himalaya Publishing House, New Delhi.

2.5 Lapage, S.P, et al. (1975), *International Code of Nomenclature of Bacteria*, International Association of Microbiological Societies, Washington, D.C.

2.6 Madigan, M.T., Martinko, J.M., and Parker, J.(1997). *Brock Biology of Microorganisms*, 8th Edn, Prentice Hall International, Inc., New Jersey.

2.7 Tortora, G.J., Funke, B.R., and Case, C.L. (1995), *Microbiology–An Introduction*, 5th Edn, The Benjamin/Cummings Publishing Company, Inc., Bridge Parkway.

2.8 Webb, F.C. (1964), *Biochemical Engineering*, D. Van Nostrand Company Limited, London, p. 3.

Review Questions

2.1 What are the components and various branches of microbiology?

2.2 How do you consider that a study of microbiology will lead to an understanding of biochemical processes?

2.3 Describe the taxonomical classification of five kingdoms as proposed by Whittaker?

2.4 Write a simplified classification of universal ancestor.

2.5 What is the chemical composition of a cell?

2.6 What are various nutritional requirements of organisms?

2.7 Describe Kreb's cycle.

2.8 Describe a procaryotic cell with its salient features.

2.9 What are the physical properties of cytoplasm?

2.10 Describe mitochondria. What is the function of mitochondria in a cell?

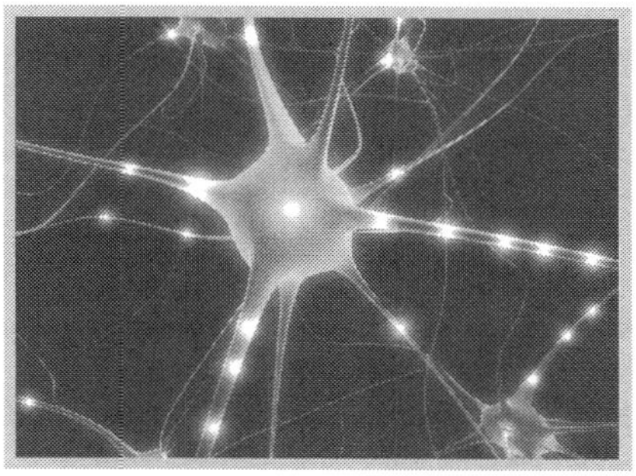

chapter 3

Introduction to Biochemistry

Biochemistry is the study of life cyclic processes in terms of chemicals. It is not the study of chemistry of living organisms. It obtains information from the basic microorganisms viz., bacteria, viruses, algae and fungi. It draws information on energitics from basic thermodynamics. It couples them to show a path how life cycle proceeds with the mutual cooperation of various activities of the living beings. Energy is required for various metabolic activities in the living organisms. This energy is released by breaking of some of the high energy storing molecules, usually phosphate containing molecules. Oxidation of NADH (nicotinamide adenine dinucleotide, reduced form) in the mitochondria is one of the main reactions.

Some of the chemical/biochemical reactions in the living organisms are facilitated by another type of compounds known as *enzymes*. The facilitation of the reaction is known as *catalysis*, and hence enzymes are known as *biocatalysts* or *biological catalysts*.

The cells themselves contain some of the enzymes. The cell regulates the type of enzyme and its number by which it regulates both the type of biochemical reaction and its rate. Genetic engineering deals with the information flow of the living organisms. Biochemistry, coupled with genetic engineering principles, plays a vital role in devising techniques for manipulating most of the biochemical transformations. Thus, of late, the theoretical knowledge of biochemistry is being diverted to applied aspects in regulating the cell metabolism.

Living organisms contain various biomolecules which are the building blocks of the cell and also help in storing and releasing energy for biotransformations, and at the same time carrying forward genetic information. They include carbohydrates, lipids, proteins, nucleic acids, vitamins, etc.

Biomolecules Living organisms contain a large number of biomolecules which are essentially composed of carbon and nitrogen. The bacteria which are the smallest cells contain a large number of (5000) different kinds of compounds including 3000 kinds of proteins, 1000 kinds of nucleic acids. These biomolecules have high molecular weights and are complex in structure. They form the basic building blocks for the cells, and serve various purposes. For example, the amino acids serve not only as building blocks of protein molecules but also as precursors of hormones, alkaloids, porphyrins, pigments etc. Similarly nucleotides serve not only as building blocks of nucleic acids but also as co-enzymes and as energy carrying molecules.

In the following sections, we shall very briefly discuss the various biomolecules, their structural formulae and the functions performed by them. A detailed discussion of these aspects is available in any standard text-book of biochemistry (Lehninger, 1983; Trehan, 1987; White et. al., 1968).

3.1 LIPIDS

Lipids are a class of compounds which are fatty/oily in nature and present in cells and tissues. They have the following distinct characteristic features.

- They are insoluble in water
- They are soluble in non-polar solvents like n-hexane, alcohol, chloroform, benzene, etc. and hence can be extracted from the organisms using these solvents
- They release a lot of energy on break-down and hence are considered as the energy-storage media
- They contain a large proportion of C-H bonds
- Upon saponification, they release fatty acids and glycerol, and hence glycerol is considered as a binding agent for fatty acids
- They are synthesized by the cells from sugars

Some lipid compounds, such as vitamins and hormones have intense biological activity. Lipids also include a heterogeneous group of structural components. In Table 3.1, they are grouped based on their chemical composition. Some lipids are combined with other classes of compounds, and they are known as:

- Lipoproteins
- Proteolipids
- Phosphatidopeptides
- Lipoaminoacids
- Lipopolysaccharides

Table 3.1 *Various types of lipids*

Fatty acids
Lipids containing glycerol Neutral Fats Phospholipids • Phosphatides • Phosphoglycerides & Phosphoinositides
Lipids not containing glycerol Sphingolipids • Ceramides • Sphingomyelins Aliphatic alcohols and waxes Terpenes Steroids

As biomolecules, they are a constituent of cell walls and form a protective coating to the cell, and encourage some species-specific reactions. They are also energy carriers, and release the energy as and when the cell requires it.

3.1.1 Fatty Acids

The fatty acids are generally available in the esterified form in cells and tissues, and are rarely available in free form. They are named on the basis of the hydrocarbon from which they are derived. These hydrocarbons may be saturated or unsaturated. Accordingly, the fatty acids are also classified as saturated and unsaturated fatty acids. In saturated fatty acids, the carbon atoms hold as many hydrogen atoms as possible in addition to the carboxylic grouping, whereas in the unsaturated form, the carbon atoms hold double and triple bonds also. Fatty acids differ from each other primarily in chain length and in the number and position of their unsaturated bonds. They are generally represented by a shorthand notation based on the length of the carbon chain and the number, position and configuration of the double bonds. For example, palmitic acid is symbolized as 16:0 and oleic acid as $18{:}1^{\Delta 9}$. Table 3.2 lists some naturally occurring fatty acids. The palmitic acid contains 16 carbon atoms without having any unsaturated double bonds, whereas oleic acid contains eighteen carbon atoms and contains one unsaturated double bond between 9th and 10th carbon atoms.

Table 3.2 Some naturally occurring fatty acids

	Saturated fatty acids	
6:0	$CH_3(CH_2)_4$ COOH	Caproic acid
8:0	$CH_3(CH_2)_6$ COOH	Caprylic acid
9:0	$CH_3(CH_2)_7$ COOH	Pelargonic acid
10:0	$CH_3(CH_2)_8$ COOH	Capric acid
12:0	$CH_3(CH_2)_{10}$ COOH	Lauric acid
14:0	$CH_3(CH_2)_{12}$ COOH	Myristic acid
16:0	$CH_3(CH_2)_{14}$ COOH	Palmitic acid
18:0	$CH_3(CH_2)_{16}$ COOH	Stearic acid
20:0	$CH_3(CH_2)_{18}$ COOH	Arachidic acid
22:0	$CH_3(CH_2)_{20}$ COOH	Behenic acid
24:0	$CH_3(CH_2)_{22}$ COOH	Lignoceric acid
	Unsaturated fatty acids	
16:1	$CH_3(CH_2)_5$ CH=CH$(CH_2)_7$ COOH	Palmitoleic acid
18:1	$CH_3(CH_2)_7$CH=CH$(CH_2)_7$ COOH	Oleic acid
18:1	$CH_3(CH_2)_7$CH=CH$(CH_2)_7$ COOH	Elaidic acid
18:2	$CH_3(CH_2)_4$CH=CHCH_2CH =CH (CH2)$_7$ COOH	Linoleic acid
18:3	CH_3CH_2CH=CHCH_2CH =CHCH_2CH=CH $(CH_2)_7$ COOH	Linolenic acid
18:3	$CH_3(CH_2)_4$CH=CHCH_2CH =CHCH_2CH=CH $(CH_2)_4$COOH	γ-Linolenic acid
18:3	$CH_3(CH_2)_3$CH=CHCH=CHCH=CH $(CH_2)_7$COOH	Eleostearic acid
20:4	$CH_3(CH_2)_4$CH=CHCH_2CH =CHCH_2CH=CH CH_2CH=CH- $(CH_2)_3$COOH	Arachidonic acid
24:1	$CH_3(CH_2)_7$ CH=CH$(CH_2)_{13}$ COOH	Nervonic acid

3.1.2 Lipids Containing Glycerol

Fatty acid esters of glycerol are called acyl-glycerols or glycerides. Depending upon the number of hydroxyl groups of the glycerol is esterified, it is known as monoacyl-, diacyl- or triacyl-glycerol. They are shown as follows.

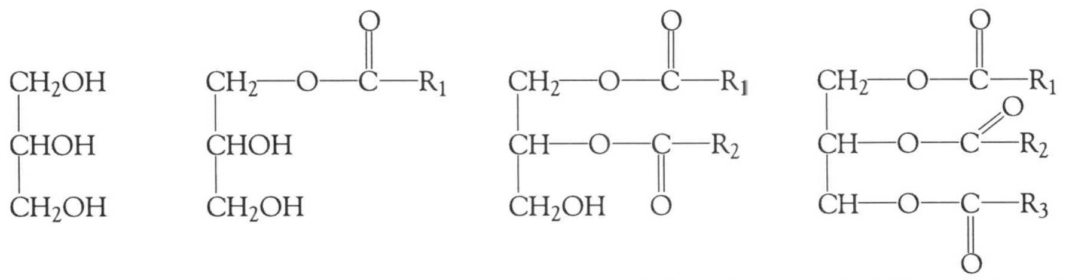

Glycerol 1-Monoacyl glycerol 1, 2-Diacyl glycerol 1, 2, 3-Triacyl glycerol

In the above R_1, R_2 and R_3 may be the same fatty acid group or different fatty acid groups. If they are all same, then the fat is known as simple triacylglycerol. If R's are stearic acid, the lipid is known as tristearoyl-glycerol or simply known as tristearin.

3.1.3 Phospholipids

Phospholipids consist of a glycerol connected to a phosphate group. They consist of two fatty acids linked to glycerol molecules. They are generally crystal white solids. They turn brown when exposed to air due to polymerization of unsaturated fatty acids. They are characterized also by the presence of nitrogen group. They generally posses a negative charge on the phosphate group. Most of the phospholipids are derivatives of L-α-phosphoric acid

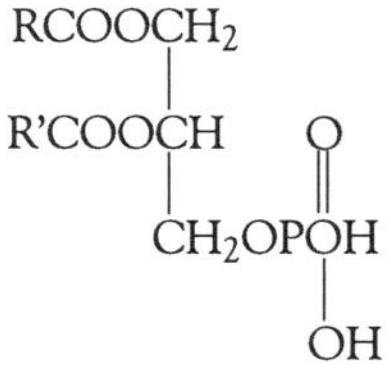

L-α-Phosphatidic acid

L-α-phosphatidic acid on hydrolysis yields one mole each of glycerol, phosphoric acid and two moles of fatty acid. In phosphatides, the phosphoric acid is bound in ester linkage to one of the following nitrogenous compounds.

Choline Ethanolamine L-Serine

3.1.4 Lipids Not Containing Glycerols

Sphingolipids Plant and animal cells contain sphingolipids as important components. The sphingolipids are present in the brain and nerve tissues in large quantities. They are lipids containing sphingosine in place of glycerol as their backbone. *Sphingomyelins* are the sphingolipids in higher animals which contain phosphoryl ethanolamine or phosphoryl choline. Sphingomyelins have the following generalized structural formula:

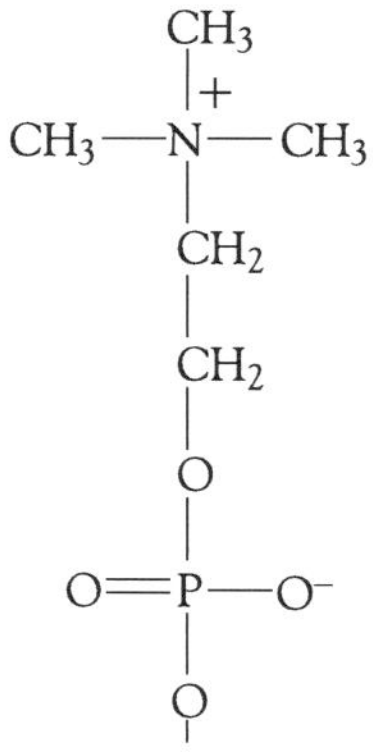

(Contd.)

(Contd.)

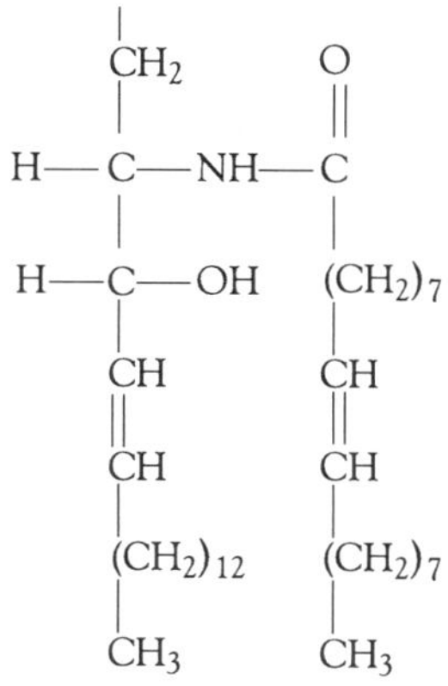

Sphingomyelin

3.1.5 Terpenes

Terpenes contain multiples of the five-carbon hydrocarbon isoprene (2-methyl-1,3 butadiene), and they are classified based on the number of isoprene units (Table 3.3).

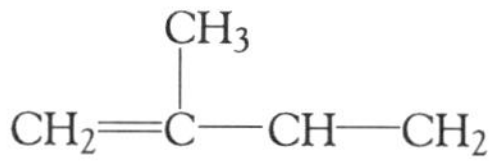

Isoprene

Table 3.3 *Classification of Terpenes*

Nomenclature of terpene	*No. of isoprene units*	*Examples*
Monoterpenes	Two	Geranoil, limonene, menthol, pinene, camphor and carvone
Sesqui-terpenes	Three	Farnesol
Di terpenes	Four	Phytol
Tri terpenes	Six	Squalene
Tetra terpenes	Eight	
Polyterpenes	More	Natural rubber

Terpenes are known for their characteristic odor and flavor, and are mostly present in the essential oils of plants.

3.1.6 Steroids

Steroids are a group of compounds derived from the saturated tetracyclic hydrocarbon perhydrocyclopentanophenanthrene. Many steroids have different functions and distinct metabolic activity. Mostly they are derived from natural sources. All steroids originate from the linear triterpenesqualene which cyclizes readily.

Lanosterol is the precursor of cholesterol, which is the most abundant steroid in animal tissues. Lanosterol and cholesterol are steroid alcohols, which occur either as free alcohols or as long-chain fatty acid esters of the hydroxyl group at carbon 3.

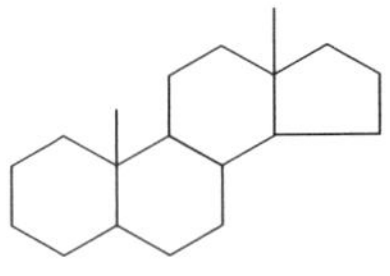

Perhydrocyclopentanophenanthrene

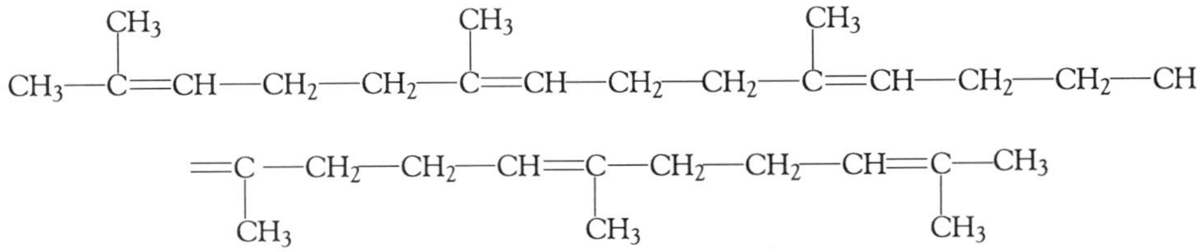

Squalene

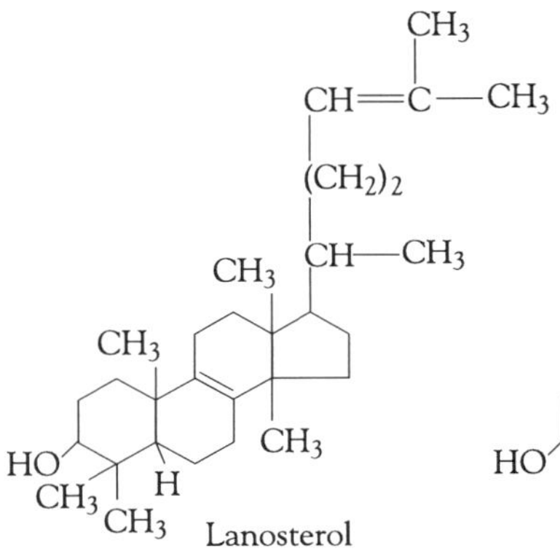

Lanosterol

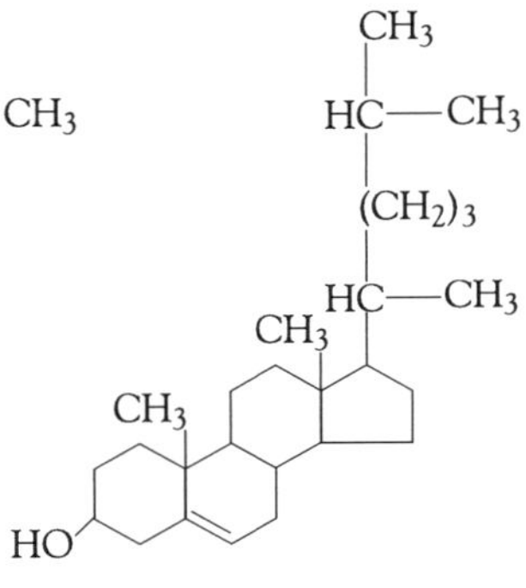

Cholesterol

3.1.7 Lipoproteins

Lipoproteins are a class of compounds formed by the association of certain lipids with proteins. Hence they exhibit properties characteristic of both lipids and proteins. They are classified into the following two major groups:

- transport lipoproteins
- membrane systems

The transport lipoproteins are so named because they carry water-insoluble lipids to various organs of the body through blood. They are also known as plasma lipoproteins and are further classified into the following three groups depending upon the type of lipid

- very low density lipoproteins (VLDL)
- low density lipoproteins (LDL)
- high density lipoproteins (HDL)

Membrane systems are the lipoproteins present in the membrane of the cell structure (Section 2.7.3). They contain about 40% lipid and 60% protein. They are mainly meant for protecting the cell from external environment. They also facilitate the transport of certain enzymes and other transport systems.

3.1.8 Lipids Combined with Other Compounds

Lipids combined with other classes of compounds, in addition to lipoproteins, are as follows:

Proteolipids: Proteolipids contain proteins associated with lipids in which lipid content is more.

Phosphatidopeptides: Lipid category of phosphatides associated with peptides are phosphatidopeptides.

Lipoamino acids: Lipids associated with amino acids are lipoamino acids.

Lipopolysaccharides: Lipids found in nature combined with polysaccharides are lipopolysaccharides.

3.2 PROTEINS

The word protein is derived from a Greek word meaning 'of the first rank'. They are considered to be the beginners and builders of biochemical reaction. Thus, they occupy a central position in the construction and functioning of living matter. All enzymes are proteins and they catalyze biochemical reactions as biocatalysts. Collagen is an elongated protein which is responsible for the tensile strength of skin and bones. In addition, proteins perform various other functions as well; viz., they help in transport of oxygen into blood, and transmission of nerve impulses, etc. Proteins constitute a major part of the animal tissues.

Proteins are a large molecules consisting of as many as twenty amino acids. Their molecular weights range from 5000 to few millions. The amino acids are linked through the peptide linkages.

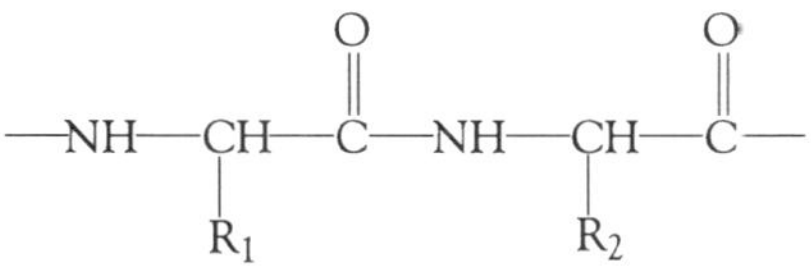

Peptide linkage

3.2.1 Amino Acids

Amino acids are the building blocks of proteins. There are twenty amino acids present in all living organisms, which is a clear manifestation of the biological unity in all living organisms. These twenty amino acids have different structural configurations and are represented by three lettered symbols (Table 3.4). They are also represented by single alphabets which help in writing the amino acid sequence in the proteins. Some proteins contain a large combination of several amino acids.

The amino acids have the structure

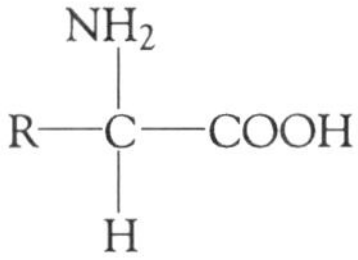

These are known as α-amino acids in which the amino group and the carboxyl group are attached to the same carbon atom. Different amino acids posses different groups in place of R. Amino acids are also classified as acidic, basic, and neutral amino acids.

The above classification is based on the number of carboxyl groups and number of amino groups present. If they are same, it is known as neutral amino acid. Structures of some amino acids are shown in Figs 3.1 to 3.4.

Table 3.4 *Amino acid symbols, chemical group and charge on the molecule*

Amino acid	*Three-Letter Symbol*	*One-Letter Symbol*	*Chemical Group*	*Remarks*
Alanine	Ala	A	Aliphatic	Non polar
Arginine	Arg	R	Aliphatic	Basic
Asparagine	Asn	N	Aliphatic	Polar
Aspartic acid	Asp	D	Aliphatic	Acidic
Asparagine/Aspartic acid	Asx	B	Aliphatic	—
Cysteine	Cys	C	Aliphatic	Polar
Glutamine	Gln	Q	—	—
Glutamic acid	Glu	E	Aliphatic	Acidic
Glutamine/Glutamic acid	Glx	Z	—	—
Glycine	Gly	G	Aliphatic	Polar
Histidine	His	H	Aliphatic	Basic
Isoleucine	Ile	I	Aliphatic	Non polar
Leucine	Leu	L	Aliphatic	Non polar
Lysine	Lys	K	Aliphatic	Basic
Methionine	Met	M	Aliphatic	Non polar
Phenylalanine	Phe	F	Aromatic	Non polar
Proline	Pro	P	Heterocyclic	Non polar
Serine	Ser	S	Aliphatic	Polar
Threonine	Thr	T	Aliphatic	Polar
Tryptophan	Trp	W	Heterocyclic	Non polar
Tyrosine	Tyr	Y	Aromatic	Polar
Valine	Val	V	Aliphatic	Non polar

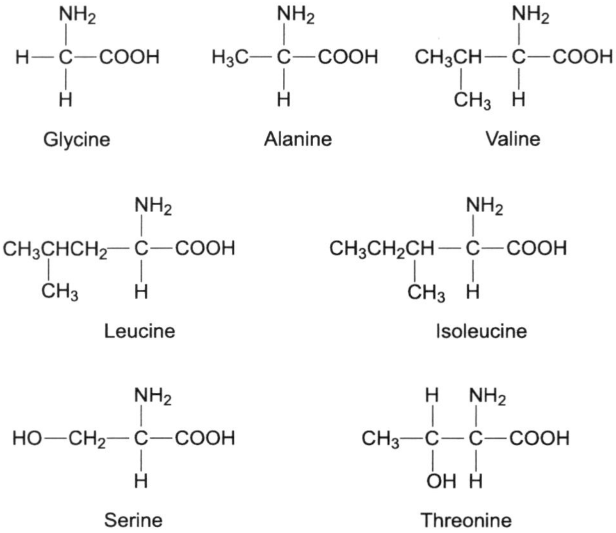

Fig. 3.1 *Structures of some neutral amino acids*

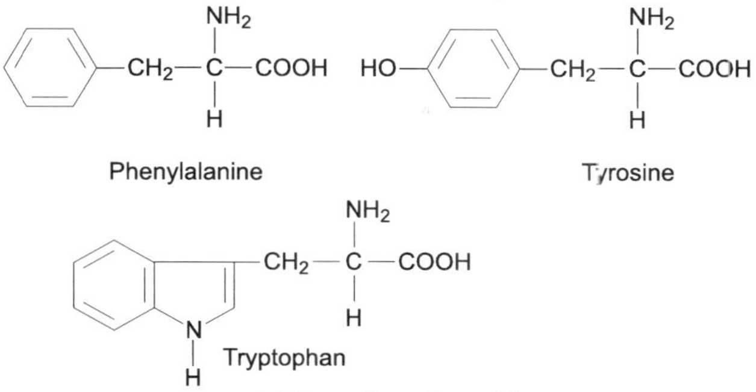

(a) Aromatic amino acids

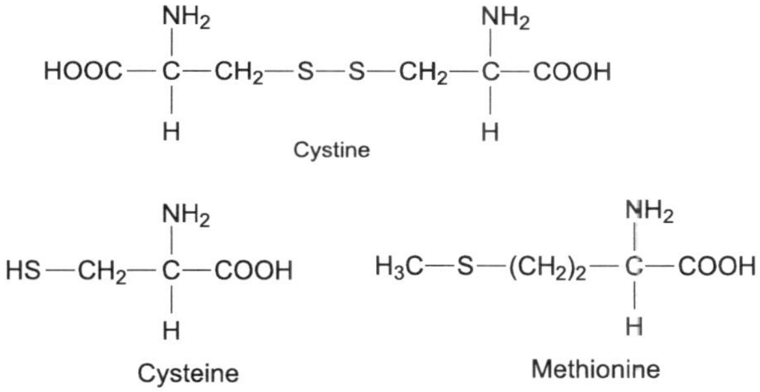

(b) S-containing amino acids

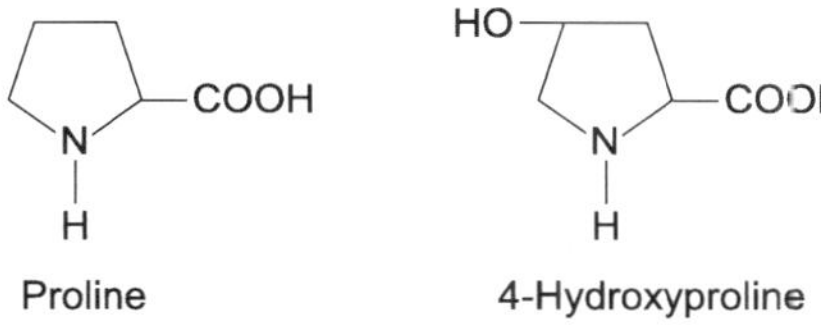

(c) Imino acids

Fig. 3.2 *Structures of some amino acids*

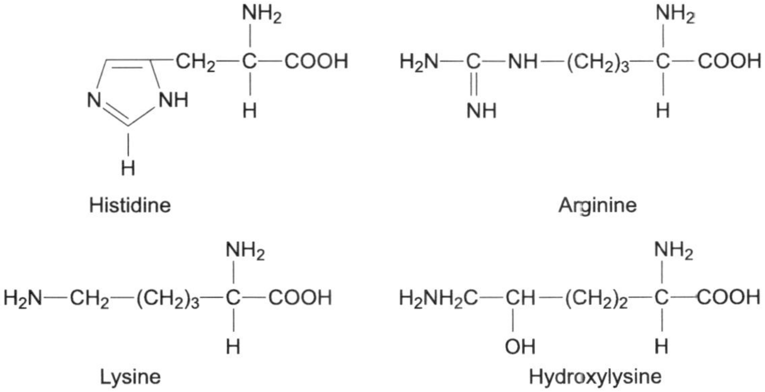

Fig. 3.3 *Structures of some basic amino acids*

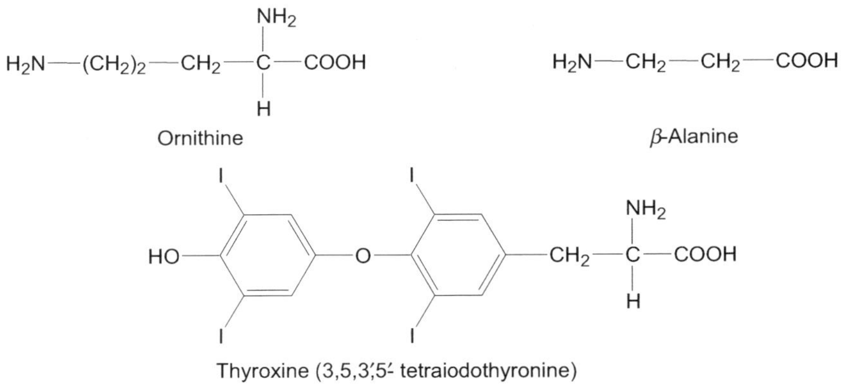

Fig. 3.4 *Structures of more amino acids (ornithine, β-alanine and thyroxine)*

3.2.2 Absorption Spectra

Spectrophotometric method is a rapid method for estimating the protein content in a solution. Tyrosine and other two amino acids namely tryptophan and phenylalanine absorb light in the ultraviolet region. Hence, the measurement of light absorption at 280 nm in a spectrophotometer is used for the estimation of proteins in a solution, since most proteins contain tyrosine.

3.2.3 Analysis of Amino Acid Mixtures

Chromatographic separation methods are useful for the separation, identification and quantitative analysis of amino acid mixtures. A few chromatographic separation methods are described as follows.

(i) Partition Chromatography The relative distribution of the solute between two immiscible phases is taken as an advantage for separation of the mixture of amino acids. It is similar to counter-current extraction principle (described in Chapter 17). When a solute is brought into contact with a two-solvent system (in which the two solvents are immiscible) the solute gets distributed between the two solvents depending upon the partition coefficient of the solute. The immiscible solvent systems could be phenol-water or n-butanol-water, etc. Subsequently, the solute is separated from the solvents.

(ii) Amino Acid Analyzer This technique of amino acid analysis was first pioneered at the Rockefeller Institute, New York in which the entire procedure of cation exchange chromatography is automated so that elution, collection of fractions, analysis of fractions and recording data are all performed automatically.

Other chromatographic methods include :
- column chromatography
- paper chromatography
- TLC
- Ion-exchange chromatography

These are described in Section 17.8.

3.2.4 Purification of Proteins

The characteristic properties of proteins which are used to separate mixtures of proteins are:
- molecular size,

- solubility,
- electric charge,
- differences in absorption characteristics, and
- biological affinity for other molecules.

The following are some methods employed to separate proteins

(i) Dialysis and Ultra-filtration Proteins can be separated from low molecular weight solutes by dialysis. This method employs a semipermeable membrane on which the bigger protein molecules are retained. Small solute molecules and water pass through the membrane. Thus separation takes place. In ultra-filtration, pressure or centrifugal force is used to filter the aqueous medium and small solute molecules through a semipermeable membrane. Cellophane and other synthetic materials are employed as membranes (also see Section 17.7).

(ii) Gel Filtration Separation is carried out based on the relative size of the proteins. Gel filtration or molecular-sieve chromatography consists of a column packed with a polysaccharide called *Sephadex*. The hydrated sephadex gel will have molecular size of 0.1 mm in diameter. Thus, it holds smaller protein molecules. The bigger protein molecules pass through the column and get separated out.

(iii) Ion-exchange Chromatography In this method, the separation is based on the charge of the proteins. The ion-exchange column is packed with negatively charged carboxylate ions. They retain the positively charged proteins at pH below 7. The negatively charged proteins will pass through the column and get separated. Subsequently, the adsorbed proteins are separated from the column by pouring a solution of different concentrations of sodium chloride.

(iv) Affinity Chromatography The method is based on the ligand-ligand interaction. The chromatographic column consists of a polysaccharide molecule (like agarose). The protein mixture is passed through the column. Those proteins, which have the property of binding non-covalently to ligand molecules of agarose, will get attached to the agarose particles. Other proteins, which do not have the affinity for such ligand interaction, will pass through the column and get separated. This method is extensively used for separation of enzymes.

(v) Electrophoresis The isoelectric point of a protein is that value of pH at which the net electric charge on the protein molecule is zero (i.e. the positive charge equals the negative charge). Hence, at isoelectric point, the mobility of the protein under an electric field will be zero. This property of a protein is used to separate a protein mixture by the electrophoresis method. The protein mixture is subjected to electric field in the retarding gel support. Proteins with different isoelectric points will have different mobilities. Thus, the separation takes place in the electrophoresis method.

This method can also be effectively used as an analytical tool for proteins.

3.3 CARBOHYDRATES

The carbohydrates or saccharides are a large group of compounds which are poly-hydroxyaldehydes or poly-hydroxyketones and their derivatives. They contain carbon, hydrogen and oxygen, with the ratio of hydrogen and oxygen being 2:1. They have a general chemical formula: $Cx(H_2O)_y$, and may be classified as :

- monosaccharides
- oligosaccharides (di- and tri- saccharides)
- poly-saccharides
- high molecular weight poly-saccharides.

3.3.1 Monosaccharides

Formaldehyde (HCHO) and hydroxy-acetaldehyde (OH—CH_2—CHO) conform to the empirical formula of carbohydrates, $(CH_2O)_n$, n being equal to 1 and 2 respectively. The smallest three-carbon sugar molecules are glyceraldehyde and dihydroxyacetone, and they are termed as trioses.

Glyceraldehyde (Aldotriose) Dihydroxyacetone (Ketotriose)

Various other monosaccharides are shown in Table 3.5, which are named based on the number of carbon atoms. All the simple monosaccharides are white crystalline solids, sweet in taste, and are freely soluble in water, but are insoluble in non-polar solvents. The carbon skeleton of the common monosaccharides is unbranched, and each carbon atom except one, contains a hydroxyl group. The carbon atom has a carbonyl oxygen which is combined in an aldehyde or ketose linkage. If the carbonyl group is at the end of the chain, the monosaccharide is an aldehyde derivative, and such monosaccharides are called aldoses. If it is at any other position, the monosaccharide is a ketone derivative, and is called as ketoses. Among monosaccharides, the hexoses are abundant. Aldopentoses are important components of nucleic acids and various polysaccharides.

Table 3.5 *Various monosaccharides*

No. of C atoms	*Generic name of monosaccharide*	*Examples*
3	trioses	glyceraldehyde, dihydroxyacetone
4	tetroses	erythrose
5	pentoses	arabinose, xylose, ribose, rhamnose
6	hexoses	glucose, galactose, fructose, mannose
7	heptoses	
8	octoses	

Epimers are defined as the sugars which have the same chemical formula and structure except around one specific carbon atom. Thus D-glucose and L-glucose are epimers with respect to carbon atom 2 (Fig. 3.5).

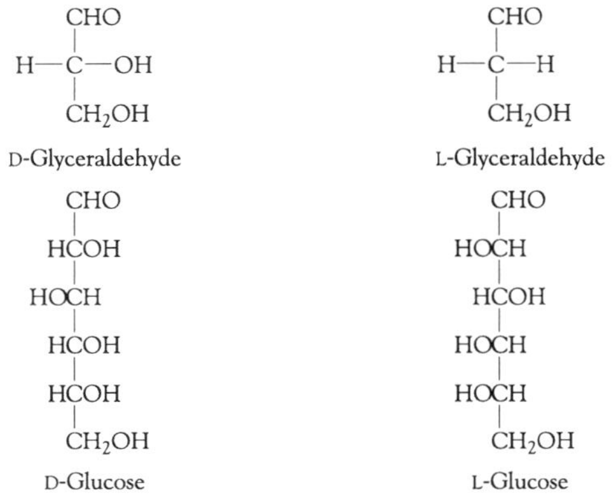

Fig. 3.5 *Structure of D and L monosaccharides*

The families of D-ketoses and D-aldoses are shown in Fig 3.6 and Fig. 3.7.

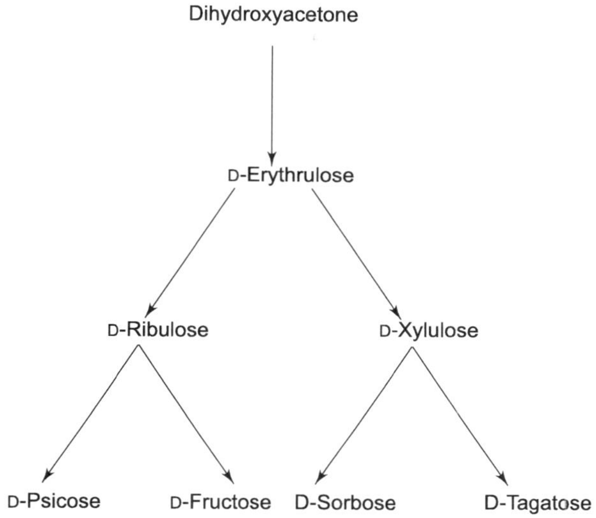

Fig. 3.6 *Family of D-ketoses*

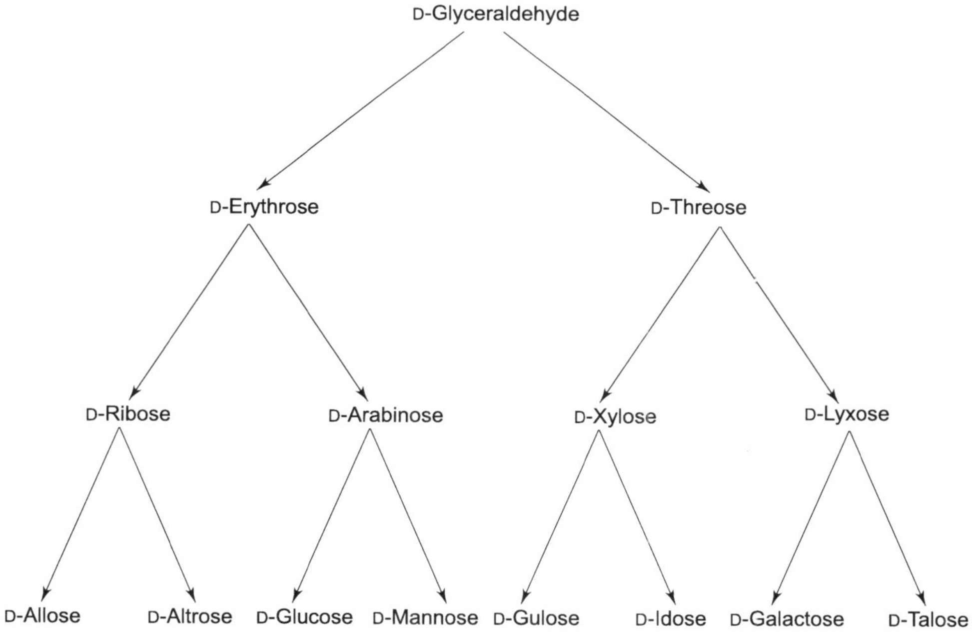

Fig. 3.7 *Family of D-aldoses*

Pyranoses are also six carbon sugars but with a ring structure. Thus α-D-glucose in the ring form is known as α-D-glucopyranose. The structural formulae of α-D-glucopyranose and α-D-glucopyranose are shown as follows

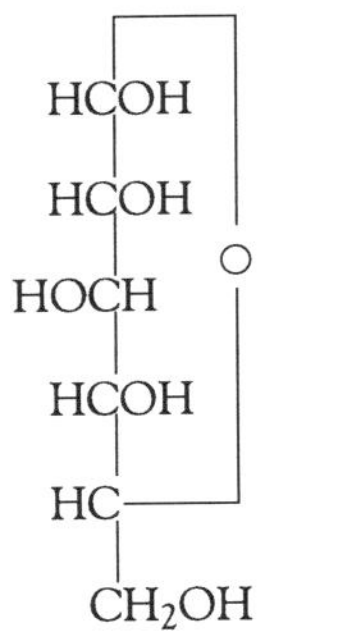

α-D-Glucopyranose

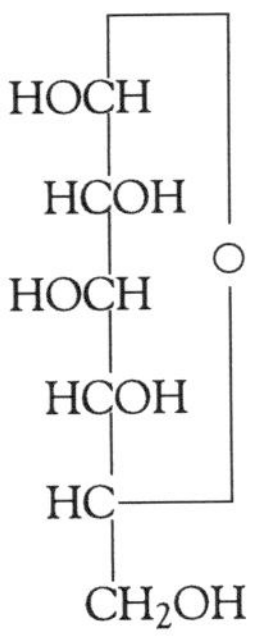

β-D-Glucopyranose

Pyranoses are also represented by two other conformations, viz. boat-form and chair-form (Fig. 3.8). The conformational forms are represented to show the sugar structures in planar form. Of the two forms, the chair form is more rigid and stable.

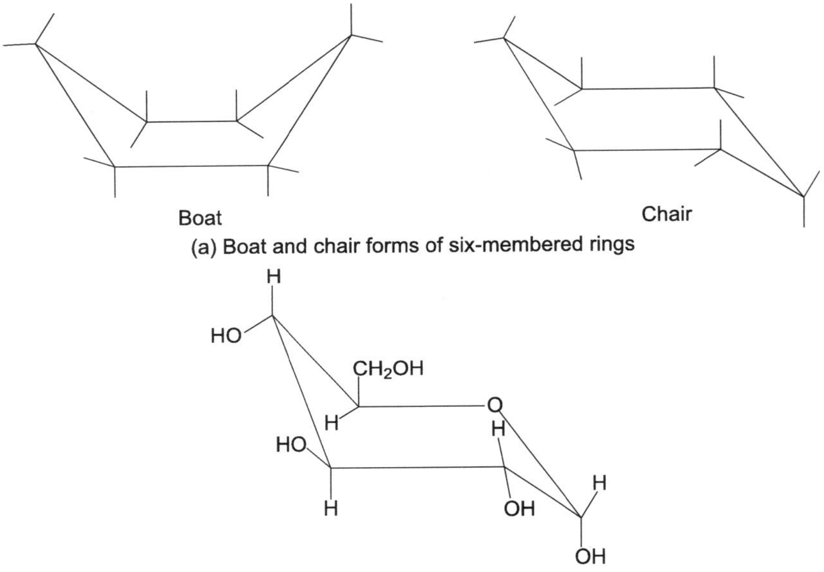

(b) Stable conformation (chair form) of α-D-glucopyranose

Fig. 3.8 *Pyranose rings (in boat and chair forms)*

3.3.2 Disaccharides

Disaccharides are anhydrides of two monosaccharides, which include sucrose, maltose, lactose and cellobiose (Fig. 3.9).

Sucrose Sucrose occurs in sugarcane or sugar beets, and it is a disaccharide of glucose and fructose (Fig. 3.9(a)). Sucrose yields an equimolar mixture of glucose and fructose on hydrolysis. The hydrolysis of sucrose is called the inversion of sucrose because during the reactions, the sign of rotation changes from *dextro* to *levo*. It is an enzymatic reaction by *invertases*. The equimolar mixture of glucose and fructose is called 'invert sugar'. It is sweeter than sucrose in taste. Hence hydrolysis of sucrose is carried out in confectionary industry. Fructose is present naturally in honey.

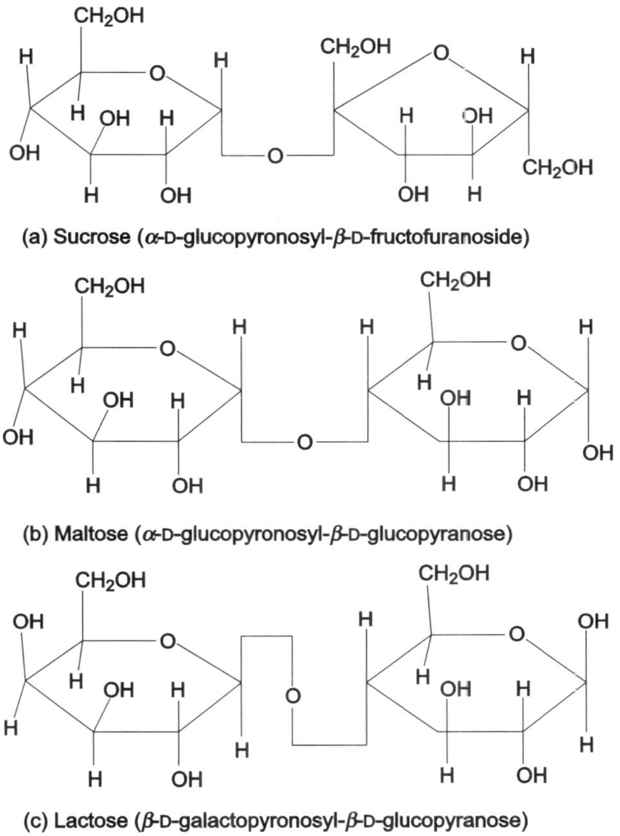

Fig. 3.9 *Structures of sucrose, maltose and lactose*

3.3.3 Oligosaccharides

Oligosaccharides are anhydrides of several monosaccharide residues. Compounds containing ten or less monosaccharide units are called oligosaccharides, while those containing more than ten are called polysaccharides.

3.3.4 Polysaccharides

Polysaccharides are also known as *glycans*. They are anhydrides of one or more monosaccharides in which a large number of units are combined. Many polysaccharides are composed of only one repeating monosaccharide, and even where a large number exists, they appear to be arranged in a systematic fashion. *Homopolysaccharides* possess one repeating unit, while *heteropolysaccharides* contain more than one monosaccharide unit. D-Glucose is the most prevalent monosaccharide unit though polysaccharides of D-mannose, D-fructose, D- and L-galactose, D-xylose and D-arabinose are common. D-glucosamine, D-galactosamine, D-glucuronic acid, N-acetylmuramic acid and N-acetylneuraminic acid are the monosaccharide derivaties which are commonly found in natural polysaccharides.

Starches: Starches are nutritional reservoirs in plants corresponding to $\alpha(1 \rightarrow 4)$ linkages. The repeating disaccharide unit in starch is maltose (Fig. 3.10). Native starches are a mixture of two compounds, viz. amylose and amylopectin. Amylose is an unbranched chain. Amylopectin is a branched chain polysac-

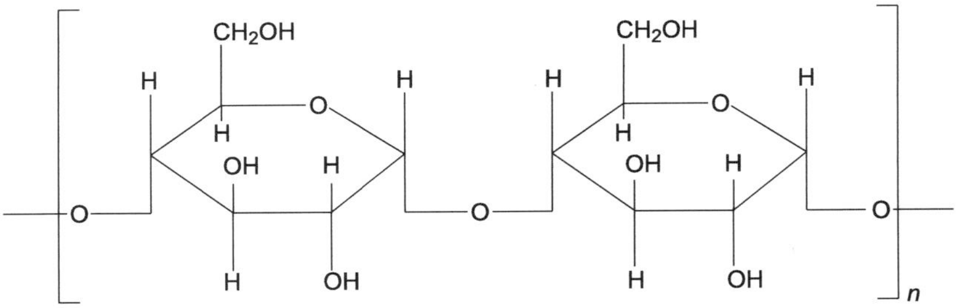

Fig. 3.10 *Repeating maltose unit of starch*

charide (Fig. 3.11). Amylose consists of long unbranched chains in which D-glucose units are bound in $\alpha(1 \rightarrow 4)$ linkages. The molecular weight of amylose varies from a few thousands to 5,00,000 whereas that of amylopectin is of the order of 10^9. It contains about 24–30 glucose residues. The backbone glycosidic linkage is $\alpha(1 \rightarrow 4)$, but the branch points are $\alpha(1 \rightarrow 6)$ linkages.

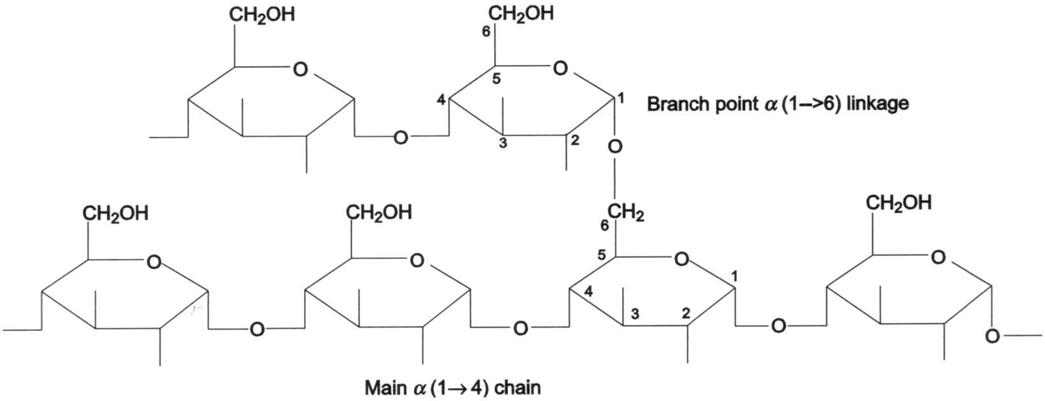

Fig. 3.11 *An $\alpha(1 \rightarrow 6)$ branch point and $\alpha(1 \rightarrow 4)$ main chain of amylopectin*

Cellulose Cellulose is the compound which gives strength to plant tissues. On acid hydrolysis of cellulose, glucose is formed in yields of 95–96%. The cellulose is abundantly available in nature with the purest source being cotton containing at least 90% pure cellulose. It is insoluble in water. Its molecular weight is in the range of 50,000 and 25,00,000 corresponding to 300–15000 glucose residues per molecule. Glucose units are joined in $\beta(1,4)$ linkages (Fig. 3.12).

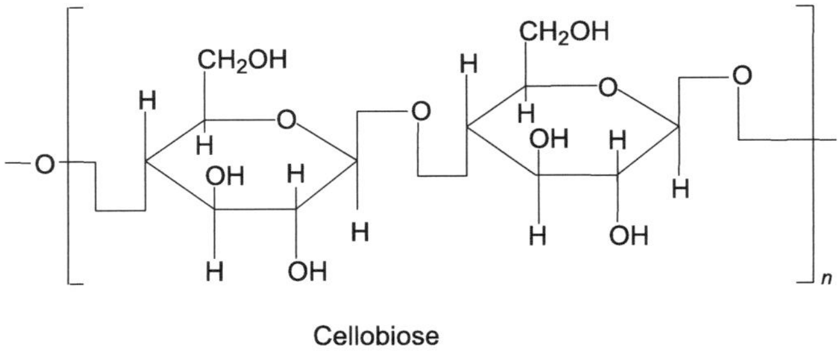

Fig. 3.12 *Repeating cellobiose units of cellulose*

On complete hydrolysis, cellulose yields D-glucose. Partial hydrolysis yields cellobiose. *Cellulase* enzyme can hydrolyze cellulose to D-glucose. But cellulase enzyme is not secreted in the digestive tract of mammals except cows.

3.4 NUCLEIC ACIDS

Nucleic acids are the polymers of nucleotides, which are the units that store genetic information in each cell. Thus, nucleic acids contain all the hereditary information. Hence they are also known as informational biopolymers. When the cells divide, a copy of the parent cell's nucleic acids is passed on to the daughter cell.

3.4.1 Nucleotides

Nucleotides are the building blocks of nucleic acids and, they are also energy carriers. They contain essentially the following three components:

- phosphoric acid,
- ribose or deoxyribose (five-carbon sugars), and
- a nitrogenous base.

The nitrogenous base is derived either from purine or pyrimidine.

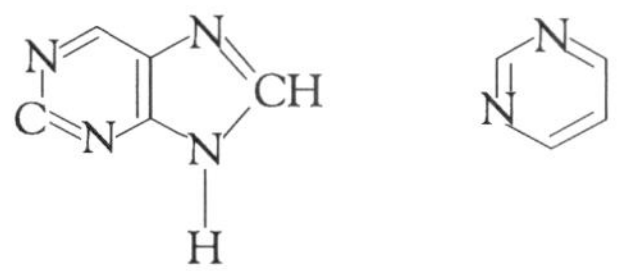

Purine Pyrimidine

The above three components of the nucleotides are joined as shown here.

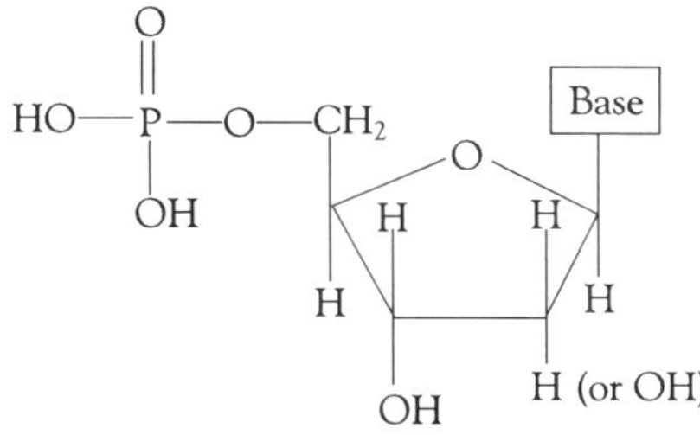

H (or OH)-Ribonucleotide

Deoxyribonucleotide

The five nitrogen bases found in DNA and RNA (shown in Fig. 3.13) are adenine, guanine (purines) and cytosine, thymine and uracil (pyrimidines). Ribonucleic acid (RNA) is a polymer of ribose containing nucleotides while deoxyribonucleic acid (DNA) is a polymer of deoxyribose containing nucleotides.

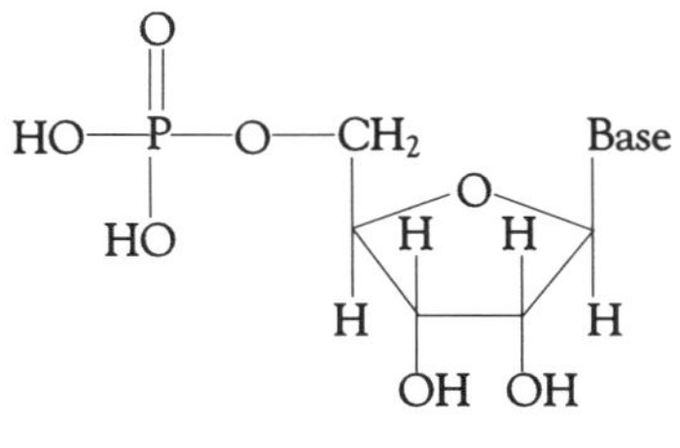

Ribonucleoside 5′– monophosphate

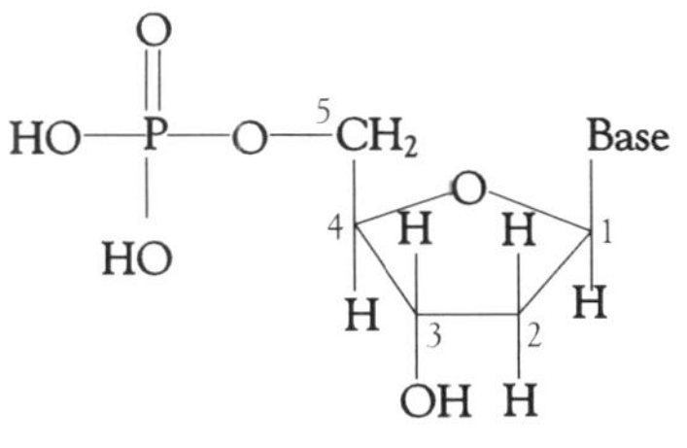

2′– Deoxyribonucleoside 5′– monophosphat

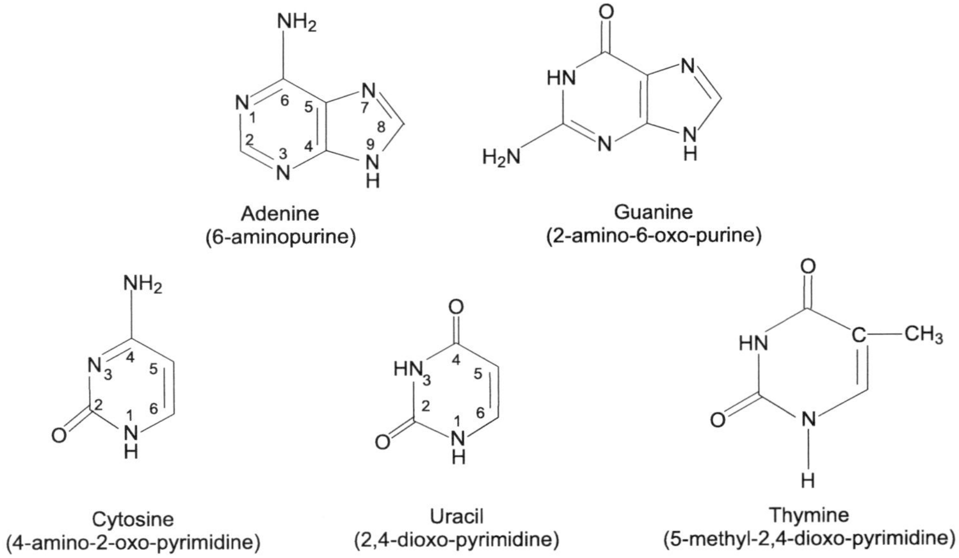

Fig. 3.13 *Five nitrogen bases found in DNA and RNA*

DNA contains 2-deoxy 2-D-ribose whereas RNA contains D-ribose sugars which are in their furanose form in nucleotides. The base in the form of purine or pyrimidine is joined to the pentose sugar by a β- N-glycosyl bond between carbon 1 of pentose and nitrogen atom of the bases. This is known as nucleoside. The phosphate group is joined to the pentose of the nucleoside in the form of the ester at the carbon 5 position. When phosphate group is added to the nucleoside, it is called nucleotide. Thus, nucleotides are the 5′-phosphates of the corresponding nucleosides.

Ribonucleotides and deoxyribonucleotides occur in free form in cells, and most of them are 5′-monophosphates. Some of them also occur as 5′-diphosphates and 5′-triphosphates. Adenosine may occur in all the three forms as adenosine 5′-monophosphate (AMP), adenosine 5′-diphosphate (ADP) and adenosine 5′-triphosphate (ATP). ATP is a carrier of phosphate and pyrophosphate, and is involved in the transfer of chemical energy. ATP and ADP are interconverted by dephosphorylation and rephosphorylation during respiration. ATP–ADP system is responsible for transferring phosphate groups in the cell. Nucleoside triphosphates (NTPs) function as energy-rich precursors of mononucleotide units in the biosynthesis of DNA and RNA.

3.4.2 DNA

Deoxyribonucleic acid (DNA) consists of deoxyribonucleotides linked with covalent bonding (Fig. 3.14). The backbone of DNA consists of alternating phosphate diesters and pentose groups with the covalent bonding in which the continuity is maintained by the phosphodiester bridges. The purines and pyrimidines constitute the side chains. Different sources of DNA molecules differ in terms of the ratio of the four major types of nucleotide monomers. Hence, their molecular weights will also differ. J.D. Watson and F.H.C. Crick postulated the DNA structure with two strands in complementary double helical arrangement in 1953 (Fig. 3.14).

DNA is present in procaryotic cells, eucaryotic cells and bacteria in different forms. In procaryotic cells, it is present in single chromosome while in eucaryotic cells, it is present in different chromosomes as different molecules. Hence its molecular weight will also change considerably. In bacteria, it is generally present

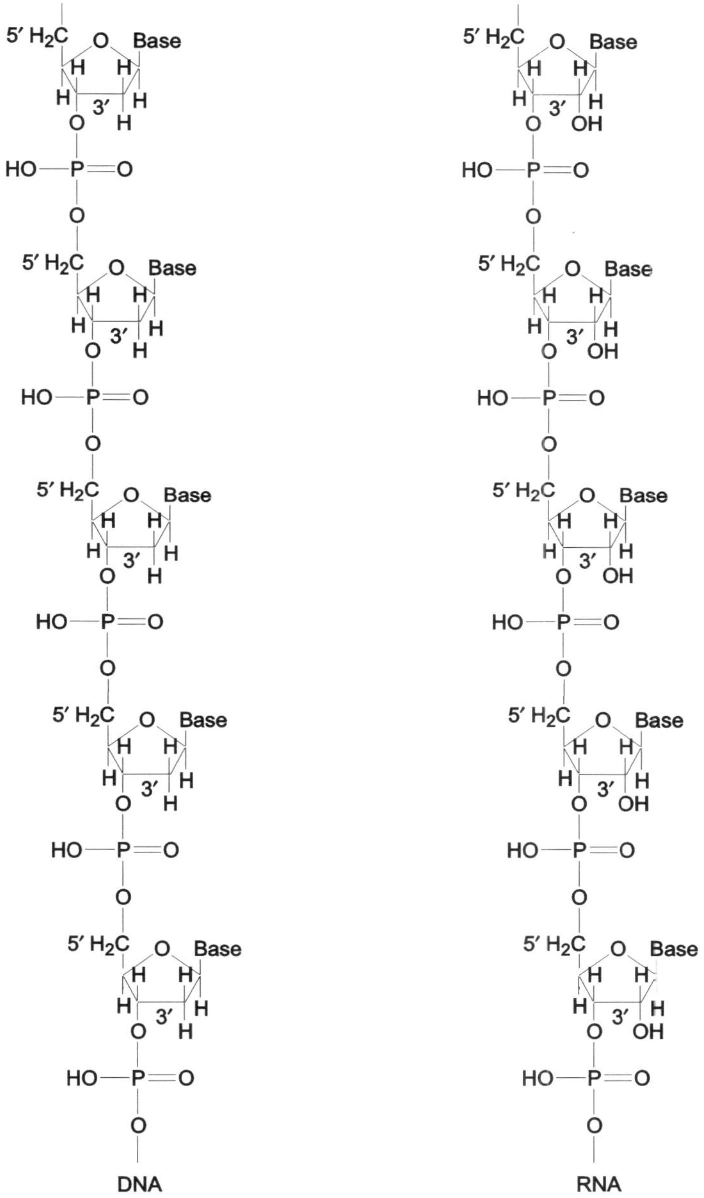

Fig. 3.14 *Structures of DNA and RNA*

in the nucleus. In some cases, it may be present in the cytoplasm in the form of extra-chromosomal DNA. In eucaryotic cells also, in some cases it may be present in the mitochondria; but in such cases its molecular weight and base composition are different from that of the DNA present in the nucleus of the cell.

3.4.3 RNA

Ribonucleic acid (RNA) was discovered in the year 1955 by Frankel-Conrat and Williams, and was assigned a single helical structure. In this respect, it differs from the DNA, which has a double helical structure. Unlike DNA, it contains ribose sugars, and the nitrogen group is provided by uracil.

There are three major types of RNAs, namely

- messenger RNA (mRNA),
- ribosomal RNA (rRNA), and
- transfer RNA (tRNA).

Each type of RNA has a characteristic range of molecular weights in between 25,000–11,00,000. Each RNA occurs in multiple molecular forms. RNA is present in the cytoplasm in bacteria. Mitochondrial rRNAs and tRNAs are different from the extramitochondrial forms.

3.5 VITAMINS

Vitamins are a group of organic compounds which are required in trace amounts to stimulate body growth and also to keep away certain types of diseases. They are also present in certain enzymes to assist them in their catalytic activity. Hence they are also known as coenzymes. Vitamins are present in living cells in addition to lipids, proteins, carbohydrates, nucleic acids, etc. Since most of the vitamins contain amines, the name vitamin was derived from *Vitamine* which indicated amines required for health. Subsequently some vitamins were found not to contain amines, but the name Vitamin was retained.

Vitamins are classified into two groups, viz. water-soluble and fat-soluble. Table 3.6 gives a list of the various types of vitamins, their sources of availability, and the diseases and other affects caused by their deficiency. Their structural formulae are presented in Figs 3.15–3.22.

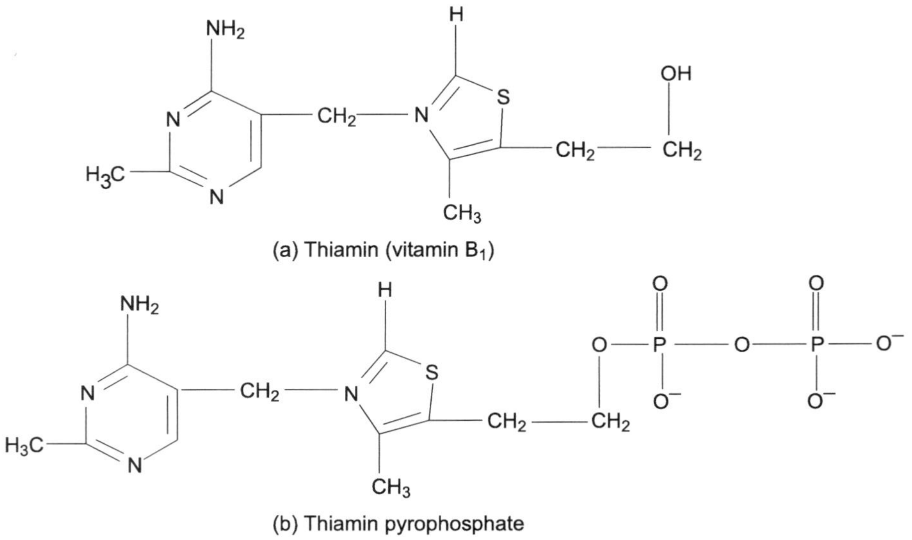

Fig. 3.15 *Structures of vitamins B_1*

Table 3.6 *Different vitamins, their sources of availability and results of their deficiency/diseases caused.*

Vitamin	*Source*	*Structural formula*	*Remarks*
WATER-SOLUBLE VITAMINS			
Ascorbic acid (Vit-C)	Citrus fruits, amla, tomatoes, several mammals	Fig. 3.22	Deficiency causes cough and cold, and the disease 'scurvy'
Biotin	Yeast cells and liver	Fig. 3.22	Important for growth, protects animals from toxicity caused by avian egg-white
Folic acid	Spinach leaves, green leafy vegetables and most plants	Fig. 3.18	Deficiency causes anemia, and impairs biosynthesis of purines and pyramidine
Lipoic acid (thioctic acid)	Liver extract, yeast cells	Fig. 3.22	An important oxidizing co-enzyme in pyruvate oxidase
Nicotinic acid (niacin)	Nicotin in tobacco, meat, eggs and milk	Fig. 3.22	Deficiency leads to disease 'pellagra'
Panthenic acid	Synthesized by plants and some microbes	Fig. 3.22	Required as a growth factor by vertebrates
Pyridoxine (Vit-B_6)	Present in several plants and animals, cereals	Fig. 3.17	Used in metabolism of various amino acids. Deficiency casues dermatitis
Riboflavin (Vit-B_2)	Present in eggs, milk and animal tissues. Synthesized by plants and several micro organisms	Fig. 3.16	Flavoenzymes function in oxidative degradation of pyruvate, fatty acids and amino acids
Thiamin (B_1)	Cereal grains, many plants, animal tissues	Fig. 3.15	Deficiency causes disease 'beriberi'
Vitamin B_{12} (Cyanocobalamine)	Human blood, liver		Deficiency causes pernicious anemia. It is essential for maturation and development of erythrocyte
Inositol	Animal tissues		Required for hair growth in mice
Carnitine	Muscle tissues		It is an essential growth factor
Vitamin-A	Widely distributed in plants, fish and some mammals	Fig. 3.19	Available in 2 forms Vitamin A1 and Vitamin A_2. Deficiency causes blindness, young animals fail to grow, development of bones and nervous system get affected, skin becomes dry, kidney and some glands degenerate or become sterile.
Vitamin-D	Fish liver oil. Commonly found in several plants, human skin, various mammals	Fig. 3.20	Deficiency results in rickets, affects bones and causes bow-legs. Excess causes bones to become fragile
Vitamin-E (tocopherol)	Wheat germ, vegetable oils	Fig. 3.22	Known for its anti-oxidant activity. Deficiency causes infertility of the liver, degeneration of kidneys, necrosis, muscular dystrophy in animals.
Vitamin-K	Green plants, several bacteria.	Fig. 3.21	Deficiency causes blood coagulation, affects blood clotting, results in failure of biosynthesis of the enzyme "proconvertin" in the liver.

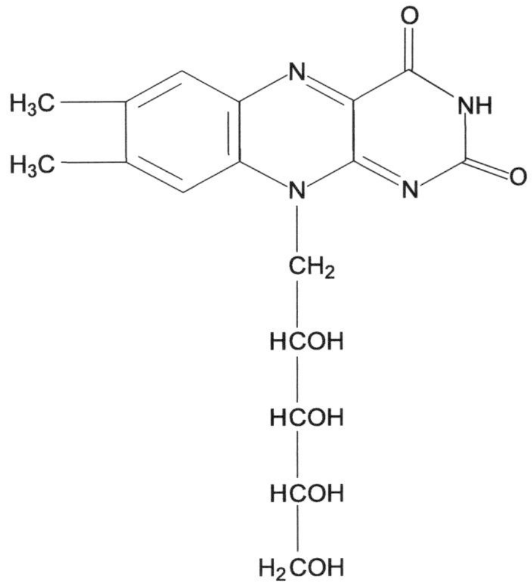

Fig. 3.16 *Structures of vitamins B_2 (Riboflavin)*

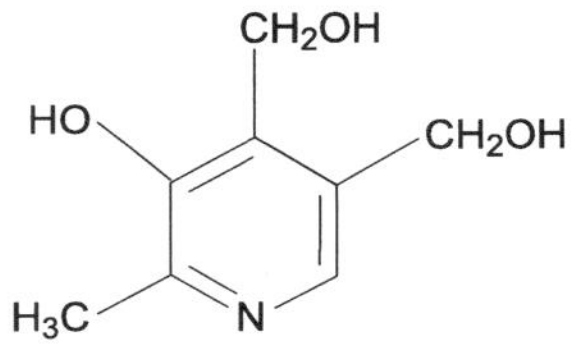

Pyridoxine

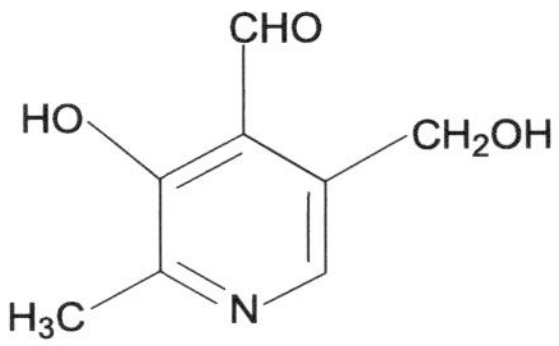

Pyridoxal

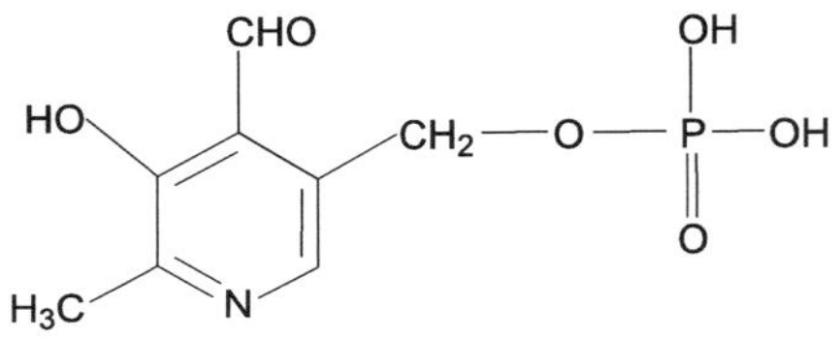

Pyridoxal phosphate

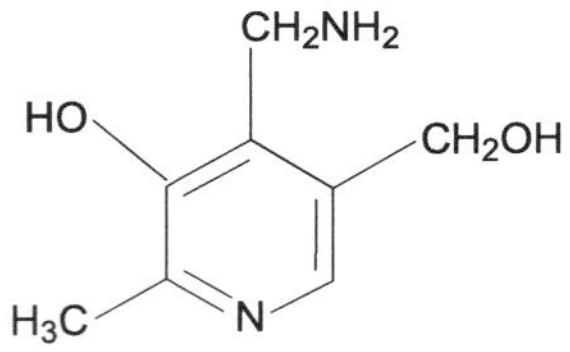

Pyridoxamine

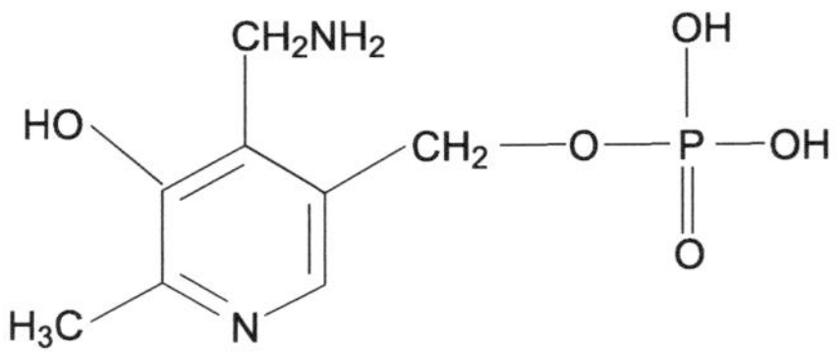

Pyridoxamine phosphate

Fig. 3.17 *Structures of pyridoxine, pyridoxal, pyridoxal phosphate, pyridoxamine and pyridoxamine phosphate*

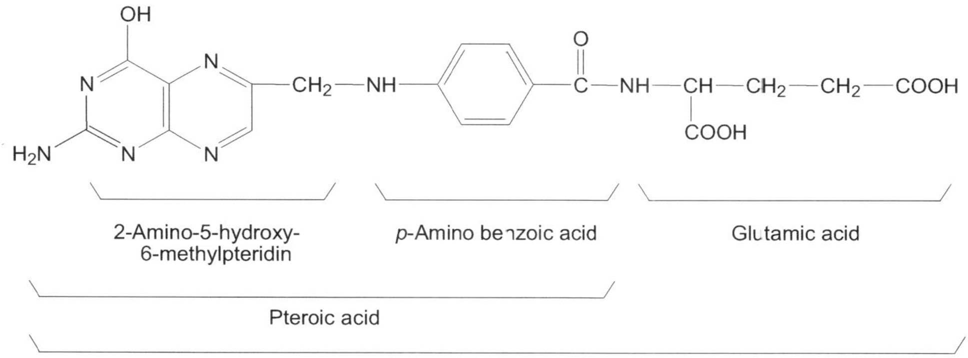

(a) Folic acid

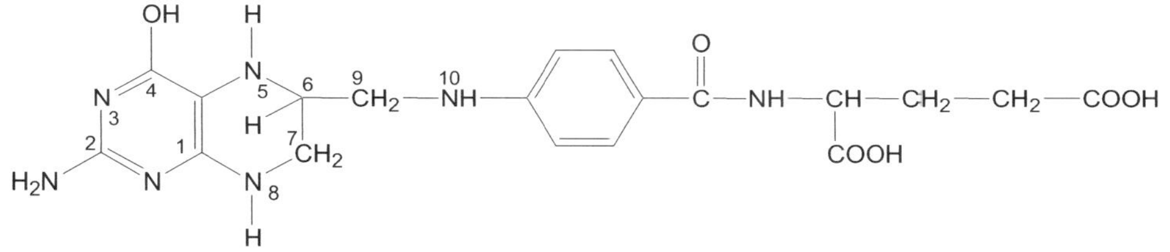

(b) Tetrahydrofolic acid

Fig. 3.18 *Structures of (a) folic acid and (b) tetrahydroflic acid*

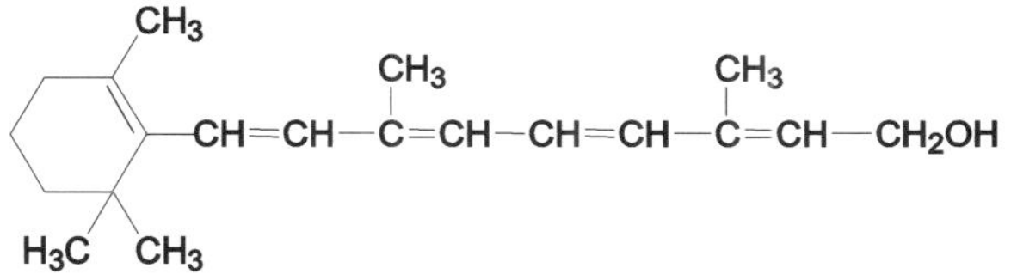

(a) Vitamin A_1 (Retinol)

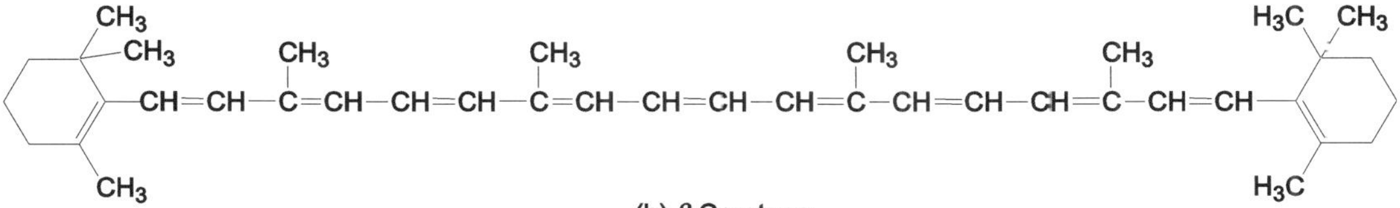

(b) β-Carotene

Fig. 3.19 *Structures of (a) vitamin A_1 (retional) and (b) β-Carotene*

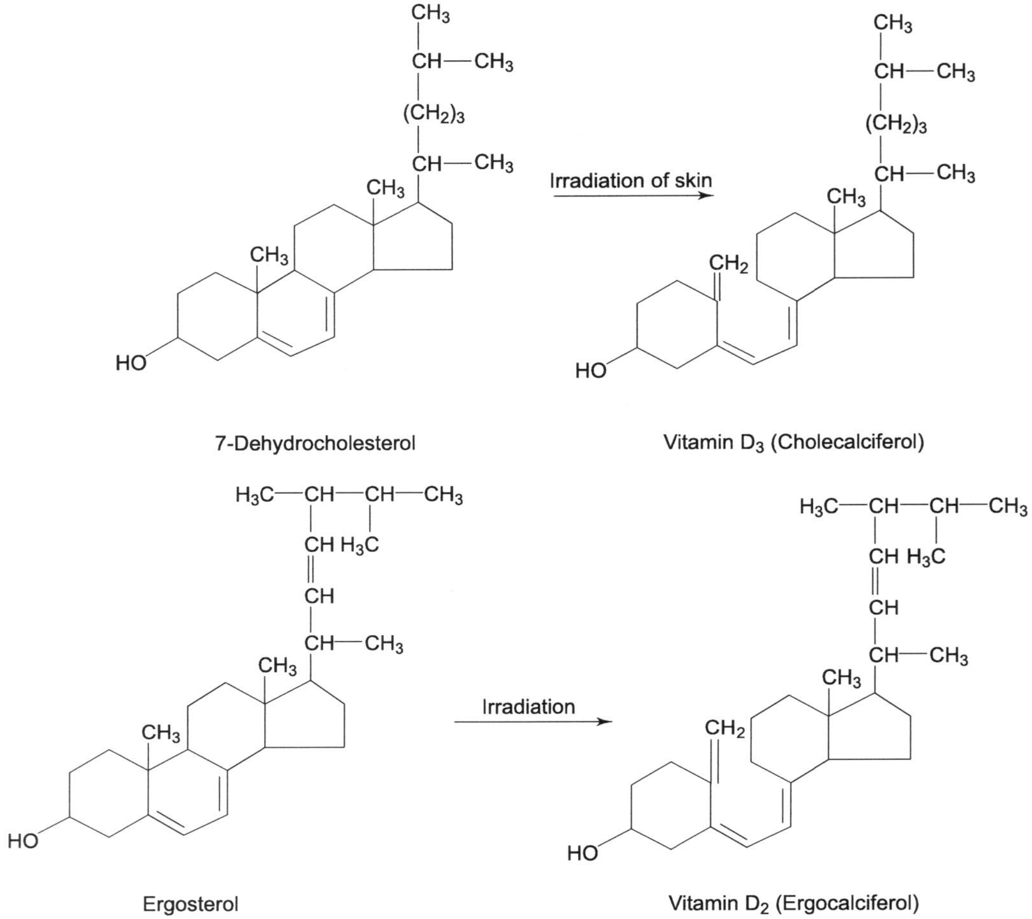

Fig. 3.20 *Structures of vitamin D_2 and D_3*

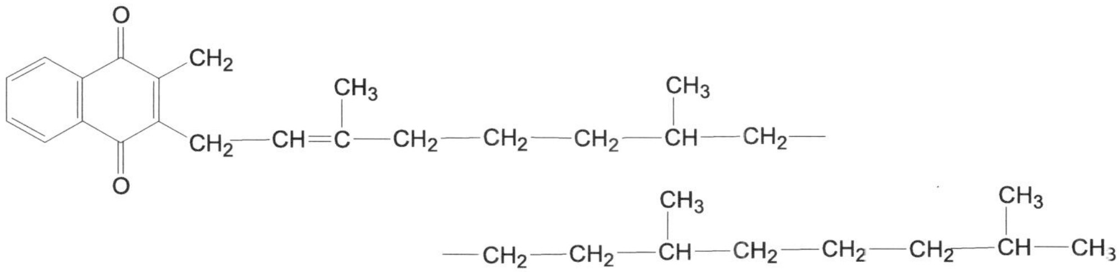

Vitamin K_1

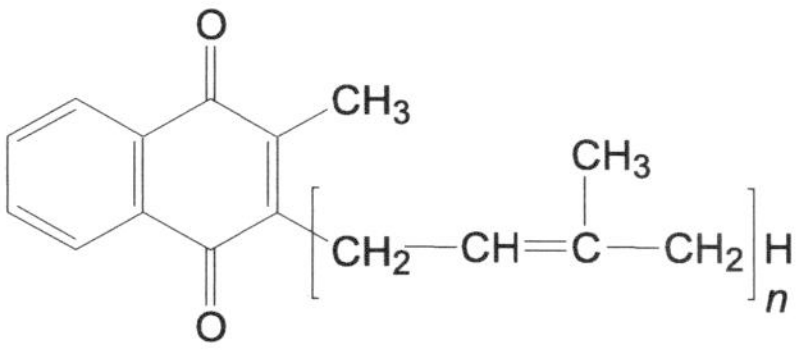

Vitamin K_2 (n = 6, 7, 8, 9, or 10)

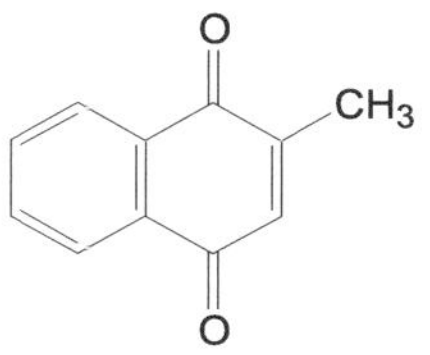

Vitamin K_3 (menadione)

Fig. 3.21 *Structures of vitamin K_1, K_2 and K_3*

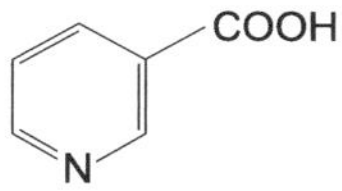

Nicotinic acid
(Niacin)

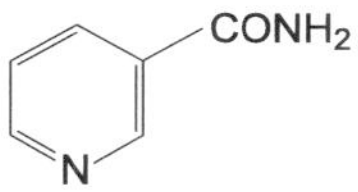

Nicotinamide

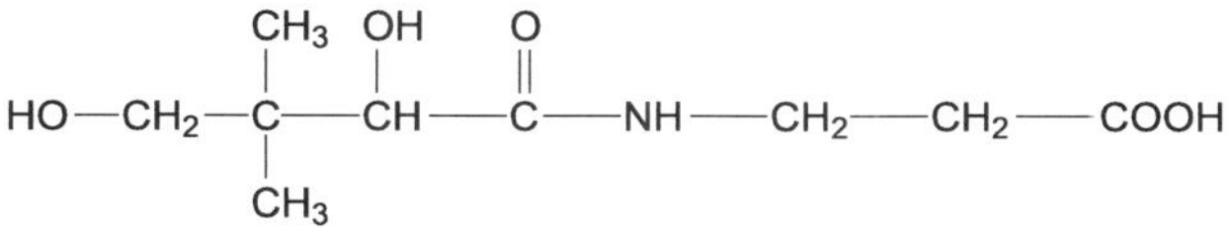

Pantothenic acid

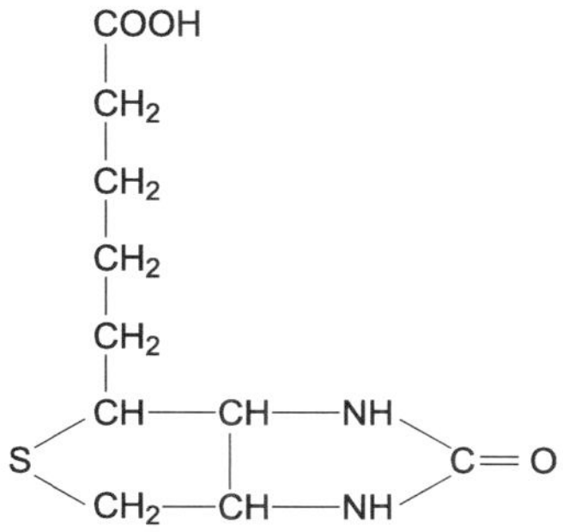

Biotin

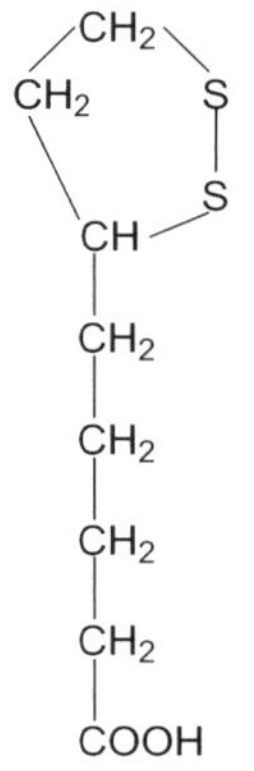

Lipoic acid

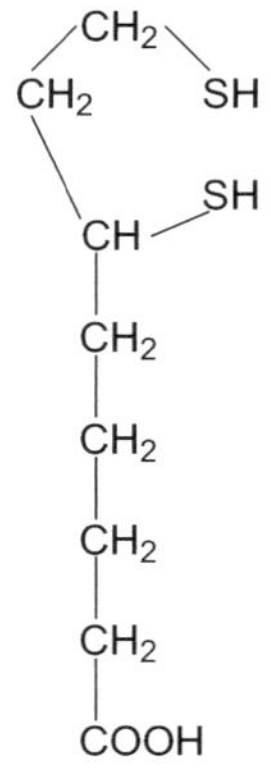

Dihydrolipoic acid

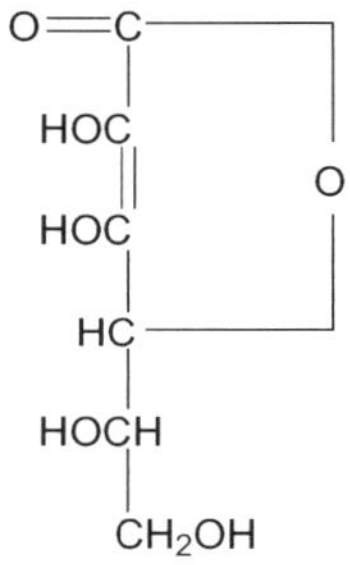

L-Ascorbic acid

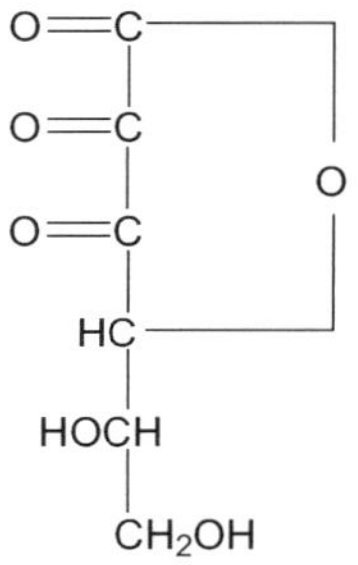

L-Dehydroascorbic acid I
(active)

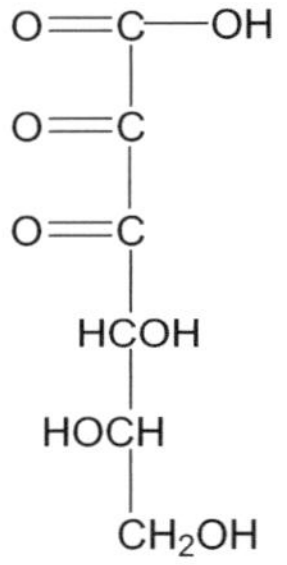

L-Diketogulonic acid
(Inactive)

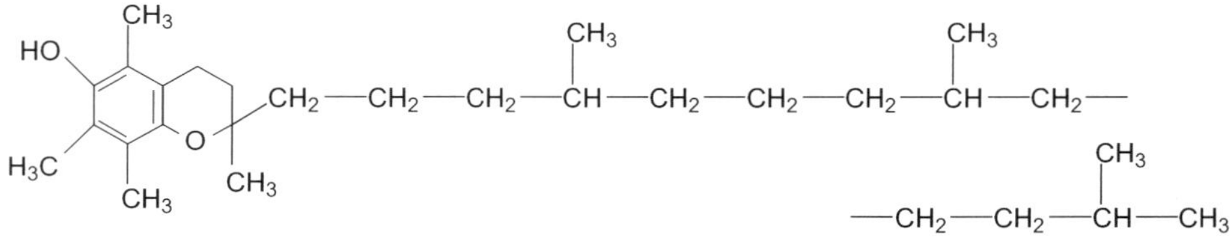

Vitamin E(α-tocopherol)

Fig. 3.22 *Structures of some vitamins and related compounds*

References

3.1 Bailey, J.E. and Ollis, D.F. (1986), *Biochemical Engineering Fundamentals*, 2nd Edn, McGraw-Hill, Inc., New York, pp.27-70.

3.2 Lehninger, A.L. (1983), *Biochemistry*, Worth Publishers, Inc., New York.

3.3 Trehan, K. (1987), *Biochemistry*, Wiley Eastern Ltd, New Delhi.

3.4 White, A., Handler, P. and Smith, E.L. (1968), Principles of Biochemistry, 4th Edn, McGraw-Hill, Inc, New York.

Review Questions

3.1 Write a short note on biomolecules.
3.2 What are various types of lipids ?
3.3 What are saturated and unsaturated fatty acids? Explain.
3.4 Write a note on phospholipids.
3.5 Explain some non-glycerol lipids.
3.6 Write a note on sphingomyelins with their structure.
3.7 Describe some terpenes.
3.8 What are steroids ?
3.9 Write a note on proteins.
3.10 What are amino acids? Describe some of them.
3.11 Describe some acid-base properties of amino acids.
3.12 Describe some methods for analysis of amino acids.
3.13 How do you separate and purify proteins ?
3.14 Describe how to separate proteins based on their isoelectric points.
3.15 How are proteins separated using selective adsorption techniques?
3.16 Write a note on nucleic acids.
3.17 Write a short note on (i) DNA and (ii) RNA.
3.18 Write a note on vitamins.
3.19 Describe water-soluble and fat-soluble vitamins.

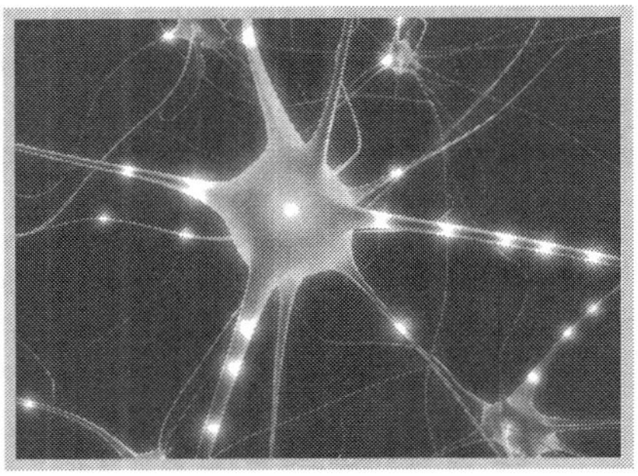

chapter 4

Enzymes

Biochemical reactions occur rapidly through the mediation of natural catalysts called enzymes. They are bio-catalysts present in living matter. A catalyst is defined as a substance that accelerates a chemical reaction but is not consumed in the overall process. The catalytic efficiency of enzymes is extremely high. Pure enzymes catalyze the transformation of 10,000 to 10,00,000 moles of substrate per minute per mole of enzyme (White, et al, 1968). Enzymes have a high degree of specificity and efficiency, which transforms the substrates into the products. Enzymes are ubiquitous, and enzyme reactions have been used by man since long ago. Making of cheese, leavening of bread, manufacture of vinegar are enzymatic reactions. The name enzyme was coined by W. Kiihne, which means "in yeast", but the word is now used to mean **biological catalyst** regardless of their origin.

Thus, enzymes are biological catalysts used to increase the rate of biochemical reactions taking place within living systems, without themselves undergoing any overall change. All enzymes are an important class of conjugated proteins. The following are a few generic terms associated with enzymology.

Cofactor: the non-protein component of an enzyme.

Coenzyme: an enzyme with organic molecule as its cofactor.

Holoenzyme: an active enzyme including cofactor.

Apoenzyme: the inactive portion of protein.

In some enzymes, metal may be conjugated to the protein part.

4.1 HISTORY OF ENZYMES

For the first time, the word 'catalyst' was coined by Berzilius in 1836 for the materials whose presence would set the reactions to go which otherwise would be dormant. He ascribed the catalytic power to be something mysterious. The concept prevailed until 1926 when Willstaetter first could achieve the isolation of pure enzyme peroxidase. Almost during the same period Sumner could successfully isolate and concentrate urease from jack beans, which had a catalytic effect in cleaving urea into carbon dioxide and ammonia.

Subsequently a large number of enzymes has been isolated from time to time. By 1985, as many as 2500 enzymes were known out of which 250 were marketable (also see Section 5.2.5).

4.2 CLASSIFICATION OF ENZYMES

Nomenclature Enzymes are usually named in terms of the reactions that are catalyzed. Customary practice is the addition of '-ase' at the end (suffix) to the major part of the name of the substrate. Enzyme that attacks urea is urease, argenine is arginase, etc. Enzymes were also classified by groups that catalyze similar chemical reactions such as proteases, lipases, oxidases, etc. The only major exception is trypsin. Lack of consistency in the nomenclature, and list of known enzymes rapidly growing up have necessitated the need for a systematic way of naming and classifying enzymes. A commission was appointed by International Union of Biochemistry which has published its report first in 1964, and later updated. All the enzyme groups are essentially classified into two groups, viz.

(i) hydrolyzing enzymes or hydrolases
(ii) other enzymes or desmolases.

Enzyme Commission's Classification The classification based on Enzyme Commission (EC) nomenclature groups enzymes into six major classes. A code number, given to each enzyme, which consists of four digits is separated by dots. First digit shows the main classes of the enzyme (Table 4.1). The second and third digits in the code further describe the kind of reaction being catalyzed. Thus enzymes will have a systematic name, a trivial name which was generally being used before 1961, and the EC's code number. For example, alcohol dehydrogenase is represented as EC.1.1.1.1.

Isoenzymes are different enzymes catalyzing identical reactions, and will have the same four figure classification. For example, lactate dehydrogenase is present in the human body in the form of five different isoenzymes, and they all will have an identical code. Thus each enzyme will have, in addition to its systematic name, an existing trivial name. This systematic name, as it is, will indicate the type of substrate it catalyses. To facilitate easy comprehension, the substrate name is included with an ending *-ase*. Thus the enzyme that catalyzases the breakdown of urea is known as *urease*. The systematic name and the enzyme commission (EC) classification number describe the reaction being catalyzed by enzyme. Generally, it is convenient to use the trivial name, but wherever there is a possibility for confusion, we may prefer to use the systematic name or the EC number. Accordingly, enzymes have been classified into six classes which is shown in Table 4.1. For more details on classification procedure, any standard textbook of enzymes may be consulted (Palmer, 2001).

4.3 ENZYMES AS BIOLOGICAL CATALYSTS

As has been already mentioned, enzymes are biological catalysts. We will see now how the biological catalysts work in making a particular biochemical reaction proceed. Let us first see how a reaction proceeds. There are two possible important theories:

(i) collision theory
(ii) transition state theory

Collision theory proposes that reactions take place by the collision of the reactant molecules. That is, more is the concentration of the reactants, more are the chances for the reactants to collide, and hence more will be the rate of reaction. But it is imperative that all collisions may not necessarily result in the reaction to proceed to produce products.

The **transition state theory** proposes that the collision of certain molecules, which have crossed certain potential energy barrier, alone will result in the reaction to take place. This potential energy barrier is known as activation energy (Fig. 4.1). The biocatalysis of the enzymatic reactions can be explained by the concept

Table 4.1 *Enzyme nomenclature and different types of reactions they catalyse*

Class No.	*Enzyme class*	*Reaction catalysed*	*2nd digit*	*Group being catalysed*	*Remarks*	*3rd digit*	*Group being catalysed*	*Remarks*
1	Oxido reductases	Oxidation/ reduction reactions	1	alcohol (>CHOH)	Hydrogen/ electron donor	1	NAD^+ or $NADP^+$	Hydrogen/ electron acceptor
			2	aldehyde/ketone (>C=O)		2	Fe^{3+}	
			3	HC=CH		3	O_2	
			4	primary amine				
			5	secondary amine				
			6	NADH or NADPH				
2	Transferases	Transfer of an atom (or) group between two molecules	1	1-carbon group	Second digit indicates type of group transferred	1	Methyl transferases	Third digit describes the group transferred
			2	>C=O		2	Hydroxy methyl transferases	
			3	acyl group—C(=O)—R		3	Carboxyl or carbamoyl transferases	
			4	glycosyl group				
			5	phosphate group				
			6	sulphur containing group				
3	Hydrolases	Hydrolysis reactions	1	ester	Hydrolytic reactions	1	Carboxylic ester	Bond being hydrolysed
			2	glycoside		2	Thiol ester	
			3	peptide—C(=O)—NH		3	Phosphoric monoester	
			4	other CN bonds		4	Phosphoric diester	
4	Layes	Removal of a group from substrate (not by hydrolysis)	1	>C=C<	Bond being broken	1	Carboxyl group	Group being removed
			2	>C=O		2	Hydroxyl	
			3	>C=N–		3	Keto acidc	
			4	C–S				
5	Isomerases	Isomerisation reactions	1	Racemisation or epimerisation	—	1	Amino acids	Type of molecule undergoing isomerisation
			2	cis-trans isomerisation		2	Hydroxyl acids	
			3	intramolecular oxido reactions		3	Carbohydrates	
6	Ligases	Synthetic joining of two molecules	1	C—O	Bond being synthesized	1	Amide	Bond being formed
			2	C—S		2	Peptide	
			3	C—N				
			4	C—C				

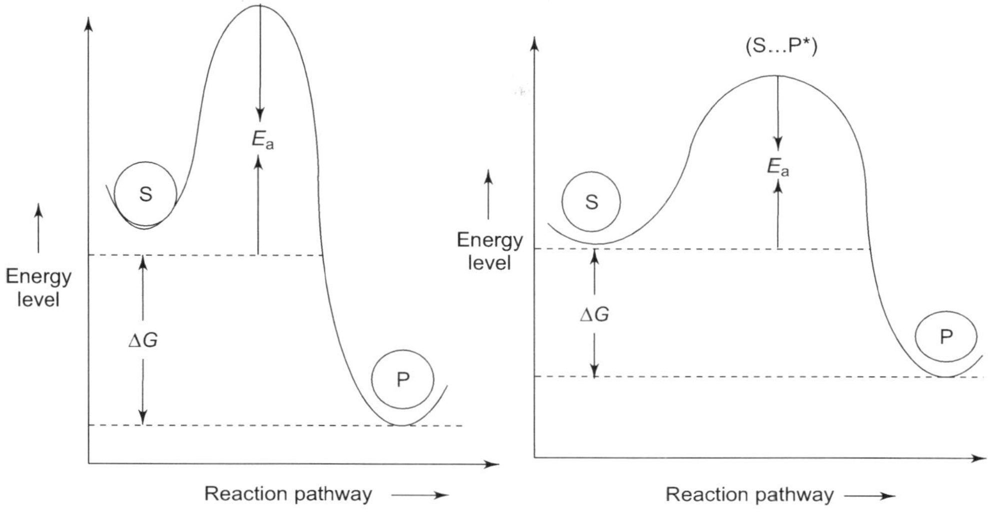

(a) Non-enzymatic process (b) Enzymatic process

ΔG and E_a (activation energy) of non-enzymatic and enzymatic reaction systems

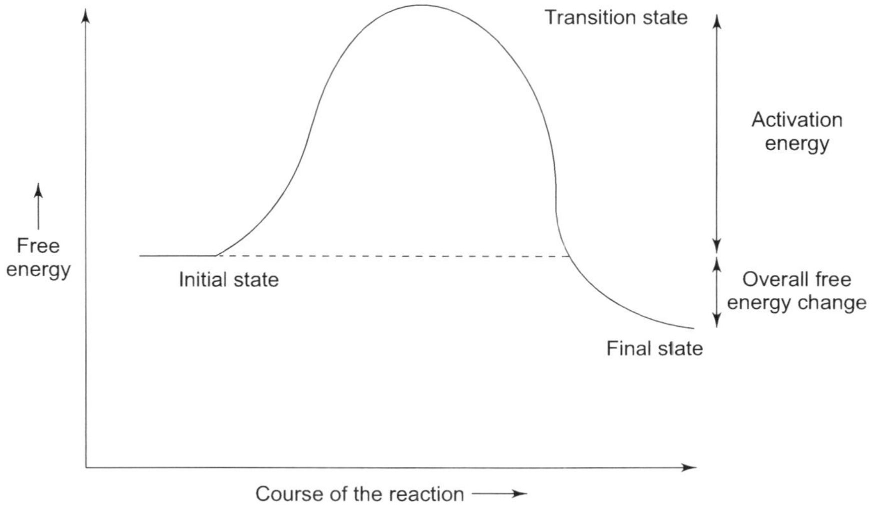

Fig. 4.1 *Free energy change for favourable reaction proceeding via formation of a transition state*

that the enzymes form with the substrate an intermediate complex which further decomposes to form the bioproducts (Fig. 4.2). For the reaction

$$S \rightarrow P$$

where S is the substrate and P is the product, the transition state proposes a halfway point and unstable intermediate complex where the bonds of S are distorted sufficiently so that conversion to P becomes possible. The rate of reaction obviously depends on the number of molecules of S that enter the transition state per unit time. Therefore the rate of reaction can be increased by

(i) raising the temperature, or
(ii) lowering the activation energy.

The enzyme complexes would selectively lower the energies of activation of vital biochemical reactions. Thus, the enzymes can carry out the reactions at ambient temperatures. This makes it possible for an enzyme-catalyzed reaction at 25°C to proceed at 10^6–10^{15} times faster as compared to an uncatalyzed reaction. The kinetics of enzymatic reactions are described in Sec. 4.5. The methodology to evaluate activation energy using Arrhenius equation was dealt in Chapter 6.

4.4 ENZYME SPECIFICITY

Enzymes as biocatalysts are particularly credited for their specificity in action. Each enzyme usually catalyzes only one particular type of reaction by selectively lowering the activation energy of only one of the several possible chemical/biochemical reactions. Thus, only stereospecific groups of the substrate alone undergo the desired reaction to form the desired product.

4.4.1 Types of Specificity

Depending upon the reaction conditions and the specific nature of the enzymes, the enzymatic catalytic process exhibit different kinds of specificity, viz.

- absolute specificity
- group specificity
- stereochemical specificity
- product specificity
- substrate specificity, etc.

Alcohol dehydrogenase is a group specific enzyme which catalyzes the oxidation of a variety of alcohols. In this case, the oxidation or dehydrogenation takes place on the alcohol group irrespective of the type of alcohol. Those enzymes which act on a particular substrate are known as 'absolute specific enzymes'. Glucokinase catalyses the transfer of phosphate from ATP to glucose. In general Fig. 4.2 shows how enzymes catalyze the conversion of substrate to product by forming an 'enzyme–substrate' complex.

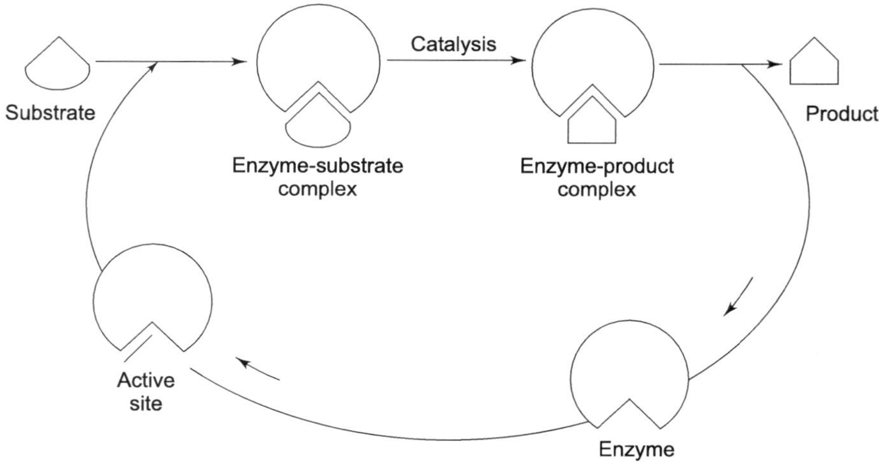

Fig. 4.2 *Enzymatic process of conversion of substrate to product*

Unlike uncatalyzed reactions, which often produce a range of products, enzyme catalyzed reactions are 'product specific' as well as 'substrate specific'. Some enzymes even exhibit stereospecificity, i.e., if a substrate exists in two stereochemical forms but are chemically identical, the enzyme will specifically catalyze only one of the isomers. A typical example is oxidation of *L*-amino acids to oxo acids which can take place by enzymatic catalytic action of *L*-amino acid oxidase; whereas oxidation of *D*-amino acids can be brought about only by the enzyme *D*-amino acid oxidase.

4.4.2 Active Site

Active site concept is sometimes useful to explain the stereochemical specificity of enzymes. An enzyme is conceived to possess two different kinds of site, viz.

(i) binding sites
(ii) catalytic sites

The substrate-enzyme interaction can be at the binding sites or catalytic sites (Fig. 4.3).

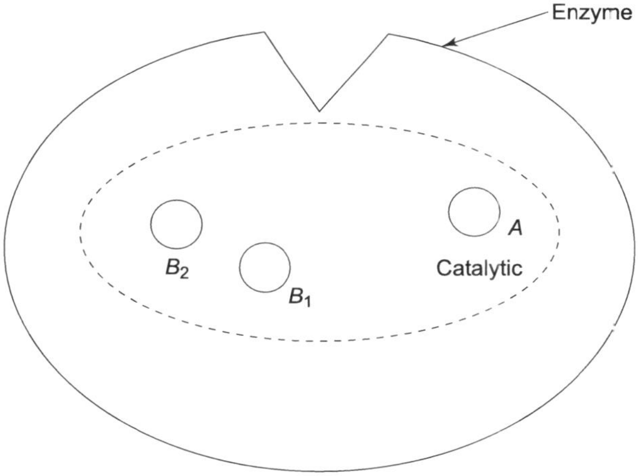

B_1 and B_2 are binding sites. A is the catalytic site

Fig. 4.3 *Active and catalytic sites in enzyme*

The active site or **active centre of the enzyme** is the region which contains the binding and catalytic sites which are only a small fraction of the total volume of the enzyme. This is situated at or near the surface of the enzyme and is thus easily accessible for substrate molecules to get attached.

The function of an enzyme depends upon the arrangement of binding and catalytic sites. The environment in which these sites occur also matters as far as enzyme interaction/specificity is concerned.

4.4.3 Enzyme Specificity Hypotheses

Several hypotheses have been proposed from time to time to explain the enzyme specificity in catalytic activity and its ability to interact with the substrates. The following are the few very prominent hypotheses (Palmer, 2001 (a)).

(i) Fischer lock-and-key hypothesis
(ii) Koshland induced fit hypothesis
(iii) Hypotheses involving strain or transition-state stabilization.

Fischer lock-and key hypothesis is probably the first theory proposed in 1890 by Fischer who conceived the concept of '**complementary structural features**' between the enzyme and the substrate. The catalytic process is brought about because the substrate fits into the complementary site on the enzyme just as a key fits into the lock. Thus, the reacting group of the substrate gets struck with the catalytic site of the enzyme. Similarly, the binding groups attach to the binding sites in the enzyme as shown in Fig. 4.4.

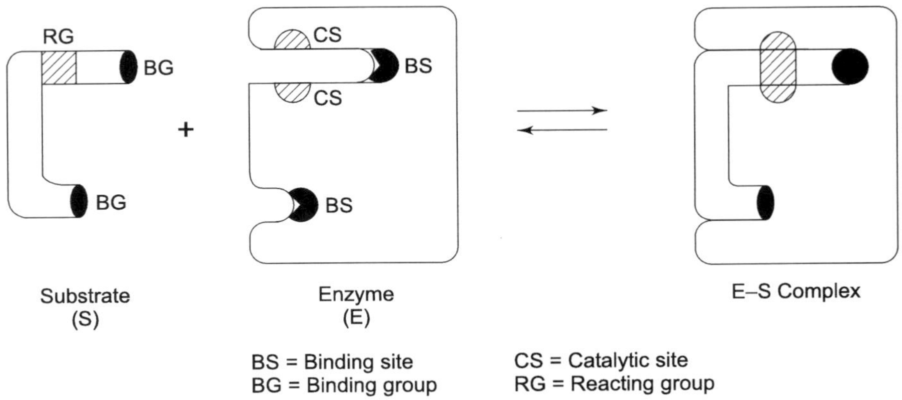

Fig. 4.4 *Interaction between an enzyme and its substrate to form an Enzyme-Substrate complex (lock-and-key hypothesis)*

The lock-and-key hypothesis has been successful in explaining many features of the enzyme specificity; but one serious drawback with this hypothesis is that it could not explain some of the conformational changes taking place in the enzymes when they come in contact with the substrate. An enzyme may not be having as an exactly complimentary feature that is compatible to the substrate, but still there are instances when reactions have taken place; particularly with proteins which are known for their flexibility. X-ray diffraction analysis and some spectroscopic analysis have shown differences in the structures of the free enzymes and substrate bound enzymes. This was explained by Koshland in 1958 with his induced fit hypothesis which is popularly known as **Koshland induced-fit hypothesis**. This hypothesis proposes that the structure of the substrate may not be complimentary to the enzyme in its native form, but it is complimentary to the active site in the substrate-enzyme complex. When the substrate and enzyme are brought together, the binding of the substrate brings in structural changes in the enzyme so that the catalytic sites in the enzyme and the reacting group in the substrate match each other (Fig. 4.5). This is explained with a hand-and-glove example. The

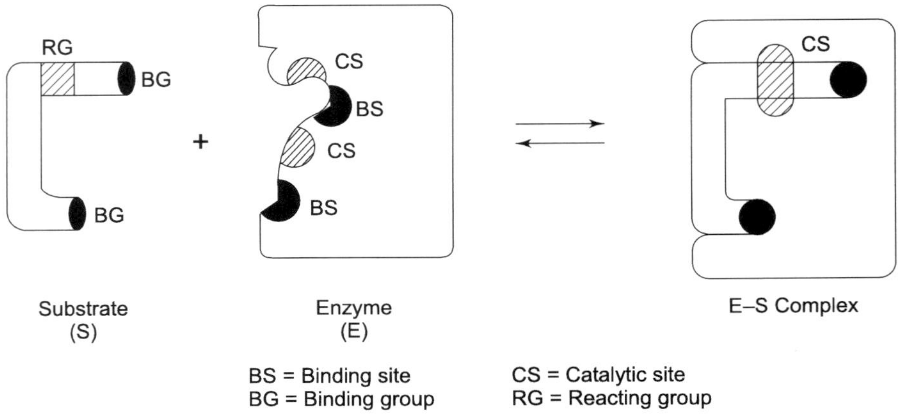

Fig. 4.5 *Interaction between an enzyme and its substrate according to the induced-fit model to form Enzyme–Substrate complex.*

substrate is like a hand which is rigid and does not change its shape, but the enzyme is like a woolen glove which can be wrapped around the hand, and just as it (woolen glove) adjusts itself to wrap the hand, so does the enzyme. The reaction,

$$\text{D-hexose} + \text{ATP} \rightleftharpoons \text{D-hexose-6-phosphate} + \text{ADP}$$

which is catalyzed by yeast hexokinase, is a typical example of induced fit hypothesis.

4.5 ENZYME KINETICS

Kinetics is the study that tells us about how fast does a reaction proceed. In this section, we study the fastness of the reactions in a homogeneous phase.

Reactions can take place in two phases:

- homogeneous phase
- heterogeneous phase.

The concept of the reaction phases is borrowed from catalysis. If the reactants and catalyst are in the same phase, it is called as 'homogeneous phase', and if the reactants and catalyst are in two different phases (viz. the reactants in the liquid or gas phase and the catalyst in the solid phase), it is called as 'heterogeneous phase'. Similarly, the **enzymatic** reactions are considered analogous to catalytic reactions. An **enzyme** is a protein or a protein-like substance with catalytic properties. A substrate, S is equivalent to the chemical reactant whose reaction rate is accelerated due to the action of enzymes. Based on the physical state of the enzyme in the fermentation, there are three possible enzymatic reactions.

Type I : Soluble enzyme–Insoluble substrate
Type II : Insoluble enzyme–Soluble substrate
Type III : Soluble enzyme–Soluble substrate

In this section, we will study only Type III reactions, which are known as **homogeneous reactions**. However, it may be noted that the equations developed for Type III reactions are also equally applicable to the other two types as well.

The rate of a reaction is normally proportional to the concentration of all the reactants or one of the reactants, as indicated by Eq. (7.2). Thus, we write the rate equation for the following reaction scheme with the first-order dependence:

$$\text{A} + \text{B} \xrightarrow{k_1} \text{C} + \text{D} \tag{4.1}$$

$$-r_\text{A} = -r_\text{B} = r_\text{C} = r_\text{D} \tag{4.2}$$

$$-r_\text{A} = k_1[\text{A}]\,[\text{B}] \tag{4.3}$$

The concentration of any component, say A, is generally expressed as [A], i.e., A in square brackets. Keeping the component in square brackets is a usual notation used for expressing its concentration, which is generally followed in physical chemistry/chemical reaction engineering. Thus, [A] indicates the concentration of A, and [B] indicates the concentration of B. Equation (4.3) can also be written as

$$-r_\text{A} = k_1\,C_\text{A}\,C_\text{B}$$

where C_A and C_B are concentrations of A and B, respectively.

4.5.1 Searching for Mechanism

In the enzymatic reactions, the enzyme E forms a complex with the substrate S and participates in the reaction. Fogler (1986) explained this with an example of catalytic activity of urease during the removal of urea from the bloodstream. The enzyme converts urea into ammonia and water. The reaction mechanism for an enzymatic reaction can be explained in the form of two steps as follows:

(i) The enzyme and substrate react to form an enzyme–substrate complex (E.S) which is a reversible reaction:

$$\mathrm{E + S} \underset{k_{-1}}{\overset{k_1}{\rightleftarrows}} \mathrm{E.S} \tag{4.4}$$

where k_1 is the forward reaction rate constant, and k_{-1} is the backward reaction rate constant, i.e. the reaction for decomposition of E.S into E and S.

(ii) The complex E.S converts to form products P.

$$\mathrm{E.S} \xrightarrow{k_2} \mathrm{E + P} \tag{4.5}$$

where k_2 is the reaction rate constant for the conversion of E.S into product P, and releasing the enzyme E.

If [C] is the concentration of the enzyme–substrate complex at any time, we can write the rate of disappearance of the substrate

$$-r_S = k_1\,[E]\,[S] - k_2\,[C] \tag{4.6}$$

The concentration of enzyme available at any time is the actual concentration of the enzyme [E] minus the concentration of the enzyme locked up in the formation of the enzyme–substrate complex E.S, which is equal to [C]. Thus, effective concentration of enzyme available at any time is equal to ([E] – [C]).

The net rate of change of concentration of enzyme–substrate complex is given by its formation (forward reaction in Eq. (4.4)), decomposition (reverse reaction in Eq. (4.4)), and its conversion to products by (Eq. (4.5)).

$$r_{E.S} = k_1\,([E] - [C])\,[S] - k_{-1}\,[C] - k_2\,[C] \tag{4.7}$$

The first term in the RHS indicates the concentration of enzyme existing at any time ([E] – [C]), after giving due consideration to its formation to complex E.S multiplied by the concentration of the substrate and rate constant, k_1. The second term indicates the consumption of E.S by decomposing back into reactants (by reverse reaction of Eq. (4.4)) multiplied by the rate constant k_{-1}. The third term indicates the conversion of E.S into products, (by Eq. (4.5)) which is a product of the concentration of E.S (i.e.,[C]) multiplied by the reaction rate constant k_2.

In the above equation, the value of [S] is assumed to be considerably more than [E]. Under steady conditions, the rate of formation of the E.S is zero, in the LHS of Eq. (4.7). Thus,

$$r_{E.S} = \frac{d[C]}{dt} = k_1[E][S] - k_1\,[C]\,[S] - k_{-1}\,[C] - k_2\,[C] = 0 \tag{4.8}$$

i.e.

$$k_1[E][S] = k_1[C][S] + [C](k_{-1} + k_2) \tag{4.9}$$

Dividing throughout by k_1

$$[E][S] = [C][S] + [C]\,\frac{(k_{-1} + k_2)}{k_1} \tag{4.10}$$

$$= [C]\left\{[S] + \left(\frac{k_{-1} + k_2}{k_1}\right)\right\}$$

Therefore,

$$[C] = \frac{[E]\,[S]}{[S] + \left(\dfrac{k_{-1} + k_2}{k_1}\right)} \tag{4.11}$$

The rate of product formation by Eq. (4.5) is

$$k_2[\mathrm{C}] = \frac{k_2\,[\mathrm{E}]\,[\mathrm{S}]}{\left\{\dfrac{k_{-1}+k_2}{k_1} + [\mathrm{S}]\right\}} \tag{4.12}$$

In the above equation, $k_2[\mathrm{E}]$ indicates the maximum production rate of the product, since all the enzyme is shown to have been forming the product, and is represented as υ_{max}, i.e. maximum rate of product formation.

Thus,

$$\upsilon = \frac{\upsilon_{max}\,[\mathrm{S}]}{K_s + \dfrac{k_2}{k_1} + [\mathrm{S}]} \tag{4.13}$$

where $K_s = k_{-1}/k_1$, which is the equilibrium constant in the dissociation of the enzyme–substrate complex. E.S. Equation (4.13) may also be written as

$$\upsilon = \frac{\upsilon_{max}\,[\mathrm{S}]}{K_m + [\mathrm{S}]} \tag{4.14}$$

where

$$K_m = K_s + \frac{k_2}{k_1} \tag{4.15}$$

If the value of k_2 is far less than k_1, which is normally the case in reality, then k_2/k_1 in Eq. (4.15) may be neglected, and hence

$$K_m = K_s \tag{4.16}$$

Equation. (4.14) is the classical **Michaelis–Menten (1913) equation**. This is similar to Monod equation[#], i.e. Eq. (7.18), for reactions brought about by microorganisms. Km represents the substrate concentration corresponding to $(\upsilon_{max})/2$, as shown in Fig. 4.6. K_m values for some enzyme–substrate systems were reported earlier (Atkinson and Mavituna, 1991). For example, the α-amylase–starch system has K_m values of 1.0 and 0.4 mM, depending upon the source of enzyme, viz. *Bacillus stearothermophilus* and *Porcine pancreas*, respectively. Similarly, glucose oxidase—D glucose system has values of 33.0 and 9.6 mM, based on whether the enzyme source is *Aspergillus niger* or *Penicillium notatum*. Some such other systems are:

alcohol dehydrogenase–ethanol	:	13.0	mM
β-amylase–amylose	:	0.07	mM
penicillinase–benzyl penicillin	:	0.049	mM
urease–urea	:	10.05	mM

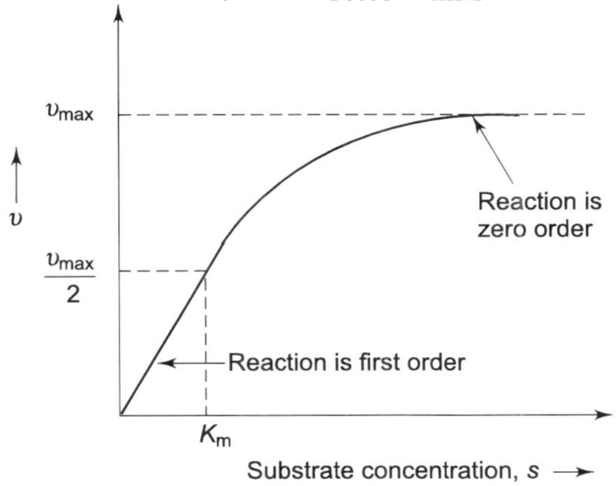

Fig. 4.6 *Michaelis–Menten plot*

[#] Monod model will be discussed in Sec. 7.3 on kinetic models for cell growth.

4.5.2 Determining Enzyme Kinetic Constants from Batch Reaction Data

The enzymatic reactions are modelled by Michaelis–Menten equation. To fully specify the growth of Michaelis–Menten kinetic expression given by Eq. (4.14), we need to evaluate υ_{max} and K_m. Unfortunately, these parameters cannot be estimated straightforward.

Usually, the first step is to calculate the data on reaction rate versus the substrate concentration, *s*, in a batch reactor. Typically, only 'initial rate data' are used. Several batch experiments are to be co ducted with different initial substrate concentrations. From each data, the reaction rate is evaluated at zero time, i.e. at initial concentration of the substrate. The data are plotted, as shown in Fig. 4.6, to find out υ_{max} and K_m. Initial rate data is usually preferred in Fig. 4.6, since the enzyme and substrate concentrations are accurately known at this point. However, the accuracy of Michaelis–Menten plot is always erroneous because of the difficulty in correctly predicting υ_{max}.

Lineweaver–Burk plot The Michaelis–Menten equation can be linearised by taking the reciprocal of Eq. (7.14):

$$\frac{1}{\upsilon} = \frac{K_m}{\upsilon_{max}}\frac{1}{(s)} + \frac{1}{\upsilon_{max}} \tag{4.17}$$

which is similar to Eq. (7.21) for microbial reactions. Equation (4.17) can be drawn in the form of a straight line by plotting $1/\upsilon$ versus 1/s, which gives as slope K_m/υ_{max} and intercept as $1/\upsilon_{max}$ (Fig. 4.7). This plot is known as **Lineweaver–Burk plot**. This method is also not free from certain inaccuracies. Particularly at low concentrations, the linearisation process distorts the experimental error in υ.

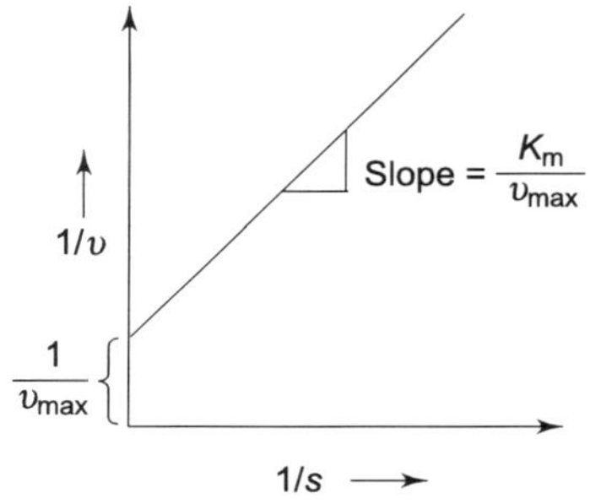

Fig. 4.7 *Lineweaver–Burk plot*

Langmuir plot Multiplying Eq. (4.17) by *s* on both sides, yields

$$\frac{s}{\upsilon} = \frac{K_m}{\upsilon_{max}} + \frac{s}{\upsilon_{max}} \tag{4.18}$$

Thus, s/υ versus *s* should give a straight line plot which is known as **Langmuir plot**. The intercept is K_m/υ_{max} and the slope is $1/\upsilon_{max}$. The distortions in experimental errors in υ are reduced considerably in this method; and hence is recommended (Moser, 1985).

Direct linear plot Eisenthal and Cornish–Bowden (1974) suggested a direct linear plot method for finding out enzyme kinetics which is different from earlier methods. The reaction rates, υ, are measured for each substrate concentration, s. The s values are plotted on abscissa in the negative scale and υ values on the coordinate, as shown in Fig. 4.8. The straight lines drawn across $-s$ and υ will reasonably meet at point A. The value for υ_{max} is read on the coordinate and K_m on the abscissa.

This method is recommended for those types of enzymes which are known to obey Michaelis–Menten kinetics (Doran, 1995).

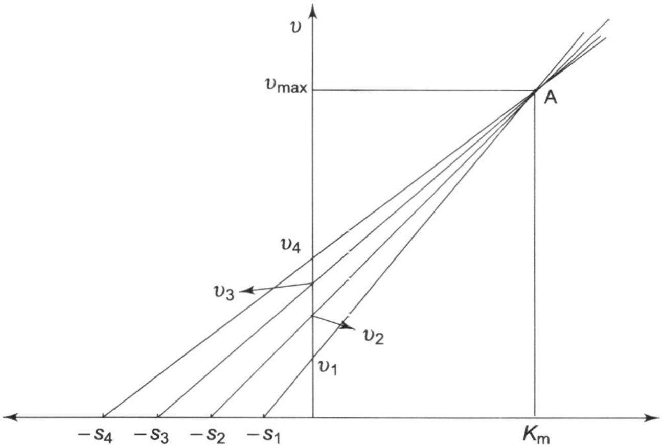

Fig. 4.8 *Direct linear plot method of Eisenthal and Cornish-Bowden for evaluating K_m and v_{max} Reproduced from "The direct linear plot : a new graphical procedure for estimating enzyme kinetic parameters", Eisenthal, R., and Cornish-Bowden, A., Biochem. J., Vol. 139, P. 716, Copyright © (1974), with permission from the Biochemical Society.*

Example 4.1

Inversion of sugar by the enzyme invertase was studied in a laboratory reactor and the following data were generated. (Table E4.1). Find the maximum reaction velocity (v_m) and Michaelis-Menten constant (K_m) by the Eisenthal and Cornish-Bowden plot.

Table E- 4.1:

Sugar concentration moles/m³	70	63	58	52	44	36	25	17	12
Reaction vel. (v_m) moles/m³ h	14.375	14.22	14.076	13.88	13.55	13.10	12.13	10.89	9.61

Solution The data in Table E4.1 is plotted on Eisenthal and Cornish – Bowden plot as shown in Fig. E4.1 and we note

$$K_m = 8 \text{ mol / m}^3$$
$$v_m = 16 \text{ mol / (m}^3 \text{ h)}$$

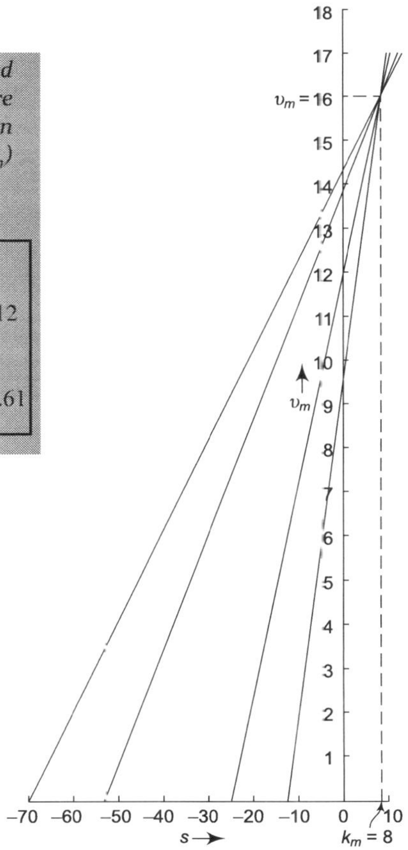

Fig. E4.1 *Eisenthol and Cornish–Bowden plot for Ex-4.1*

4.6 IMMOBILIZATION OF ENZYMES

The movement of enzymes in a fixed location is restricted by the technique known as **immobilization**. Immobilized enzymes are attached to an insoluble support medium or enclosed by the support medium which is also known as a **carrier**. In some cases, the enzyme molecules are cross-linked to each other so that their movement is restricted but their catalytic activities are retained. Thus, immobilization restricts only the movement of enzymes either completely or to some region.

Since the enzymes are proteins and they are soluble in water, it is difficult to separate them once the reaction is over for reuse in a batch process. To overcome this difficulty, enzymes are immobilized on the carrier materials or inside an insoluble matrix by various physical or chemical methods.

Immobilization has the following distinct advantages:

(i) component enzymes can be reused
(ii) the cell density or enzyme density in a location can be increased
(iii) continuous operations are possible
(iv) enzyme stability is improved considerably
(v) product is enzyme-free, and hence is particularly useful with intracellular particles
(vi) since the immobilized enzymatic processes are continuous, it is easy to incorporate process control techniques
(vii) catalytic process route can be controlled more accurately
(viii) allows development of multi-enzyme reaction systems
(ix) effluent disposal problems are considerably reduced

In view of the above advantages, immobilized enzymes have become a potential source for industrial and medical use.

4.6.1 Immobilization Methods

Immobilization techniques can be divided into four main groups as shown in Fig. 4.9. The division is based on whether the enzymes are entrapped in a limited space or bound to a carrier material.

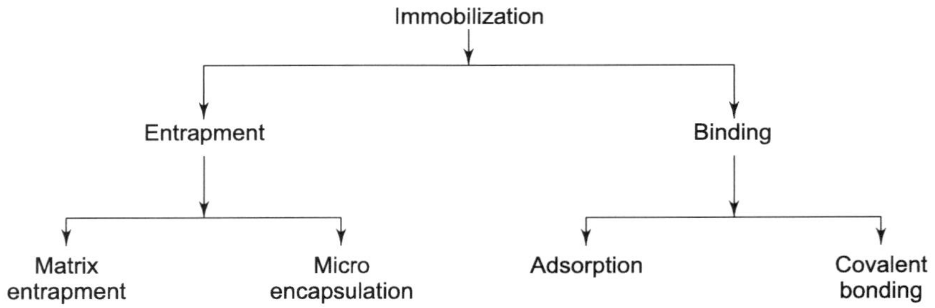

Fig. 4.9 *Various immobilization method*

(i) Entrapment or Occlusion It is a simple physical process of immobilization. The active enzyme is entrapped in the matrix of a polymerized gel. Polyacrylamide is generally the most preferred gel. This method is very simple, and can be applied for most enzymes. The advantages of this method are:

(i) enzymes are not chemically modified
(ii) enzyme properties are not altered.

There are certain disadvantages with this method. They are as follows:

(i) the deactivation of enzyme may take place during gel formation
(ii) enzyme leakage may take place continuously depending upon the pore size of the gel.
(iii) diffusional limitations may pose reduced accessibility for the substrate.

(ii) Micro Encapsulation In this method, the enzyme is entrapped within semipermeable membrane in the form of microscopic hollow spheres. The entrapping methods do not affect the activity of the enzyme.

However, the diffusional limitations may restrict the movement of the substrate to the active site. Thus, this method of immobilization may not be suitable for proteolytic enzymes, or for macro molecular substrates. Even though the micro encapsulation method is not free from some degree of leakage of enzyme, the best advantage of this method is that each enzyme is in much closer contact with the substrate in the surrounding solution.

(iii) Covalent Attachment In this method, the enzyme is attached to the surface by covalent bond formation via certain functional groups. The most commonly used functional groupings of the support material are:

- amino groups
- carboxyl groups
- hydroxyl groups or
- sulphydryl groups

The only restriction imposed by this method is that these functional groups should not be in the active sites of the enzyme. Certain chemical reagents are used for activating the functional grouping. They are:

- cyanogen bromide
- carbodiimide and
- glutaraldehyde.

The covalent attachments of enzyme molecules can be simply described as follows.

$$\vdash X + E \rightarrow \vdash E + X$$

The co-polymerization of enzyme with a reactive monomer of the support material is also followed for some immobilization by covalent attachment.

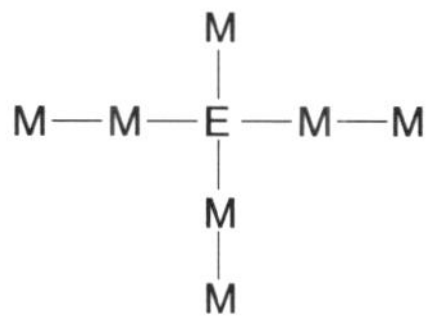

$$n\text{M} + \text{E} \rightarrow \text{M}_n\text{E}$$

Some of the water-insoluble support materials are:

Synthetic support materials →
- Acrylamide-based polymers
- Maleic anhydride polymers
- Styrene-based polymers, Polypeptides

Natural support materials →
- Agarose (sepharose)
- Cellulose
- Dextran
- Glass and Starch

Covalent bonding method provides more permanent linkage between the enzyme and the support material. Covalent bonds can be formed under mild conditions, and the active site of enzyme must remain free from covalent attachments. There is still some possibility for loss of activity of the enzyme during bond formation mainly because of chemical reaction.

(iv) Adsorption One of the simplest methods for enzyme immobilization is by adsorption. Enzymes can be adsorbed physically on a surface-active adsorbent by weak physical forces such as van der Waals' forces or dispersion forces. Commonly used adsorbents are: alumina, clay, silica, anion-exchange resins,

$CaCO_3$, C, cation exchange resins, collagen and glass plates. For effective immobilization, the surfaces of these support materials may have to be chemically or physically pretreated. Ion exchange resins DEAE–Sephadex and carboxymethylcellulose (CMC) can also be used as support media. One of the drawbacks with the adsorption procedure is that since adsorption is a non-specific process, many other substances may also be attached to the carrier in addition to the immobilized enzyme. Another disadvantage of this method is that the loading of enzyme on a unit amount of surface is always very low, and the bonding strength is very weak. Still this method is followed for the following distinct advantages:

(i) the immobilization procedure is easy and simple
(ii) the adsorption process is reversible
(iii) enzymes are not deactivated by adsorption

4.6.2 Properties of Immobilized Enzymes

Enzymes are usually immobilized in particle or pellet form; but enzymes may be attached to, or entrapped within carriers in the form of membranes, tubes or fibers, based on the requirements of a given application. In view of this, an immobilized enzyme may have different properties as compared to the same enzyme in a free solution form. The method of immobilization and nature of insoluble carrier may have influence on the enzyme properties. The specific activity may reduce in the immobilized enzyme, particularly if a chemical process is involved in the immobilization method. The enzyme stability may vary on heating or storage. The pH optimum can change by as much as two pH units for the immobilized enzyme, mainly because of the new microenvironment as compared to the pure enzyme.

The internal and external mass transfer limitations are imposed on the immobilized systems because of diffusional limitations. They will reduce biochemical reaction rate considerably making the effectiveness factor less than one. These aspects, including the mathematical treatment of the diffusional process in immobilized enzyme systems are separately dealt in detail in Chapter 15.

4.6.3 Reactors for Immobilized Enzyme Systems

As has been mentioned earlier, the immobilized enzyme systems are particularly credited for their

(i) amenability to use in continuous systems
(ii) ease in separation of the immobilized pellets for reuse or recycling
(iii) separation of the enzyme-free or cell-free products at the end of the reaction

Thus, immobilized enzyme systems are ideal for use in continuously-operated processes, and hence continuously operated reactors can be used with advantage. Some of the continuously operated flow reactors in chemical/biochemical systems are:

(i) continuous stirred tank reactors
(ii) plug-flow reactors
(iii) packed-bed reactors
(iv) fluidized-bed reactors

Typical of them are shown in Fig 4.10. Their detailed description is given in subsequent chapters (Ch. 5, 16 and 19).

All the above reactors have got their own advantages and operational features. The continuous stirred tank reactors (CSTR) have the advantage of providing fully mixed conditions so that every time the enzyme pellet can have an opportunity to have a new environment of the substrate concentration. Except for the disadvantage of the shearing, which may be detrimental to the shear-sensitive particles, these reactors are very ideal.

The packed-beds and plug-flow reactors almost operate under similar flow conditions. In either case there is no movement of the bed, and hence are ideal for any systems except those in which heat generation and dissipation are a problem to reckon with. The fluidized-bed bioreactors incorporate some of the beneficial features of both stirred-tank and packed-bed reactors. In this reactor the upward flow system is ideally suitable for processes in which the substrate solution is highly viscous.

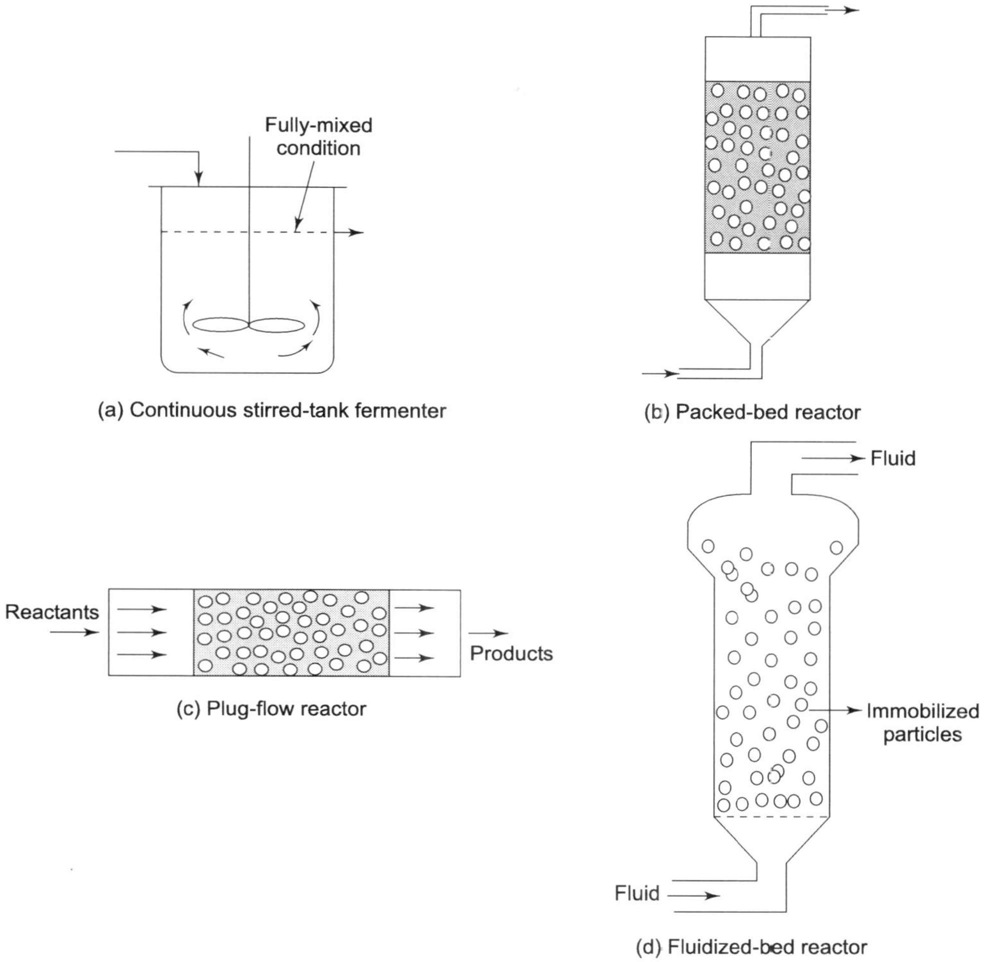

Fig. 4.10 *Immobilized bioreactors*

4.7 INDUSTRIAL APPLICATIONS OF ENZYMES

Enzymes find innumerable applications in process industries. The list is increasing day by day because of the newer applications coming up with recombinant DNA techniques. One of the major reasons for their wider applications in industry is their ability to selectively catalyze the reactions as biocatalysts. The first enzyme produced industrially was fungal *Amylase takadiastase* in the US in 1964. By 1985, about 2500 enzymes were known out of which only 250 were marketable. Table 4.2 lists some of the industrial enzymes available by 1993 (Nielsen, et al, 1994).

The prospects of enzyme applications are mainly due to:

(i) higher yields and selectivity obtained by genetic manipulation
(ii) mild fermentation conditions which result in low cost of nutrients, optimal utilization of components in nutrient solution and amenability to fed-batch fermentation conditions
(iii) continuous methods for production of stereospecific enzymes in special reactors

Table 4. 2 *Available industrial enzymes*

Year launched	*Enzyme type*	*Brand names*[a]	*Company*
1950	Bacterial α-amylase	BAN Canalpha Rapidase	Novo Nordisk Solvay Enzymes IBIS
1963	Protease with a low alkaline pH optimum	Alcalase Maxatase Optimase Superase Milezyme	Novo Nordisk IBIS Solvay Enzymes Pfizer Solvay Enzymes
1965	Amyloglucosidase	AMG Amigase Optidex Optilase Diazyme Spezyme	Novo Nordisk IBIS Solvay Enzymes Solvay Enzymes Solvay Enzymes Genencor International
1973	Thermostable α-amylase	Termamyl Maxamyl Takatherm Optitherm Kleistase	Novo Nordisk IBIS Solvay Enzymes Solvay Enzymes Daiwa Kasei
1973	Microbial rennet (chymosin)	Rennilase	Novo Nordisk
1974	Protease with high alkaline pH optimum	Esperase	Novo Nordisk
1976	Glucose isomerase	Sweetzyme Maxazym Optisweet Spezyme Spezyme Takasweet	Novo Nordisk IBIS Solvay Enzymes Pfizer Genencor International Solvay Enzymes
1982	Low-temperature protease with high alkaline pH optimum	Savinase Maxacal Opticlean	Novo Nordisk IBIS Solvay Enzymes
1982	Pullulanase	Thermozyme	Novo Nordisk
1986	Cellulase for alkaline conditions	Celluzyme	Novo Nordisk
1988	Lipase for alkaline conditions made by DNA technology	Lipolase	Novo Nordisk
1991	Bleach-stable protease with high pH optimum made by protein engineering	Maxapem Durazym	IBIS Novo Nordisk

[a] First product on market is listed first for each enzyme type.

(Reprinted from "Encyclopaedia of Chemical Technology", Nielsen, et al., Kroschwitz, J.I., (Ed.), 4th Edn., Vol. 9, p. 570, Copyright © (1994), with permission form John Wiley & Sons, Inc.)

(iv) development of process for immobilization of enzymes and for their recycling

(v) developments in isolation methods for intracellular products

(vi) a number of downstream processes such as counter-current extraction, ion-exchange chromatography, affinity chromatography and precipitation techniques which have made the isolation of product easier.

Industrial processes were generally getting restricted as long as the enzymes in aqueous solution were being used. Subsequently efforts were made for utilization of enzymes which function in non-aqueous media. We will now mention just a few of the many applications of enzymes in industry (also see Section 5.2.5).

Enzyme applications in the industry may be broadly classified into the following major areas:

(i) food and beverage industry including baking and brewery
(ii) pharmaceuticals
(iii) medical applications
(iv) analytical purposes

Applications of enzymes in food industry are wide and have been in use from time immemorial; the most classical application being making of cheese using rennet.

α-Amylase and proteases are used in baking industry to convert starch into dextrin and sugars. They also help in hydrolysis of gluten which is a major flour protein. These characteristic changes in the dough will help in leavening of bread by improved fermentation conditions. The raising quality of bread is caused by this treatment.

Fruit juices and wines, which have a haze because of the presence of pectin, are treated with pectolytic enzymes for clarifying the fruit juices. Some of the fruit pulps, like banana or guava, need pectolytic enzyme treatment for breaking of pectin to release the free-flowing fruit juice from the fibrous materials. This process is known as **clarification of fruit juices** for making RTS (Ready to Serve) beverages. Similarly wines made out of the fruit juices will also need clarification for getting a crystal-clear wine. Pectolytic enzymes (viz., Pectinol, Mylase) are used for such treatments.

For brewing of beer, amylase is used which otherwise should have come by malting of barley. Chillproofing of beer is another very useful industrial application. Beers, since they are brewed from cereal grains, contain a certain amount of haze (cloudiness) which needs to be removed. This is know as **Chillproofing of beer** and is done by an enzyme papain (extracted from the latex of raw papaya fruits).

Tenderization of meat is desired if the meat is obtained from aged animals, which is normally the case in India. Such meats need to be tenderized for improving the texture. It is done by using certain pectolytic enzymes like papain or bromelin.

In the confectionary industry, the enzyme *invertase* is used for inversion of sucrose to a mixture of glucose and fructose. This also helps in avoiding the problem of crystallization of sucrose. Immobilized glucose isomerase is used for conversion of glucose to fructose which is about 1.7 times more sweet than glucose, and hence is used in the soft drinks as sweetener.

The enzyme lactase is used for hydrolysis of lactose present in skim milk and whey. This step will facilitate use of lactose products in the preparation of ice-creams without the possible crystallization of lactose during cooling process. Aspartame is a low-calorie sweetener, and is produced by reaction of L-phenylalanine with aspartate which in turn is formed by conversion of fumerate in a packed tower using immobilized *aspartase* enzyme.

Liquid glucose or corn syrup is made industrially by hydrolysis of starch. Starch is first solubilised and hydrolysed using amylo glucosidase. The enzyme cleaves the glucosidic bonds in starch to give good yields of liquid glucose. Liquid glucose, because of its high consistency and viscosity, is extensively used in confectionery industries (Fig. 4.11).

Enzymes are also used in the food industry for analytical purposes. The activity of certain enzymes is a measure of the degree of sterility the material has attained. For example, in the pasteurization of milk, the activity of alkaline phosphatase and invertase which should not be present in the sterile milk, gives an indication of the degree of pasteurization the milk has attained by heating. Similarly the bacterial contamination of food can be assessed by an assay of the microbial enzymes which should not normally be present.

Fig. 4.11 ***Flash Cooker for enzymatic conversion of liquid starch slurry to maltodextrin using α-amylase.*** *(Courtesy: M/s. Sukhjit Starch Mills, Nizamabad, AP).*

High amounts of α-amylase present in wheat flour is an indication that the flour is made out of the germinated (spoiled due to long storage) wheat grains. The germination process breaks down the starch and protein (particularly gluten) which makes the flour unsuitable for baking. Thus, estimation of α-amylase in the wheat flour helps know whether it can be used for baking. Freshness of meat can be measured by the presence of amines which are formed due to degradation. The amines are assessed with the enzyme monoamine oxidase.

Some of the organo-phosphorous pesticides present in the foods can be assessed by the Cholineesterase-catalyzed procedure.

Some of the important applications of enzymes in medicine are their utilization for diagnosis, therapy and treatment purposes. Their application in medicine and pharmaceutical industry is increasing at a rapid rate. Chirally- pure compounds from the isomers in recemic mixtures is the need of the hour in the pharma industry. Ability of the enzymes to recognize the chiral isomers and reacting with them selectively is exploited in the pharmaceutical industry. In the antibiotic industry, Penicillin acylase is used for conversion of Penicillin-G to 6-amino penicillinic acid (6-APA). This is a precursor for making a wide variety of penicillin derivatives.

Enzymes in immobilized form find use in biosensors. Biosensors are instruments used to sense some of the biological materials (substrate or product) either qualitatively or quantitatively or both. Generally they are used for qualitative analysis. Enzymes like glucose oxidase or urease are used as sensing enzymes in the immobilized form of the bio-sensor membranes for sensing glucose and urea respectively. (Also see Sec. 5.4.3).

Another important industrial application of enzymes is their utilization in washing powders. Bacterial proteases have been incorporated into washing powders, and are available for several years. Subsequently the fears about the possible effect of enzyme dust on the respiratory system had greatly retarded the commercial importance. Later, such fears were set aside by using the enzymes in the form of granules which rupture only on contact with water. The enzyme-incorporated washing powders have become more popular for removal of blood and other protein stains which was made possible by using subtilisins from *Bacillus subtilis* mutants which are stable to alkali, high-temperature detergents and bleaches. Bacterial proteases are also very popular for loosening hair or wool and separating it from hides, and hence are used in the leather and textile industries.

Table 4.3 lists some of the industrial applications of enzymes.

Table 4.3 *Enzymes in industrial use**

Enzyme	*Application*
Trypsin	Anti inflammatory agent
Lysozyme	Anti bacterial agent
Streptokinase	Anti inflammatory agent useful for dissoluting blood clots
Urokinase	Prevention of blood clots
Aspuraginase	Anti cancer agent
Glucose oxidase	Oxidation of glucose to gluconic acid
Penicillinase	Treatment of allergies against penicillin
Papain	Tenderization of meat
	Chillproofing of beer
Pectinase	Clarification of fruit juices
	Clarification of wine
Immobilized amino acylase	Separation of D- and L- amino acids
Chymosin	Cheese production
Penicillin acylase	Used in antibiotic industry for conversion of Penicillin-G to 6-amino penicillinic acid (6-APA)
Urease	Conversion of urea into CO_2 and NH_4^+ in dialysis process
Pseudomonas melanogenum	For ampicillin synthesis

*The list is not exhaustive

Symbols

E	: activation energy (kJ/k mol)
[A]	: concentration of component A (kmol/m^3)
[B]	: concentration of component B (kmol/m^3)
[C]	: enzyme–substrate concentration
C	: concentration (g/l or mol/m^3)
[E]	: concentration of enzyme
E.S	: enzyme–substrate complex
k_2, k_{-1}	
k_n, k_1,	: rate constants
K_m	: Michaelis – Menten–constant, i.e. substrate concentration when $\upsilon = \upsilon_{max}/2$ [Eq. (7.15)]
K_s	: k_{-1}/k_1
r	: rate of reaction (mol/(m^3.s))
s	: substrate concentration
[S]	: concentration of substrate
t	: time (s)

Subscripts

0	: initial conditions
A	: component A
B	: component B
E.S	: enzyme–substrate complex
i	: component i
max	: maximum
s	: substrate

Greek letters

υ	: rate of enzyme reaction

References

4.1 Atkinson, B., and Mavituna, F. (1991), *Biochemical Engineering and Biotechnology Hand Book*, 2nd Edn., Macmillan, Basingstoke.

4.2 Doran, P.M. (1995), *Bioprocess Engineering Principle*, Academic Press, London, 271–272, 277.

4.3 Eisenthal, R., and Cornish-Bowden, A. (1974), "The direct linear plot: a new graphical procedure for estimating enzyme kinetic parameters", Biochem. J., 139, 715–720.

4.4 Fogler, H.S. (1986), *Elements of Chemical Reaction Engineering*, Prentice-Hall, New Jersey, 328–332.

4.5 Michaelis, L., and Menten, M.L. (1913), "The kinetics of invertin working", Biochem. Z. (in German), 49 , 333.

4.6 Moser, A. (1985), "Rate equations for enzyme kinetics", *Biotechnology*, Rehm, H. –J., and Reed, G. (Eds.), Vol. 2, VCH, Weinheim, pp. 199–226.

4.7 Nielson, P.H., et al. (1994), *Enzyme applications (Industrial), Kirk–Othmer ECT*, Kroschwitz, J.I. (Executive Ed.), 4th Edn, Vol. 9, John Wiley, New York, p.164.

4.8 Palmer, T., (2001), *Enzymes: Biochemistry, Biotechnology, Clinical Chemistry*, Horwood Pub. Co., Chichester. (a) pp 70–74

4.9 Wang, D.I.C, Cooney, C.L., Demain, A.L., Dunnill, P., Humphrey, A.E., and Lilly, M.D., (1979), *Fermentation and Enzyme Technology*, John Wiley & Sons, New York, p. 319.

4.10 White, A., Handler, P., and Smith, E.L., (1968), *Principles of Biochemistry*, 4th Edn, McGraw-Hill Inc., New York, p. 209.

General

4.11 Boyer, P.D. (Ed.) 1970–1974. *The Enzymes*, 3rd Edn, Vol. 1–10, Academic Press, New York.

4.12 Jencks, W.P., (1969), *Catalysis in Chemistry and Enzymology*, McGraw-Hill, Inc., New York.

Review Questions

4.1 Describe how enzymes are classified.

4.2 What are various chemical theories that describe the functioning of enzymes as biocatalysts?

4.3 What is meant by enzyme specificity? What are various types of enzyme specificity?

4.4 Describe the Fischer lock-and-key hypothesis for enzyme specificity.

4.5 Describe the Koshland induced-fit hypothesis for enzyme specificity.
4.6 What is meant by homogeneous reactions?
4.7 What is Michaelis–Menten equation?
4.8 Explain how Michaelis–Menten equation can be derived for enzymatic kinetics from first principles.
4.9 What is meant by immobilization of enzymes? Describe different methods of immobilization.
4.10 How do you determine the enzyme kinetic constants from batch reactor data?
4.11 What is Lineweaver–Burk plot? How can it be used to calculate Michaelis–Menten constants?
4.12 Explain the Langmuir plot and how it can be used to calculate Michaelis–Menten constants?
4.13 Explain the direct linear plot method to calculate Michaelis-Menten constants. What are its merits over other methods?
4.14 What type of bioreactors do you suggest for immobilized enzyme systems?
4.15 What are various industrial applications of enzymes?
4.16 What are the applications of enzymes in food and beverage industries?
4.17 What are the medical applications of enzymes?

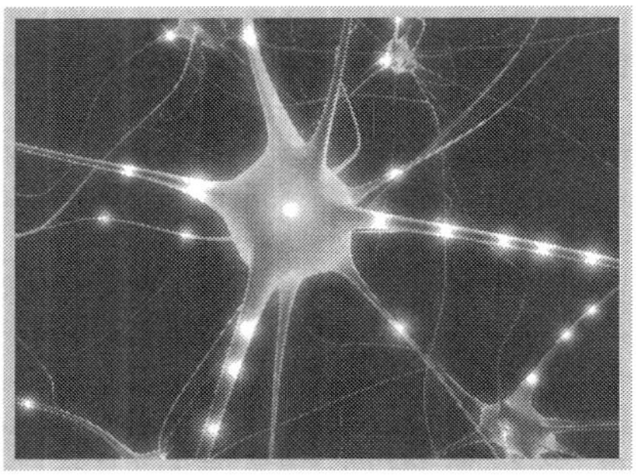

chapter 5

Fermentation

Fermentation is a word derived from the Latin verb *FERVERE* (to boil), which describes the evolution of carbon dioxide bubbles in anaerobic conditions by the action of yeast on fruit juices. Since the carbon dioxide evolution gave the impression that the fruit juice was boiling, the process was equated to boiling, and was termed as **fermentation**.

5.1 GENERAL REQUIREMENTS OF FERMENTATION PROCESS

Since fermentation is a biochemical process brought about by the intervention of living organisms, it is essential that any fermentation process should have:

(i) a microorganism for carrying out the bioconversion
(ii) a substrate to get converted to useful products
(iii) maintenance of fermentation conditions
(iv) a provision for recovery and purification of the products
(v) effluent-treatment section
(vi) facilities for packaging and marketing.

Figure 5.1 highlights them in a nutshell. Stanbury and Whitaker (1993) classified them into six basic components, viz.

- Isolation and production of pure culture.
- medium formulation
- Sterilization
- fermentation in the production fermenter
- product recovery
- effluent treatment

These steps are shown in Fig. 5.2. The requirements/components of fermentation process show the true interdisciplinary nature of bioprocessing. These requirements also manifest the need for Research and Development (R&D) in the areas of both biological sciences and engineering sciences, in the absence of which, the development of a successful fermentation process cannot be realized.

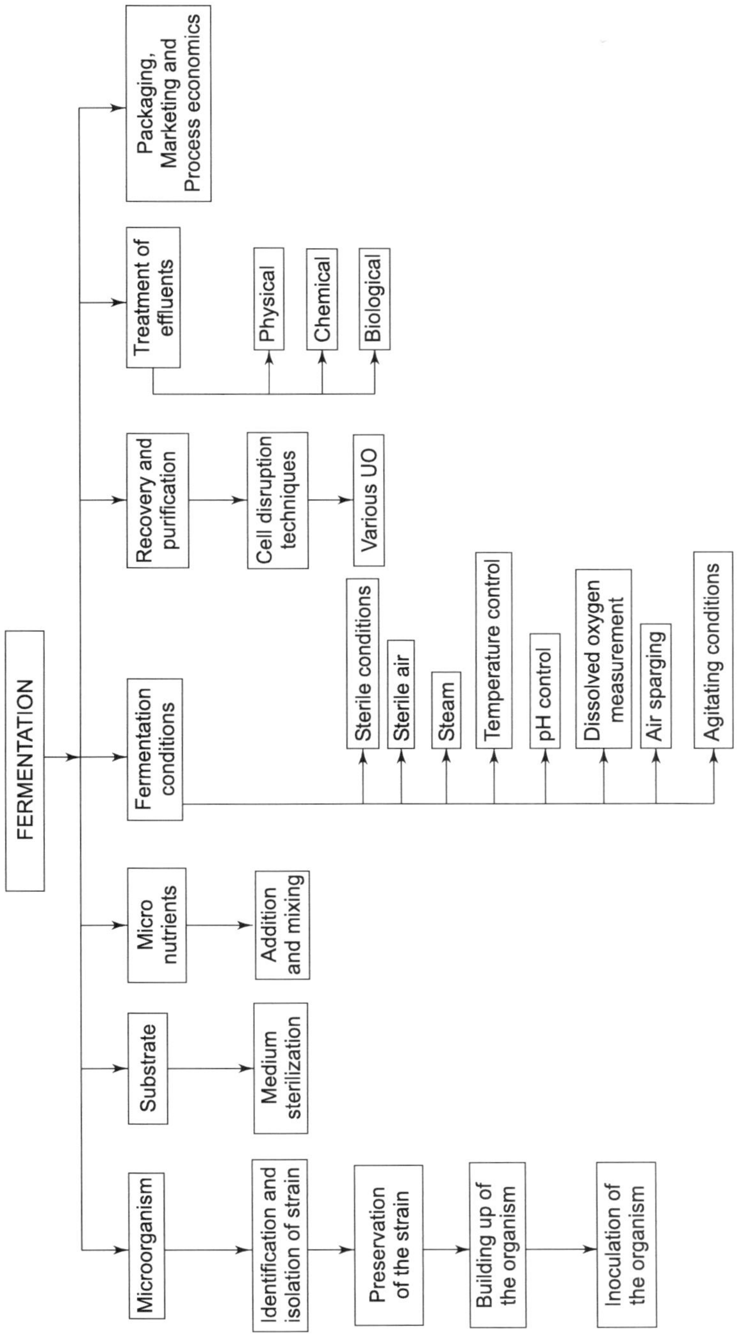

Fig. 5.1 *General requirements of a fermentation process*

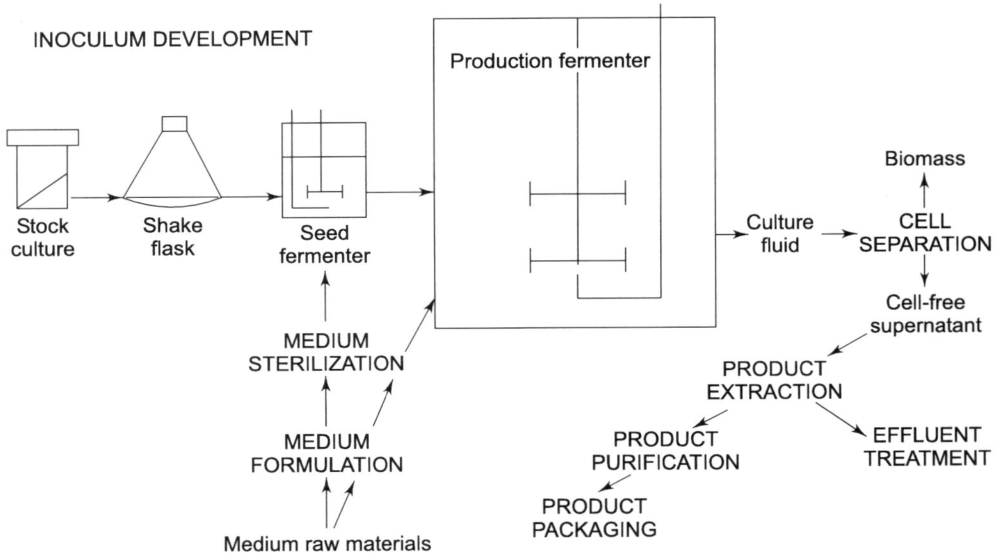

Fig. 5.2 *Generalized schematic representation of a typical fermentation process (Reprinted from "Principles of Fermentation Technology", Stanbury , P.F., and Whitaker, A., p. 9, Copyright © (1993), with permission from Elsevier.)*

5.2 RANGE OF FERMENTATION PROCESS

From biochemical point of view, "fermentation" is the name usually given to the general class of chemical changes produced in organic compounds (substrates) through the activity of living organisms (Atkinson, 1974). Fermentation processes have made deep inroads into human life from time immemorial. Production of wine from fruit juices since prehistoric age, to the production of antibiotics and vaccines in the modern age, to the genetically modified foods and gene cloning in the present days are a saga of miracles bestowed by biotechnology on mankind. The list is endless and can be catalogued as follows (Webb, 1964,(a)):

- food and beverages
- industrial solvents
- antibiotics
- vaccines
- enzymes
- organic acids
- animal feeds
- effluent disposal
- agricultural and related areas

Of the above, products like antibiotics, vaccines, enzymes and effluent disposal techniques are the exclusive domain of fermentation. Table 5.1 highlights some of the major products of bioprocessing (Doran, 1995). The products/processes can be classified into the following five categories (Sengha, 1993):

(i) Production of biomass, viz. baker's yeast, single cell protein (SCP), animal feed formulation.
(ii) Production of some primary metabolites, viz. amino acids, organic acids, vitamins, etc.

Fermenters for production of ethanol from molasses using yeast (Sachromises cervacia). The fermenters have working volume of 100 m³, diameter 2.5 m and height 6.5 m *(Courtesy: M/s. Nav Bharat Ferro Alloys Ltd., Deccan Sugars, Samalkot, AP.).*

(iii) Production of bioconversion products or bio-transformation products, viz. steroid conversion, etc.
(iv) Production of enzymes, viz. amylase, lipase, etc.
(v) Production, by genetically engineered microbes, of therapeutic proteins, polypeptides, hormones, etc. which are not normally synthesized, viz. insulin, etc.

5.2.1 Food and Beverages

The use of fermentation processes in foods and beverages is very extensive. It is difficult even to guess an approximate value of market potentiality. Most of the traditional foods in India are fermented foods. They are in the unorganized sector. The fermentation is known to impart organoleptic qualities to the foods, besides improving their nutritional value. For example, it was reported that the fermented and cooked traditional food 'ragi ambali' made out of 'ragi' (finger millet) flour improved the thymine content by 50% and niacine content by 37% as compared to the raw material (Rajyalakshmi and Geervani, 1990). Only the alcoholic beverage industry is in the organized sector.

5.2.2 Industrial Solvents

Industrial solvents such as acetone, alcohol, butanol, vinegar, etc. were being extensively produced by fermentation route. But subsequently during the Second World War, the requirement of these solvents was beyond the production capacity by the of fermentation route, and hence they were produced by synthetic routes using petrochemical sources. The same trend continues even today also; but most of the potable alcohol which is used for human consumption or for use in the pharmaceutical industry is made by fermentation process.

Table 5.1 *Major products of biological processing*

Fermentation product	*Typical organism used*	*Approximate world market size (kg per year)*
Bulk organics		
Ethanol (non-beverage)	*Saccharomyces cerevisiae*	2×10^{10}
Acetone/butanol	*Clostridium acetobutylicum*	2×10^{6} (butanol)
Biomass		
Starter cultures and yeasts for food and agriculture	Lactic acid bacteria or Bakers' yeast	5×10^{8}
Single-cell protein	*Pseudomonas methylotrophus* or *Candida utilis*	$0.5–1 \times 10^{8}$
Organic acids		
Citric acid	*Aspergillus niger*	$2–3 \times 10^{8}$
Gluconic acid	*Aspergillus niger*	5×10^{7}
Lactic acid	*Lactobacillus delbrueckii*	2×10^{7}
Itaconic acid	*Aspergillus itaconicus*	
Amino acids		
L-glutamic acid	*Corynebacterium glutamicum*	3×10^{8}
L-lysine	*Brevibacterium flavum*	3×10^{7}
L-phenylalanine	*Corynebacterium glutamicum*	2×10^{6}
L-arginine	*Brevibacterium flavum*	2×10^{6}
Others	*Corynebacterium* spp.	1×10^{6}
Microbial transformations		
Steroids	*Rhizopus arrhizus*	
D-sorbitol to L-sorbitol (in Vitamin C production)	*Acetobacter suboxydans*	4×10^{7}
Antibiotics		
Penicillins	*Penicillium chrysogenum*	$3–4 \times 10^{7}$
Cephalosporins	*Cephalosporium acremonium*	1×10^{7}
Tetracyclines (e.g. 7-chlortetracycline)	*Streptomyces aureofaciens*	1×10^{7}
Macrolide antibiotics (e.g. erythromycin)	*Streptomyces erythreus*	2×10^{6}
Polypeptide antibiotics (e.g. gramicidin)	*Bacillus brevis*	1×10^{6}
Aminoglycoside antibiotics (e.g. streptomycin)	*Streptomyces griseus*	
Aromatic antibiotics (e.g. griseofulvin)	*Penicillium griseofulvum*	
Extracellular polysaccharides		
Xanthan gum	*Xanthomonas campestris*	5×10^{6}
Dextran	*Leuconostoc mesenteroides*	Small
Nucleotides		
5'-Guanosine monophosphate	*Brevibacterium ammoniagenes*	1×10^{5}

(Contd.)

(*Contd.*)

Enzymes		
Proteases	*Bacillus* spp.	6×10^5
α-amylase	*Bacillus amyloliquefaciens*	4×10^5
Glucoamylase	*Aspergillus niger*	4×10^5
Glucose isomerase	*Bacillus coagulans*	1×10^4
Pectinase	*Aspergillus niger*	1×10^4
Rennin	*Mucor miehei* or recombinant yeast	1×10^4
All others		5×10^4
Vitamins		
B_{12}	*Propionibacterium shermanii* or *Pseudomonas denitrificans*	1×10^4
Riboflavin	*Eremothecium ashbyii*	
Ergot alkaloids	*Claviceps paspali*	5×10^3
Pigments		
Shikonin	*Lithospermum erythrorhizon* (Plant-cell culture)	60
β-carotene	*Blakeslea trispora*	
Vaccines		
Diphtheria	*Corynebacterium diphtheriae*	< 50
Tetanus	*Clostridium tetani*	
Pertussis (whooping cough)	*Bordetella pertussis*	
Poliomyelitis virus	Live attenuated viruses grown in monkey kidney or human diploid cells	
Rubella	Live attenuated viruses grown in baby-hamster kidney cells	
Hepatitis B	Surface antigen expressed in recombinant yeast	
Therapeutic proteins		< 20
Insulin	Recombinant *Escherichia coli*	
Growth hormone	Recombinant *Escherichia coli* or recombinant mammalian cells	
Erythropoietin	Recombinant mammalian cells	
Factor VIII-C	Recombinant mammalian cells	
Tissue plasminogen activator	Recombinant mammalian cells	
Interferon α_2	Recombinant *Escherichia coli*	
Monoclonal antibodies	Hybridoma cells	< 20
Insecticides		
Bacterial spores	*Bacillus thuringiensis*	
Fungal spores	*Hirsutella thompsonii*	

(Reprinted from *Bioprocess Engineering Principles*, Doran, P.M., p. 4., Copyright © (1995), with permission from Elsevier.)

5.2.3 Antibiotics

Antibiotics production is one of the areas that falls in the exclusive domain of biotechnology. They are considered as the chemical substances which are produced by living organisms or microbes as products of secondary metabolites, and are capable of inhibiting the growth of bacteria. The history of antibiotics can be traced back to the classical work of Louis Pasteur in 1877, who reported that common bacteria inhibited the growth of anthrax culture. Subsequently, a number of other observations followed until the pioneering work of Alexander Fleming in 1928, which had revolutionized the whole antibiotic field. Now it has becomes the second largest pharmaceutical sector, holding 9% of the total world pharma market. The world antibiotic production in 1989 was of the order of $12.3 billion (Borders, 1992). Till now, more than 10,000 antibiotics and similar bioactive substances have been made by microorganisms, but out of them only a 100 products have found applications. Kirk–Othmer Encyclopedia of Chemical Technology (Borders, 1992) highlighted, in a tabular form, some 35 antibiotics with their year of discovery or year of introduction in the market. Table 5.2 shows some important antibiotics made by fermentation process (Aiba, et al., 1965; Atkinson, 1974).

Table 5.2 *Biochemical products of industrial fermentation processes*

Antibiotics		
Aterrimin	Gramicidin	Polymyxin
Amphotericin	Griseofulvin	Ristocetin
Bacitracin	Hygromycin	Spiramycin
Blasticidin-S	Kanamycin	Streptomycin
Colistin	Leucomycin	Tetracyclines
Chloramphenicol	Lincomycin	Thiostrepton
Cycloheximide	Neomycin	Tricomycin
Cycloserine	Novobiocin	Tyrocidine
Erythromycin	Nystatin	Tyrothricin
Fucidin	Oleandomycin	Tylosin
Fumagillin	Paromomycin	Vancomycin
Gentamicin	Penicillins	Viomycin
Steroids		
Cortisone	9α-Fluoro-16α-methylprednisolone	
Hydrocortisone	9α-Fluoro-16β-methylprednisolone	
Prednisolone	6α-Methylprednisolone	
Enzymes		
γ-Aminobutyric transaminase	Hemicellulase	
Amylases	Invertase	
Catalase	Lipase	
Cellulase	Pectinase	
Collagenase	Penicillinase	
Glucose oxidases	Proteases	
Glutamic acid decarboxylase	Streptokinase-streptodornase	
Organic acids		
Acetic (vinegar)	Glutamic	Lactic
Citric	Fumaric	Lysine
Gibberellic	Isoascorbic	5-Nucleotides
Gluconic	Itaconic	

(Reprinted from "Biochemical Engineering", Aiba, S., Humphrey, A.E., and Millis, N.F., pp. 11–12., Copyright © (1965), with permission from Elsevier.)

In addition to the usual medical applications, some antibiotics found use in feed preparations, where a small quantity of addition of it (approximately 50g in 1 ton) will help reduce the incidence of diseases and also help increase the growth. That is, in other words, these antibiotics help increase the conversion ratio of food to flesh.

5.2.4 Vaccines

The vaccines function in a peculiar way. They help prevent certain infectious diseases by alerting the host and induce immunity against such infectious diseases. The origin of vaccines can be traced back to 1796, when Edward Jenner performed the first human immunization against small pox by vaccination. Still, the vaccination technology did not receive much attention for the next 100 years, until Louis Pasteur proposed and demonstrated his classical theory that most of the diseases were caused by microorganisms. However during early 1900's, vaccination became very popular for preventing many diseases like tetanus, tuberculosis, etc.

Subsequently, during the 1950's with the mass production of antibiotics, however, the interest in vaccination research diminished. But still, even till today, vaccination is resorted to as a preventive approach. Mass vaccination programme by various governments is taken up as a major public health campaign, particularly for eradication of polio by 2000 as per the campaign of WHO. The world market for vaccines is estimated to be approximately \$3 billion, with about 35% of it accounting for pediatric portion (Hsiah and Ritchey, 1997).

Vaccination has also become very popular in its application for animals and poultry. Now poultry industry has become one of the biggest consumers of vaccines. Presently, the vaccine production is more based on the factory-type operations in which bacteria and virus are grown by tissue culture techniques or on semi-synthetic media, than the older techniques in which the living animals were used for production of vaccines.

Fermenters for fermentation of genetically modified Pichia pastoris yeast to produce intracellular Hepatitis-B antigen by methanol induction. The fermenter has a bottom driven agitation with double mechanical seal and has a compact control unit with software for display of process values, calibration and control of parameters. *(Courtesy: M/s. Shantha Biotechnics Pvt. Ltd, Hyderabad, A.P.)*

5.2.5 Enzymes

The word 'enzyme' was first proposed in 1878, which means "in Yeast" in Greek. Enzymes were known to be in use as early as 800 BC for production of cheese. However, the modern enzyme technology has its origin in 1874, from the time the Danish scientist Christian Hansen produced the first industrial enzyme *chymosin* (rennet) by extracting it from the dried calves' stomachs. Enzymes are used for many industrial applications for bringing about a bioconversion or changing the functional characteristics. Because of their very special feature in the industrial applications, enzymes indeed open up an important branch of biotechnology (also see Chapter 4). They help in making industrial chemicals via alternate routes, which bring down the extreme reaction conditions to virtually atmospheric conditions. They even make conversions, which are otherwise not possible; the classical example being the special syrups made from starch, which are not possible by chemical hydrolysis. Generally, enzymes are used in very small quantities in the applications, except some of the food enzymes like amylase, pectinase, etc. Various industrial enzymes that are being produced commercially were compiled by Aiba, et al. (1965) and are listed in Table 5.2. Nielsen, et al, (1994) reported the worldwide production of industrial enzymes is of the order of \$720 million in 1990 which has come up from \$450 million in 1987.

The various uses to which enzymes are put, are classified as:

- Enzymes used in the industry (industrial enzymes)
- Enzymes used for analytical purposes
- Enzymes used in medicine

Table 5.3 highlights various features of these enzymes. Detergent industry is probably the largest user of industrial enzymes. Approximately 34–35% of the industrial enzymes are used in detergent industry and starch industry each, whereas dairy, distilling and fruit juice industries use about 10% each with the brewing and baking industry consuming approximately 7–8%. The enzyme applications are increasing day by day with the improvement in their production methodology and purification steps. Nielsen et al. (1994) listed the available industrial enzymes by 1993 (Table 4.2).

Table 5.3 *Various features of enzymes and their qualities*

Type of enzyme	*Application*	*Various enzymes*	*Origin and production features*
Industrial enzymes	Used for industrial operations	Amylases, catalases, isomerases, papain, penicillin acylases, proteases,	Cost of production is low, generally produced in tonnes. Their origin of production is microbial and usually obtained in extracellular form. The degree of purification is crude and hence secondary activities are present.
Analytical enzymes	Used for analytical purposes	Alcohol dehydrogenase, cholesterol oxidase, galactose oxidase, hexokinase, muvamidase	Cost of production is medium to high. Generally produced in grams in pure crystalline form. The secondary activities are generally nil, and if any, they will be defined. They are produced from microbial, animal and plant sources in intracellular form.
Clinical enzymes	Used in medicine	Asparginase, lipase, protease, streptokinase	Cost of production is high since they are produced in pure crystalline form in milligrams levels. They are produced from microbial, animal and plant sources, usually in intracellular form.

5.2.6 Organic Acids

A good number of organic acids is produced by fermentation route, out of which some were subsequently produced by chemical methods and were found to be cheaper, viz. acetic acid. But the production of citric acid by mould fermentation process is competitive to its manufacture from citrus juice. Even the production of lysine, glutamic acid, gluconic acid and isoascorbic acid by fermentation route was not challenged by chemical routes. Table 5.2 lists industrially important organic acids.

5.2.7 Animal Feeds

Most of the animal feeds are made by a combination of grains (usually minor millets) and oil-cake residues. They are supplemented with the wastes of fermentation industry. Molasses is used in most of the cattle feeds not only as a source of energy, but also to enhance the flavour and taste.

5.3 DESIGN AND CONSTRUCTION OF FERMENTER

The basic purpose of design of a reactor or bioreactor or fermenter is to visualize the size of the unit to deliver the product both qualitatively and quantitatively. The size is based on the quantitative output, and the implicit quantitative input to achieve the desired output which is constrained by the yield factors. After the size is designed, next task is to achieve the transport properties, viz.

- fluid mechanics
- heat transfer
- mass transfer effects.

In fact, Charles (1985) notes that in spite of the fact that the transport properties are all governed largely by rheology of fermenter, adequate importance has not been given to this important aspect. Hence, rheology (which speaks about the flow behaviour of the fermentation broth) should be considered before going for any rational design. He also feels that far more attention must be devoted to heat transfer aspects of the fermenter, since the heat transfer effects, which are not very relevant in the laboratory-scale fermenter in view of its high surface-to-volume ratio, play a dominant role in scale-up, and are difficult to achieve. Once the transport processes are decided, their monitoring and control will be the next task. The last activity which is the nightmare of any designer is to visualize how easy or how difficult it is to scale-up the size of the fermenter.

Mechanical design, body construction, and materials of construction are the associated tasks of design of fermenters. The structure of a reactor design, which was originally made for chemical reactors, and is equally applicable to fermenters also, is shown in Fig. 5.3 (Bisio and Kabel, 1985).

A further dimension in the case of design of fermenters (which is not typical of a design of a chemical reactor) is to achieve the aseptic conditions or to maintain asepsis. This most important aspect, which makes or mars the technical viability of a biochemical process, is seldom given adequate importance at the design stage. On the contrary, the subject of asepsis is often left for mechanical fabrication, and is considered as nothing to do with heat transfer and mass transfer.

The following points need to be considered in designing and constructing a fermenter (Stanbury and Whitaker, 1993 (a)):

- operating under aseptic conditions continuously for long periods of time
- interplay of the transport parameters
- ability to control various fermentation process parameters, viz. temperature, pH, etc.
- supply of adequate quantity of oxygen so that the cells do not suffer from inadequacy of oxygen supply
- power requirement for mixing should be minimum
- adequate amount of mixing should be ensured without causing damage to the cells
- flexibility in operation of the fermenter for various purposes, so that the vessel should be suitable for a range of processes
- the vessel geometry should be such that it should facilitate scale-up
- the operating costs should be minimum, particularly for labour in operation, harvesting, cleaning and maintenance.

From the above points, it is clear that the design of a fermenter requires the expertise of various faculties, viz. chemical engineering, microbiology, biochemistry, mechanical engineering and costing.

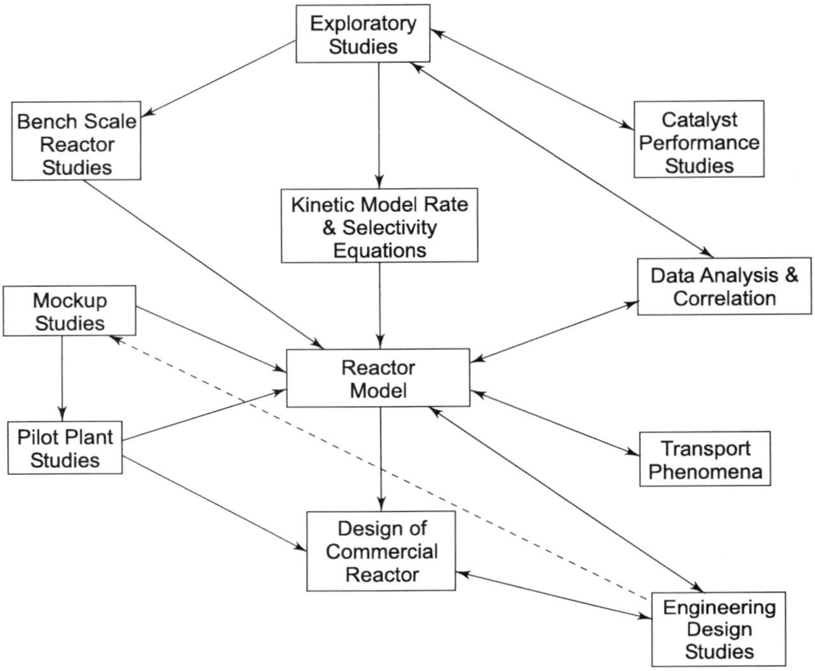

Fig. 5.3 *Scale-up of chemical processes (Reprinted from "Scale-up of Chemical Process", Bisio, A., and Kabel, R.L., p. 4, Copyright © (1985), with permission form John Wiley & Sons, Inc.)*

Bartholomew, *et al.* (1950) described the design and operation of a laboratory fermenter, which is shown in Fig. 5.4.

Construction of a Fermenter The construction details of a fermenter comprise

- materials of construction
- fabrication
- maintenance of asepsis
- piping and instrumentation
- grouting and fixtures.

Since asepsis is an important aspect, Charles (1985) suggested that the design and construction of fermenters should be given to such engineering firms which have adequate expertise in the construction of fermenters, rather than going to the usual fabricators of chemical reactors, since the latter normally overlook the aseptic conditions part, and asepsis becomes the casualty. It is also cautioned that errors in faulty construction are very difficult to locate, and sometimes impossible to correct.

A general representation of a jacketed fermenter is given in Fig. 5.5. (Sengha,1993).

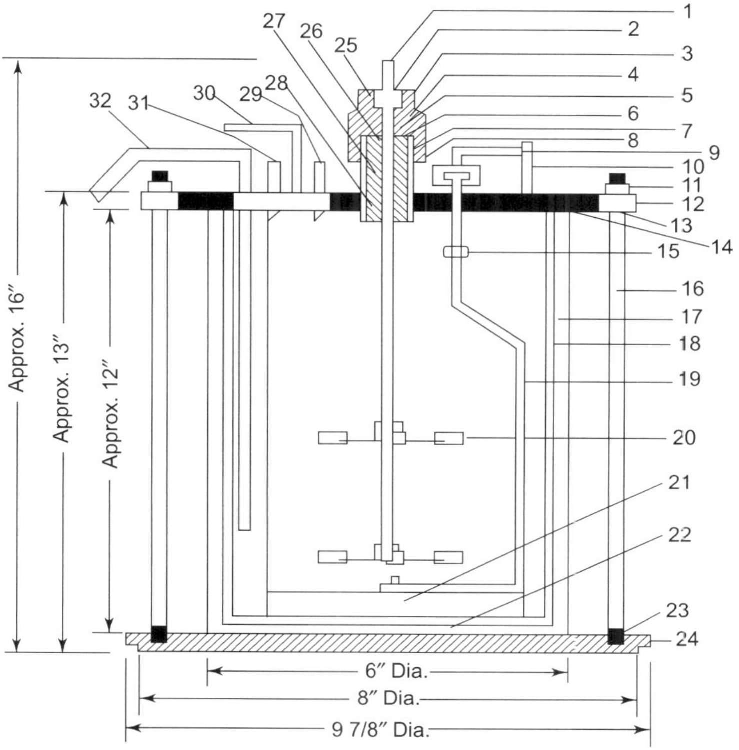

1. Agitator shaft, 15″ long, 0.3125″ dia.
2. Agitator shoulder, 0.5″ long, 0.5″ dia.
3. Upper graphitar thrust bearing, 0.25″ long, 0.8125″ OD, 33/64″ ID
4. Thrust water 0.125″ fluid, 0.75″ dia. force fit to shaft
5. Lower Graphitar thrust bearing, 0.25″ long, 0.8128″ OD, 0.3125 ID
6. Packing gland 0.5″ long, 25/32″ OD, 21/64″ ID, 1/16″ collar, 7/8″ dia.
7. Stuffing box head made from hex stock and turned to fit jam
8. Jam nut hex stock
9. Air line, 1/8″ standard pipe
10. Thermometer well, 7/16″ OD, 5/16″ ID, extended 9″ into fermenter
11. Lining nut
12. Fermenter head plate, 0.5″ plate, 8″ dia.
13. Rubber gasket
14. Inoculation port with cap
15. Union
16. Stud
17. Cylindrical pyrex museum jar, 6″ dia. 12″ tall
18. Baffles (4), 5/8″ wide, 11.5″ long No. 16 gauge
19. Sparger
20. Four-blade turbine impeller, 3″ overall dia.
21. Tie ring
22. Rubber pad
23. Fermenter base plate, 0.5″ plate, 8″ dia.
24. Positioning leg
25. Snap ring
26. Packing
27. Graphitar main bearing, 1.25″ long 0.8128″ OD, 0.3125" ID
28. Stuffing box fixed section, 2.395″ long, 1.25″ outside dia. 0.8125″ ID
29. Water addition line, 1/8″ dia. standard pipe
30. Gas outlet line 1/8″ dia. standard pipe
31. Antifoam addition line, 1/8″ dia. standard pipe
32. Sampling line, 1/8″ dia. standard pipe

Fig. 5.4 *Detail diagram of a laboratory fermenter (Reprinted from Bartholomew, W.H., Karow, E.O., and Sfat, M.R., Ind. Eng. Chem., Vol. 42(9), 1827, Copyright © (1950), with permission from American Chemical Society.)*

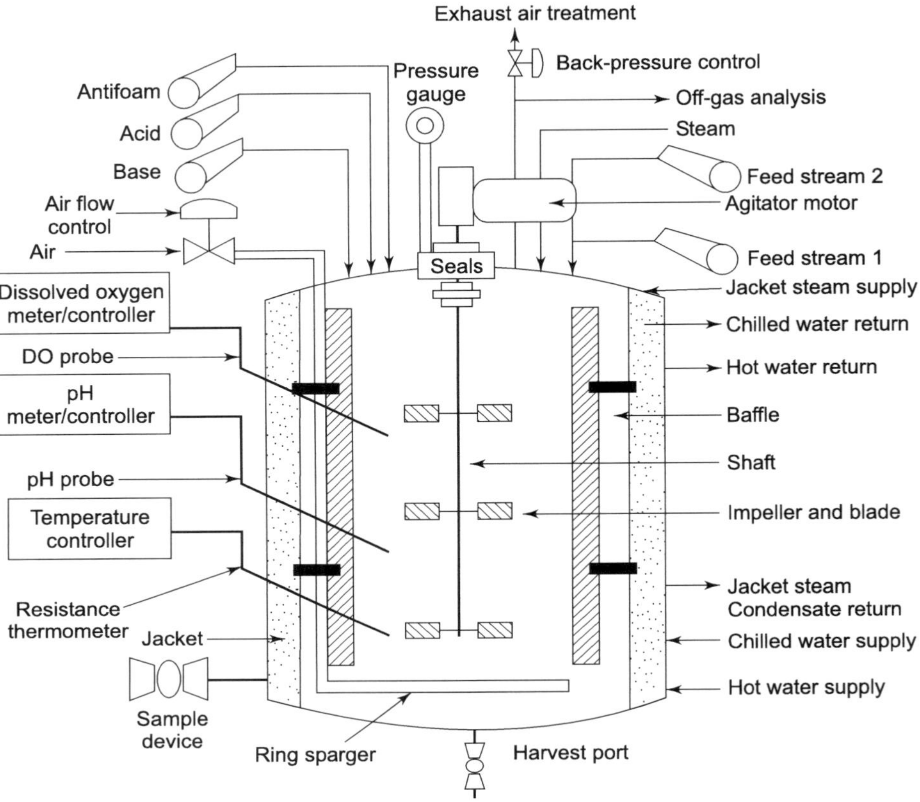

Fig. 5.5 *Representation of a jacketed fermenter. (Reprinted from "Encyclopaedia of Chemical Technology", Sengha, S.S.; Kroschwitz, J.I. (Ed.), 4th Edn, Vol. 10, p. 373, Copyright © (1993), with permission from John Wiley & Sons, Inc.*

5.4 MEDIA DESIGN FOR FERMENTATION

Any fermentation process proceeds through the action of microorganisms which perform in the presence of a medium. Hence, proper design of the medium is an essential component in the design of a fermentation process. Thus, detailed investigations are needed to identify the most suitable medium for any fermentation process to proceed.

5.4.1 Medium Requirements

Since the medium is desired to support the functioning of microorganisms, the requirements of the medium are decided by those of the microorganisms. They are:

- carbon
- nitrogen
- energy source
- minerals
- other nutrients like vitamins, etc.
- oxygen/air for aerobic processes.
- water.

The medium used in a laboratory-scale process, or for that matter even in the pilot plant-scale level, can be reasonably composed of pure components; but such a luxury is not affordable in the case of commercial production, where the cost of production rules the economic viability, and hence the commercial viability of the process. Thus, for large-scale productions, we look for a medium, which has the following attributes:

- It should be cheap, and easily available at a consistent cost and quality.
- It should have higher productivity, i.e. it should produce maximum amount of product per unit of the substrate consumed.
- The rate of formation of the product should be high.
- It should minimize the formation of undesired products.

Considering the above points, we look for the materials which supply the requirements of the medium. Some examples of fermentation media were compiled by Stanbury and Whitaker (1993) and are presented in Table 5.4.

Table 5.4 *Some examples of fermentation media*

Itaconic acid		**Gibberellic acid**	
Cane molasses (as sugar)	150 g/dm^3	Glucose monohydrate	20 g/dm^3
$ZnSO_4$	1.0 g/dm^3	$MgSO_4$	1 g/dm^3
$MgSO_4 \cdot 7H_2O$	3.0 g/dm^3	NH_4NO_3	1 g/dm^3
$CuSO_4 \cdot 5H_2O$	0.01 g/dm^3	KH_2PO_4	5 g/dm^3
		$FeSO_4 \cdot 7H_2O$	0.01 g/dm^3
		$MnSO_4 \cdot 4H_2O$	0.01 g/dm^3
		$ZnSO_4 \cdot 7H_2O$	0.01 g/dm^3
		$CuSO_4 \cdot 5H_2O$	0.01 g/dm^3
		Corn steep liquor (as dry solids)	7.5 g/dm^3
Amylase		**Glutamic acid**	
Ground soybean meal	1.85%	Dextrose $NH_4H_2PO_4$	270 g/dm^3
Autolysed brewers			2 g/dm^3
yeast fractions	1.50%	$(NH_4)_2HPO_4$	2 g/dm^3
Distillers dried solubles	0.76%	K_2SO_4	2 g/dm^3
NZ-amine (enzymatic		$MgSO_4 \cdot 7H_2O$	0.5 g/dm^3
casein hydrolysate)	0.65%	$MnSO_4 \cdot H_2O$	0.04 g/dm^3
Lactose	4.75%	$FeSO_4 \cdot 7H_2O$	0.02 g/dm^3
$MgSO_4 \cdot 7H_2O$	0.04%	Polyglycol 2000	0.3 g/dm^3
Hodag KG-I antifoam	0.05%	Biotin	12 μg/dm^3
		Penicillin	11 μg/dm^3
Riboflavin		**Penicillin**	
Soybean oil	20 cm^3/dm^3	Glucose or molasses	
Glycerol	20 cm^3/dm^3	(by continuous feed)	10% of total
Technical grade glucose	20 g/dm^3	Corn-steep liquor	4–5% of total
Corn-steep liquor	12 cm^3/dm^3	Phenylacetic acid	0.5 – 0.8% of total
Casein	12 g/dm^3	(by continuous feed)	
KH_2PO_4	1 g/dm^3	Lard oil (or vegetable oil) antifoam by continuous addition pH to 6.5–7.5 by acid or alkali addition	0.5% of total

(Reprinted from "Principles of Fermentation Technology", Stanbury, P.F., and Whitaker, A., p. 75, Copyright © (1993), with permission from Elsevier.)

Carbon Source The carbon requirement for the medium is normally provided by molasses, starches, some cereal grains like maize, tubers like potatoes and cassava, sucrose, glucose, and lactose. The carbon source also provides the energy requirements for the medium. By and large, one of the cheapest sources of carbohydrates which provide the carbon requirement for the medium is molasses, a byproduct of the sugar industry. Another important source is corn starch. Malt extract, which is an aqueous extract of the malted barley, is also an excellent substrate for many fungi, yeast and actinomycetes. The selection of the carbon source plays a major role in the economics of the processes, particularly in those cases where the raw material constitutes 60–75% of the cost of production of the product.

Nitrogen Source Microorganisms can utilize the nitrogen from organic or inorganic sources. Ammonia and salts of ammonia (like ammonium sulphate or nitrate) are common inorganic sources of nitrogen salts which provide both acidic and basic environments, depending upon the type of the salt. When once the ammonium ion is consumed, ammonium sulphate creates acidic environment, whereas ammonium nitrate provides basic environment.

Amino acids, proteins or urea serve as organic sources of nitrogen. Microorganisms are able to grow faster in the presence of organic nitrogen, and some of them will have an absolute need for amino acids. Using pure amino acids is often expensive. We may prefer to use some precursor to the amino acid. For example, metheonine and threonine are obtained from soyabean hydrolysate, which are in turn used in the production of lysine. The following are a few nitrogen sources through proteins or amino acids:

- corn steep liquor
- soya meal
- soya beans
- groundnut meal
- cotton seed meal
- fish meal
- casein hydrolysate
- slaughter house wastes
- yeast extract, etc.

Energy Source The energy required for the biochemical reaction comes either from the oxidation of the medium constituents or from the light. Photosynthetic bacteria and algae avail light energy, but most of the microorganisms are **chemo-organotrophs** which get the energy from carbon sources. Carbohydrates, lipids and proteins are the most common sources to get the energy. Adenosine triphosphate (ATP) is the most important compound in energy transformations in cells (Aiba, et al., 1965). It enables some of the most unfavourable reactions to proceed at a considerable rate. The oxidation of inorganic compounds in the medium enables the autotrophic bacteria to generate ATP, whereas ATP is formed in fungi, yeast and heterotrophic bacteria while oxidizing organic compounds.

Minerals Minerals like magnesium and phosphorous are important in the medium, since they are concerned with all energy transport reactions involving ATP. Other minerals like potassium, sulphur and calcium are also important in the medium. They are also found in significant quantities in most of the cells. Some other minerals in trace quantities, viz. iron, cobalt, copper, zinc, manganese and molybdenum are essential, but they are available in the medium ingredients as impurities. These micronutrients are required by the cells (see Table 2.7).

Other Nutrients Other nutrients generally required in the medium are some of the vitamins. Some or most of the vitamins are generally present in the carbon or nitrogen sources. If any specific vitamin deficiency is felt in the fermentation process, it can be separately added by some other outside sources. If only one vitamin is required, it is economical to add it to the medium in the pure form rather than adding a mixture of vitamins from a cheaper source. For example, calcium pantothenate, which is required in the production of vinegar, was added in one medium formulation (Beaman, 1967).

Oxygen/Air Oxygen, which is required in the aerobic fermentation processes, is supplied from outside by sparging air. Even though the desired active component is oxygen, it is rarely supplied in the pure form. The reasons are both economical and technical. Air, which is a cheaper source of oxygen, is generally sparged through the fermenter. This is very economical compared to using pure oxygen (which approximately costs Rs 8000 per ton). Technically, pure oxygen causes a local rise in the partial pressure of oxygen, and this causes excess growth of organisms and excess release of heat which raises the temperature locally and takes the reaction away from control.

Air is generally sparged into the fermenter through sparger of various designs, and the air is dispersed into the fermentation broth through different types of impellers. The impeller design is based on the physico-chemical and thermal properties of the liquid. The following are a few common types of impellers (Doran, 1995):

- anchor
- propeller
- flat blade disc turbine
- paddle type
- helical ribbon.

The basic purpose of designs of impellers (as shown above) is to disperse air in the medium. A good number of studies was reported on the mixing pattern and power requirements for various impellers, etc.; the details on the above are beyond the scope of the book (also see Fig. 11.13 and 11.14).

Water Water is an important requirement in any fermentation process, particularly for most of the submerged fermentation processes. It is a major component in the preparation of the medium. It is also used for most of the peripheral services like cleaning, etc. Thus, a good source of quality water is an important aspect in the selection of the location of the plant.

The water quality is determined by

- pH
- dissolved salts
- minerals
- microbial contamination
- contamination by effluents
- chlorine content in case of treated (municipal) water supply.

The water available from the natural sources is cleaned by means of deionization technique or some water softening methods before being used in the fermentation process. If the water requirement is high, like in algae production units, the used water is recycled after certain amount of fortification or adjustment to meet the requirements.

5.4.2 Medium Formulation

Formulation of medium is one of the important parameters in successful fermentation processing. It needs to cater to the requirements of:

- optimal growth of the cells
- product formation,

irrespective of the scale of processing; may it be at laboratory scale, pilot plant scale or at industrial scale. More stress needs to be given for industrial processing because the cost of the medium plays a vital role in deciding the economic viability of the fermentation process; whereas at the laboratory scale formulation, the stress is more to elicit the information on the technical feasibility. It also needs to supply the energy requirement component for cell maintenance and for biosynthesis. Obviously, the medium formulation has to meet all the requirements mentioned in Section 5.4.1. The equation, based on the stoichiometry for product formation and growth of cells, can be given (Stanbury and Whitaker, 1993) as:

carbon and energy source		cell biomass
+		+
nitrogen source	$\longrightarrow$	products
+		+
other requirements		$CO_2 + H_2O$ + Energy

The above equation reasonably gives information for media formulation. For example, the requirements of the nutrients can be calculated or predicted from the amount of biomass produced. If the stoichiometry of the fermentation process is known, it is possible to know how much of the substrate is required to produce a desired amount or yield of the biosynthetic product. The product yield should give us information to find the cell density required and hence the cell growth requirements. Based on the cellular 'yield coefficient' (which is defined as the ratio of the cell dry matter produced to the quantity of the carbon substrate utilized), it should be possible for us to calculate the quantity of the substrate required, and hence the formulation of the medium to meet the demand for the carbon source.

5.4.3 Simple and Complex Media

Depending on the type of microorganism used (whether autotrophic or heterotrophic) the medium could be simple or complex (Casida, 1996). Autotrophs depend on carbon dioxide as a source of carbon. Thus, autotrophs will have improved growth in an environment rich in carbon dioxide; whereas, heterotrophs usually fix carbon dioxide during growth; and atmospheric air generally supplies adequate quantity of carbon dioxide for this purpose.

Autotrophs generally require simple inorganic media. They require:

- inorganic salts
- water
- nitrogen source
- carbon source (usually carbon dioxide from air).

They synthesize, from the inorganic nutrients, the complex organic compounds required for growth and survival of the cells. Their total energy requirements are met from the oxidation of the inorganic salts in the medium.

A diametrically opposite class of microorganisms are the highly fastidious class which cannot synthesize many of their sustenance and growth requirements. Hence, the media designed for them should contain all types of simple to complex preformed nutrients. A classical example is some of the lactic acid bacteria.

These two types of organisms are two extremes, and most of the organisms function in between these two extremes. However, in addition to the above considerations, fermentation growth conditions also impose a certain amount of "metabolic stress" on the organisms. We generally employ high aeration rates in aerobic fermentation processes, and also provide high substrate concentrations, which demand higher amount of nutrients and growth factors in commercial fermentation, which is not normally desired in laboratory-scale processes. This situation calls for complex media, and hence are normally employed.

Media Design and Usage A further classification of media is:

- synthetic medium
- crude medium.

A **synthetic medium** is one in which all constituents are specifically defined and added more or less in pure form. Their composition and concentrations are pre-defined. Thus, a typical synthetic medium containing inorganic salts, water, purified sugar and magnesium sulphate, etc. is given in Table 5.5, which is used for the production of thermostable α-amylase in the laboratory. Composition of various other laboratory media are available in any manual on microbiology (Gunasekaran, 1995; Refai, 1979).

Table 5.5 *Minimal medium composition for production of α-amylase in laboratory*

Component	*Amount*
Soluble starch (g/l)	20.0
Glucose (g/l)	5.0
Peptone (g/l)	5.0
Yeast extract (g/l)	2.5
K_2HPO_4 (g/l)	2.0
$MgSO_4 \cdot 7H_2O$ (g/l)	1.0
pH	7.0
Distilled water (ml)	100.0

Synthetic medium has some specific advantages. Since the chemical structure and composition of various components of the medium are known, their effect on the product yield can be assessed by specifically altering one of their compositions. Thus, the composition of one or several components can be varied to assess the product yield. This will help in re-designing the medium for maximum product yield. These studies also help in finding out the metabolic process route leading to the formation of the desired product. Radioactive isotopes can be incorporated into one of the components of the medium, which help us to study the effect of it on the product yield by identifying whether the radioactive isotopic component is present in the product or not, and if so, to what extent. Based on these studies, the medium can be designed. The synthetic media have got various other advantages given as follows:

- It will help in reproducibility of the results from time to time, because the composition of the medium is well-defined.
- The problem of foaming is almost nil since the foaming agents like proteins and high molecular weight peptides are absent.
- The recovery and purification of the fermentation products are much easier, since the medium can be designed by avoiding the interfering components.

Sometimes, all the above advantages will be counter-balanced by the single disadvantage of higher cost of the medium, which destabilizes the economic viability of the process. However, the synthetic media are generally used in the developmental stage, i.e. at the R&D stage.

Another source of medium is the **crude medium**, which does not have any of the advantages of the synthetic medium. Their compositions are not defined. They are available as they are, e.g. soya bean meal, molasses, corn steep liquor, ammonium sulphate, etc. The main feature of the crude medium is its economic advantage. It also contains the sources of carbon and nitrogen in a form that can be easily consumed by the microorganisms. The crude media are generally used in the industrial scale of production after perfecting the composition of the medium at the R&D stage. Some of the sources of the crude media are shown in Table 5.6.

Table 5.6 *Sources for crude medium*

Carbon sources			*Nitrogen sources*	*Nutritive sources*
Sugars	*Carbohydrates*	*Cellulose by-products*		
Beet molasses	Corn	Wood waste	Salts	Molasses
Cane molasses	Wheat	Oat hulls	Nitrate	Corn steep liquor
Whey	Rhy	Corn cobs	Urea	Sulphite waste liquor
Sulphite waste liquor	Milo	Straw	Corn steep liquor	
Cull fruits	Rice		Casein	
Cannery wastes	Potatoes		Cereal grains	
	Sweet potato		Peptones	
			Fish meal	
			Stick liquor	
			Meat scraps	
			Soya bean meal	
			Yeast extracts	
			Cotton seed meal	
			Linseed oil meal	
			Peanut oil meal	

Applications for Industrial Fermentations The requirements of the medium and the sources of supply have already been discussed earlier. We have also seen what are the attributes of the medium; the basic attribute being its cost, availability and ability to yield the desired product. We now discuss some of the frequently used substrates which are the sources of carbon, nitrogen, etc. for the medium and their applications are given in Table 5.7 (Crueger and Crueger, 1989). Obviously, most of them are for the crude media, since the synthetic media are mostly used in the developmental stage.

Table 5.7 *Industrial applications of some naturally available substrates*

Substrate	*Source of C/N2*	*Industrial application*
Malt	—	Brewing of beer
Beet/Cane molasses	—	Many fermentation processes
Barley malt	—	Scotch whisky
Molasses	Carbon	Alcohol, butanol, acetic acid (vinegar), acetone, citric acid, etc.
Starch and dextrins	Carbon	For ethanol fermentation
Sulphite waste liquor	Carbon	Cultivation of yeast
Cellulose	Carbon	On hydrolysis, cellylose yields sugar syrup which is used for alcohol fermentation, fermetative production of butanol, acetone and isopropanol
Whey	Carbon	The lactose is used for the production of ethanol, SCP, xanthan gum, vitamin B_{12}, 2,3-butanediol, lactic acid and gibberellic acid
Vegetable-based fats like soya, palm oil, olive oil	Carbon	Antibiotic fermentation
Methanol	Carbon	SCP, glutamic acid, serine and vitamin B_{12}
Corn steep liquor	Nitrogen	Production of lactic acid
Soya meal	Nitrogen	Antibiotic fermentation

5.5 AEROBIC AND ANAEROBIC FERMENTATION PROCESSES

Fermentation carried out in the presence of air/oxygen is known as aerobic fermentation, whereas that in the absence of air is called anaerobic fermentation. Industrially, a few processes demand total absence of oxygen. Many processes require a nominal amount of oxygen or at least they are not affected by oxygen.

5.5.1 Anaerobic Fermentation

Yeast fermentation process to produce alcohol requires a small amount of aeration for the cells to multiply. Afterwards, no air is required. On the contrary, air is detrimental for the process, which will otherwise oxidise the substrate. Most of the anaerobic fermentations produce carbon dioxide gas. Many times, the gas covers the surface and acts as a blanket to prevent the effect of oxygen. The evolved carbon dioxide will also help in better mixing conditions, which is more evident in large industrial tanks because of longer pathways for the gas bubbles to go before they leave the fermenter. Mechanical agitation, though improves the degree of mixing, does not seem to have any significant effect (Webb, 1964(b)).

The anaerobic digestion of the sludge is also one classical process of anaerobic processing. Wet sludge or effluent with solids is produced in large quantities in most of the processing operations. For example, mango processing produces 50% of effluents which amounts to 9.2 million tonnes per annum (Mokshapathy, 1988). These wet sludges need to be processed and reduced in volumes before they are let into the outside environment. It can be done by sludge digestion in an anaerobic fermenter to produce huge quantities of methane and carbon dioxide, using anaerobic bacteria, in large tanks which may be of capacity up to 12,000 m^3.

5.5.2 Aerobic Fermentation

Sparging air/oxygen is a very common phenomenon in fermentation processes to supply oxygen for the cells to meet their specific oxygen demands. Such fermentation processes which are associated with the bubbling of oxygen are termed as aerobic fermentation. There are normally three categories of aerobic fermentation (Webb, 1964(b)):

- For growing organisms on solid or semi-solid medium in the presence of air, though air is not specifically blown. This is also known as **solid state fermentation.**
- For growing organism in liquid medium, air is not specifically blown or sparged, but air helps the growth of organisms. The air enters into the system through a cotton plug kept on the top, either in static condition or in shake flask experiments.
- Air is dispersed in the form of fine bubbles through a sparger or some other kind of bubbling mechanism, so that the oxygen gets dissolved in the liquid medium and is available for the cells or microorganisms.

Industrial production of antibiotics has given the main boost for the dispersion of air in the liquid medium, and the whole subject of "air sparging" has gained significance in biochemical engineering. In place of air, pure oxygen can also be dispersed, but some times pure oxygen is detrimental to the microorganisms, mainly because of generation of excess heat locally. Hence, unless and otherwise it is specially demanded, pure oxygen is not used in aerobic fermentation. In addition to this, air, which is a mixture of oxygen and nitrogen, contributes to the escape of undissolved and unutilised nitrogen, which in turn helps in the mixing and desorption of product gases like carbon dioxide.

'Effluent treatment' is one very important industrial application of aerobic processes. Most of the industrial effluents, before they are let out into the environment, are treated for reducing the BOD. In most of the effluent-treatment plants, the effluents are treated by sparging air/oxygen with a heavy amount of agitation using propeller or vaned disc impellers. The high-speed agitation gives a vigorous mixing, so that the effluent will be able to absorb atmospheric oxygen (Fig. 18.1).

5.6 SOLID STATE AND SUBMERGED FERMENTATION AND THEIR APPLICATIONS

5.6.1 Solid State Fermentation

Solid state fermentation (SSF) is a method of growing microorganisms in an environment of limited moisture without having free flowing water. That is, microorganisms grow on a solid surface which is moistened and which has also got free access to air. Thus, it may be viewed that the growth environment is similar to natural processes in which ideal conditions are provided for the growth of certain organisms for the production of certain metabolites. Thus, SSF is a process which has been in vogue from time immemorial, but has been neglected as a science subsequently. To some extent, SSF is being practised in oriental countries including China, Japan, Korea, etc., where it is referred to as "koji" fermentation, for the production of soya products such as tempeh, soya sauce, etc. However, in the last two decades, there has been some renewed interest on the SSF processing in view of its some obvious advantages such as lower manufacturing costs because of the use of crude solid agro wastes like wheat bran. The solid surface directly comes in contact with the air and hence aeration costs are avoided. Lonsane (1994) summarised some of the industrial and economic advantages of the solid state fermentation process. They are:

- low capital investment and recurring expenditures
- low water utilisation and hence negligible outflow of waste water
- low energy requirements for the fermentation process because of absence of agitation
- absence of foam formation because of absence of excess water
- high reproducibility of the results
- simple fermentation media
- less fermentation space, and less complex plant and machinery
- absence of rigorous control techniques
- any level of scale of operation
- absence of elaborate aeration requirements
- ease in controlling bacterial contamination
- facility of using wet and dry fermented solids directly
- ease in induction and suppression of spores
- lower costs of downstream processing.

Some of the biochemical engineering aspects of solid state fermentation were discussed by Ramanamurthy et al. (1993) in sufficient detail.

These include mainly:

- the interaction of transport processes, viz. heat and mass transfer aspects, with the bioreaction kinetics
- the inter-particle and intra-particle diffusional limitations of oxygen on the growth of microorganisms, etc.

5.6.2 Submerged Fermentation

In the case of submerged fermentation (SmF) the microorganisms and the substrate are present in the submerged state in the liquid medium, where a large quantity in the form of solvent is present. This has many advantages over SSF, and hence became more popular. Since the contents are in submerged state in the liquid medium, the transfer of heat and mass is more efficient, and is amenable for modeling the process. The scaling-up of the process is very easy. Some of the essential features and differences in SSF and SmF are shown in Table 5.8.

Table 5.8 *Essential differences between SSF and SmF*

Characteristic feature	*SSF*	*SmF*
Condition of microorganisms and substrate	Static	Agitated
Status of substrate	Crude	Refined
Nature of microorganism	Fungal systems	—
Availability of water	Limited	High
Supply of oxygen	By diffusion	By bubbling/sparging
Contact with oxygen	Direct	Dissolved oxygen
Requirement of fermentation medium	Small	Huge
Energy requirements	Low	High
Study of kinetics	Complex	Easy
Temperature and concentration gradients	Steep	Smooth
Controlling of reaction	Difficult	Easy
Chances of bacterial contamination	Negligible	High
Quantity of liquids to be disposed	Low	High
Pollution problems	Low	High

5.6.3 Applications

- Citric acid can be produced by both SSF and SmF. Generally, the latter technique is used industrially. SSF method has not yet become a commercial success because of its labour intensity (Marison, 1988).
- Soya-based oriental foods like tempeh and soya sauce are produced by SSF alone.
- The leavening of bread by yeast is a SSF process.
- Production of Roquefort cheese from sheep milk is a classical example of SSF. This process is in existence in Southern France for many thousands of years.
- Mushroom cultivation is another example of the growth of fungus on solid medium like paddy straw. Mushroom is directly consumed as a vegetable.
- Fish and meat products are preserved in the form of sausages as fermented foods.

5.7 VARIOUS TYPES OF BIOREACTORS

A biochemical reactor is a vessel in which substrate (reactant) undergoes a biochemical transformation by the action of living cells, microorganisms or cell-free enzyme systems. In biochemical engineering, we use terms like

- biochemical reactor
- biological reactor
- bioreactor
- fermenter
- microbial reactor

which are all synonymous. Thus, essentially a bioreactor is for carrying out

- microbial fermentations
- enzyme (cell-free) reactions.

Depending on process requirements, different types of bioreactors are used, and are operated as

- aerobic
- anaerobic
- solid state
- enzymatic → immobilised / soluble
- well-mixed/stratified.

A generalised outline of the biochemical process is illustrated as follows (Moo-Young and Blanch, 1981), where the central importance of a bioreactor is highlighted.

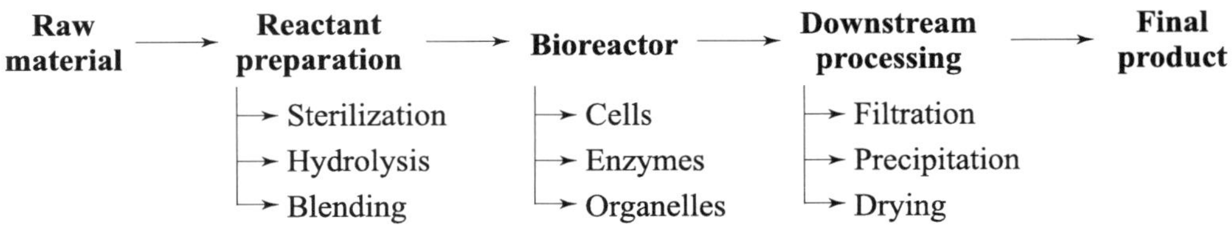

Fermentations give rise to a variety of products based on the type of cell and its growth pattern. They are:

- primary metabolites such as alcohol, citric acid
- biomass such as Baker's yeast, single cell protein (SCP)
- transformed substrates such as steroids
- purified solvents as in the case of water treatment.

Most of the biochemical reactions are accompanied by interphase mass and heat transfer for bringing out a desired biochemical change with the inter-play of fluid dynamics and mixing of various components, which are all complicated because of changes in physico-chemical characteristics of the fermentation broth with the progress of biochemical conversion. Hence, these operations will have to handle:

- fluid–fluid systems
- fluid–solid systems
- fluid–fluid–solid systems.

The biochemical systems are characterised by the fact that the biochemical reactions involve a living organism (a cell) which grows and proliferates during the course of the reaction. Thus, the kinetics of the cell growth is superimposed on the kinetics of the product formation. That is, the substrate is partly consumed for:

- cell growth
- product formation.

The cells perform both bioconversion and biotransformation depending on the environmental conditions. Another characteristic feature of the biochemical system is that the microbial growth rates are considerably lower than that of the chemical pathways, and hence they require relatively larger reactor volumes with high residence times.

Enzymatic reactions are yet another important class of biochemical reactions, which can be operated both in soluble and immobilised (insoluble) form. Their behaviour is analogous to homogeneous and heterogeneous catalytic chemical reactions, in which principles of reaction and diffusion play a major role. In order to meet the requirements of such a wide spectrum of situations, various types of bioreactors are in vogue. Some of the principle conventional reactor types are:

- stirred tanks
- bubble columns
- packed beds
- fluidised beds
- trickle beds
- air lift towers, etc.

Analogous to chemical reactor theory (Resnick, 1981), all the above types of bioreactors can be classified, based on the feeding mechanism, into:

- batch fermenters
- fed-batch (semi-batch) fermenters
- continuous stirred tank fermenters.

They will be discussed in detail in the subsequent sections.

Based on the state of operation, i.e. steady state or unsteady state, and the flow characteristics (or the mixing mechanism) in the bioreactors, they can be classified into three types of flow regimes which are shown in Fig. 5.6. In an 'unsteady' state operation, the concentration of reactants and products change with time, whereas in the 'steady' state operation, the concentration does not change with time. However, the concentration may or may not change with the position in the reactor. The ideal unsteady state operated reactor is a batch reactor in which the reactants are fed into the reactor. The inlet and outlet of the reactor are subsequently closed (Fig. 5.6(a)). The contents are allowed to react. Fully mixed conditions prevail in the reactor, i.e. the concentration at any position in the reactor is same; but the concentration changes with time. As time progresses, the concentration of the reactant falls and that of the product builds up, which is a characteristic feature of unsteady state operation.

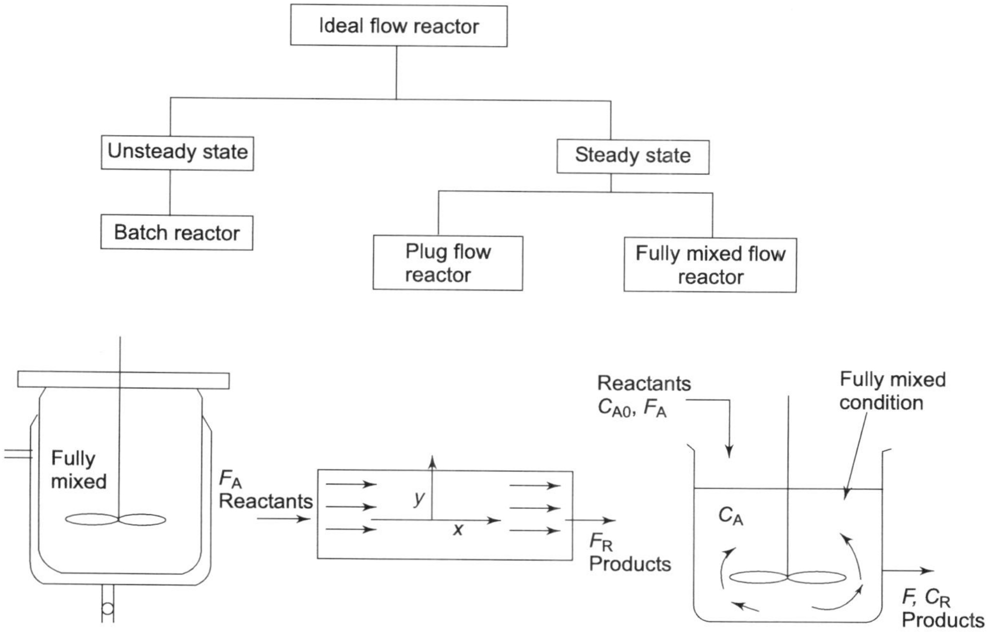

(a) Batch reactor **(b) Plug flow reactor** **(c) CSTR (fully mixed flow reactor)**

Fig. 5.6 *Classification of various types of ideal flow reactors*

Under steady state conditions, we have two situations. The flow in the reactor could be like a piston flow in which the reactants do not mix at all, and we call this as '**plug flow**'; or the reactants fully mix the moment they enter the reactor, and the concentration is uniform throughout the reactor (like a batch reactor), which we call as 'continuous stirred tank reactor' (CSTR). In the plug flow conditions, shown in Fig. 5.6(b), the reactants enter the reactor from one side, reaction takes place along the length of the reactor, and the products leave the reactor from the other side. Inside the reactor, the reaction progresses progressively, and the concentration will be changing as we move forward, i.e. the change in concentration takes place progressively in the axial direction. There will not be any change in concentration in the radial direction. The reactants takes place in an orderly direction as if they are like a plug/piston. The reaction takes place along the axis, and hence the concentration of the reactants falls and that of the products increases in the axial direction (i.e. in the X-axis); however, there will not be any changes in the radial direction (i.e. in the Y-axis). There also will not be any accumulation of reactants or products which is a typical character of 'steady state flow' conditions.

The third type of ideal reactor flow conditions is a fully mixed flow condition (Fig. 5.6(c)). In this, the reactant with initial concentration C_{A0} enters the reactor continuously at the rate of F_A. The contents in the reactor are perfectly mixed. Hence, the concentration of the reactant falls, the moment it enters the reactor to C_A instantaneously which is the concentration prevailing in the reactor. The contents are allowed to react in the reactor, and the products form. The products leave the reactor at a flow rate of F_R and a concentration of C_R continuously. These conditions prevail in the reactor as long as the external and internal conditions do not change in the reactor; the volume remains constant and there will not be any accumulation of reactants or products in the reactor. Hence, the reactor is typically called as continuous stirred tank reactor, which is a case of existence of perfect mixing conditions to prevail in the reactor.

Thus, the two ideal continuous flow conditions are: (i) total plug flow (non-mixing) conditions, and (ii) total mixing conditions. However, in reality, realization of either is impossible. Most of the reactors operate in between these two extreme conditions (which we call as non-ideal flow, and will be discussed in detail in Chapter 10). Mathematical equations are developed to model these two ideal flow regimes, which are helpful in the design of reactors/bioreactors.

5.7.1 Classification Based on Feeding Mechanism

Based on the way the reactants are fed to the reactor, the reactor operations can be classified as:

(i) batch
(ii) continuous
(iii) semi-batch/semi-continuous/fed-batch

They are discussed briefly here.

(i) Batch reactor As has already been mentioned, the batch reactor is fed with all the reactants and the contents are sealed (Fig. 5.6(a)). There will not be any further in-flow or out-flow of materials. The products are withdrawn after the reaction time is over. The batch reactors are ideally suited for:

- small-scale operations
- for testing new processes at R&D stage
- for making some of the expensive products
- for making various types of products on shift basis with ease and flexibility
- for processes that are difficult to be translated into continuous operations.

However, the batch reactors are not free from disadvantages. For example, the operating costs are high, and hence the large-scale production in batch processes is not economical.

(ii) Continuous Reactors The two types of continuous reactors have already been discussed previously (Fig. 5.6(b) & (c)). By and large, we refer to the CSTR as a continuous reactor. Since the contents are thoroughly mixed with a stirrer, good mixing conditions can be achieved with good maintenance of uniform temperature. Such configurations for cultivation of cells are generally referred to as **chemostats** in biochemical terminology.

The typical disadvantage of CSTR is that the conversion per unit reactor volume is the lowest, which results in higher reactor volumes for the given desired conversion/yield (Fogler, 1986).

(iii) Fed-batch/Semi-batch Reactors Semi-batch or semi-continuous reactors are very common in biochemical operations, and they are also popularly referred to as 'fed-batch reactors'. Their operation is intermediate to batch operation and continuous operation, and hence the name. There are four types of feeding methods which are possible, as shown in Fig. 5.7. The first three types of feeding methods were shown by Levenspiel (1974).

Suppose there are two reactants, viz. A and B:

(a) One possibility is that one reactant A is taken in the reactor as a batch, and rectant B is added continuously, with the result that both the volume and composition of the contents in the reactor change.

(b) Another possibility is that both the reactants A and B are continuously added, but the products are not withdrawn, with the result that the volume alone changes, but the composition does not change.

(c) A third possibility is that the reactant A is fed initially as batch, but the reactant B is added continuously, and the product is withdrawn continuously, with the result that the volume of the contents in the reactor does not change, but their composition changes continuously.

(d) A fourth possibility is that if B is a gaseous reactant, it enters continuously and the unreacted gas leaves continuously, but the product does not leave continuously. Thus, the volume of the reactor contents does not change appreciably, but the composition changes. This is a typical case of an aerobic fermentation process, in which the air is bubbled continuously, the oxygen is absorbed by the fermentation broth, and the unreactable nitrogen leaves continuously. The most common example is production of antibiotics.

It is difficult to analyse the data of the semi-batch reactor operations compared to the batch and continuous systems, though it is flexible in operation (Levenspial, 1974). It offers a good control over the reaction speed (progress), as one of the reactants is in limited supply. Control of reaction temperature and pressure is possible. Unwanted side reactions can be avoided or restricted. Most of the gas–liquid reactions are carried out in semi-batch reactors, in which the gas is continuously bubbled through the reactor into the liquid reactant.

5.7.2 Various Other Types of Reactors

A large number of types of bioreactors are used in fermentation processes. Some of them are very common, and some are less common. There is a good review on less common types of bioreactors by Schugerl (1982), which have different modes of aeration for use in aerobic fermenters. However, in the present text, we restrict ourselves to some of the very common types of bioreactors. The following are a few:

(i) fluidised bed reactor
(ii) immobilised cell reactor
(iii) air-lift reactor.

A brief description on the above is presented here (also see section. 19.3).

(i) Fluidized Bed Reactor The fluidised bed reactors are used generally for intimate solid–fluid contacting. The bed of the solids is taken in a reactor with perforated bottom, and has facility for pumping the fluid from bottom. Initially, the solids are stationary in the form of a fixed bed of particles, and the fluid is pumped at a lower velocity. The fluid escapes through the crevices of the voids in the solid bed. If the

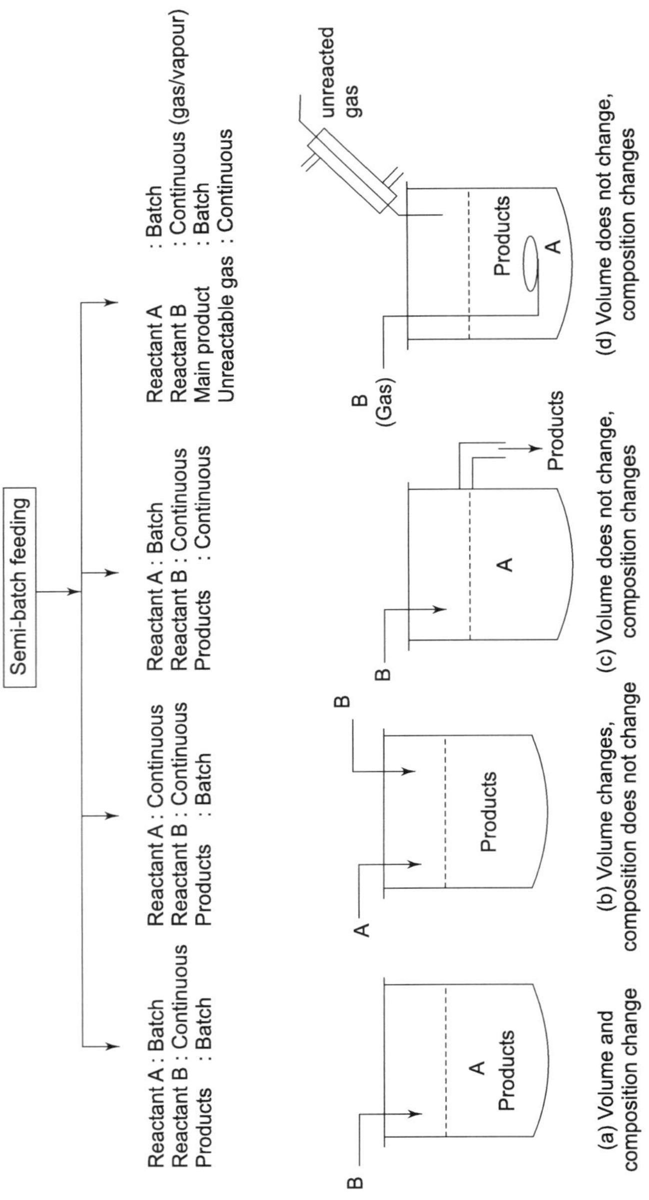

Fig. 5.7 *Various methods of semi-batch feeding of reactants A and B*

velocity is further increased, the solid particles detach from other solid particles, and start rising up. At this stage, the gravitational force of the particle, acting downwards, is countered by the buoyancy forces by the fluid going up. Any further increase in velocity of the fluid pushes the particles; they move as if they are fluids. Hence, the bed is called as 'fluidised bed' (Fig. 5.8). This velocity of the fluid at which the solid particles just start moving is known as 'minimum fluidisation velocity', and the bed is under incipient fluidization.

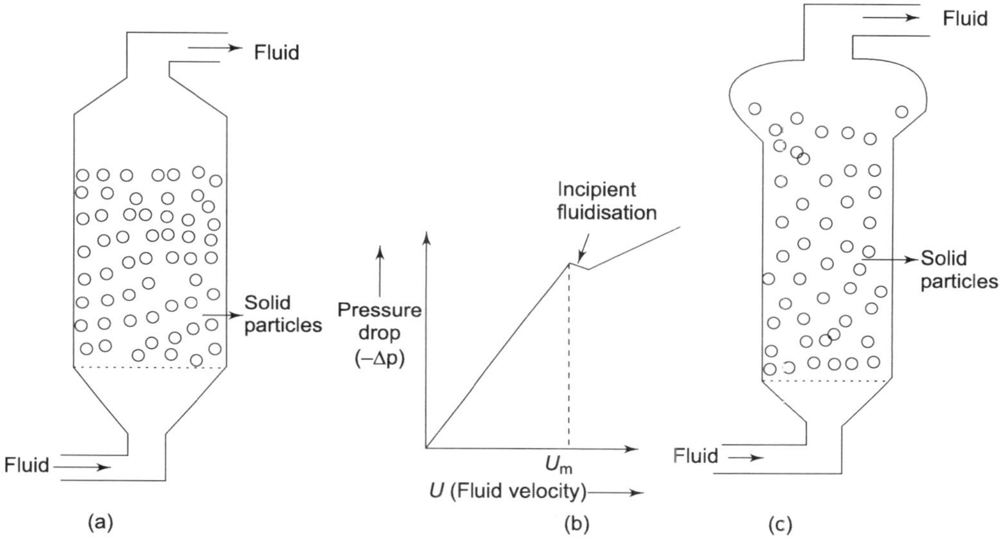

Fig. 5.8 *Fluidised bed reactors*

When once the bed is under fluidisation, all solid particles come in close contact with the fluid, and hence effective utilisation of the entire surface area of the particles is realised as they swirl and move. Thus, any surface-active or surface-catalysed reactions are effectively promoted by fluidised beds. There will be a better control of temperature by effective heat dissipation.

Some of the highly exothermic reactions can be carried out in fluidised bed reactors only. In biotechnology, the immobilised enzyme reactions are effectively carried out in fluidised bed reactors.

(ii) Immobilised Cell Bioreactors The cells used in any fermentation process are normally in the form of suspension. When the products form, they sometimes become 'intracellular' (the product is held up inside the cells). Hence, some of the cell disruption techniques are to be resorted to (Stanbury and Whitaker, 1993(c)), which are time-consuming and are not economical. To overcome these difficulties, one of the classical techniques is to immobilise the cells or enzymes. Mostly, the enzymes are immobilised. 'Immobilisation' of enzymes is a technique of entrapping of the enzymes on an inert immobilised material like calcium alginate. There are various methods of immobilisation which were described earlier in Section 4.6 (also see Bailey and Ollis, 1986). Once the enzymes are immobilised in the form of pellets, the system becomes akin to a fluid–solid system, and the fermenter operates in the form of a packed bed or fluidised bed (Fig. 5.8).

Some of the advantage of immobilisation are:

- the cell density in a local place can be increased
- the product is free of the cells and hence the product recovery is easy
- the immobilised cells can be recycled\reused

- certain types of cells, viz., the mammalian cells, grow only if they are attached to a surface; in such cases immobilisation is a necessity rather than an option
- cell morphology and viscosity of the broth can be controlled.

Once the immobilisation of the enzymes is done satisfactorily, the fermenter is similar to a fluid–solid contacting system, and the same design principles will be applicable.

(iii) Air-lift Reactor The air-lift reactor is a typical motionless reactor where the internal circulation and mixing are achieved by the bubbling air. A good number of the configurations of the air-lift reactors have been described by Schugerl (1982). A schematic representation of the air-lift reactors is shown in Fig. 5.9.

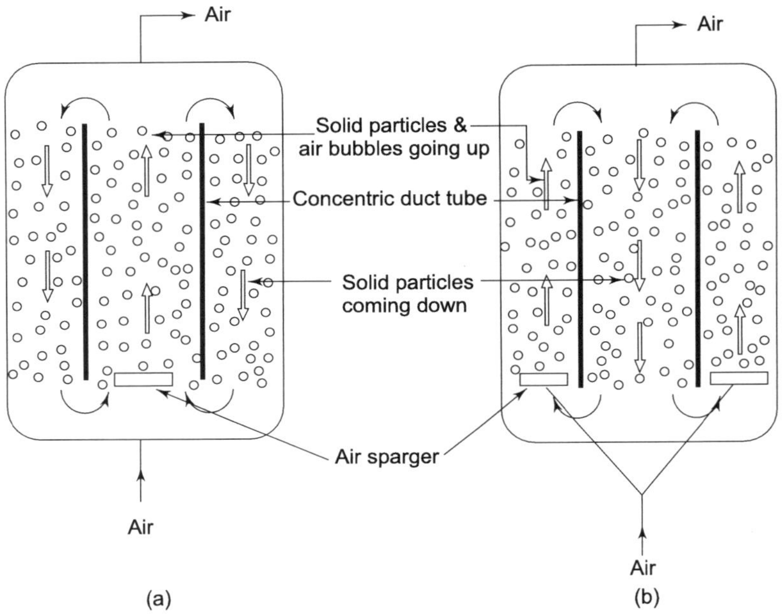

Fig. 5.9 *Air lift reactors*

The reactor vessel contains the fermentation broth, with or without biocatalyst pellets. Air, at high velocities, is admitted through the sparger either at the centre (Fig. 5.9(a)) or at the sides outside the draft tube (Fig. 5.9(b)). Because of the presence of the air, the local zone becomes less dense and rises up. When once the air bubbles along with the solid biocatalysts and the fermentation broth reach to the top, they get themselves disengaged and go into the outlet air stream. The fermentation broth, along with the solids, reaches the bottom through the down comer zone. The up-flow and down-flow movements keep the surface-active biocatalysts distributed throughout the broth (Erickson and Stephanopoulos, 1987).

Air-lift reactors have some of the following advantages:

- mixing is achieved through the air draft
- absence of moving parts for mixing, which in turn eliminates the maintenance problems associated with moving parts
- the shearing action is less on the solid particles compared to stirred tanks, which helps in reducing the disintegration of the solid structures of the particles.

Chisti (1989) gave a detailed description of air-lift bioreactors. The hydrodynamic models provide the basic design equations for air-lift reactors.

References

5.1 Aiba, S., Humphrey, A.E, and Millis, N.F. (1965), *Biochemical Engineering*, Academic Press, New York, pp. 4–15.

5.2 Atkinson, B. (1974), *Biochemical Reactors*, Pion Ltd., London, p. 4.

5.3 Bailey, J.E., and Ollis, D.F. (1986), *Biochemical Engineering Fundamentals*, 2nd Edn., McGraw-Hill, Inc., New York, pp. 595–606.

5.4 Bartholomew, W.H., Karow, E.O., and Sfat, M.R. (1950). "Design and operation of a laboratory fermenter", *I&EC*, 42, 1827–1830.

5.5 Beaman, R.G. (1967), "Vinegar fermentation", *Microbial Technology*, Papper, H.J. (Ed.), Reinhold, New York, pp. 344–359.

5.6 Bisio, A., and Kabel, R.L. (1985), *Scale-up of Chemical Processes*, John Wiley & Sons, New York.

5.7 Borders, D. (1992), "Antibiotics survey", *Kirk–Othmer ECT*,. Kroschwitz, J.I.(Executive Ed.), 4th Edn., Vol. 2, pp. 893–904.

5.8 Casida Jr., L.E. (1996), *Industrial Microbiology*, New Age International (P) Ltd., New Delhi, Ch. 7, pp. 117–135.

5.9 Charles, M. (1985). "Fermenter design and scale-up", *Comprehensive Biotechnology*, Moo-Young, M. (Chief Ed.) Vol. 2, Pergamon Press, pp. 57–75.

5.10 Chisti, M.Y. (1989), *Air-lift Bioreactors*, Elsevier Applied science, Barking.

5.11 Coughanowr, D.R (1991), *Process Systems Analysis and Control*, McGraw-Hill, Inc., New York, pp. 111–163.

5.12 Crueger, W., and Crueger, A. (1989), Biotechnology: A *Textbook of Industrial Microbiology*, 2nd Edn, Sinauer Association, Inc., Sunderland, pp. 59–63.

5.13 Doran, P.M. (1995), *Bioprocess Engineering Principles;* Academic. Press, London, pp. 4–5, 143

5.14 Erickson, L.E., and Stephanopoulos, G. (1987), *Chemical Reaction and Reactor Engineering*, Carberry, J.J., and Varma, A.(Eds), Marcel Dekker, Inc., New York, ch. 12, p. 799–838.

5.15 Fogler, H.S. (1986), *Elements of Chemical Reaction Engineering*, Prentice Hall, Englewood Cliffs, NJ, pp. 15–18.

5.16 Gunasekaran, P. (1995), *Laboratory Manual in Microbiology*, New Age International (P) Ltd., New Delhi, pp. 125–134.

5.17 Hall, E.A.H. (1990), *Biosensors*, Open University Press, Buckingham, pp. 2–29.

5.18 Hampel, W. (1979), "Applications of microcomputers in the study of microbial processes", *Advances in Biochemical Engineering*, Ghose, T.K., Fiechter, A, and Blakebrough, N. (Ed.), Springer Verlag, Berlin, pp. 1–33.

5.19 Hsiah, C.L., and Ritchey, M.B. (1997), "Vaccine technology", *Kirk–Othmer ECT*, Kroschwitz, J.I.(Executive Ed.), 4th Edn, Vol. 24 John Wiley, New York, pp. 727–750.

5.20 Karube, I., and Sode, K. (1989), "Biosensors for fermentation process control", *Bioprocess Engineering: The First Generation*, Ghose, T.K. (Ed.), Ellis Horwood, Chichester, pp. 207–219.

5.21 Levenspiel, O. (1974), *Chemical Reaction Engineering*, Wiley Eastern Pvt. Ltd., New Delhi, pp. 94–95.

5.22 Lonsane, B.K. (1994), "Resurgence of interest in solid-state fermentation", *Solid State Fermentation*, Pandey, A. (Ed.), Wiley Eastern Pvt. Ltd., New Delhi, pp. 11–20.

5.23 Marison, I.W. (1988). "Citric acid production", *Biotechnology for Engineers*, Scragg, A.H. (Ed.), Ellis Horwood Ltd., Chichester, UK, p. 324.

5.24 Mokshapathy, S. (1988), *Handbook of Complete Statistics on Processed Fruits and Vegetables Exported from India (1977-1985)*, CFTRI Press, Mysore.
5.25 Moo-Young, M., and Blanch, H.Y. (1981), "Design of biochemical reactors: Mass transfer criteria for simple and complex systems", *Advances in . Biochemical Engineering*, Fiechter, A.(Ed.), Vol. 19, Springer-Verlag, Berlin, pp. 1–69.
5.26 Nielsen, P.H., et al. (1994), "Enzyme applications (Industrial)", in *Kirk–Othmer ECT*, Kroschwitz, J.I.(Executive Ed.), 4th Edn., Vol. 9, John Wiley, New York, p. 164.
5.27 Rajyalakshmi, P. and Geervani, P. (1990), "Studies on tribal foods of South India: Effect of processing methods on the vitamin and in vitro protein digestibility (IVPD) of cereals/millets and legumes", *J. Food Sci. Technol.*, 27(5), 260–263.
5.28 Ramanamurthy, M.V., Karanth, N.G., and Raghava Rao, K.S.M.S. (1993), "Biochemical engineering aspects of solid state fermentation", *Advance in Applied Appl. Microbiology*, Vol. 38, Academic Press, Inc., San Diego, pp. 99–147.
5.29 Refai, M.K. (1979). *Manuals of food quality control: 4. Microbiological analysis*, FAO, Rome, pp. F1–F8.
5.30 Resnick, W. (1981), *Process Analysis and Design for Chemical Engineers*, McGraw-Hill, Inc., New York, pp. 111–113.
5.31 Schugerl, K. (1982), "New bioreactors for aerobic processes", Int. Chem. Eng., 22(4), 591–610.
5.32 Schugerl, K. (2000), "Development of bioreaction engineering", *Advances in Biochemical Engineering and Biotechnology*., Vol. 70, Fiechter, A.(Ed.), Springer-Verlag, Berlin, pp. 49–50.
5.33 Sengha, S.S. (1993), "Fermentation", *Kirk–Othmer ECT*, Kroschwitz, J.I.(Executive Ed.), 4th Edn., Vol. 10, John Wiley, New York, pp. 361–381.
5.34 Stanbury, P.F., and Whitaker, A. (1993), *Principles of Fermentation Technology*, Pergamon Press, Oxford, (a) p. 121, (b) pp. 146–150, (c) 206–208.
5.35 Webb, F.C. (1964), *Biochemical Engineering*, D. Van Nostrand Co. Ltd. London, (a) pp. 4–9 (b) pp. 219–220.

Review Questions

5.1 What is the basic difference between fermentation and chemical reaction?
5.2 What are the general requirements of a fermentation process?
5.3 Enumerate the various fermentation processes that you come across in process industries.
5.4 Write a note on antibiotics production.
5.5 Write a note on vaccines.
5.6 Classify the enzymes based on their uses, and write a note on the enzymes production.
5.7 Describe how microorganisms can be used for effluent treatment.
5.8 What are the essential attributes of an industrial medium in the fermentation process?
5.9 What are the requirements of the medium in a bioprocessing operation?
5.10 Describe simple and complex media with examples.
5.11 Write a note on synthetic medium and crude medium.
5.12 Describe the medium composition for the production of α-amylase in laboratory scale.
5.13 Describe briefly the concept of design of a fermenter. What factors do you consider as essential for a successful design and operation of a fermenter?
5.14 Distinguish between aerobic and anaerobic fermentations, and describe them briefly.
5.15 Distinguish between solid state fermentation and submerged fermentation, and tabulate their essential features.

5.16 What is meant by solid state fermentation and what are its applications?

5.17 What are the various types of bioreactors? Bring out their features.

5.18 Is batch fermentation a steady state process or unsteady state process? Justify your answer.

5.19 What are the various ideal flow bioreactors? Explain them briefly.

5.20 What are the various methods of fed-batch/semi-batch fermentation process? Explain with their feeding methods.

5.21 Describe the working of a fluidized bed bioreactor. What are its specific advantages over other types of bioreactors?

5.22 What are the advantages of immobilisation?

5.23 Describe an immobilisedcell bioreactor.

5.24 Describe an air-lift bioreactor with a neat diagram. What are its advantages over other conventional bioreactors?

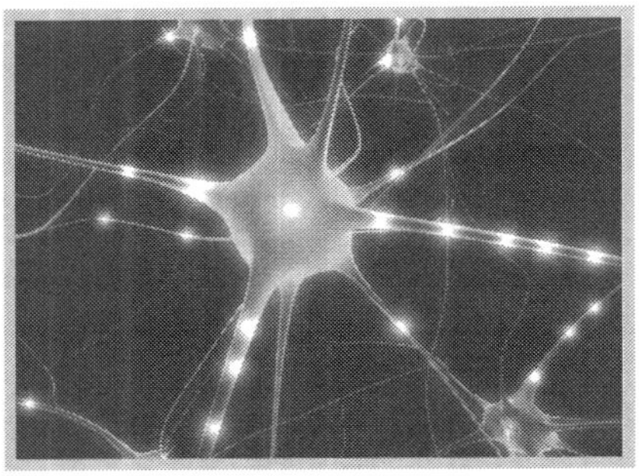

chapter 6

Sterilization

6.1 GENERAL INTRODUCTION

Sterilization is probably one of the most important operations which differentiates a biochemical process from a chemical process. The role of an engineer (or more precisely, biochemical engineer) in the overall process is probably to provide a 'contamination-free' environment (Aiba et al., 1965(a)).

Earlier, most of the processes which have fared well in terms of technical feasibility at laboratory-scale processing faced the formidable problem of contamination when the scale of operation increased. One of the most classical examples is penicillin fermentation. The technical and operational problem faced by the engineer in penicillin fermentation was to design and operate a deep aerobic fermenter under absolutely contamination-free environment with pure culture, since the environment in the fermenter was also conducive for the contaminants to grow. In fact, this has ushered in the new era of biochemical engineering as a distinct discipline in the field of engineering and technology. As has already been mentioned in Chapter 1, even the mechanical fabrication of the fermenter should preferably be kept in the hands of engineers who are well-versed with design and fabrication of fermenters, since otherwise a mechanical engineer, who is versatile in fabrication of chemical equipments, is likely to overlook the subtle point of asepsis, which is so vital for biochemical engineering operations, and not so important for a typical chemical equipment.

Fermentation is a biochemical process of producing metabolic products by the action of a microorganism or a group of microorganisms on a substrate, in the presence of nutrients in the medium. Obviously, fermentation proceeds with the involvement of

- a microorganism
- a medium
- a fermenter
- nutrients/other additives
- air in the case of aerobic processes.

Thus, a sterile environment is needed for all the above. Otherwise, the contaminants would dominate the whole scene, and virtually no product will form, or a contaminated product forms with toxicants. Invasion by a foreign (or contaminant) organism may result in the following ills of fermentation:

- The foreign organism may contaminate the final product.
- The medium would be consumed unnecessarily to support the growth of contaminating organisms.

- The contaminated product may outweigh the desired product; particularly, it is more so in the case of continuous fermentation processes.
- The contaminated product may interfere with the recovery of the desired product, and may render it inaccessible.
- Unsterile air in aerobic fermentations (antibiotic fermentations) may result in the spoilage of the fermentation product.

To overcome the above difficulties, the only alternative is to have a contamination-free environment which can be achieved by sterilizing all the involved parameters as shown in Table 6.1.

Table 6.1 *Methods to avoid contamination in a fermentation process*

1.	Sterilization of the medium
2.	Employing as pure an inoculum as possible
3.	Sterilization of the fermenter
4.	Sterilizing the pipes, valves, bends etc. which come in contact with the fermentation process
5.	Sterilization of all materials to be added to the fermenter
6.	Sterilization of air
7.	Disinfecting the fermenter and contact parts with a non-toxic disinfectant
8.	Maintaining aseptic conditions in the fermenters during fermentation
9.	Maintaining optimum or desired pH which discourages the growth of certain contaminants or undesired organisms or undesired products

The sterilization of a biochemical process can be achieved by the application of some lethal agent, viz.

- heat in the form of steam
- radiation
- a chemical,

or by restricting the undesired organisms or spores by some physical unit operation such as filtration.

6.2 STERILIZATION OF MEDIUM

Media used in industrial fermentations need to be sterilized before being used, to avoid the contaminating microorganisms which may

- use the nutrients in the medium
- change the chemical structure of the nutrients
- change the pH
- create more foam in the fermentation which affects the aeration
- produce metabolic products which affect the growth of fermentation microorganisms
- alter the oxidation–reduction potential of the medium
- further convert, degrade or destroy the desired fermentation product.

Sterilization is normally done by

- boiling in water
- passing live steam
- autoclaving (subjecting the medium to steam under pressure in a pressure cooker).

All the above methods are based on thermal treatment. Various techniques were developed to achieve thermal sterilization of the medium. It should be a batch process or continuous process or a HTST (high temperature, short time) process. Sometimes, the cost of sterilization would be prohibitive and destabilize the process economics. In such situations, the industrial fermentations are carried out by employing other means, viz.

- low pH
- contamination inhibitors (such as lactic acid).

Synthetic media do not require much sterilization efforts as compared to crude media. The synthetic media may require a small amount of heating for sterilization. The crude media are likely to contain some of the heat-resistant bacterial spores, and require prolonged heating. However, care must be exercised to see that the medium does not get cooked due to over-heating, but still adequate amount of heating must be given to see that the medium is thoroughly sterilized. This requires a bit of experience. Thermal death studies need to be made 'a priori' to evaluate the sterilization time (they will be discussed in the next section). Excessive heating may denature the proteins or caramalize the sugars or encourage thermal degradation of the components in the nutrient medium by inter-reactions (like sugars reacting with phosphates). If pH is a critical factor (whether low or high) during sterilization, it is advisable to adjust the pH to neutrality, sterilize the medium and then bring the pH of the medium to the original value after the sterilization by addition of pre-sterilized acid or alkali. If some of the enzymes and nutrients like vitamins are sensitive to heat sterilization, they are separated initially by passing through a bacteriological filter; sterilization is carried out, and the separated (filtered) enzymes/nutrients are added back to the medium. Similarly, the volatile components in the medium may be separated by bacteriological filtration to avoid/reduce their loss due to thermal sterilization.

Thus, various types of methods/alternatives need to be devised to see that the active components/nutrients are not lost during heat sterilization, but at the same time, the contaminating organisms are destroyed to the maximum possible extent.

6.3 THERMAL DEATH KINETICS OF MICROORGANISMS

The rate of destruction of microorganisms by steam or moist heat is known as thermal death kinetics of microorganisms. The destruction kinetics may be described as a first-order chemical reaction (Stanbury and Whitaker, 1993) and may be written as

$$-\frac{dN}{dt} = kN \tag{6.1}$$

or

$$\text{In}\frac{N_0}{N_t} = \int_0^t k dt$$

where N is the number of live organisms present
t is the sterilization time period
k is the first-order thermal death kinetic rate constant.

The negative sign in the above equation indicates that as t increases, N decreases. On integration of the above equation, with the boundary conditions that N_0 is the initial number of organisms and N_t is the number of organisms still present after a time period of t,

$$\frac{N_t}{N_0} = e^{-kt} \tag{6.2}$$

or

$$\text{In}\frac{N_t}{N_0} = -kt \tag{6.3}$$

or

$$\text{In}\frac{N_0}{N_t} = kt \tag{6.4}$$

The ratio given in Eq. (6.4) is known as **inactivation factor**, whereas N_t/N_0 given by Eq. (6.3) is the **survival factor**. N_0/N_t is the design criterion which indicates the contamination level of the medium, which is initially present as N_0 and needs to be brought to a sterility level of N_t. Equations (6.1) to (6.3) can be shown graphically in Fig. 6.1. Figure 6.1(a) shows that as time of sterilization increases, the number of surviving spores decreases. Figure 6.1(b) shows the decline of ratio N_t/N_0 in an exponential fashion as the time proceeds. Plot of ln N_t/N_0 versus time, in Fig. 6.1(c), gives the value of $-k$ as its slope. Equation (6.3) indicates that complete 100% sterilization (i.e. $N_t = 0$) is never possible: or in other words, it takes an infinite amount of time to get $N_t = 0$. If we get the values of N_t as a fraction of 1, i.e. $N_t = 0.1$, it looks to be practically non-viable. However, a value of N_t less than 1 or a fraction of 1 may be viewed as a probability of an organism surviving the treatment (Stanbury and Whitaker, 1993). It may better be explained with an example. If $N_t = 0.1$, it indicates that 1 organism is surviving the treatment out of 10. In other words, this means in ten batches, one batch of treatment will be contaminated.

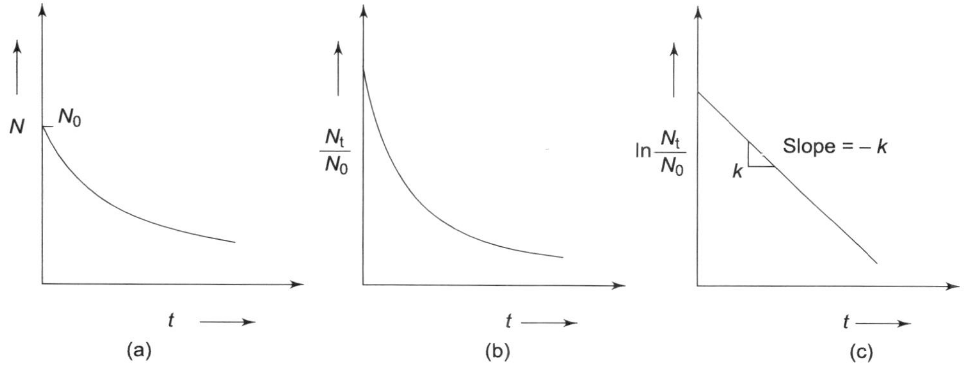

Fig. 6.1 *Plot of number of cells versus time, subjected to thermal sterilization over a period of time*

The value of k, which is obtained as the negative slope of Fig. 6.1(c), is not only dependent on the type of cells (i.e. species of the cells), but also on the physiological form of the cells. Figure 6.2 shows the logarithmic and non-logarithmic death rates (Crueger and Crueger, 1989). The death kinetics of vegetative cells of *E. coli* follows a logarithmic pattern, whereas that of *Bacillus stearothermophilus* follows a non-logarithmic path (Fig. 6.2(b)).

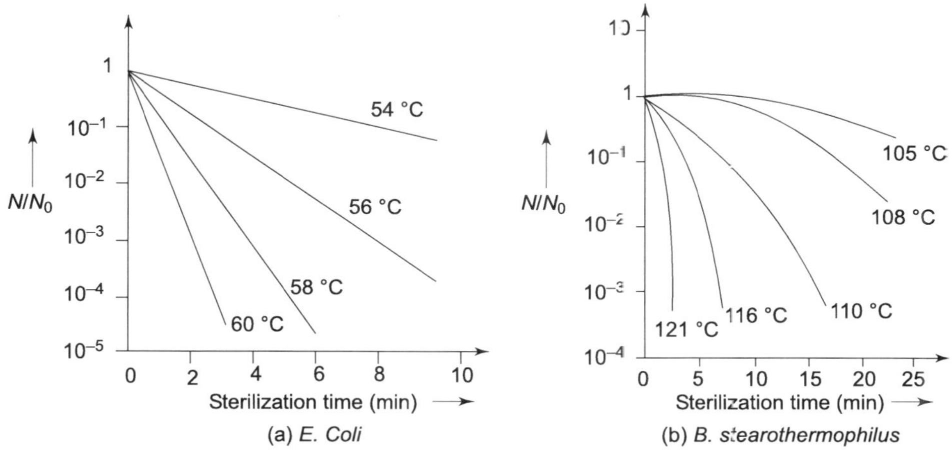

Fig. 6.2 *Non-logarithmic and logarithmic death rates of B. stearothermophilus and E. coli (Reprinted from "Biotechnology: A Textbook of Industrial Microbiology", Crueger, W., and Crueger, A; Brock, T.D. (Ed.), 2nd Edn (English), p. 97, Copyright © (1989), with permission from Oldenbourg, Industrieverlag, GmbH, Munchen)*

Example 6.1

A fermentation medium contains an initial spores concentration of 8.5 × 10¹⁰. The medium is sterilised thermally at 120°C, and the spore density was noted with the progress of time. The data are as follows:

Table E6.1

Time (min)	0	5	10	15	20	30
Spore density (m^{-3})	8.5×10^{10}	4.23×10^9	6.2×10^7	1.8×10^6	4.5×10^4	32.5

(i) Find the thermal death kinetic rate constant in s^{-1}.
(ii) With the above data, calculate the inactivation factor at 40 min.

Solution It is a straight application of Eq. 6.2. We prepare the data from Table E6.1 to draw the plot between ln (N_t/N_o) and t as follows:

Table E6.2 *Data for drawing Fig. E6.1*

t (min)	N_t	N_t/N_o	$\ln (N_t/N_o)$
0	8.5×10^{10}	1	0
5	4.23×10^9	0.05	–3.0
10	6.2×10^7	7.23×10^{-4}	–7.22
15	1.8×10^6	2.11×10^{-5}	–10.76
20	4.5×10^4	5.3×10^{-7}	–14.45
30	32.5	3.82×10^{-10}	–21.7

Fig. E6.1 is drawn using the data in the first and last columns in Table E6.2.

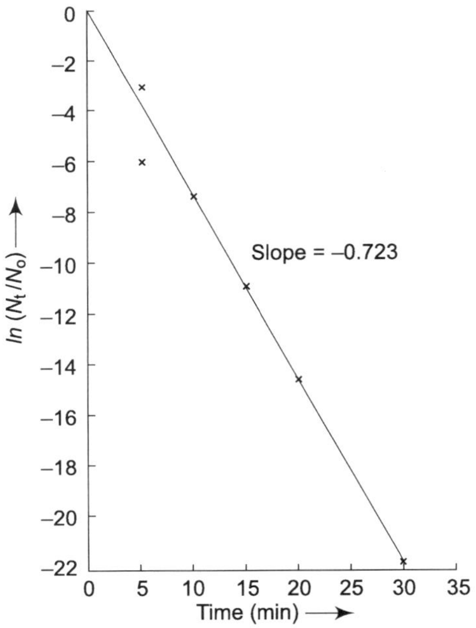

Fig. E6.1 *Data for Example 6.1*

From Fig. E6.1, the slope of the line gives the value for k_d in the units of min^{-1}.

$$\text{Slope} = -k_d = -0.723\ min^{-1}$$

$$\therefore \quad k_d = 0.72\ min^{-1} = 0.012\ s^{-1}$$

(ii) Inactivation factor at 40 min

$$N_t = N_o\, e^{-0.723 \times 40} = 8.5 \times 10^{10} \times 2.75 \times 10^{-13}$$
$$= 0.023$$

6.3.1 Effect of Sterilization on Quality of Nutrients

The sterilization time and temperature may have the following adverse effects on the nutrient quality:

Interaction between nutrient components As has been mentioned earlier, the interaction between various nutrients sometimes causes deleterious effects on the fermentation medium. Discolouration of the media due to Milliard-type browning is a common occurrence. The reducing sugars have carboxyl groupings, which generally react with the amino groups of the amino acids, etc. In such cases, it is suggested that the carbohydrate component of the medium should be separated from the remainder of the medium, and sterilized separately.

6.3.2 Sterilization of Culture Media

Nutrient media contain different vegetative cells and spores. Some of them are more heat labile and some are less. Temperature of sterilization can be used as a useful proposition to selectively destroy some.

Degradation of heat labile compounds Different compounds will have different sterilization periods and they will also be affected by the temperature of sterilization. The thermal destruction of various organisms may be given by first-order reaction kinetics and may be represented by equation similar to Eq. (6.2), and the temperature effect can be assessed by Arrhenius equation.

For simplicity of explanation, let us assume only two organisms, A and B, are present in the culture which is being subjected to sterilization. Let organism A have higher activation energy E_A as compared to that of B which has activation energy E_B. If we draw a plot of ln k versus $1/T$ for A and B (Fig. 6.3), the slope of the graph will give activation energy of destruction. From the figure it is evident that any increase in temperature will have a greater effect on A as compared to B from the steepening of plot for A relative to that of B. That is, increase in temperature will destroy more A compared to B. For achieving higher destruction of A, the medium can be heated to a higher temperature for shorter periods, which is known as HTST technique. How to achieve HTST in a large batch of fermentation is an engineer's nightmare; it is easier said than done. One of the methods to achieve HTST is to use continuous sterilization process, which will be discussed in later sections.

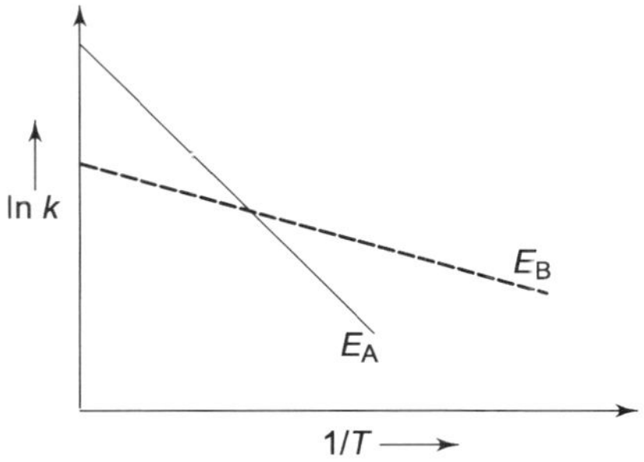

Fig. 6.3 *Effect of activation energy on viability of two different types of spores*

The sterilization of mixed cultures having different sterilization times will leave to different extents the viable cells. Let us call the mixed culture as C which contains A and B. A plot of number of survivors on a semi-log scale versus the sterilization time is shown in Fig. 6.4. From Fig. 6.4(a), it is clear that the more sensitive organism will be getting destroyed in the early stages, which is evident from the fact that the curve C is following line A. When once, more amount of A is destroyed, subsequently the curve follows the path of

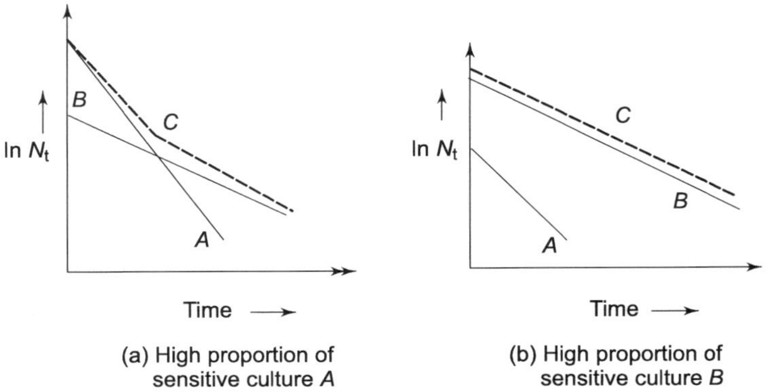

Fig. 6.4 *Sterilization effect on a mixed culture C, containing two cultures A and B*

B. In the case of Fig. 6.4(b), the situation is different. Since the more sensitive organism is less in number, it gets destroyed quickly, and hence the total number of cells will be always equal to the number of cells of the relatively resistant organism i.e. B. Hence, the curve for C almost follows that of B.

6.4 BATCH STERILIZATION

Having understood the principles of sterilization, let us see the two distinct methods of thermal sterilization. They are:

- batch sterilization
- continuous sterilization.

As has already been mentioned for batch processes, batch sterilization process is one in which all the contents are loaded into the sterilizer, steam is injected to see that the sterilization takes place as per the programme, and later the contents are discharged for further processing or transferred directly into the fermenter. The batch sterilization process has some inherent disadvantages. This procedure is less successful in avoiding the risk of destruction of nutrients while destroying the contaminants as compared to the continuous sterilization. However, it has the following advantages:

- The initial costs of investment are lower.
- The chances of contamination after the sterilization is over are less because the sterilization can be done in the same vessel or in the fermenter itself.
- Manual control is normally done, and hence the chances of mechanical failure are less. Media containing high proportion of solids can be handled easily.

6.4.1 Design Aspects

In Eq. (6.1), the death kinetic rate constant k is similar to chemical reaction rate constant for a first-order reaction, and its dependence on temperature can be conveniently represented by Arrhenius equation

$$\frac{d(\ln k)}{dT} = \frac{E}{RT^2} \tag{6.5}$$

where E is the activation energy (in J/mol) which is a specific constant for the population;
R is universal gas constant (8.27 J/(mol K)),
T is temperature (in K).

Integration of Eq. (6.5) yields

$$k = k_0 e^{-E/RT} \tag{6.6}$$

where k_0 is Arrhenius constant.

In the above equation, the constants k_0 and E/R can be calculated by the Arrhenius plot drawn between ($\ln k$) and ($1/T$), as shown in Fig. 6.5. For this purpose, a set of experiments is conducted at one temperature, and the values of k are evaluated by Eq. (6.3) by measuring N_t at different time intervals. The procedure is repeated at different other temperatures, and the Arrhenius plot is made as given in Fig. 6.5. The slope of the line gives ($-E/R$) and the intercept is equal to $\ln (k_0)$. Thus, the temperature effect on N_t is evaluated and incorporated into Eq. (6.4) by combining it with Eq. (6.6):

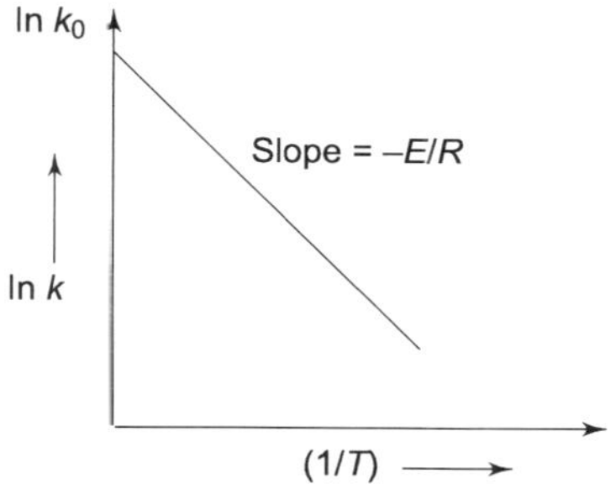

Fig. 6.5 *Arrhenius plot*

$$\ln\left(\frac{N_0}{N_t}\right) = k_0\, t\, e^{-E/RT} \tag{6.7}$$

Equation (6.7) acts as the design criterion which is called as "Del factor" (∇), Nabla factor, or sterilization criterion represented by ∇ (Deindoerfer and Humphery, 1959). The Del factor can be defined as the measure of fractional reduction in living organisms count over the initial number present, and is produced by a certain heat and time regime.

Thus,

$$\nabla = \ln \frac{N_0}{N_t} \tag{6.8}$$

Eq. (6.7) and (6.8) combine to give

$$\nabla = k_0\, t\, e^{-E/RT} \tag{6.9}$$

which on rearrangement yields

$$\ln t = \left(\frac{E}{R}\right)\left(\frac{1}{T}\right) + \ln \frac{\nabla}{k_0}$$

Figure 6.6 shows plots of $\ln(t)$ versus ($1/T$) (Stanbury and Whitaker, 1993).

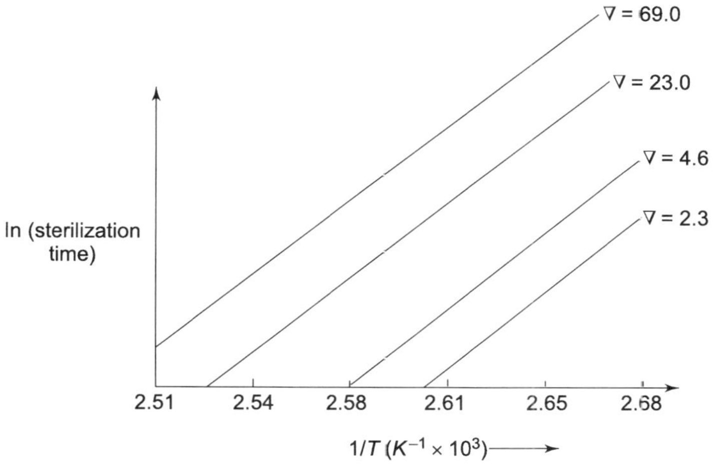

Fig. 6.6 *Effect of sterilization time and temperatures on Del factors.*
(Reprinted from "Principles of Fermentation Technology", Stanbury, P.F., and Whitaker, A., p. 94. Copyright © (1993), with permission from Elsevier.)

Deindoerfer and Humphrey (1959) suggested that a risk factor of one in thousand being contaminated could be considered as reasonable for all practical purposes. However, to have a complete profile of all the contaminating spores is seldom possible. Hence, it is assumed that all the contaminants can be clubbed into a single category of spores of *Bacillus stearothermophilus* which is one of the most heat-resistant microbial type. Deindoerfer and Humphrey (1959) reported for *B. stearothermophilus* spores,

$$\text{Activation energy} = E = 283 \text{ kJ/mol}$$

$$\text{Arrhenius constant} = k_0 = 1 \times 10^{36.2} \text{ s}^{-1}$$

The idea of choosing *B. stearothermophilus* is that, since it is one of the most heat-resistant organisms, the design would automatically take care of inclusion of a good safety factor in calculations.

From the above discussion, it is evident that a frequently adopted assumption of a risk factor of one in thousand is justified. Hence,

$$N_t = 10^{-3}$$

The Del factor ∇ was defined as

$$\nabla = \ln \frac{N_0}{N_t}$$

The design procedure for calculation of ∇ factor can be explained with an example (Stanbury and Whitaker, 1993). If the unsterile bath contains initially 10^{11} viable cells,

$$N_0 = 10^{11}$$

Hence,

$$\nabla = \ln\left[\frac{10^{11}}{10^{-3}}\right] = \ln\left(10^{14}\right) = 32.2$$

This value is considered as the overall Del factor. Normally the destruction of cell takes place at 121 °C (corresponding to a steam pressure of 15 psig, which is equivalent to 0.1 MN/m^2 gauge). This period is known as "holding period" since the medium is held at 121 °C. However, some number of cells would be destroyed during the temperature build-up to 121 °C which is known as "heating period". Similarly, some cells will also be destroyed during "cooling period" from 121 °C to room temperature. Thus, the overall period is a sum of heating period, holding period and cooling period. In other words,

$$\nabla_{\text{overall}} = \nabla_{\text{heating}} + \nabla_{\text{holding}} + \nabla_{\text{cooling}} \tag{6.10}$$

Example 6.2

The thermal death kinetic data of Bacillus stearothermophilus are as follows at three different temperatures.

Temp, °C	115	120	125
k_d, min^{-1}	0.035	0.112	0.347

(i) Calculate the activation energy and Arrhenius constant for sterilisation
(ii) Find k_d at 130°C

Solution This is a direct application of Arrhenius equation (Eq. (6.6)). The data in the above is reformatted to draw the Arrhenius plot and is shown in Table E6.3.

Table E6.3 *Data to draw Arrhenius plot for Ex. 6.2*

Temp, °C	*k_d, min^{-1}*	*Temp, K*	*1 / T*	*ln (k_d)*
115	0.035	288	3.47×10^{-3}	–3.35
120	0.112	293	3.41×10^{-3}	–2.189
125	0.347	298	3.36×10^{-3}	–1.058

From the data in Table E6.3, the Arrhenius plot is drawn and is shown in Fig. E6.2. The slope of the Arrhenius plot is measured.

$$\text{Slope} = \frac{-(3.4-(-0.9))}{(3.474-3.35)\times 10^{-3}} = -2.016\times 10^4$$
$$= -E/R$$

$\therefore$ $$E = 2.016\times 10^4 \times 8.314 = 1.67\times 10^5$$

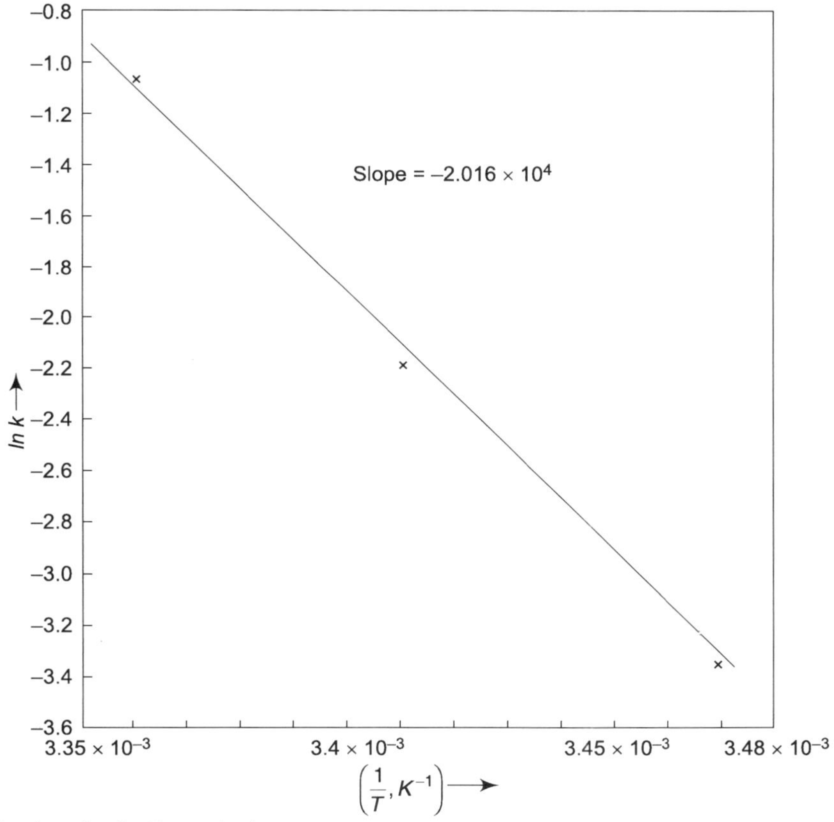

Fig. E6.2 *Arrhenius plot for Example 6.2*

Since it is difficult to draw the Arrhenius plot in the entire range of scale, k_0 will be calculated using Eq. (6.6).

$$k_d = k_0\, e^{-(E/RT)}$$

in which $k_d = 0.347\ \text{min}^{-1}$

$T = 398$ K

$R = 8.314$

$E = 1.67\times 10^5$

$$k_0 = \frac{0.347}{\exp(-(1.67\times 10^5)/(8.314\times 398))} = 2.875\times 10^{21}\ \text{min}^{-1}$$

At 130°C, $$k_d = 2.875\times 10^{21}\times \exp((-2.016\times 10^4)/(273+130))$$
$$= 2.875\times 10^{21}\times 1.88\times 10^{-22} = 0.541\ \text{min}^{-1}$$

Example 6.3

The specific death constants of heating and cooling during sterilisation of a medium at 121°C are 0.1 min^{-1} and 0.2 min^{-1} respectively.

$t_{heating} = 20$ min

$t_{holding} = 30$ min

$t_{cooling} = 30$ min

The decimal reduction time during holding is 2 min. The initial batch contains 6 × 10^{15} organism at 30 °C. find the sterilisation effect.

Solution Given:

$$N_o = 6 \times 10^{15}$$

During heating period

$$k_d = 0.1 \text{ min}^{-1}$$
$$T = 20 \text{ min}$$

During holding at 121°C

$$t = 30 \text{ min}$$
$$\text{decimal reduction time} = 2 \text{ min}$$

During cooling

$$k_d = 0.2 \text{ min}^{-1}$$
$$t = 30 \text{ min}$$

From Eq. (6.2), we have

$$\frac{N_t}{N_o} = e^{-kt}$$

During heating

$$\frac{N_{heating}}{6 \times 10^{15}} = e^{-0.1 \text{ X } 20} = 0.8187$$

∴ $$N_{heating} = 6 \times 10^{15} \times 0.135 = 8.12 \times 10^{14}$$

During heating, N_t will reduce by 10 times in 2 minutes

i.e., $$\frac{N_t}{N_0} = 0.1 = e^{-k_d(2)}$$

$$-2\,(k_d) = \ln(0.1) = -2.303$$

∴ $$k_d = 2.303 / 2 = 1.152 \text{ min}^{-1}$$

$$(N_0)_{holding} = 8.12 \times 10^{14}$$

$$\frac{N_{holding}}{(N_0)_{holding}} = e^{-1.152\,(30)} = 9.79 \times 10^{-16}$$

∴ $$N_{holding} = 8.12 \times 10^{14} \times 9.79 \times 10^{-16} = 0.795$$

During cooling

$$(N_0)_{cooling} = 0.795$$

$$\frac{N_{cooling}}{(N_0)_{cooling}} = e^{-(0.2 \text{ X } 30)} = 2.479 \times 10^{-3}$$

∴ $$N_{cooling} = 0.795 \times 2.479 \times 10^{-3} = 1.97 \times 10^{-3}$$

i.e., N_t at the end of cooling period $= 1.97 \times 10^{-3}$

Example 6.4 *In the above ecamples, if the power fails after 15 min of holding time (i.e. after it reaches 121 °C), what will happen to the sterilisation if the cooling process is not affected by it?*

Solution During holding time, the power went off after 15 minutes and so heating stopped and cooling started.

$\therefore \quad t_{holding} = 30 - 15 = 15$ min

and cooling time increases by 15 min

$$\therefore \quad t_{cooling} = 30 + 15 = 45 \text{ min}$$

$$(N_0)_{holding} = 8.12 \times 10^{14}$$

$$(N_t)_{holding} = (N_0)_{holding}\, e^{-k_d t}$$

$$= 8.12 \times 10^{14}\, e^{-(1.152 \times 15)}$$

$$= 8.12 \times 10^{14} \times 3.129 \times 10^{-8} = 2.54 \times 10^7$$

Hence $\quad (N_0)_{cooling} = 2.54 \times 10^7$

$$(N_t)_{cooling} = (N_0)_{cooling} \times e^{-(0.2\,(30+15))}$$

$$= 2.54 \times 10^7 \times 1.234 \times 10^{-4} = 3.135 \times 10^3$$

The number of cells surviving after the sterilisation = 3.13×10^3

Hence the sterilisation is not complete.

6.4.2 Del Factors during Heating and Cooling

The effect of temperature on ∇ factor was shown by Eq. (6.9). Thus, as temperature increases, ∇ factor also increases. Based on this, it will be possible to calculate ∇ at any one temperature, which is the case for holding period. However, the temperature is not constant during the heating-up or cooling-down periods. Hence, Eq. (6.9) can be written in integral form by combining with Eq. (6.1) as follows:

$$\nabla = \ln \frac{N_0}{N_t} = \int_0^t k\, dt \tag{6.11}$$

It is possible to integrate the above equation and find ∇ factor during heating or cooling, provided we know the functionality of temperature and time. Deindoerfer and Humphrey (1959) have given analytical expressions for the integration of Eq. (6.11) for various functions of t and T, viz. linear, hyperbolic or exponential. However, experience shows that it is rarely possible to describe the functionality of t and T in any of the forms such as linear, hyperbolic or exponential, etc. In such cases, graphical integration is possible by experimentally noting the data. The method is described as follows (Richards, 1968).

The time–temperature profile can be experimentally recorded, and drawn in the form of a plot (Fig. 6.7). The X-axis is divided into a large number small equal intervals like t_1, t_2, t_3, etc. Hence, $t_1 = t_2 = t_3 = \ldots = t$. The temperature T_1 during t_1 period is noted as the average temperature at the beginning of t1 and ending of t_1, i.e. T_1 is noted at the midpoint of t_1. The accuracy of the method depends upon how small is t. Usually it is suggested that 30 is a reasonable number of divisions for total time. Having known the temperature T and time t, ∇ can be calculated by Eq. (6.9) for *B. stearothermophilus* using the values for activation energy and Arrhenius constant as follows:

$$E = 283 \text{ kJ/mol}$$

$$k_0 = 1 \times 10^{36.2} \text{ s}^{-1}$$

Thus,

$$\nabla_1 = k_1 t$$

$$\nabla_2 = k_2 t$$

$$\nabla_3 = k_3 t \text{ etc.}$$

The overall ∇ is a summation of individual ∇'s.

Hence, $$(\nabla) = \nabla_1 + \nabla_2 + \nabla_3 + \dots$$

$$\nabla_{\text{heating}} = \sum_{i=1}^{n} \nabla_i \tag{6.12}$$

Alternatively, ∇ at every time period is calculated based on the temperature which is measured and plotted in Fig. 6.7. That is, the right hand side part of Eq. (6.11) is plotted on Y-axis versus the time intervals on X-axis, as shown in Fig. 6.8. The time intervals are equally placed. Integration of Eq. (6.11) is evaluated by graphical method by measuring the area under the curve ABCD. Thus, the ∇ for heating period (∇_{heating}) can be calculated by Eq. (6.11). The method is described with the following numerical example:

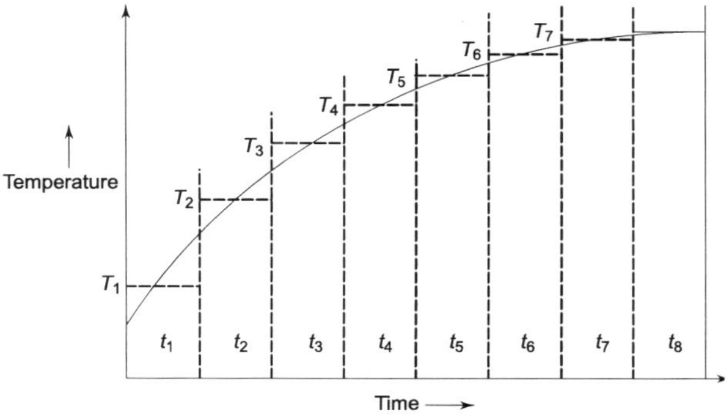

Fig. 6.7 *Graphical integration method for evaluating ∇ during heating period*

The time–temperature profile during thermal sterilization of a medium is measured in the laboratory, and the data are shown in Table 6.2. The calculated values of ∇_i are also shown in Table 6.2, and plotted in Fig. 6.8 versus the time intervals. The area under the curve ABCD is measured to be 8.7.

Table 6.2 *The time–temperature data of batch sterilization of a medium*

t (min)	T_i (°C)	$\nabla_i = k_o(e^{-E/RT_i})^*$
1	100.0	0.0227
2	101.5	0.0327
3	103.0	0.047
4	104.5	0.0674
5	106.0	0.0963
6	107.5	0.137
7	109.0	0.195
8	110.5	0.276
9	112.0	0.39
10	113.5	0.55
11	115.0	0.772
12	116.5	1.082
13	118.0	1.513
14	119.5	2.109
15	121.0	2.839

* $E/R = 34043\ \text{K}^{-1}$

$k_0 = 9.5 \times 10^{37}\ \text{min}^{-1}$

$\therefore$ ∇ for the process during heating period from 100–121°C in time of 1–15 min = 8.7

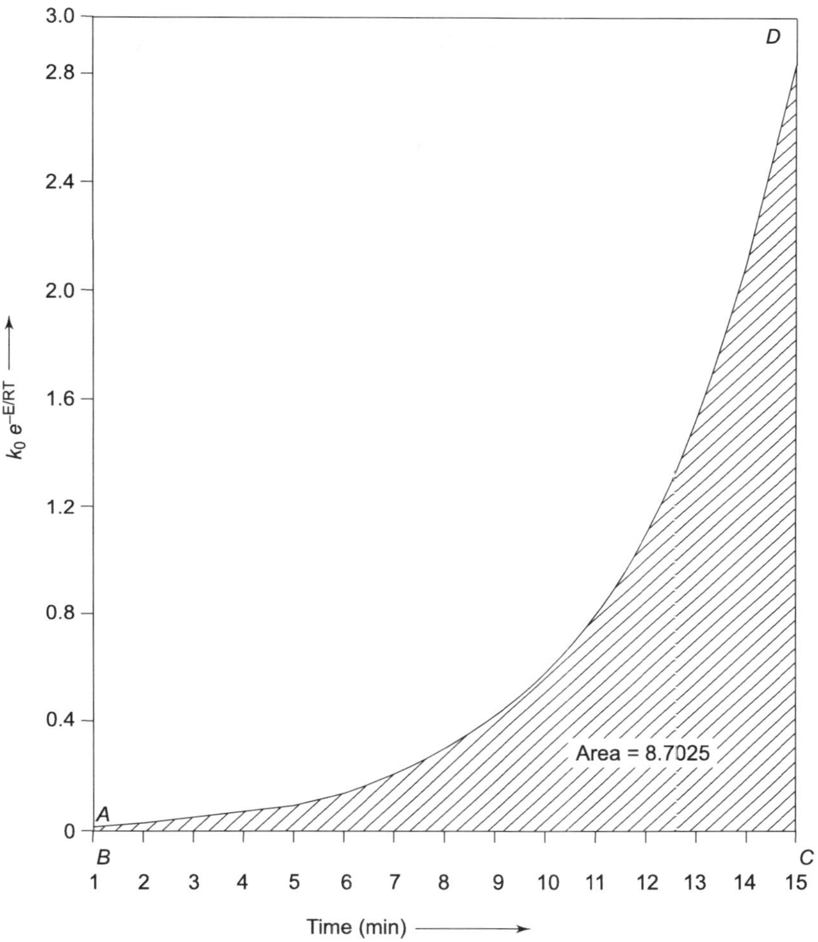

Fig. 6.8 *Data for carrying out graphical integration of Eq. (6.11)*

Similarly, ∇_{cooling} can be calculated by having the time–temperature profile during cooling process of sterilization.

Holding Time at 121°C After having calculated ∇_{heating} and ∇_{cooling} by graphical integration method, ∇_{holding} can be calculated by Eq. (6.10), provided ∇_{overall} is known. ∇_{overall} can be arrived at as explained earlier for

$$N_t = 10^{-3} \quad \text{and} \quad N_0 = 10^{11} \text{ cells.}$$

$$\nabla_{\text{overall}} = 32.2$$

The calculation of ∇_{holding} will enable us to determine the holding time, assuming that the holding temperature is 121°C. This can be explained with an example. If the ∇ factor for heating is 9 and ∇ factor for cooling is 10,

$$\nabla_{\text{holding}} = 32.2 - 10 - 9 = 13.2$$

The specific death rate for *B. stearothermophilus* at 121°C was reported to be 2.54 min^{-1} (Deindoerfer and Humphery, 1961).

$$t = \frac{\nabla}{k} = \frac{13.2}{2.54} = 5.2 \text{ min}$$

Example 6.5 *A certain fermentation medium was sterilised in a fermenter. the heating up took 30 min from 100° to 121°C and the cooling took 15 min from 121° to 100°C. The ∇ factor value is $V_{1°/min\ 121°C} = 12.55$. The thermal death constant $k_{121°C}$ is 2.54 min^{-1}. The ∇ for sterilization is 43.5. Find $\nabla^{121°C}_{holding}$ and holding period.*

Solution Given:

100 to 121°C

$$\nabla_{1°/min} = 12.55, \quad (k_d)_{holding} = 2.54 \text{ min}^{-1}$$

Heating time taken is 30 min instead of 21 min.
Similarly, cooling time taken is 15 min instead of 21 min.

$$\nabla_{total} = 43.5$$

$$\nabla_{heating} = 12.55 \times \frac{30}{21} = 17.91$$

$$\nabla_{cooling} = 12.55 \times \frac{15}{21} = 8.96$$

By Eq. (6.10)

$$\nabla_{overall} = \nabla_{heating} + \nabla_{holding} + \nabla_{cooling}$$

∴ $$\nabla_{holding} = 43.5 - 17.91 - 8.96 = 16.63$$

By Eq. (6.11), we have

$$\nabla_{holding} = k_d \times t_{holding}$$

$$16.63 = 2.54 \times t_{holding}$$

∴ $$t_{holding} = 16.63 / 2.54 = 6.55 \text{ min}$$

6.4.3 Methods for Evaluating Del Factor

There are two methods of calculation of Del factors during heating and cooling. The graphical integration method is one which we have just seen in Section 6.4.2. There is another method which was proposed by Richards (1965) and is called as **Richard's rapid calculation method for heat sterilizations.** It does away with all the tedious graphs and calculations as described earlier. Richards rapid method basically depends upon the fact that the destruction of the spores will be considerable only above 100°C, i.e. the heating period will be effective only beyond 100 °C and up to the holding temperature of 121°C, and similarly during cooling period also, the destruction of the spores will be taking place until the temperature drops up to 100°C; below which it will not be significant. Accordingly he presented a table for temperature (T), the specific constant (k) and the ∇ factor in the temperature range of 100–130°C with a temperature raise of 1°C/min (Table 6.3) for *B.stearothermophilus,* using Eq. (6.6), in which k_0 and E were reported by Deindoerfer and Humphrey (1959):

Table 6.3 *k and Del values calculated in the temperature range 100 – 130 °C, using Eqs (6.6) and Eq. (6.9)*

T (°C)	*k* (min^{-1})	∇
100	0.015	—
101	0.019	0.033
102	0.024	0.057
103	0.031	0.088
104	0.039	0.126
105	0.049	0.176
106	0.063	0.239
107	0.080	0.318
108	0.100	0.419
109	0.127	0.547
110	0.161	0.708
111	0.203	0.911
112	0.256	1.167
113	0.322	1.491
114	0.405	1.894
115	0.509	2.403
116	0.638	3.042
117	0.800	3.841
118	1.000	4.842
119	1.250	6.093
120	1.562	7.654
121	1.947	9.601
122	2.426	12.027
123	3.019	15.046
124	3.752	18.800
125	4.658	23.456
126	5.777	29.233
127	7.158	36.391
128	8.858	45.249
129	10.951	56.200
130	13.525	69.725

$$k_0 = 9.51 \times 10^{37}\ \text{min}^{-1}$$
$$E = 283\ \text{kJ/mol.}$$

The ∇ value at 100 °C is neglected, and the ∇ value at any other temperature is based on the ∇ value at the previous temperature which has occurred 1 min earlier, since the temperature raise is 1 °C/min. Thus,

$$\nabla_{(T)} = \nabla_{(T-1)} + k_{(T)} \times 1.0$$

The second term in the RHS indicates the thermal sterilization that occurred at temperature T in 1 min (Eq. (6.4)).

Even if the temperature rise is different from 1 °C/min, the table can be used with the simple assumption that the heating and cooling cycles above 100 °C are linear. Thus, the rapid method relies on 'two assumptions' (Richards, 1965):

(i) the spores destruction takes place only beyond 100 °C.
(ii) the cooling and heating cycles are linear.

These two assumptions are reasonably valid and do not sacrifise the accuracy of calculations. The second assumption is useful to calculate the ∇ factor at any other rate of temperature rise. For example, if the temperature during heating rises in t_1 min (for example, in 25 min) from 100 °C to 121 °C, instead of 21 min, ∇_{heating} will proportionately change.

$$\nabla_{\text{heating}} = 9.601 \times \frac{t}{21}$$

$$\nabla_{\text{heating}} = 9.601 \times \frac{25}{21} = 11.43$$

Similarly, if the cooling takes place rapidly in 15 min instead of 21 min.

$$\nabla_{\text{cooling}} = 9.601 \times \frac{15}{21} = 6.86$$

Once the Del factor for heating and cooling are calculated, ∇_{holding}, and hence holding time, can be calculated by Eq. (6.10), as in the earlier case.

6.5 CONTINUOUS STERILIZATION

In the continuous sterilization process, there will be continuous inflow and outflow of material. Thus, it is best suited if the capacity of operation is high. The continuous sterilization has many advantages vis-à-vis the batch sterilization.

(i) the throughputs (capacity) can be higher
(ii) the medium quality can be maintained better
(iii) the system can be automated for control; hence it avoids human error during operation
(iv) the running costs are less
(v) the sterilization times can be shorter
(vi) this is the only option left out for HTST operation
(vii) the holding capacities of steam can be less
(viii) ease in scaling-up of the process.

The major disadvantage with the continuous operation is the initial capital investment. It necessitates the use of aseptic transfer system for the sterile broth to be transported to the sterile vessel.

However, if the medium is to be exposed to high temperatures for a short time (HTST process) to avoid denaturation of proteins or to avoid destruction of some of the enzymes, etc., it is not possible in commercial-scale operations in batch sterilization to quickly heat large volumes of the medium in a short time and to cool it also in short times. HTST is possible only in continuous sterilization.

Basically there are two methods of continuous sterilization

(i) continuous plate heat exchange sterilization
(ii) continuous steam injection and flash cooling sterilization.

Figure 6.9 shows the two types of continuous sterilization systems. In a plate heat exchanger, the incoming unsterile medium is preheated by heat exchange with the outgoing sterile medium. Subsequently, it is heated with steam in a heat exchanger. It is then passed through the holding section. The residence time in the holding section is decided by the holding time requirement, flow rate of the heated medium and length of the holding section. The typical time–temperature profiles are shown in Fig. 6.10 (Aiba, et al., 1965(b)). Heating time and cooling time are considerable, before and after the holding time.

In case of steam jet sterilization, steam is directly injected along with the medium continuously. Hence, the heating time and the heating section are negligible as shown in Fig. 6.10. The holding time is based on the length of the holding pipe. Sterilization takes place during the holding period. The steam and sterilized

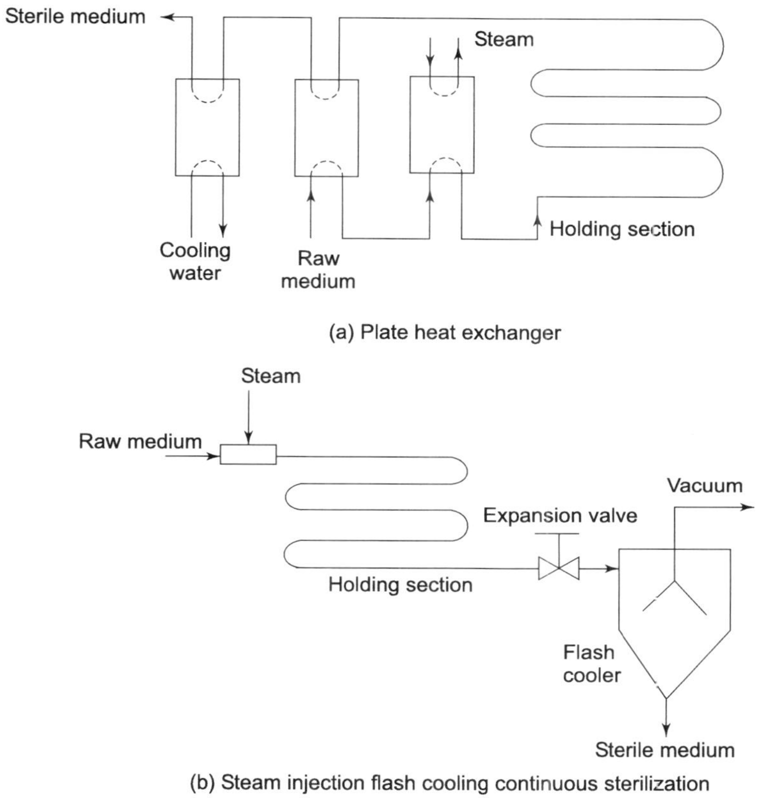

Fig. 6.9 *Continuous sterilization methods (Reprinted from "Biochemical Engineering," Aiba, S., Humphrey, A.E., and Millis, N.F., p. 207, Copyright © (1965), with permission from Elsevier.)*

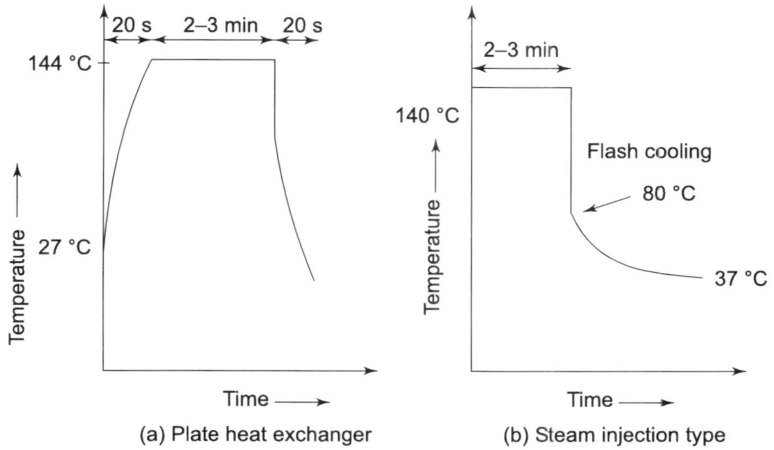

Fig. 6.10 *Typical time–temperature profiles of continuous sterilization (Reprinted from "Biochemical Engineering," Aiba, S., Humphrey, A.E., and Millis, N.F., p. 208, Copyright © (1965), with permission from Elsevier.)*

medium under pressure are passed through the expansion valve into the vacuum chamber. Steam is removed out under vacuum. The sterile medium passes into the preheat exchanger zone and gives out some heat to the incoming unsterile medium through the heat exchanger (not shown in Fig. 6.9). Finally, the sterile medium passes through the cooling zone. Hence, it will have a considerable cooling zone as shown in Fig. 6.10; but the temperature is much below 100 °C and hence it may be neglected for the purpose of thermal death kinetics. The advantages and disadvantages of the steam injection system are summarized in Table 6.4 (Banks, 1979).

Table 6.4 *Advantages and disadvantages of steam injection continuous sterilization system*

Advantages	*Disadvantages*
Media containing solids can be sterilized better.	Direct contact of the steam with the medium makes it necessary that the steam should be clean and free of any anti-corrosive agents.
Capital investment is low.	Foaming may occur during both heating and cooling.
It is easy to clean and maintain the system.	
Heating and cooling periods are shorter.	
Steam efficiency is very high as live steam is directly injected into the medium.	

6.6 STERILIZATION OF AIR

For aerobic fermentation processes on industrial scale, a huge quantity of air is required, and it is needless to say that the air should be clean and sterile. Achieving sterility for air is a stupendous task for biochemical engineers.

There is an excellent coverage on air sterilization (Aiba et al., 1965 (c)) which mentions various species and numbers of air-borne microbes. Table 6.5 shows various representative species of air-borne bacteria and bacterial spores. A cubic metre of air contains approximately 4×10^3 to 20×10^3 particles with a maximum possibility of 12×10^3. However, on an average, we can assume that air contains 10^3 to 10^4 particles/m^3. The average size of the dust particle is approximately 0.6 microns.

Table 6.5 *Representative species of air-borne bacteria.*

Species	*Width (microns)*	*Length (microns)*
Aerobacter aerogenes	1.0 – 1.5	1.0 – 2.5
Bacillus cereus	1.3 – 2.0	8.1 – 25.8
Bacillus licheniformis	0.5 – 0.7	1.8 – 3.3
Bacillus megaterium	0.9 – 2.1	2.0 – 10.0
Bacillus mycoides	0.6 – 1.6	1.6 – 13.6
Bacillus subtilis	0.5 – 1.1	1.6 – 4.8
Micrococcus aureus	0.5 – 1.0	0.5 – 1.0
Proteus vulgaris	0.5 – 1.0	1.0 – 3.0
(spores)		
Bacillus megaterium	0.6 – 1.2	0.9 – 1.7
Bacillus mycoides	0.8 – 1.2	0.8 – 1.8
Bacillus subtilis	0.5 – 1.0	0.9 – 1.8

(Reprinted from "Biochemical Engineering," Aiba, S., Humphrey, A.E., and Millis, N.F., p. 218, Copyright © (1965), with permission from Elsevier.)

6.6.1 Methods of Air Sterilization

There are various possible ways to sterilize the air:
(i) sterilization by heating
(ii) use of ultraviolet rays and other electromagnetic waves
(iii) use of germicidal sprays
(iv) sterilization by filtration.

Even though it is possible to heat the air and sterilize it, normally it is not economical to heat the air because of its poor thermophysical properties resulting in lower heat transfer coefficient. Ultraviolet rays in the wavelength range 2265–3287 Å are effective in killing air-borne microbes. This is one of the oldest known techniques. Hospital rooms, food factories and fermentation rooms are equipped with lamps emitting UV rays. However, their reliability to sterilize large volumes of air in factories is doubtful, even though they are good to sterilize the air to a reasonable extent in rooms or in small containers.

There will be a significant reduction in air-borne bacteria by spraying of small amounts of germicides, viz. phenol, ethylene oxide or formalin (formaldehye). This is also an effective method of disinfecting the air in a room to a good extent. This method also cannot ensure 100% sterility of air. The disinfectants are sprayed into the room along with the air recirculation systems like air-conditioners or pedestal fans, etc.

Thus, it can be seen that amongst the various options given above, none of them was very effective or economical or practicable. Hence, normally air is sterilized in the process industries by passing it through a filter bed which separates the spores present in the air. The air filters used for the removal of microorganisms or spores from the air are of two types:
(i) those filters whose pores are smaller than the size of the microorganisms to be removed
(ii) those in which the pore size is bigger than the size of the microorganisms.

In the former category, virtually all the organisms are filtered off and hence they are called as absolute filters. In the latter category of filters, fibrous materials such as cotton, glass-wool, slag or steel-wool are used to make filter beds with bigger gaps or pore openings. They are called as fibrous-type air filters. Obviously, they cannot assure 100% sterilization of air, no matter whatever be the length of the filter bed. Table 6.6 shows comparison of these two types of filters.

Table 6.6 *Comparison of absolute air filters and fibrous-type air filters for sterilization of air*

Absolute air filter	*Fibrous-type air filter*
The pore size is smaller than the particle size.	The gaps between the fibrous materials are bigger than the particle size.
Sterilization efficiency is 100%.	Sterilization cannot be 100%.
Pressure drops are high.	Pressure drops are not that high.
Filter thickness is small.	Filter thickness is more.
Cost of initial investment is high.	Initial investment costs are low.
Choking/clogging takes place very fast.	Choking and clogging do not take place that fast.
Cleaning and replacement are difficult.	Cleaning and replacement are easy.

6.6.2 Theory of Fibrous-type Filters

The mechanisms by which microorganisms are removed from air by fibrous-type air filters may be (Aiba et al., 1965(c)):
- inertial impaction
- interception

- diffusion
- settling by gravitational force
- electrostatic force.

A detailed description on the above mechanisms is beyond the scope of the book. A simpler description of it may be given here which is sufficient to understand the concept of fibrous filters.

An air-borne particle will escape the filtration process only if it does not hit any fibrous material on its passage through the filter bed, as normally the passage gap is more than the particle size. In other words, a particle which touches the fibrous media would get attached to them and hence will be detained by the filter bed to make the passing air cleaner to that extent. Thus, there is a possibility to have a uniform concentration of particles at any given depth in the filter. We may also assume here that each unit layer of thickness will retain the population of solids to the same extent. These assumptions would enable us to formulate the following expression (Stanbury and Whitaker, 1993):

$$\frac{dN}{dX} = -kN \tag{6.13}$$

where N = number of particles at a depth of X in the filter
k = constant.

Integration of the above equation gives

$$\frac{N}{N_0} = e^{-kX} \tag{6.14}$$

where N_0 is number of particles at the entrance of the filter bed
i.e. at $X = 0$

Equation (6.14) can be written as

$$\ln \frac{N}{N_0} = -kX \tag{6.15}$$

We define the efficiency of the filter as the ratio of the number of particles removed to the initial number of particles, which can be mathematically expressed as

$$E_f = \frac{N_0 - N}{N_0} = 1 - \frac{N}{N_0} \tag{6.16}$$

Thus, $$E_f = 1 - e^{-kX} \tag{6.17}$$

Humphrey and Gaden (1955) have used the concept of X_{90}, which was defined as *the depth of the air filter required to remove 90% of the solids entering the fibrous filter,* i.e. $E_f = 0.9$.

Combining Eqs (6.15) and (6.17), we have

$$X_{90} = \frac{2.303}{k} \tag{6.18}$$

in which k is a function of the filter medium and obviously air velocity. If the air velocity is high, then k will increase initially, which may be due to the fact that higher air velocities will result in higher chances of impact of the particles with the fibrous filter media and hence higher k.

The design of the filter beds involves evaluating the filter bed thickness (or depth) to get a desired decrease in the proportion of particles in the outgoing sterilized air. This is possible only with a knowledge of k, which has to be calculated by Eq. (6.15) after experimentally measuring N at a distance of X. The values of N and N_0 are to be measured for different bed depths for a given fibrous material.

The filter design details have been explained by Stanbury and Whitaker (1993) with a numerical example.

6.7 STERILIZATION OF FERMENTERS

Sterilization of the fermenter is done separately if the medium is sterilized outside and introduced into the fermenter. Any joints, crevices, pits or flange joints are potential hazardous points where the nutrients and microorganisms stay, and hence contamination can take place. Sterilization is done by heating. Steam at 15 psig (0.1 MN/m^2 gauge) is admitted and kept inside for approximately 20 minutes for sterilization to complete. All entry and exit points, like those of air, medium, substrate, etc. are to be kept sterile. All the condensate exit points must have slopes towards the discharging end to avoid any steam condensate forming and staying in the pipelines. The condensers, thermowells, pressure transducers, analytical probe points, etc. are likely to have pits and crevices which harbour contaminant microorganisms.

After the sterilization is complete, the fermenter should be flushed with sterile air kept under positive pressure; lest on cooling, the fermenter develops vacuum and sucks in atmospheric air which is not sterile. If the medium is sterilized in a different vessel, sterile medium should be admitted into the sterilized fermenter under aseptic conditions. The transport lines, etc. should be maintained under aseptic conditions. Any failure in this regard will result in contamination which is difficult to notice subsequently.

The inside of the fermenter should have minimum number of joints. Wherever the joints are made with welding, the joint points should be made as smooth as possible to avoid them from becoming potential sources of contamination. After the fermentation process is completed, all the contents in the fermenter should be discharged and cleaned thoroughly.

If the medium is also sterilized inside the fermenter, the medium in hot condition should be admitted into the fermenter. Otherwise, the medium absorbs the heat from the steam and condenses it. Thus, the temperature is brought down and the condensate remains with the medium. This may be a potential source for contamination.

Thus, the sources for contamination at every stage are to be looked into and checked to plug the loop-holes of contamination.

Symbols

E	: activation energy (kJ/k mol)
E_A, E_B	: E for organism A and B
E_f	: efficiency of filter [in Eq. (6.16)]
k	: first-order thermal death kinetic rate constant [in Eq.(6.1)]; a constant [in Eq. (6.13)] (s^{-1})
k_0	: Arrhenius constant [in Eq. (6.6)]
k_1, k_2, k_3	: first-order thermal death constants at times, t_1, t_2, t_3
N	: number of cells present
R	: universal gas constant = 8.27 (kJ/(kmol K))
t	: time (s)
t_1, t_2, t_3	: time intervals
T	: temperature (K)
X	: depth of the filter (m)
X_{90}	: depth of the filter to remove 90% of spores (m)

Subscripts

0	: initial conditions
i	: condition at i

T	: at temperature T
t	: conditions at any time t

Greek symbols

∇	: Del factor or Nabla factor [in Eq. (6.8)]

References

6.1 Aiba, S., Humphrey, S.E., and Millis, N.F. (1965), *Biochemical Engineering*, Academic Press, New York, (a) p. 7, (b) pp. 207–208, (c) Ch. 9, pp. 217–244.

6.2 Banks, G.T. (1979), "Scale up of fermentation processes", *Topics in Enzyme and Fermentation Biotechnology*, Wiseman, A. (Ed.), Vol. 3, Ellis Horwood, Chichester.

6.3 Crueger, W., and Crueger, A. (1989), Biotechnology: A Textbook of Industrial Microbiology, 2nd Edn., Sinauer Associates, Inc., Sunderland, pp. 96–97.

6.4 Deindoerfer, F.H., and Humphrey, A.E. (1959), "Analytical method for calculating heat sterilization times," *Appl. Microbiol*, 7, 256–264.

6.5 Deindoerfer, F.W., and Humphrey, A.E. (1961). "Scale-up of heat sterilization operations," *Appl. Micro.*, 9, 134–145.

6.6 Humphrey, A.E., and Gaden, E.L. (1955), *Air sterilization by fibrous media*, I&EC, 47, 924–930.

6.7 Richards, J.W. (1965), *Rapid calculations for heat sterilizations*, *British Chem. Eng.*, 10(3), 166–169.

6.8 Richards, J.W. (1968), *Introduction to Industrial Sterilization*, Academic Press, London.

6.9 Stanbury, P.F., and Whitaker, A. (1993), *Principles of Fermentation Technology*, Pergamon Press, Oxford, pp. 92–106.

Review Questions

6.1 What is the importance of sterilization in bioprocessing?
6.2 What are the various methods of sterilization?
6.3 What is meant by "inactivation factor"? Give an expression for it.
6.4 Describe thermal sterilization methods for heat labile components in the culture media.
6.5 Describe the process of batch sterilization.
6.6 How do you describe the effect of temperature on sterilization?
6.7 What is meant by Del factor?
6.8 Describe a method for calculation of Del factor.
6.9 Describe the process of continuous sterilization.
6.10 Compare and contrast the batch and continuous sterilization processes.
6.11 What are the different methods of air sterilization?
6.12 Define the efficiency of air filter in sterilization of air.
6.13 What is X_{90}? Write an expression for it.
6.14 Write a note on the sterilization of the fermenter.

Problems

6.1 A liquid medium initially contains bacterial spores at a concentration of $9.2 \times 10^{11}/m^3$. The medium is sterilized thermally at 125°C, and spores density is recorded. The data are as follows.

Table 6.7

Time (min)	0	5	10	20	30	35
Spore density (m^{-3})	9.2×10^{11}	2×10^{10}	3.32×10^{9}	2.3×10^{5}	115	3

(i) Find the thermal death kinetic rate constant in s^{-1}.

(ii) Find the inactivation factor after 40 min.

(Ans: $k_d = 0.0127\ s^{-1}$, $N_t = 0.06$)

6.2 The thermal death kinetic data for *Bacillus stearothermophilus* are collected in laboratory at different temperatures. The data are presented in Table 6.8.

Table 6.8 *Time versus spore density at different temperatures*

Time (min)	*No. of spores (m^{-3})* 110 °C	115 °C	120 °C
0	9.2×10^{11}	9.2×10^{11}	9.2×10^{11}
3	4.5×10^{11}	9.43×10^{10}	9.2×10^{8}
5	2.73×10^{11}	2.0×10^{10}	9.32×10^{6}
10	8.34×10^{10}	3.32×10^{9}	94
15	2.5×10^{10}	1.02×10^{7}	1.0×10^{-3}
20	7.6×10^{9}	2.3×10^{5}	
30	6.8×10^{8}	115	

The effect of temperature on the kinetic rate constant can be represented by Arrhenius equation. Find the Arrhenius constant and activation energy for thermal sterilization.

(Ans: $k_0 = 9.5 \times 10^{37}\ min^{-1}$, E = 2.8×10^{5} J/(g mol))

6.3 An unsterile bath contains initially 1.0×10^{10} viable cells. Find the Del factor of sterilization ($N_t = 10^{-3}$).

(Ans: $\nabla = 29.93$)

6.4 The time–temperature profile during heating of a medium for thermal sterilization is shown in Table 6.9. Calculate $\nabla_{heating}$ for the spores *Bacillus stearothermophilus* using graphical integration method.

Table 6.9 *Time-temperature profile*

Time (min)	1.0	1.5	2.0	2.5	3.0	3.5	4.0	4.5	5.0
Temperature (°C)	100.0	103.4	106.7	109.5	111.6	113.8	116.4	118.7	121.0

(Ans: $\nabla_{heating} = 2.8$)

6.5 Derive an expression for the efficiency of fibrous filter for sterilising air. Calculate the filter depth for 90% efficiency of removal of spores from the air (X_{90}) when filter constant (k) is 0.42 cm^{-1}.

(Ans: X_{90} = 5.48 cm)

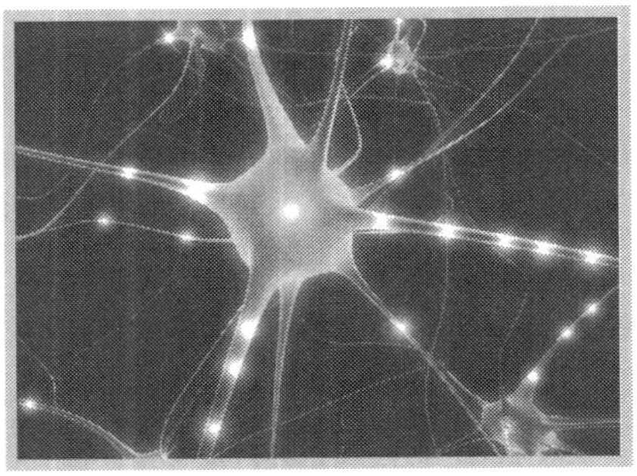

chapter 7

Kinetics of Microbial Growth and Biochemical Reactors

Kinetics is the study that deals with speed. The kinetics of a reaction, in general parlance, means how fast a reaction is progressing. The chemical kinetics deals with the subject matter that how the rate of reaction is dependent upon the concentration of the reactant(s). Thus, we can write for the chemical reaction

$$A \rightarrow \text{Products} \tag{7.1}$$

$$(-r_A) = k\,(C_A)^n \tag{7.2}$$

where r_A = is the rate of reaction of component A
k = reaction rate constant or kinetic constant
C_A = concentration of the reactant A
n = exponential power of C_A

In other words, n is also called as the 'order of the reaction' with respect to the component A.

However, in the case of biochemical reactions, further complications are added to the kinetics. A biochemical reaction proceeds with the intervention of living organisms or cells in the presence of nutrients of the medium under specific conditions of pH, temperature, etc. All the microorganisms are not added in the desired quantities (unlike in the case of chemical reactions). The cells consume a part of the substrate (reactant) and essential nutrients from the medium of fermentation. The cells initially multiply and grow. Depending upon the type of the cells, whether they are unicellular or moulds, the growth pattern varies. For example, the unicellular organisms, which divide when they grow, will increase in the number of cells or increase the biomass. As they increase in number, they consume more and more of substrate. The other type of cells, viz. moulds, will not increase in number, but they increase in size and hence density of the cells, which also result in increase in the viscosity of the broth. This in turn manifests itself in the form of changes in various other physico-chemical and thermal parameters of the broth.

The process does not end here. Associated with the cell growth, there are two other processes also. They are:

(i) uptake of some material from the cell's environment
(ii) convert it into some metabolic end products.

This starts destabilizing the cell growth and the availability of substrate for competitive processes, viz. cell growth and bioconversion of the substrate into metabolic products. This indeed complicates the situation for applying directly Eq. (7.2) because of the division of substrate for cell morphology and bioconversion.

Thus, the complexities in the kinetic study of biochemical reactions/processes can be summarized as follows:

(i) The type of the cells, viz. unicellular or moulds, affects the cell growth and physico-chemical characteristics of the fermentation broth.

(ii) Nutrients and substrates are consumed concurrently and competitively for increasing the production of cell mass and bioconversion to metabolic products.

(iii) Cell-to-cell heterogeneity is a common problem in the growing cells.

(iv) Cells of different ages often manifest several metabolic functions and activities.

(v) Characterization of biochemical pathway by a simple equation like Eq. (7.2) is seldom possible.

Thus, it is almost impossible to try to formulate a simple kinetic model which includes all the complexities mentioned above.

7.1 PHASES OF CELL GROWTH

The cell growth has several phases in a batch culture. They are shown typically in Fig. 7.1. Initially, when the cells are inoculated into the medium, they take some time to adjust to the new environment, which we call as **lag phase** (ab). Subsequently, they start consuming the nutrients in the medium and multiply. Once the growth process starts, the cells multiply in an exponential order. The growth is very fast. This is known as growth phase or exponential growth phase or typically **log phase** (bc). This phase is very important and all the kinetic growth parameters are studied in this phase. Since the growth is exponential in nature, when the cell concentration is plotted versus time on a semi-log plot (Fig. 7.1), the log phase results in a straight line. When once the cell numbers reach a maximum at point c, they stop growing any more. Hence, the growth becomes stagnant. This phase continues up to point d, which we call as **stationary phase** (cd). It is probably in this period, that they start converting the substrate into metabolic products. More and more amount of substrate goes for biochemical reaction to yield some useful products. After reaching point d, the cells start decaying, what we call as decay phase or **death phase** (de). After point d, probably the cells face shortage of substrate for their consumption. Hence they start decaying.

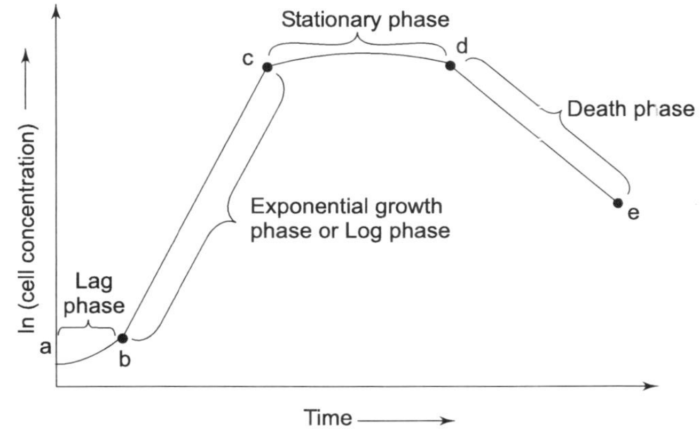

Fig. 7.1 *Growth curve of cells in a batch process*

7.2 BATCH REACTOR DATA ANALYSIS

The batch reactor was discussed in Section 5.7 and shown in Fig. 5.6(a). Batch reactor is frequently used to collect the kinetic data in view of its ease of construction and operation, and versatility. That is, the same batch reactor can be used for various reactions. The data are collected on concentration versus time, and processed to evaluate the reaction rate constant given in Eq. (7.2). A material balance of the reactants fed and the products formed gives us an expression for reaction time. If we assume the system to be multi- component (containing more than one reactant), material balance on any one component, *i*, is given by

$$\frac{d}{dt}\left[\text{reactor volume }(C_i)\right] = \frac{(\text{reactor volume})\text{ moles of } i \text{ formed by reactior}}{(\text{unit reactor volume})(\text{unit time})} \tag{7.3}$$

where C_i is the molar concentration of component *i* in mol/vol.

$$\frac{d}{dt}(V C_i) = V r_i \tag{7.4}$$

Since, in biochemical operations, V can be taken as the culture volume which remains reasonably constant with time, we can write

$$\frac{d C_i}{dt} = r_i \tag{7.5}$$

But

$$r_i = \frac{d C_i}{dt} = k_1 C_i \tag{7.6}$$

where k_1 is the reaction rate constant, assuming that the reaction is a first-order type, i.e. the rate of reaction is proportional to the first power of the concentration (i.e. r_i is proportional to concentration).

Initially, the concentration versus time data are determined experimentally by noting down the C_i with time. The same is plotted in the form of a graph; which is typically shown in Fig. 7.2(a). Slopes are drawn at different points, say (C_1, t_1), (C_2, t_2), (C_3, t_3), etc, to the curve in Fig. 7.2(b). The rate of reaction (or conversion) for the reactant *i* which is equal to

$$-r_i = \frac{-d C_i}{dt} = k_1 C_i \tag{7.7}$$

is obtained by measuring the slopes. Equation (7.7) on integration yields $-\ln(C_i/C_{i0}) = k_1 t$ for first-order kinetics. The reaction rate constant k_1 is measured by plotting $-\ln(C_i/C_{i0})$ against time t in Fig. 7.3(a). The plot is a straight line passing through origin with the slope being equal to k_1.

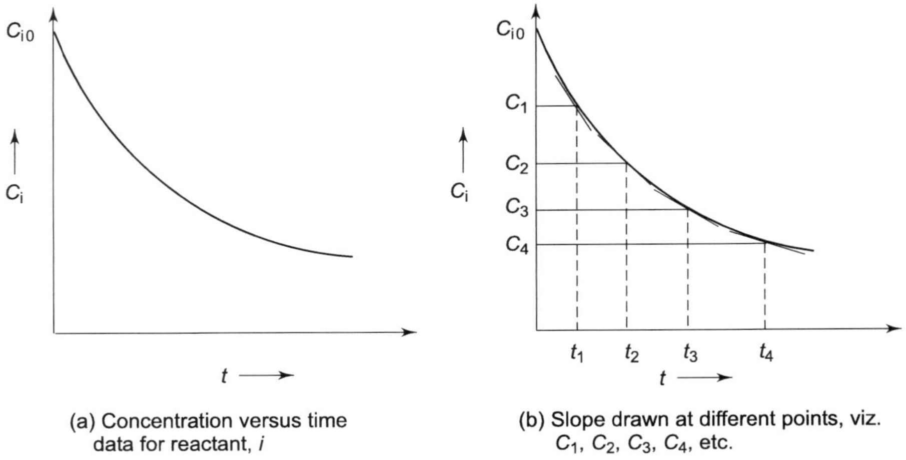

Fig. 7.2 *Time–Concentration data for reactant i with C_{io} being initial concentration*

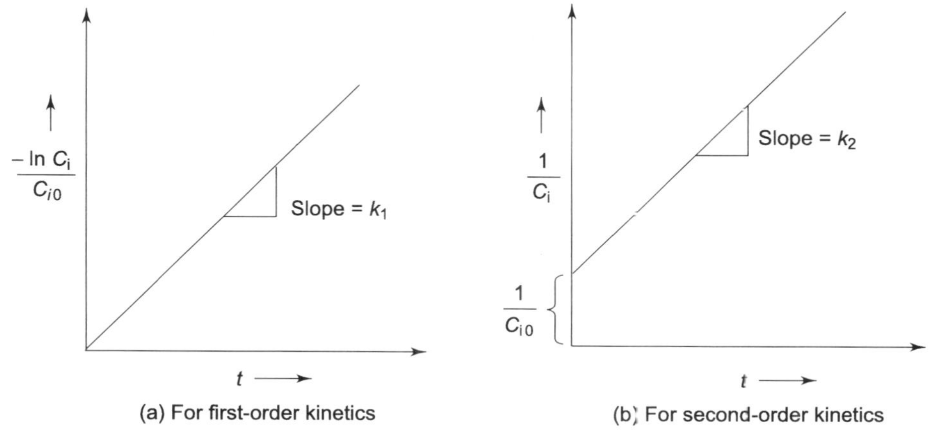

Fig. 7.3 *Procedure to find out the reaction rate constants for first and second-order kinetics from batch reactor data*

For a second-order kinetics, Eq. (7.7) becomes

$$\frac{-\mathrm{d}\, C_i}{\mathrm{d}t} = k_2\, C_i^2 \tag{7.8}$$

which on integration yields,

$$\frac{1}{C_i} - \frac{1}{C_{i0}} = k_2\, t \tag{7.9}$$

where C_{i0} is the initial concentration of component *i*. Plotting $1/C_i$ versus *t* yields a straight line, as shown in Fig. 7.3(b), with the intercept on *y*-axis being equal to $1/C_{i0}$ and slope equal to k_2.

For a reaction of *n*-order kinetics, Eq. (7.7) becomes

$$\frac{-\mathrm{d}\, C_i}{\mathrm{d}t} = -r_i = k_n\, C_i^n \tag{7.10}$$

The above equation can be linearised by taking logarithms on both sides

$$\ln \frac{(-\mathrm{d}\, C_i)}{\mathrm{d}t} = \ln k_n + n \ln C_i \tag{7.11}$$

The rate versus concentration data are obtained by drawing slopes to the time concentration data, as shown in Fig. 7.2(b). The logarithm of the reaction rate [$\ln(-\mathrm{d}C_i/\mathrm{d}t)$] is plotted against $\ln C_i$, which gives a straight line with the intercept being equal to $\ln(kn)$ and the slope equal to *n*, which is the order of the reaction (Fig. 7.4).

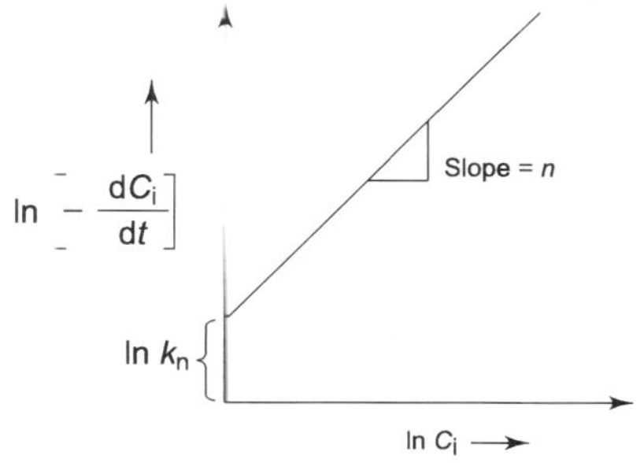

Fig. 7.4 *A generalized plot of reaction rate vs concentration for the nth order reaction.*

Example 7.1

The enzyme ATPase catalyzes the reaction ATP + ADP → P_i. In the process the enzyme gets deactivated from an initial concentration of 10 mg/l. The data on the concentration of of enzyme versus time, at 30°C, are presented in Table E7.1. Find the deactivation kinetics and calculate the deactivation constant (k_d).

Table E7.1 *Concentration of ATPase vs time data at 30 °C.*

time, min	0	5	10	15	20	25	30	45	60
Concentration 10^3 kg/m³	10	6.077	3.694	2.245	1.364	0.829	0.504	0.113	0.025

Solution With the data in Table E7.1, we draw a plot of concentration vs time data on Fig. E7.1. Tangents are drawn at different points on the curve to find

$$-\frac{dC_A}{dt} = -r_A$$

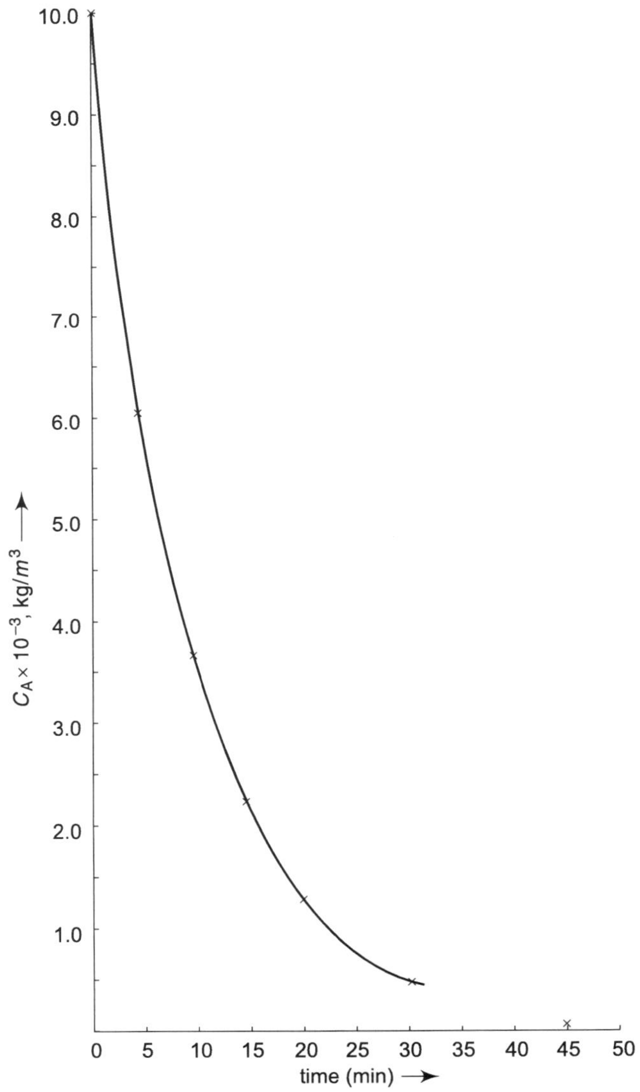

Fig. E7.1 *Conc. vs time data for Example 7.1*

The data on ($-r_A$) are shown in Table E7.2

Table E7.2 *Data for drawing Fig. E7.2*

t, min	$C_A \times 10^3$ kg/m³	$(-r_A) \times 10^3$ kg/(m³ h)	ln C_A	ln $(-r_A)$
0	10	$\frac{10}{3} = 3.333$	–4.605	–5.704
5	6.077	$\frac{8.8}{15} = 0.587$	–5.1	–7.44
10	3.694	$\frac{7}{21} = 0.333$	–5.6	–8.0
15	2.245	$\frac{5.5}{25} = 0.22$	–6.1	–8.42
20	1.364	$\frac{4.15}{29} = 0.143$	–6.6	–8.85
25	0.829	$\frac{2.5}{36} = 0.069$	–7.1	–9.58
30	0.504	$\frac{1.8}{41} = 0.044$	–7.6	–10.03
45	0.113			
60	0.025			

From the data in last two columns of Table E7.2, ln C_A vs ln $(-r_A)$ were plotted in Fig E7.2. It resulted in the form of a straight line indicating that the data obey the rate equation of the type given by Eq. (7.10).

$$\text{Slope of the line} = \frac{10.1 - 6.7}{7.7 - 4.5} = \frac{3.4}{3.2} = 1.065$$

Here, n is taken as $\cong 1.0$

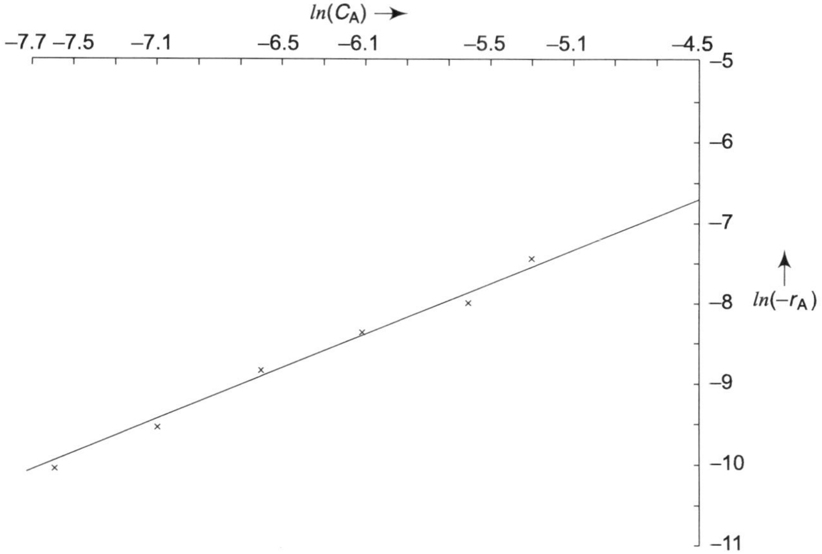

Fig. E7.2 *ln(C_A) vs ln($-r_A$) data for Example 7.1*

Hence the reaction is considered as the first-order reaction.

$$\therefore \qquad (-r_A) = -\frac{dC_A}{dt} = k_d\, C_A$$

k_d can be obtained by noting the intercept corresponding to ln $C_A = 0$. Since it is not possible to find such a long graph we shall use an alternate approach to find ln (k_d),

$$y_2 - y_1 = m\,(x_2 - x_1)$$

in which
$$y_2 = -10.1$$
$$y_1 = ?$$
$$m = 1$$
$$x_2 = -7.7$$
$$x_1 = 0$$

i.e., $-10.1 - y_1 = 1\,(-7.7 - 0)$

i.e $-y_1 = -7.7 + 10.1 = 2.4$

i.e $-\ln(k_d) = 2.4$

$\therefore$ $k_d = e^{-2.4} = 0.0907$ min

Example 7.2

For the above example, the deactivation rate constant was determined at two other temperatures. The results are

Temp °C	*k_d, min^{-1}*
25	0.0825
35	0.212

Determine the activation energy and Arrhenius constant for deactivation kinetics.

Solution The Arrhenius equation can be written as

$$k_d = k_{do}\, e^{-(E/RT)}$$

where E is activation energy,
R is universal gas constant and
T is degrees in Kelvin

Arrhenius equation is linearised by taking logarithms on both sides of the equation

$$\ln(k_d) = \ln(k_{do}) - (E/R)(1/T)$$

From the given data, we prepare Table E7.3

Table E7.3 *Data of ln k_d vs (1/T)*

Temp, °C	*k_d, min^{-1}*	*T, °K*	*1/T*	*ln k_d*
25	0.0825	298	3.356×10^{-3}	– 2.495
30	0.0907	303	3.3×10^{-3}	– 2.40
35	0.1	308	3.247×10^{-3}	– 2.302

We draw or plot last two columns on Fig. E7.3

$$\text{Slope} = \frac{-2.495 - (-2.302)}{(3.356 - 3.247) \times 10^{-3}} = -1771$$

$$= -\frac{E}{R}$$

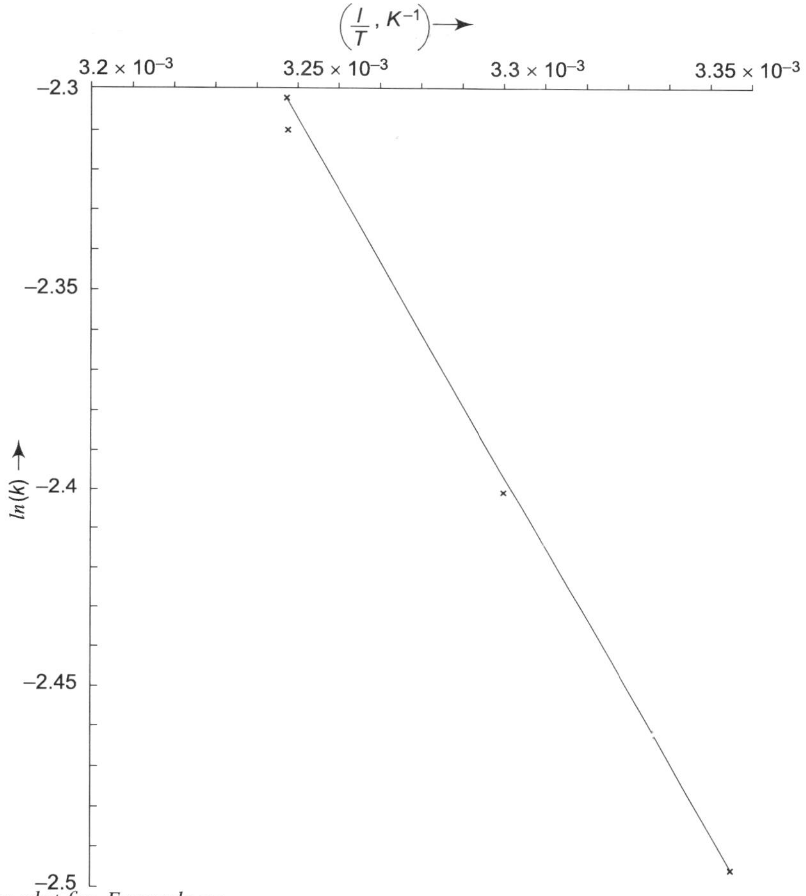

Fig. E7.3 *Arrhenius plot for Example 7.2*

$$\therefore \quad E = 1771 \times R = 1771 \times 8.303 = 1.47 \times 10^4 \text{ Joules / mole}$$

$$k_d = k_{do}\, e^{-E/RT}$$

at 308 K,

$$k_d = 0.1 \text{ min}^{-1}$$

$$\therefore \quad 0.1 = k_{do}\, e^{-(1771/308)} = k_{do} \times 3.183 \times 10^{-3}$$

$$k_{do} = 0.1/3.183 \times 10^{-3} = 31.42$$

Example 7.3

The concentration of glucose being fermented by yeast is measured as the time progresses. The data are presented in Table E7.4. Find whether the data can be represented by rate expression of the form $-r_A = k\, C_A^{\,n}$ *where* C_A *is the concentration of glucose and* r_A *is the rate of conversion of glucose. If so, find k and n. If not, explain the reasons thereof.*

Table E7.4 *Concentration of glucose vs time data*

t, hours	0	2	4	6	8	10	12	14
C_A, *kgmoles/m*^{3}	70	51	41	33	27	23	20	18

Solution We draw a graph between C_A vs t on Fig. E7.4. From the curve, by drawing tangents at various points, we find the $(-r_A)$. Later we draw a plot on Fig. E7.5 between ln C_A vs ln $(-r_A)$ to find k and n by Eq. (7.11). The data required for Fig. E7.5 is shown in Table E7.5. Since, the plot is not a straight line as expected in Fig. 7.5, the data cannot be represented by rate equation similar to Eq. (7.10).

Reasons: Initially, the consumption of glucose is fast since it is being consumed for the growth of cells. Later the consumption slows down. At the end again, the rate of consumption of glucose is faster since the cell density is high, and the rate of formation of primary metabolic products is high when the cell density is high.

Table E7.5 *Data for drawing Fig. E7.5*

t, h	C_A *kg moles/m*^{3}	$-r_A$ *kg moles/(m*^{3}*-h)*	*ln* C_A	*ln* $(-r_A)$
0	70	$\frac{70}{2.4} = 29.17$	4.225	3.37
2	51	$\frac{64.5}{9.5} = 6.8$	3.93	1.92
4	41	$\frac{56.5}{14.5} = 3.9$	3.71	1.36
6	33	$\frac{49}{18} = 2.72$	3.5	1.2
8	27	$\frac{46.5}{18.3} = 2.54$	3.29	1.0
10	23	$\frac{43-5}{18.7} = 2.03$	3.14	0.71
12	20	$\frac{34-15}{15.9} = 1.19$	3.0	0.17
14	18			

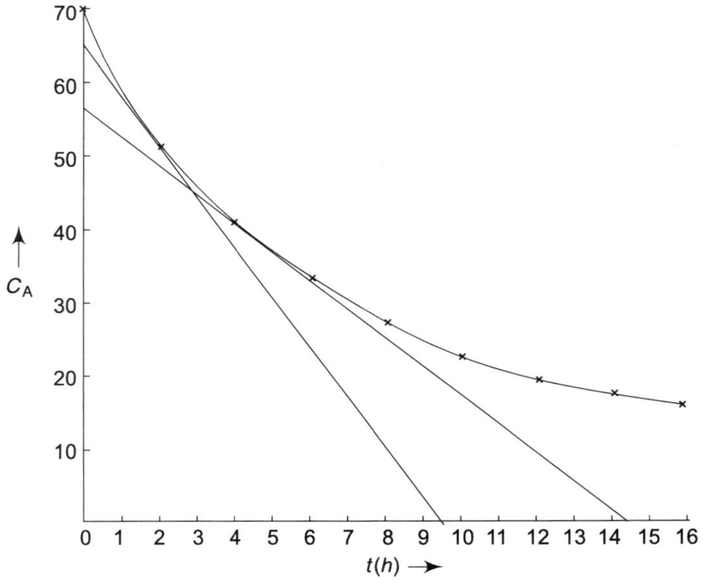

Fig. E7.4 *Conc. vs time data for Example 7.3*

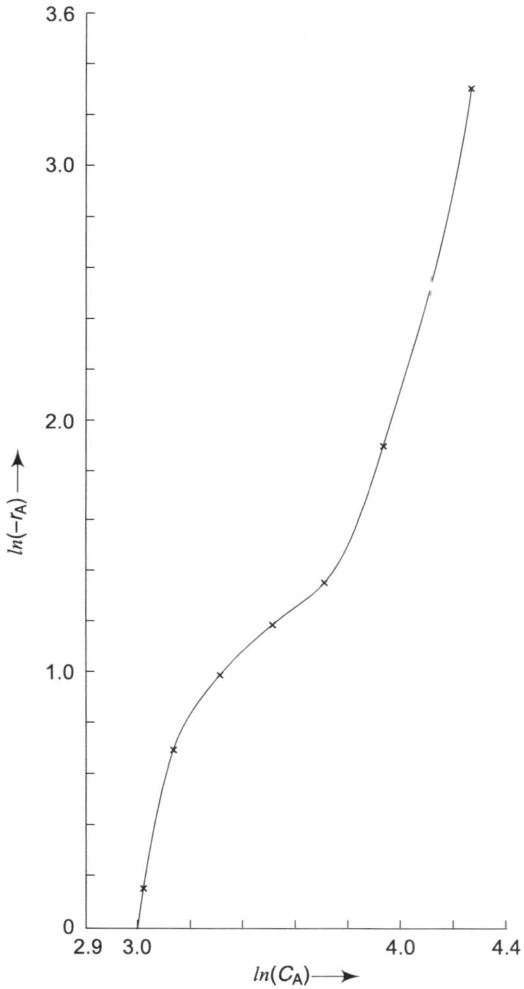

Fig. E7.5 *$ln(C_A)$ vs $ln(-\gamma_A)$ data for Example 7.3*

7.3 KINETIC MODELS FOR CELL GROWTH

After having seen that the growth of the cells in the log phase follows an exponential path in Fig. 7.1, let us try to understand the cell growth by describing it in the form of a mathematical equation. It is generally believed that the growth rate of the cells at any time in the growth phase is proportional to the number of the cells present at that time, which is popularly described by **Malthus' law**

$$\frac{dx}{dt} = \mu x \tag{7.12}$$

where x is the mass of cells per unit volume, μ the proportionality constant known as 'specific growth rate' in $(\text{hour})^{-1}$ and t is the time in hours.

Equation (7.12) is similar to Eq. (7.1), except with a difference of –ve sign. In Eq. (7.12), as t increases, x also will be increasing.

On integration of Eq. (7.12), we get

$$\ln \frac{x}{x_0} = \mu t \tag{7.13}$$

or
$$\ln x = \ln x_0 + \mu t$$

Thus, Eq. (7.13) shows that a plot of $\ln x$ versus time gives a straight-line plot with $\ln x_0$ as its intercept and μ as the slope. Or in other words, Eq. (7.13) enables us to calculate the **specific growth rate** μ with the knowledge of x versus t data.

The biomass produced from the growth of the cells is always due to a change in concentration of the substrate. Thus, we can mathematically express

$$x = Y_{x/s}(S_R - s) \tag{7.14}$$

where x is the biomass concentration which is produced by the depletion of the substrate with its original concentration of S_R, and s is the residual concentration of the substrate. $Y_{x/s}$ is some yield parameter which relates the conversion of substrate to biomass, and is dimensionless. The biomass yield or the cell growth is dependent upon the residual concentration of the substrate.

The stationary phase and death phase of the cells, as observed in Fig. 7.1, are caused by the depletion of one of the components in the medium, which we call as the 'limiting substrate'. Thus, limiting substrate is that substrate in the medium, the concentration of which affects the cell growth and changes the growth phase from the log phase to the stationary phase and finally to the death phase.

The residual limiting substrate concentration also affects the specific growth rate. If it (s) is more, the specific growth rate (μ) also increases up to a certain extent and ultimately reaches a plateau as shown in Fig. 7.5. This indicates that more the substrate, more is the specific growth rate up to a certain extent, beyond which it does not have any effect. The plateau value which μ reaches is called as 'maximum specific growth rate constant' (μ_m).

Here we introduce the concept of 'doubling time' (t_d) for cell growth rates, which is defined as the time required to double the number of cells, i.e. $x = 2x_0$. From Eq. (7.13), we have

$$\ln \frac{x}{x_0} = \ln 2 = \mu t_d$$

or
$$t_d = \frac{\ln 2}{\mu} = \frac{0.693}{\mu} \tag{7.15}$$

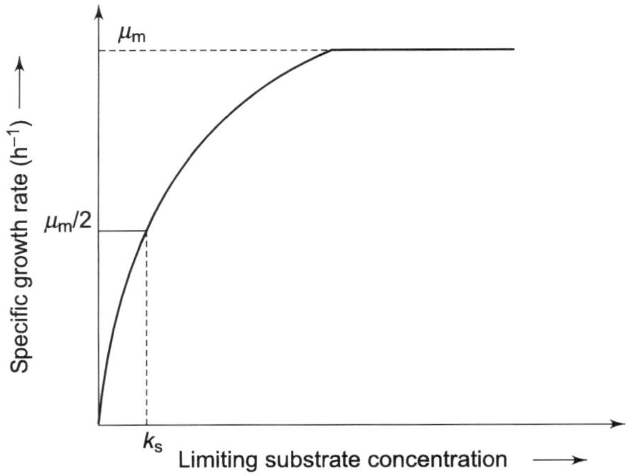

Fig. 7.5 *Dependence of μ on the concentration of the limiting substrate*

where μ is the 'specific growth' for cells, which is defined as the rate of increase in cell concentration per unit of cell concentration (h^{-1}). From Eq. (7.12), we can write

$$\frac{1}{x}\frac{dx}{dt} = \mu$$

The above equation is frequently referred to as the **growth equation**.

7.3.1 Monod Model

As can be seen from Fig. 7.1, the growth rate passes through various phases, viz. high growth phase, low growth phase and finally cessation. In other words, the specific growth rate varies with the residual concentration of the limiting substrate. The relationship is well explained by an empirical equation proposed by Monod (1949).

$$\mu = \frac{\mu_m s}{K_s + s} \tag{7.16}$$

where μ_m is the maximum growth rate achievable, and K_s the limiting substrate concentration when the specific growth rate is equal to half the maximum specific growth rate, i.e. when

$$\mu = \mu_m/2 \text{ (as shown in Fig. 7.5)}$$

which is evident by substituting $\mu_m/2$ for μ in Eq. (7.15), i.e.

$$\frac{\mu_m}{2} = \frac{\mu_m s}{K_s + s} \tag{7.17}$$

by cancelling μ_m on both sides and on rearrangement, $K_s = s$.

Marison (1988) reported some typical values of K_s.

E. coli grown on substrates glucose, mannitol, lactose and phosphate has K_s values like 0.068, 2.0, 20.0 and 1.6 mg/l, respectively. However, the autotrophic forms of *E. coli* and *A. niger* on substrates tryptophan and arginine have less K_s values, viz. 1.1×10^{-3} and 0.5 mg/l. *S. cerevisiae* on substrate glucose has a K_s value of 25.0 mg/l.

In Eq. (7.16), if $s >> K_s$, the denominator may be equated to simply s, in which case

$$\mu = \frac{\mu_m s}{K_s + s} \simeq \frac{\mu_m s}{s} \simeq \mu_m \tag{7.18}$$

In other words, the above equation predicts the percentage increase in μ_m as compared to μ, when s is very large.

Since, it is difficult to always measure μ_m, it can be found by linearizing the Monod equation

$$\frac{1}{\mu} = \frac{K_s}{\mu_m}\left(\frac{1}{s}\right) + \frac{1}{\mu_m} \tag{7.19}$$

By plotting $(1/s)$ versus $(1/\mu)$ (as shown in Fig. 7.6), we can calculate μ_m and K_s.

Though Monod's equation is a simple expression which reasonably predicts the kinetics of the cell growth, it is felt that it (Monod's equation) is an over-simplification of the process which was made evident from some studies of the cell's biochemistry. However, this is generally the case with most of the areas of engineering studies in which either

(i) a very complicated model is proposed and it is further simplified with a good number of assumptions and approximations to make it a workable model

or

(ii) a very simple model is proposed and it is further upgraded (complicated?) by adding various other parameters to reasonably predict the performance of the system.

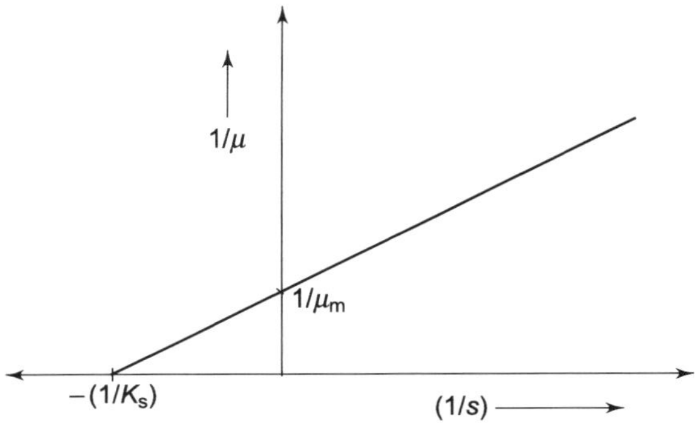

Fig. 7.6 *Plot of linearized Monod equation [Eq, (7.19)]*

The Monod equation thus depicts the general variation of μ with s except when $s >> K_s$. In such cases (i.e. when $s >> K_s$), the equation predicts the percentage increase in μ_m as compared to μ.

Example 7.4 *The substrate concentration versus specific growth rate data were collected for growth of S.Cerevisiae on glucose in a fermenter, and presented as follows.*

Table E7.6 *Glucose concentration (s) vs specific growth rate (μ) data*

s, (g/l)	15	12	9	6	2.5	1.7
μ, (h^{-1})	0.34	0.33	0.32	0.3	0.22	0.18

Calculate Monod constants.

Solution We calculate the Monod constants by drawing a plot of linearized Monod equation (Eq. (7.19))

$$\frac{1}{\mu} = \frac{K_S}{\mu_m}\frac{(1)}{s} + \frac{1}{\mu_m}$$

For this we need to calculate $1/\mu$ and $1/s$, which are tabulated in Table E7.7

Table E7.7 *Data for drawing linearised Monod plot*

S (g/l)	*μ (h^{-1})*	*1/s (l/g)*	*1/μ (h)*
15	0.34	0.067	2.94
12	0.33	0.083	3
9	0.32	0.111	3.125
6	0.3	0.167	3.33
2.5	0.22	0.4	4.54
1.7	0.18	0.588	5.55

The last two columns in Table E7.7 are drawn in Fig. E7.6. From the figure, we note

$$\frac{1}{\mu_m} = 2.5$$

and $$\frac{-1}{k_s} = -0.48$$

$\therefore$ $$\mu_m = 1/2.5 = 0.4\ h^{-1}$$

$\therefore$ $$K_s = 1/0.48 = 2.08\ (g/l)$$

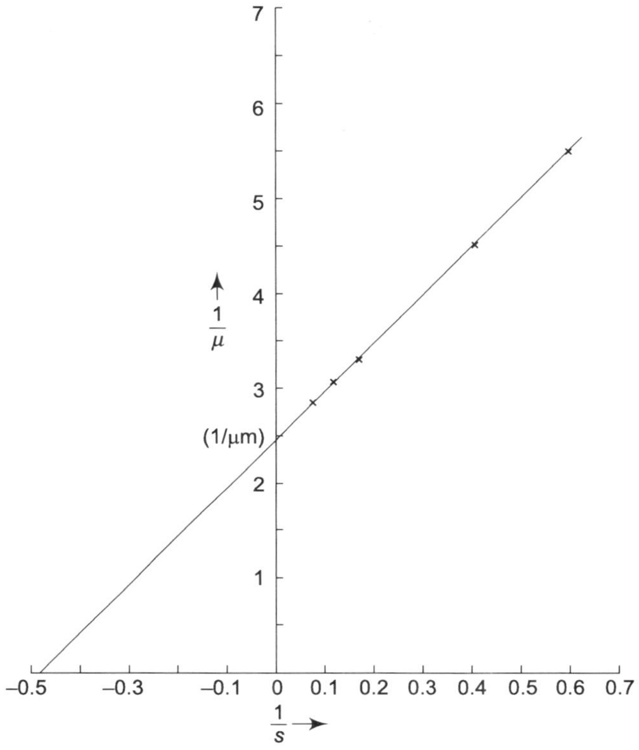

Fig. E7.6 *Linearized Monod Plot for Example 7.4*

7.3.2 Rate Equations

To overcome some of the limitations of the Monod model, Moser (1958) included an additional empirical feature. Tessier (1942) has used the concept of growth deficiency. This has resulted in a semi-logarithmic relationship between the growth rate and concentration.

Table 7.1 lists some of the rate equations (Atkinson, 1974).

7.4 GROWTH OF FILAMENTOUS ORGANISMS

The growth of filamentous organisms like moulds, fungi, etc., does not necessarily increase the number of cells (i.e. biomass), but increase them in length, and hence there will be changes in physical properties like density of the cell mass and viscosity of the broth.

Production of penicillin as a secondary metabolite in batch fermentation is one classical example of non-growth associated kinetic models. In such cases, the rate of product formation is proportional to the cell concentration rather than the growth rate.

The product formation by the organisms may be classified as:

(i) growth associated
(ii) non-growth associated.

The above classification is based on the fact whether the product formation is proportional to the cell growth or otherwise; in other words, whether the product is formed as a result of the primary metabolic function of the cell or whether product formation is due to secondary metabolism.

Table 7.1 *Suggested forms of rate equations for microbial growth*

Equation	*Reference*
$\mu = \mu_m s / \{K_s + s\}$	Monod (1949)
$\mu = \mu_m (ks)^{Bi} / \{1 + (ks)^{Bi}\}$	Moser (1958)
$\mu = \mu_m \dfrac{1}{\left[1 + \dfrac{K_k\ x}{s}\right]}$	Contois (1959)
$N = \{k_1 / k_2\}\ s$	Atkinson et al. (1967)
$\mu = \mu_m \dfrac{s}{K_s + s + (s^2/K_1)}$	Edwards (1970)
$\mu = \mu_m \left[\dfrac{s}{K_s + s}\right]\left[\dfrac{1}{1 + (P/K_p)}\right]$	Aiba, et al. (1965)

B_i = an empirical coefficient

k_1 = biological rate equation coefficient, T^{-1}

k_2 = biological rate equation coefficient, L^{-1}

k = biological rate equation coefficient, $M^{-1} L^3$

K_I = inhibition constant

K_k = an empirical constant in Contois equation

K_p = product inhibition constant

N = flux of the substrate at the interface between microbial mass and adjacent solution, $ML^{-2} T^{-1}$

P = concentration of product, which inhibits the growth

s = substrate concentration

x = mass of cells per unit volume of batch culture

7.4.1 Growth Associated

Figure 7.7 shows a schematic representation of the growth of the microorganism and product formation during the log phase. The substrate concentration continuously falls as fermentation progresses.

In case of a 'growth-associated' product formation, the rate of production of the product parallels the rate of growth of cell population. Typical examples are production of alcohol by the anaerobic fermentation of glucose by yeast, or production of gluconic acid from glucose by *Gluconobacter*. Figure 7.8 shows a representation of growth-associated product formation, in which the product forms parallel to cell growth. This is a typical representation of growth-associated culture according to the classification of fermentation processes as suggested by Gaden (1959).

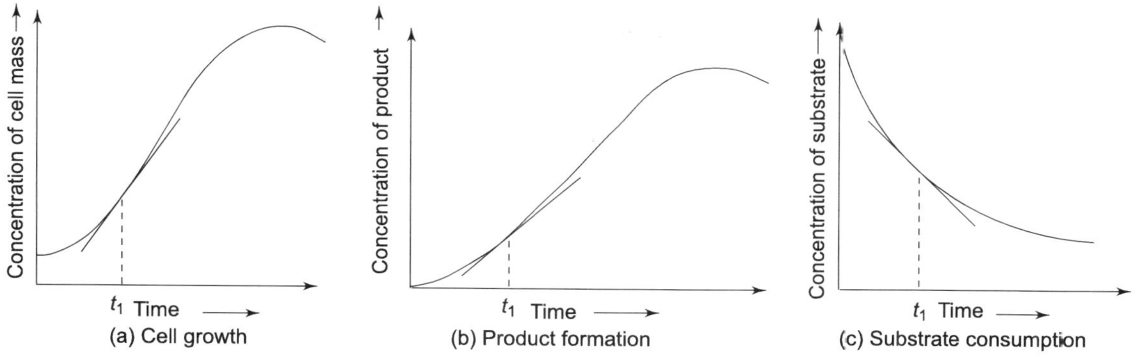

Fig. 7.7 *Growth-associated kinetic pattern*

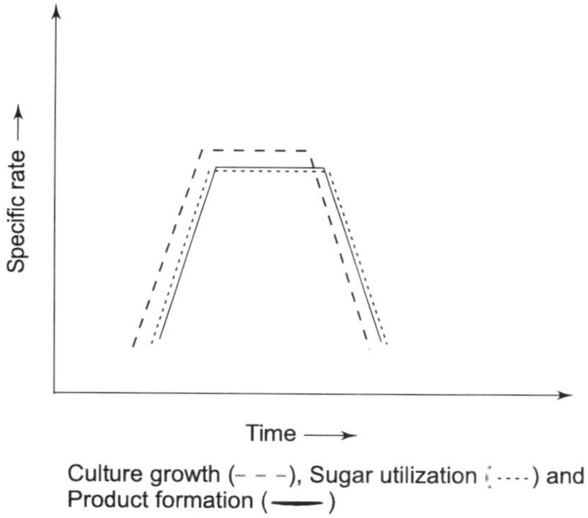

Fig. 7.8 *Representation of growth-associated culture.*
(Reprinted from "Fermentation process kinetics", Gaden, Jr., J.Biochem. Microbial. Tech. & Eng,. p.422, Copyright © (1959),. with permission from John Wiley & Sons, Inc.)

7.4.2 Non-growth Associated

Products such as antibiotics and vitamins are produced in batch fermentations at the end of the log phase or only after entering into the stationary phase. Such secondary metabolites are known as 'non-growth-associated products'. Figure 7.9 shows the concentration of the product formed, concentration of the micelia, and the substrate consumed in g/l. Since the representation of the substrate is in terms of g/l consumed with time, the shape of the curve is that of (c) shown in Fig. 7.9, with an increasing trend with time. Had it been the concentration of the substrate, it also would have taken shape similar to (c) in Fig. 7.7. The penicillin fermentation is a typical example of such non-growth-associated behaviour of product formation in which the product (penicillin) starts forming almost after the completion of micelial growth or utilization of the carbohydrate (Fig. 7.10). This is a typical representation of the non-growth-associated culture according to the classification of fermentation processes as suggested by Gaden (1959).

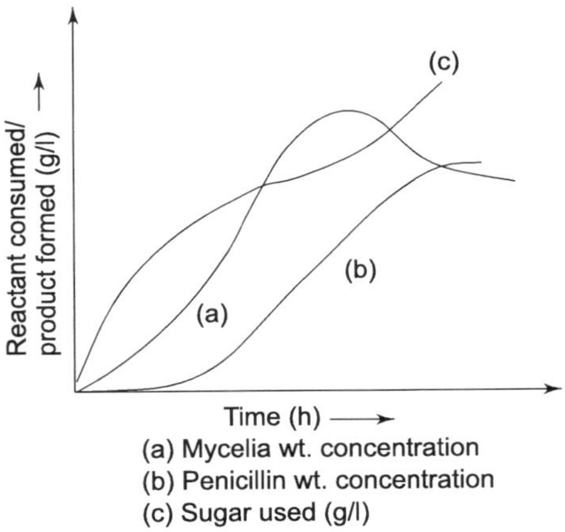

Fig. 7.9 *Non-growth-associated kinetic pattern of product, viz. penicillin.*

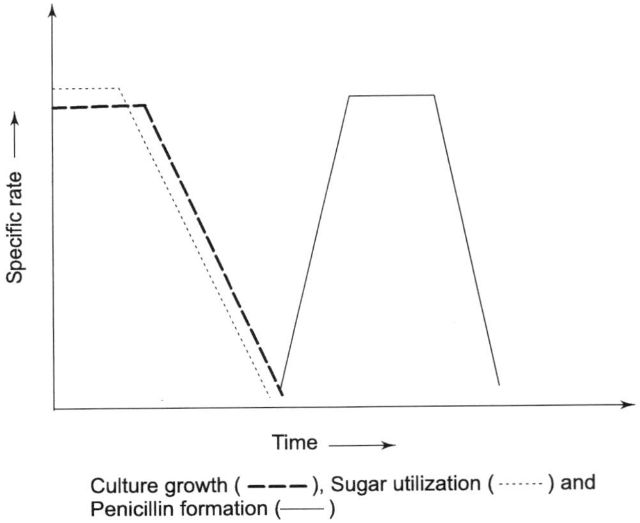

Fig. 7.10 *Representation of non-growth associated culture. (Reprinted from Fermentation process kinetics, Gaden, Jr., E.L,- J .Biochem. Microbial. Tech. & Eng. p. 422. Copyright © (1959), with permission from John Wiley & Sons, Inc.)*

7.5 SUBSTRATE AND PRODUCT INHIBITION ON CELL GROWTH

Inhibition is a process in which some substances tend to decrease the rate process, could it be the growth of cells. In microbial processes, some higher concentrations of the substrate or the product inhibit the growth of the microorganisms, due to which the growth rate falls (Fig. 7.11). Edwards (1970) used an equation originally proposed for substrate-inhibited enzyme reactions to describe the growth of microorganisms, and how the substrate concentrations affect the growth rate of the cells. The specific growth rate can be represented as follows, which is a modified version of the Monod equation

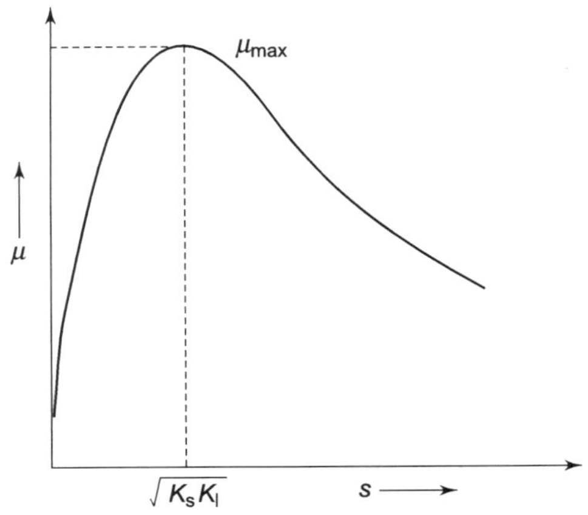

Fig. 7.11 *Influence of substrate concentration on the specific growth rate*

$$\mu = \frac{\mu_m \ s}{K_s + s + \left[s^2 / K_I \right]} \tag{7.20}$$

where K_I is the inhibition constant.

The form of the above relationship is shown in Fig. 7.11 (Lovitt and Jones, 1994). Initially, as the substrate concentration increases, μ will increase up to a certain extent. The maximum value reaches at μ_{max}. Subsequently, further increase in substrate concentration results in decreasing μ. The maximum value of specific growth rate, μ_{max}, can be found by differentiating Eq. (7.20) with respect to s and equating it to zero. Thus, the value of s corresponding to μ_{max} is given as

$$s = (K_s \ K_I)^{1/2} \tag{7.21}$$

which is shown in Fig. 7.11. Substitution of Eq. (7.21) into Eq. (7.20) results in

$$\mu_{max} = \frac{\mu_m}{2(K_s / K_I)^{1/2} + 1} \tag{7.22}$$

The growth of microbes on acetate is an example of the use of Eq. (7.20) and shows the inhibitory effect of the substrate concentration on the specific growth rate of microorganisms.

In certain cases, the product also inhibits the growth of microorganisms, which is called as 'product inhibition' of cell growth. A classical example for this is the effect of build-up of ethanol concentration in fermentation of glucose by yeast. As the alcohol concentration builds up to 8–9%, it will have an inhibitory effect on the growth of organisms. The yeast cells will not be able to survive any more build-up of alcohol concentration. In such cases, Monod equation does not give an adequate estimate of the specific growth rate. One method of mathematically representing the situation is to incorporate a term into the Monod equation which will depress the value of μ with increase in product concentration. Aiba et al. (1965) have proposed one such equation

$$\mu = \frac{\mu_m \ s}{(K_s + s)} \frac{K_p}{(K_p + P)} \tag{7.23}$$

where K_p is a constant representing the inhibitory effect due to the product, and P the concentration of the product. As the value of P increases, μ will decrease. If $P = 0$, Eq. (7.23) reduces to Monod equation, i.e, Eq. (7.16).

Example 7.5

A particular organism follows substrate-inhibition kinetic growth equation of Edwards (1970). The kinetic parameters are (μ_m) = 0.25 h^{-1}, Ks = 2g/l, and K_I = 1.35g/l. Determine the value of substrate concentration at which the specific growth rate is maximum.

Solution Given:

$$\mu_m = 0.25 \text{ h}^{-1}$$

$$K_s = 2 \text{ g/l}$$

$$K_I = 1.35 \text{ g/l}$$

It is a straight application of Eq. 7.21

$$s \text{ for maximum } \mu = (K_s \ K_I)^{1/2}$$

$$= \sqrt{(2 \times 1.35)} = 1.64 \text{ g/l}$$

7.6 STRUCTURED MODELS

So far, when we have been studying the growth of microbial cultures, we have considered that the entire biomass is a homogeneous material. Typically we carry out material balance for (i) biomass and (ii) substrate. In this, we implicitly imply that the conditions in the biomass are the same, i.e. all cells in the culture experience homogeneous conditions, and hence they respond to the stimulus also in the same way. This is over-simplification of the situation, and hence such models will not be able to predict the performance in situations other than that in the log phase.

If we try to take a realistic situation and incorporate the nature of reactions occurring within the cell, it gives rise to the concept of 'structure'. The structured models do not consider the cell as a single entity which simply reacts or interacts with its surrounding environment. These models consider the individual reactions or groups of reactions occurring within the cell.

One of the earliest structured models is the one that was proposed by Williams (1967), which considered the cell to be consisting of two components: (i) active component and (ii) structural component. According to this model, all the cells in the fermentation broth were initially identical, and when once the substrate was incorporated into the active component, the structural component of the cell becomes active. The structural component controls the cell growth.

Product Formation Chemically structured product formation kinetics is described here by considering an interesting secondary metabolite production process. This serves a useful purpose of describing the various other antibiotic fermentations. Phosphate is an important nutrient for growth in fermentation processes; but high concentrations of phosphate inhibit the synthesis of several microbial antibiotics and other secondary metabolites. Hence, it is necessary to find out the optimum value of phosphate concentration. It is described here with an example during the production of alkaloids by *Claviceps purpurea* (Fig. 7.12).

The experimental data is reproduced from the original text (Pazoutova et al., 1981), and the kinetic models for the batch growth of product formation are:

Biomass growth

$$\frac{dx}{dt} = k_1[1 - \exp(-K_1 p_i)]x - k_2 x^2 \tag{7.24}$$

Product formation

$$\frac{dP}{dt} = k_4 \frac{K_3\ x}{K_3 + p_i^2} \tag{7.25}$$

where p_i is the intracellular phosphate concentration (g of KH_2PO_4/g of biomass) and P is the concentration of alkaloid (product). k_1, k_2, k_4, K_1, K_3 are all kinetic constants.

Extracellular phosphate:

$$\frac{dp}{dt} = -k_3 \frac{p}{p + K_2} x + (Y_{p/x} + p_i)\, k_2\, x^2 \tag{7.26}$$

where p is the extracellular phosphate concentration in g/l of culture, k_3 and k_2 are the kinetic constants, and $Y_{p/x}$ is a yield parameter of product based on cell growth.

Intracellular phosphate:

$$\frac{dp_i}{dt} = -\frac{1}{x}\frac{dp}{dt} - (Y_{p/x} + p_i)\frac{1}{x}\frac{dx}{dt} \tag{7.27}$$

in which dp/dt and dx/dt are replaced by Eqs. (7.26) and (7.24), respectively.

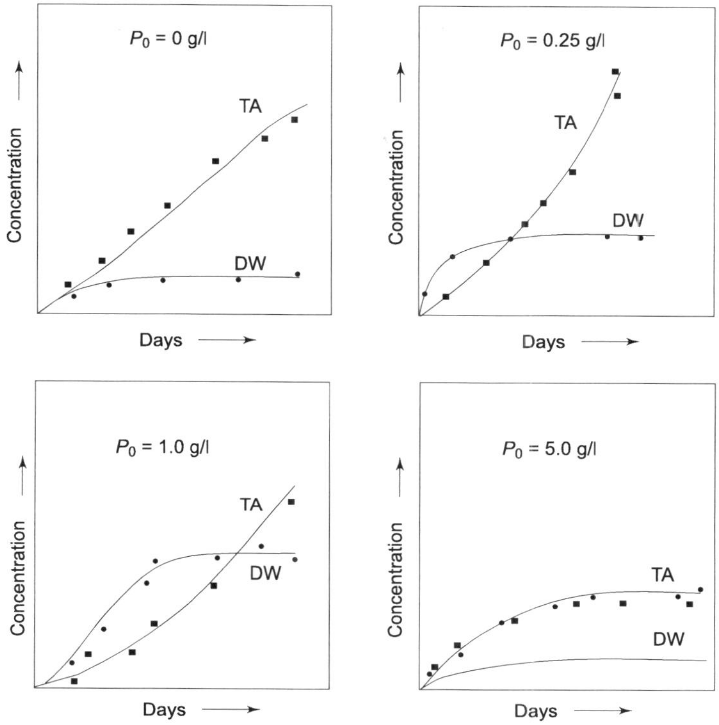

Fig. 7.12 *Batch cultivation data showing C. purpurea mass density in gram dry wt./l (DW●) and total alkaloids (TA■) in mg/l.*
(Reprinted from "A mathematical model for growth and alkaloid production in the submerged culture of Claviceps purpurea", Pazoutova, S., Votruba, J,. and Rehacek, z. from Biotechnol. Bioeng, vol. 23, p. 2843, Copyright © (1981), with permission from John Wiley & Sons, Inc.)

The experimental points in Fig. 7.12 were all well represented by the model equations, i.e. Eqs. (7.24) to (7.27). The numerical values of various coefficients used are given in Table 7.2.

Table 7.2 *Kinetic parameter values for the structured model of alkaloids production*

Parameter	*Dimensions*	*Value*
k_1	day^{-1}	0.5
k_2	l /(g day)	0.016
k_3	day^{-1}	0.0575
k_4	mg /(g day)	6.028
$Y_{p/x}$	–	0.0025
K_1	–	1.87×10^{-4}
K_2	g / l	4.29×10^{-4}
K_3	–	4.65×10^{-4}

(Reprinted from "A mathematical model for growth and alkaloid production in the submerged culture of Claviceps purpurea", Pazoutova, S., Votruba, J., and Rehacek, Z., Biotechnol. Bioeng., vol. 23, p. 2842, Copyright © (1981), with permission from John Wiley & Sons, Inc)

At high concentration of phosphate, the cell growth rate (DW) is not well represented by the model. However, the maximum accumulation of total alkaloids are well represented by the structured model. The above model predicts an optimum phosphate concentration of 0.17 g/l for maximum production of alkaloids (Pazontova, 1981).

7.7 DESIGN EQUATIONS BASED ON BIOCHEMICAL REACTIONS

The design equations for bioreactors are essentially meant for finding out the reaction time or reactor volume, and this needs to be done by coupling the mass balance equations for conversion of substrate, for the growth of the microorganisms and formation of products. The conversion of the substrate (reactant) results in the formation of the products or in the growth of cells or in both.

For a batch reactor

(i) Growth of the cells

$$\frac{dx}{dt}(V) = r_x V \tag{7.28}$$

(ii) Consumption of substrate

$$\frac{-ds}{dt}(V) = r_s V \tag{7.29}$$

(iii) Product formation

$$\frac{dP}{dt}(V) = r_p V \tag{7.30}$$

where V is the volume of the fermentation broth; and r_x, r_s and r_p are rates of cell growth, consumption of the substrate and formation of product, respectively.

For a CSTR A cell separator unit, followed by a provision for recycle of streams containing concentrated cells, is usually incorporated in CSTR to increase the biomass generation and to improve the product yield. It is typically shown in Fig. 7.13 (Bailey and Ollis, 1986). Material balance on the cells in the inlet and outlet streams, with a part of the product stream being recycled after concentrating the cells, shows that

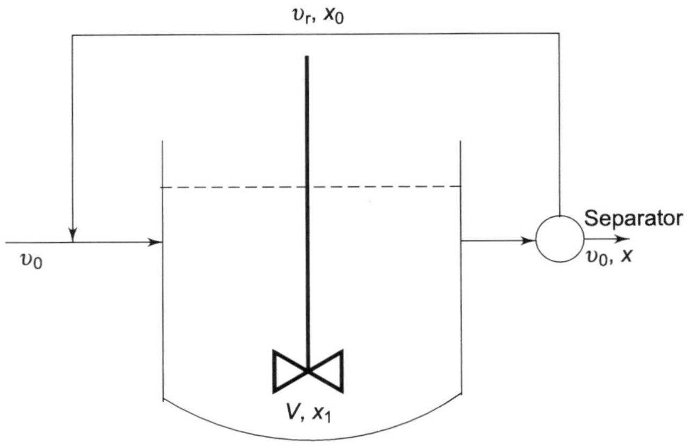

Fig. 7.13 *Schematic representation of CSTR with a recycle stream*

$$\upsilon_r x_0 + \mu x_1 V = (\upsilon_0 + \upsilon_r)x_1 \tag{7.31}$$

where υ_0 and υ_r are feed inlet volumetric rate and product recycle volumetric flow rate; x_0 and x_1 are the biomass concentrations of the recycle-stream and in the reactor, respectively. The LHS terms of the above equation indicate the quantity of the cells entering and consumed in the CSTR, whereas the RHS terms indicate those leaving with no accumulation in the CSTR.

The 'dilution rate' D is defined as

$$D = \frac{\upsilon_0}{V} \tag{7.32}$$

which indicates the number of volumes of the reactor processed in a unit time and will have units of (time)$^{-1}$.

Eq. (7.31) can be written as

$$\frac{\upsilon_r}{V}\frac{x_0}{x_1} + \mu = \frac{\upsilon_0}{V} + \frac{\upsilon_r}{V}$$

$$\mu = \frac{\upsilon_0}{V} + \frac{\upsilon_r}{V}\left[1 - \frac{x_0}{x_1}\right]$$

$$= D + \frac{\upsilon_r}{\upsilon_0} * \frac{\upsilon_0}{V}\ (1 - b)$$

$$= D + a\,D(1 - b)$$

$$= D(1 + a(1 - b))$$

which on rearrangement yields

$$D = \frac{\mu}{1 - a(b - 1)} \tag{7.33}$$

where $a = \upsilon_r/\upsilon_0$, and $b = x_0/x_1$

Similarly, assuming a constant yield factor, $Y_{x/s}$, material balance on substrate yields

$$\upsilon_0\,(s_0 - s) = \frac{\mu x_1}{Y_{x/s}} V$$

The LHS indicates the difference between the inlet flow and outlet flow of the substrate per unit time, whereas the RHS indicates the substrate converted to biomass per unit time in the reactor volume V. Replacing υ_0/V with D from Eq. (7.32), we get

$$D\,(s_0 - s) = \frac{\mu x_1}{Y_{x/s}} \tag{7.34}$$

where s_0 is the initial substrate concentration.

Combining (Eqs. 7.33) and (7.34), we get

$$\mu x_1 = \frac{\mu Y_{x/s}\,(s_0 - s)}{1 - a(b - 1)} \tag{7.35}$$

Equation (7.35) indicates the beneficial effect of recycling, which shows that the non-recyclable rate increases by a factor of $[1 - a(b - 1)]^{-1}$ due to recycling, which is always greater than 1, as b is always greater than 1, and hence $[1 - a(b - 1)] < 1$.

For a plug flow reactor Assuming that Monod's law applies for growth kinetics, the mass balance on cells yields

$$\frac{dx}{dt} = \mu x = \frac{\mu_{max}\, x s}{(s + K_s)} \tag{7.36}$$

Mass balance on substrate gives

$$\frac{ds}{dt} = \frac{-1}{Y_{x/s}} \left[\frac{\mu_{max}}{(s + K_s)} \right] (x s) \tag{7.37}$$

with initial conditions

$$x = x_0 \quad \text{and} \quad s = s_0, \text{ when } t = 0 \tag{7.38}$$

We can also write mass balance on cells as follows:

$$x + (Y_{x/s})(s) = x_0 + (Y_{x/s})(s_0) \tag{7.39}$$

Using Eq. (7.39) in Eq. (7.37), we get

$$\frac{ds}{dt} = \frac{\mu_{max}}{Y_{x/s}} \frac{\left[x_0 + Y_{x/s}(s_0 - s)\right] s}{(s + K_s)} \tag{7.40}$$

which on integration yields (Bailey and Ollis, 1986):

$$x_0 + Y_{x/s}(s_0 + K_s) \ln \left[\frac{x_0 + Y_{x/s}(s - s_0)}{x_0} \right] - K_s\, Y_{x/s} \ln \left[\frac{s}{s_0} \right] = \mu_{max}\, t\, (x_0 + Y_{x/s}\, s_0) \tag{7.41}$$

where

$$t = AL/\upsilon_0 \tag{7.42}$$

in which L and A are the length and cross-sectional area of the plug flow reactor, and υ_0 is the volumetric inlet flow rate.

Symbols

a	: ratio used in Eq. (7.33)
A	: cross-sectional area of plug flow reactor (m^2)
b	: ratio used in Eq. (7.33)
C	: concentration (kg/m^3 or $kg\ mol/m^3$)
D	: dilution rate in Eq. (7.32) (s^{-1})
k	: reaction rate constant in Eq. (7.2)
k_1, k_2, k_3, k_4	: kinetic constants in Eqs. (7.24)–(7.26)
K_1, K_2, K_3	: kinetic constants in Eqs. (7.24)–(7.26)
K_1	: inhibition constant in Eq. (7.20)
K_p	: product inhibition constant in Eq. (7.23)
K_s	: Monod constant in Eq. (7.16)
L	: length of plug flow reactor (m)
n	: order of reaction in Eq. (7.2)
p	: extracellular phosphate concentration (g/l)
p_i	: intracellular phosphate concentration in Eq. (7.27) (g/g of dry mass)
P	: concentration of the product (kg/m^3 or $kg\ mol/m^3$)

r	: rate of reaction or rate of growth
s	: substrate concentration (kg/m^3)
S_R	: original or initial substrate concentration
t	: time (s or h)
t_d	: doubling time (s)
υ	: volumetric feed rate (m^3/s)
V	: volume of reactor or volume of fermentation broth (m^3)
x	: mass of cells per unit volume (kg/m^3)
x_1	: biomass concentration in the reactor in Eq. (7.31)
$Y_{p/x}$	: yield factor of product based on x
$Y_{x/s}$	: yield factor of x based on s

Subscripts

0	: for initial values
A	: component
p	: product
r	: recycle
s	: substrate
x	: cells

Greek letters

μ	: specific growth rate (h^{-1})
μ_m	: maximum specific growth rate (h^{-1})

References

7.1 Aiba, S., Humphrey, A.E., and Millis, N.F. (1965), *Biochemical Engineering,* Academic Press, New York, p. 85.

7.2 Atkinson, B., Swilley, E.L., Busch, A.W., and Williams, D.A. (1967), "Kinetics, mass transfer and organism growth in a biological film reactor", *Trans. Inst. Chem. Eng.* 45, 257–264.

7.3 Atkinson, B. (1974), *Biochemical Reactors*, Pion Ltd., London, p. 77.

7.4 Bailey, J.E., and Ollis, D.F. (1986), *Biochemical Engineering Fundamentals*, 2nd Edn., McGraw-Hill Inc., New York, pp. 539–543.

7.5 Contois, D.E. (1959), "Kinetics of microbial growth: Relationship between population density and specific growth rate of continuous culture", *J. Gen. Microbiol.*, 21, 40.

7.6 Edwards, V.H. (1970), "The influence of high substrate concentration on microbial kinetics", *Biotechnol. Bioeng.*, 12, 679.

7.7 Gaden Jr., E.L. (1959), "Fermentation process kinetics", *J. Biochem. Microbial. Tech. Eng.*, 1, 413.

7.8 Lovitt, R., and Jones, M. (1994), "Biochemical reaction engineering", *Coulson & Richardson's Chemical Engineering*, Richardson, J.F., and Peacock, D.G. (Eds.), 3rd Edn, Vol. 3, Pergamon Press, London, pp. 346–348.

7.9 Marison, I.W. (1988), "Growth Kinetics", *Biotechnology for Engineers,* Scragg, A.H. (Ed.), Ellis Horwood Ltd., Chichester, p. 192.

7.10 Monod, J. (1949), "The growth of bacterial cultures", *Ann. Rev. Microbiol.*, 3, 371–394.

7.11 Moser, H. (1958), "The dynamics of bacterial population in the chemostat", *Carnegie Inst.*, Pub. No. 614, Washington.
7.12 Pazoutova, S., Votruba, J., and Rehacek, Z. (1981), "A mathematical model of growth and alkaloid production in the submerged culture of *Claviceps purpurea*", *Biotechnol. Bioeng.*, 23, 2837–2849.
7.13 Tessier, G. (1942), "Croissance des populations bacteriennes at quantite d'aliment disponible", *Rev. Sci. Paris,* 80, 209.
7.14 Williams, F.M. (1967), "A model of cell growth dynamics", *J. Theoret. Biol.*, 15, 190–207.

Review Questions

7.1 How are biochemical reactions different from chemical reactions?
7.2 What are the various complexities one comes across in the case of biochemical reactions/processes?
7.3 Describe the various phases of cell growth in a batch culture, with a neat diagram.
7.4 What is Malthus' law?
7.5 Describe the kinetic equations for cell growth.
7.6 How do you calculate the specific growth rate (μ)?
7.7 Define the yield factor.
7.8 What is meant by 'limiting substrate concentration'?
7.9 Describe the Monod model for specific growth rate.
7.10 Explain how do you find out the μ_m and K_s in Monod model by linearizing it.
7.11 Describe the growth-associated and non-growth-associated product formation in fermentation process.
7.12 Describe a substrate-inhibited cell growth.
7.13 Describe a product-inhibited cell growth.
7.14 Describe the concept of structured models.
7.15 Define and explain the term "dilution rate".
7.16 Write the material balance equation for CSTR and explain it with a neat schematic diagram.
7.17 In a CSTR (chemostat), show how recycling improves the production rate.
7.18 Explain how do you determine the rate of a reaction.
7.19 Explain how do you calculate the first-order reaction rate constant.
7.20 Describe how do you find out the order of a reaction.

Problems

7.1 A batch fermenter was operated for the production of alcohol from glucose using yeast. The following data were obtained in the fermenter.

Time (h)	*0*	*2*	*4*	*6*	*8*	*10*	*12*	*14*
Concentration of glucose (kg mol/m³)	*70*	*68*	*66*	*63*	*58*	*52*	*44*	*36*
Yeast count (×10⁶ cell/ml)	–	*6*	*10*	*20*	*38*	*80*	*117*	*140*
Alcohol production (kg mol/m³)	0	1	5	11	20	30	41	55

Find out whether the above rate data could be interpreted by a simple kinetic equation of the type

$$-r_A = k_n\, C_A^{\,n}$$

where C_A stands for substrate concentration. If so, find k_n and n. If not, indicate the reasons thereof.

7.2 During the growth of *Saccharomyces cerevisiae* on glucose in a 20 l fermenter, the following data were observed on the glucose concentration (s) versus the specific growth (μ), and presented in Table 7.3. The growth can be reasonably represented by Monod growth kinetics. Find the maximum specific growth rate (μ_m) and Monod constant (K_s), following the linearized Monod equation procedure.

Table 7.3 *Glucose concentration (s) versus specific growth rate (μ) data*

(s), (g/l)	10	5	3.3	2.25	2.0	1.7	1.4
μ (h^{-1})	0.4	0.36	0.3	0.27	0.25	0.22	0.2

(Ans: $\mu_m = 0.5\ h^{-1}$, $K_s = 2$ g/l)

7.3 Edwards equation (Edwards, 1970) originally proposed for substrate-inhibited enzyme reactions can be written as

$$\mu = \frac{\mu_m\ s}{K_s + s + (s^2/K_I)}$$

where K_I is the inhibition constant. Show that the maximum specific growth rate ($\neq\mu_m$) occurs for the substrate concentration

$$s = \sqrt{K_s\ K_I}$$

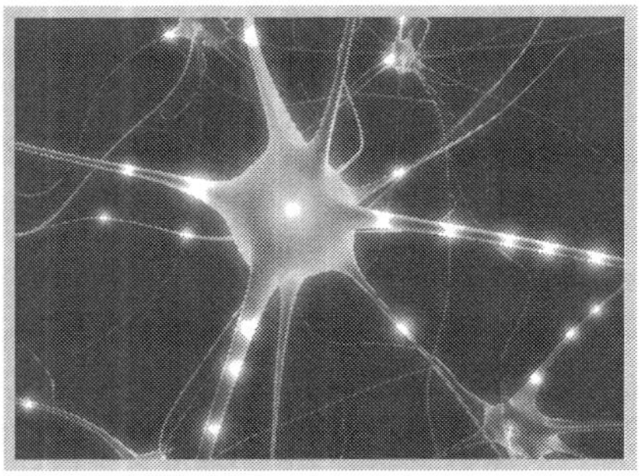

chapter 8

Ideal Reactors

A reactor is a vessel in which chemical reaction takes place to convert the reactants into products, with or without the influence of a foreign material called as catalyst. A catalyst is one which indeed takes part in a reaction, but does not get converted to products. Thus, the catalyst concentration remains virtually constant. It may only get inhibited or deactivated during the course of reaction, in which case, we may have to replenish or reactivate or regenerate the catalyst. If the conversion takes place with the intervention of living microorganisms, we call the reaction a 'biochemical reaction', and the reactor a 'fermenter'.

An ideal reactor is one in which the ideal conditions prevail. The ideal conditions refer to:

- mixing conditions, in case of multiphase reactions in a CSTR
- total non-mixing conditions, in case of a PFR
- ideal flow conditions
- uniform temperature at different zones
- different temperatures at different times as programmed.

In reality, the above ideal conditions do not prevail, which we call as nonideal state, and will be discussed in detail in Chapter 10.

From the kinetic studies, we arrive at rate equations which tell us how fast a reaction proceeds by converting the reactants to produce the products, and the dependence of the rate of reaction on the concentration of the reactants (Eq. (7.2)). A differential equation for rate expression can be expressed as

$$r_i = \frac{1}{V}\left(\frac{dN_i}{dt}\right) \tag{8.1}$$

where N_i is the number of moles of component i, dN_i/dt indicates the rate of change of N_i with time, in a reactor of volume V. Thus, r_i will have units of mol/(unit vol.)/(unit time).

8.1 DESIGN OF IDEAL REACTORS

In reactor design, our primary aim is to know what should be the type and size of the reactor, and what methods of operation are best suited for a given job. The objective behind the reactor design is to design an optimum size of the reactor for getting maximum yield of the product with highest selectivity. The conditions in the reactor may vary either with time or position. The conditions generally include composition and temperature. In addition to these, the geometry of the reactor will determine the path of the flow of the fluids in the reactor, which in turn determines the mixing pattern in the reactor.

The 'design of bioreactors' (fermenters) adds an additional feature of providing a conducive environment for the sustenance and growth of microbial population, and the maintenance of aseptic conditions. Thus, the design of a bioreactor involves the following salient features (Erickson and Stephanopoulos, 1987):

- type of mechanical agitation
- agitation rate
- gas flow rates
- liquid volume
- operating pressure (generally bioreactors are operated at atmospheric pressure)
- nutrient composition of the feed
- sterilisation method
- maintaining sterile conditions
- pH
- temperature
- liquid level
- foam control methodology
- mode of operation (batch, fed-batch or continuous).

Thus, the design features also call for the type of ideal reactor, viz.

- batch
- semi-batch (fed-batch)
- continuous
- CSTR
- PFR.

A general description about various types of the above ideal reactors was already given in Chapter 5 (Section 5.7, Figs 5.6 and 5.7).

The starting point for all design equations is invariably the material balance equation expressed for the reactant or product. It is shown in Fig. 8.1, for a reactant in unit time.

$$\begin{pmatrix}\text{component}\\\text{inlet flow}\end{pmatrix}=\begin{pmatrix}\text{component}\\\text{outlet flow}\end{pmatrix}+\begin{pmatrix}\text{disappearance}\\\text{of component}\\\text{by reaction}\end{pmatrix}+\begin{pmatrix}\text{accumulation of}\\\text{component in the}\\\text{reactor volume}\end{pmatrix} \tag{8.2}$$

If any component is forming in the reactor, it is taken in place of the second term in the RHS with a –ve sign.

If the situation in the reactor is non-isothermal (i.e. the temperature changes with time), we may need to incorporate the energy (heat) balance also, which is similar to Eq. (8.2)

$$\begin{pmatrix}\text{rate of energy}\\\text{inflow from}\\\text{reactor}\end{pmatrix}=\begin{pmatrix}\text{rate of energy}\\\text{outflow from}\\\text{reactor}\end{pmatrix}+\begin{pmatrix}\text{rate of consumption}\\\text{(disappearance) of}\\\text{energy in the reactor}\end{pmatrix}+\begin{pmatrix}\text{rate of accumulation}\\\text{of energy in the}\\\text{reactor}\end{pmatrix} \tag{8.3}$$

If any energy is generated in the reactor, it is taken in place of the second term in the RHS with a –ve sign.

The material and energy balance equations (Eqs. (8.2) and (8.3)) are tied together by the disappearance term or the formation term which is caused by the reaction. The above two equations are the starting design

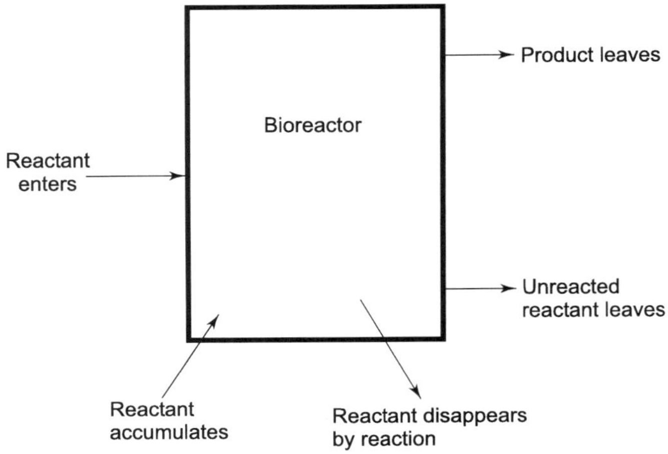

{reactant flow rate in inlet} = {reactant flow rate in outlet} + {rate of loss of reactant due to chemical reaction} + {rate of accumulation of reactant}

Fig. 8.1 *Material balance in a bioreactor*

equations for any reactor design procedure. They need to be integrated for a variety of real situations with a certain degree of complexity, because real situations are never ideal. We also should be able to predict the performance of the reactor in case of change of temperature, pressure, pH, etc. We may also like to look for various design alternatives (viz. the type of reactor, etc.). Then we formulate the design equation with all degrees of complexity. Subsequently, some simplifying assumptions are made to make a working model (see Section 7.3.1). Some shortcuts are made based on certain assumptions, and ignoring some parameters. The reliability and completeness of the observed experimental laboratory data also matter in the rational design of reactors. Comprehending all the above parameters, Levenspiel (1974) observes that "*Good engineering judgment, which only comes with experience, will suggest which course of action to take.*"

Before we proceed with the design procedure for ideal reactors, it is worthwhile to understand certain terms in reaction engineering, which we frequently come across viz.:

(i) conversion
(ii) yield
(iii) selectivity.

Let us consider the reaction represented by Eq. (4.1) and enlarge it by incorporating the stoichiometric coefficients also, as follows:

$$aA + bB \rightarrow cC + dD \tag{8.4}$$

Equation (8.4) shows that a moles of reactant A react with b moles of reactant B to form c moles of product C and d moles of product D. Thus, we have incorporated molar quantities into Eq. (4.1) to make Eq. (8.4). Dividing throughout by a, we have

$$A + (b/a)B \rightarrow (c/a)C + (d/a)D \tag{8.5}$$

It is necessary to note at this point that all reactions need not necessarily go to completion as shown by Eq. (8.4). That is, if we put simply a moles of A and b moles of B in a reactor and provide some optimal reaction conditions (viz. temperature, pressure, pH, etc.), it does not mean that all A reacts with all B to form C and D. Here we need to introduce the concept of **conversion**. Even after providing all the best reaction conditions, all A may not react, which is dictated by **thermodynamic considerations**. That is, the conversion of A is not 100%. Let us say, only 0.8 mol of A reacts by Eq. (8.5) and the remaining 0.2 mol are going with the products or remain unconverted in the reactor which need to be separated subsequently for recycling. In such situations, we call the conversion of A as X_A, and this is defined as the ratio of number of moles of A reacted or converted to the total number of moles of A originally fed. We can write that

$$X_A = \frac{\text{moles of A converted}}{\text{moles of A fed}} \tag{8.6}$$

It may be noted here that Eq. (8.6) only indicates how much of A is converted, but it does not guarantee that all A that has been converted has gone to form products C and D. It could have converted to form various other products also. Thus, Eq. (8.6) only talks about the conversion of reactants, and remains silent as far as formation of products is concerned.

Yield is defined as the ratio of mass or moles of product formed to the mass or moles of reactants consumed. Thus, yield essentially refers to the product in relation to the amount of reactant converted. The yield of product C in the Eq. (8.4) with reference to the reactant A is given by:

$$Y_{C/A} = \frac{\text{moles of C formed}}{\text{moles of A converted}} \tag{8.7}$$

Sometimes, for easy comprehension, the moles are replaced by grams or kilograms, i.e. we generally say that how many kg of raw material yields on conversion how many kg of product, based on which the costs can be worked out. However, in scientific terminology, we always go on molar basis only. So, even if the conversion is low, it does not matter much because we are concerned only with how much of the product is being formed out of the converted raw material. The unconverted reactants, it is presumed, will be recovered and recycled into the reaction stream or into the reactor. Thus, the process economics are worked out.

Selectivity Now we introduce one more new term called 'selectivity,' which is solely concerned with the product formation. It is defined as *the ratio of the amount of product formed to the amount of product that would have actually formed if all the feed material was converted to that product.*

$$\text{Selectivity} = \frac{\text{moles of product formed}}{\text{moles of product that should have actually formed}} \tag{8.8}$$

Thus, if a small amount of reactant is converted to a small amount of product, but if the whole amount of the converted reactant has yielded the desired product only, then obviously the conversion is low but the selectivity is 100%. In any case, what is always desired is that the selectivity should be 100%, yield may or may not be 100%, depending upon the stoichiometry of the reaction. Let us see the example of microbial conversion of glucose to glutamic acid:

$$\underset{\text{(glucose)}}{C_6H_{12}O_6} + \underset{\text{(ammonia)}}{NH_3} + \frac{3}{2}\,O_2 \rightarrow \underset{\text{(glutamic acid)}}{C_6H_9NO_4} + CO_2 + 3H_2O \tag{8.9}$$

Out of 1 mol of glucose fed to the fermenter, let us say, only 0.1 mol of glucose react to form 0.1 mol of glutamic acid.

In this case,

$X_{(\text{glucose})} = 10\%$

Selectivity to glutamic acid = 100%

$Y_{(\text{glutamic acid/glucose})} = 100\%$

Let us see another example for the conversion of glucose in molasses to ethanol.

$$\underset{\text{(glucose)}}{C_6H_{12}O_6} \rightarrow \underset{\text{(ethanol)}}{2C_2H_5OH} + 2CO_2 \tag{8.10}$$

In the above example, let us say, we have fed 1 mol of glucose, out of which 0.1 mol of glucose has reacted and produced 0.2 mol of ethanol.

$$X_{(\text{glucose})} = 10\%$$
$$Y_{(\text{ethanol/glucose})} = 200\%$$
$$\text{Selectivity to ethanol} = 100\%$$

8.2 SINGLE REACTOR

Under this category, we consider all the ideal reactors, viz.

(i) Batch reactor

(ii) Flow reactor — CSTR / PFR

We try to develop the performance equations for a single fluid reacting in three types of ideal reactors. We start with the material balance equations for these reactors.

8.2.1 Batch Reactor

For a batch reactor, the first two terms in Eq. (8.2) are zero. Hence,

$$-\begin{pmatrix}\text{rate of disappearance} \\ \text{of A by reaction}\end{pmatrix} = \begin{pmatrix}\text{rate of accumulation} \\ \text{of A in reactor}\end{pmatrix} \tag{8.11}$$

If N_{A0} is the initial number of moles of A, N_A the number of moles of A present at any time, t, V the volume of the reactor, and C_A is the concentration of A, then we can write

$$C_A = C_{A0}(1 - X_A) \tag{8.12}$$

and

$$N_A = N_{A0}(1 - X_A) \tag{8.13}$$

Equation (8.11) becomes

$$(-r_A)\, V = -\frac{dN_A}{dt} \tag{8.14}$$

$$N_{A0}\frac{dX_A}{dt} = (-r_A)\, V \tag{8.15}$$

$(-r_A)$ indicates the rate of conversion or depletion of A.

Equation (8.15) is called the **design equation in differential form for a batch reactor.** In an integral form, it can be written as

$$t = N_{A0}\int_0^{X_A} \frac{dX_A}{(-r_A)\,V} \tag{8.16}$$

Equation (8.16) shows the time required to achieve the conversion X_A.

If the density of the fluid remains constant, or if V is constant, it can be taken out of the integral sign, and we can write Eq. (8.16) as

$$t = C_{A0}\int_0^{X_A} \frac{dX_A}{(-r_A)} = -\int_{C_{A0}}^{C_A} \frac{dC_A}{(-r_A)} \tag{8.17}$$

Equation (8.17) gives the time required for a certain amount of conversion to take place. Normally, in a batch reactor, the reaction proceeds as time increases, or until the equilibrium is reached.

Example 8.1 *A batch fermenter was operated for the production of alcohol from glucose using yeast. The rate of glucose conversion is measured, and the data are presented in Table E8.1. Find the time taken in the batch fermenter when the concentration falls down to 12 kgmol/m³, when C_{Ao} = 70 kgmol/m³.*

Table E8.1 *Concentration vs. rate of consumption of glucose.*

Concentration (C_A) kg-mol/m³	68	54.6	33.0	20.0	12.2	7.4	4.5	2.7
Rate of reaction ($-r_A$) kg-mol/(m³h)	2.82	4.71	4.31	3.74	3.16	2.46	1.91	1.4

Solution It is a direct application of Eq. (8.17). The data of C_A vs ($1/-r_A$) are shown in Table E8.2 and drawn in Fig E 8.1.

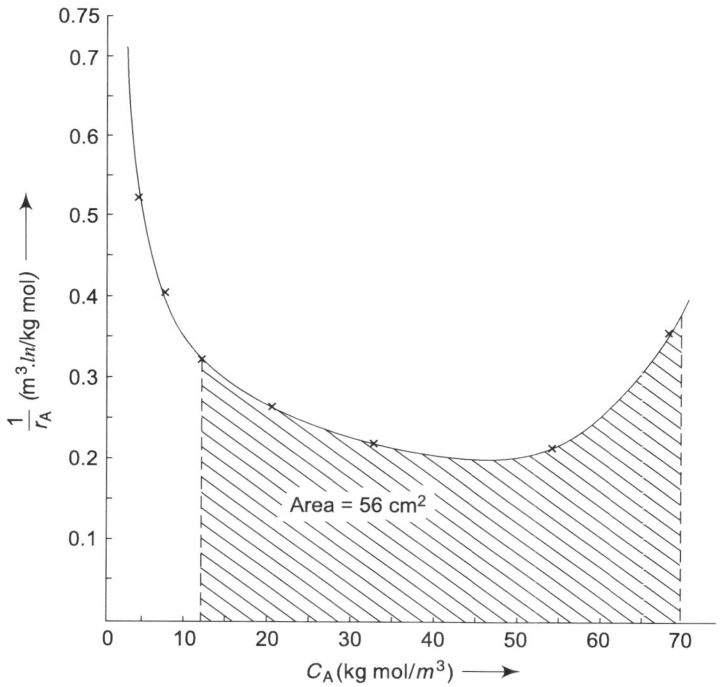

Fig. E8.1 *Data on $C_A$vs ($1/r_A$) for Example 8.1*

The last two columns in Table E8.2 are used to draw Fig E8.1. Area of the graph under the curve between the abscissa C_{A0} = 70 to C_{Af} = 12 kg-mol/m³

$$= 56 \times 0.05 \times 5 = 14 \text{ h}$$

∴ time taken for the concentration to fall down to 12 kg-mol/m³ = 14 h

Table E8.2 *Data for drawing Fig E8.1*

$(-r_A)$ (kg-moles/m^{3}h)	C_A (kg-mol/m^3)	$1/(-r_A)$ m^{3}h/kg-mol
2.82♦	68	0.354
4.71	54.6	0.212
4.31	33.0	0.232
3.14	20.0	0.267
3.16	12.2	0.316
2.46	7.4	0.406
1.91	4.5	0.523
1.4	2.7	0.714

8.2.2 Flow Reactor

In the case of flow reactors, we have continuous inflow of reactants and continuous outflow of products after reaction is over. The flow reaction system calls for the need for defining terms

- space time
- space velocity.

The reaction time t is the time taken in a batch reactor for conversion of reactants into products [Eq. (8.16) or Eq. (8.17)]. Similar concept is used in flow systems also for finding out the reaction time which is defined as follows (Levenspiel, 1974)

$$\tau = \begin{pmatrix} \text{time required to process one reactor volume of feed} \\ \text{measured at specific conditions} \end{pmatrix} \tag{8.18}$$

Then,

$$\tau = \frac{V}{v} = \frac{\text{volume of reactor}}{\text{volumetric feed rate}} \tag{8.19}$$

$$= \frac{C_{A0}\,V}{F_{A0}} \tag{8.20}$$

where F_{A0} is molar feed rate of A.

Space velocity is defined as *the number of reactor volumes of feed at specified conditions which can be treated in a unit time.*

$$\text{Space velocity} = s = \frac{1}{\tau} \tag{8.21}$$

CSTR or fully mixed reactor Since there will not be any accumulation of reactants or products in the continuous stirred tank reactor (CSTR), the fourth term in Eq. (8.2) will become zero. Hence, the material balance equation becomes

$$F_{A0} = F_A + F_{A0}X_A \tag{8.22}$$

$$\therefore \quad F_A = F_{A0}(1 - X_A) \tag{8.23}$$

Conversion of A by reaction (in mol/time)

$$= (-r_A)\,V \tag{8.24}$$

Thus,

$$F_{A0}X_A = (-r_A)\,V \tag{8.25}$$

♦The low reaction rate may be due to the fact that the microorganism build-up is the least.

or
$$\frac{V}{F_{A0}} = \frac{X_A}{(-r_A)}$$

$$\tau = \frac{V}{\upsilon} = \frac{C_{A0}\,V}{F_{A0}} = C_{A0}\frac{X_A}{(-r_A)}$$
$$= \bar{t}_m \tag{8.26}$$

where $\bar{t}_m$ is average residence time in a fully mixed flow reactor.

In the above equation, generally, r_A stands for reaction rate at the exit conditions, since the exit composition from the CSTR is the same as that in the reactor.

$$\tau = \frac{C_{A0} - C_A}{(-r_A)} \tag{8.27}$$

or
$$V = \frac{F_{A0}\,X_A}{(-r_A)_{exist}} \tag{8.28}$$

Either of the above equations is considered as the design equation for the CSTR.

Example 8.2 *If the fermentation mentioned in Example 8.1 were carried out in a CSTF, find the space time, and volume of the fermenter if the volumetric flow is 5.5 × 10⁻³ m³/h*

Solution It is a direct application of Eq. (8.27)

$$C_{A0} = 70 \text{ kg-mol/m}^3$$
$$C_A = 12 \text{ kg-mol/m}^3$$
$$1/(-r_A) = 0.315 \text{ (from Fig. E8.1) when } C_A = 12 \text{ kg-mol/m}^3$$

∴ $$r_A = 3.17 \text{ kg-mol/m}^3\text{h when } C_A = 12 \text{ kg-mol/m}^3$$

By Eq. (8.27), $$\tau = \frac{C_{A0} - C_A}{(-r_A)} = \frac{70 - 12}{3.17} = 18.27 \text{ h}$$

The volume of the CSTF is calculated by Eq. (8.19) when $\upsilon = 5.5 \times 10^{-3}$ m³/h

$$\tau = v / \upsilon$$

∴ $$v = \tau \times \upsilon = 18.27 \times 5.5 \times 10^{-3} = 0.1 \text{ m}^3$$

Plug-flow reactor (PFR) The plug flow reactor or the tubular flow reactor of Fig. 5.6(b) is shown in Fig. 8.2 with the notation of flow parameters into and out of a small differential element of the PFR (Levenspiel, 1974). Let us consider the first three terms in Eq. (8.2) for the elemental volume of reactor, dV.

Molar flow rate of A into the element = F_A

Molar flow rate of A out of the element = $F_A + dF_A$

Disappearance of A in the element by reaction = $(-r_A)(dV)$

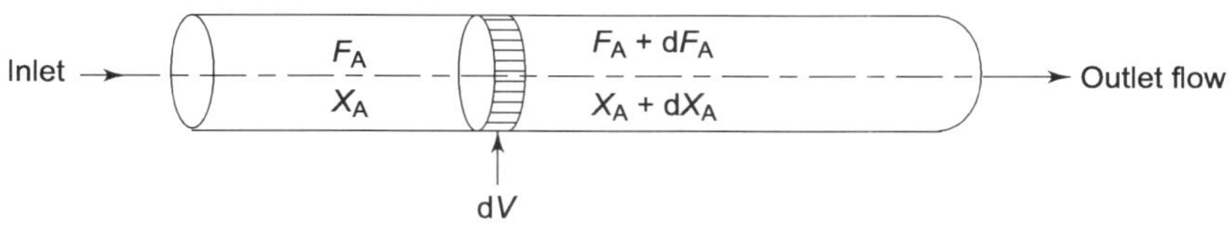

Fig. 8.2 *Schematic representation of a PFR*

Therefore, Eq. (8.2) becomes

$$F_A = (F_A + dF_A) + (-r_A)(dV) \tag{8.29}$$

On simplification,

$$-dF_A = (-r_A)dV \tag{8.30}$$

$$dV = \frac{-dF_A}{(-r_A)} \tag{8.31}$$

From Eq. (8.20), we have

$$\frac{V}{F_{A0}} = \frac{\tau}{C_{A0}} \text{ or } \frac{V}{F} = \frac{\tau}{C_A} \tag{8.32}$$

$$\frac{V}{F_A} = \frac{V}{\upsilon}\frac{1}{C_A}$$

or $\upsilon C_A = F_A$

Hence, Eq. (8.31) becomes

$$dV = -\upsilon\frac{dC_A}{(-r_A)}$$

which on integration yields

$$V = -\upsilon \int_{C_{A0}}^{C_A} \frac{dC_A}{(-r_A)}$$

or

$$\frac{V}{\upsilon} = \tau = -\int_{C_{A0}}^{C_A} \frac{dC_A}{(-r_A)} \tag{8.32(a)}$$

Conversion $X_A = \dfrac{C_{A0} - C_A}{C_{A0}} = 1 - \dfrac{C_A}{C_{A0}}$

or $\dfrac{C_A}{C_{A0}} = 1 - X_A$

which on differentiation gives $dC_A = C_{A0}(-dX_A)$

Hence, the above equation (Eq. 8.32(a)), on replacement, becomes

$$\frac{\tau}{C_{A0}} = \int_0^{X_A} \frac{dX_A}{(-r_A)} \tag{8.33}$$

or

$$\tau = C_{A0}\int_0^{X_A} \frac{dX_A}{(-r_A)} \tag{8.34}$$

Combining Eqs. (8.32) and (8.33) yields

$$V = F_{A0}\int_0^{X_A} \frac{dX_A}{(-r_A)} \tag{8.35}$$

Equation (8.34) or (8.35) is the design equation for a PFR.

Comparing the Volumes of CSTR and PFR For a given duty (or performance), the volumes of a PFR and CSTR are compared and explained with a numerical example by Fogler (1986). The volume of a CSTR is given by Eq. (8.28)

$$\left(\frac{V}{F_{A0}}\right)_{CSTR} = \frac{X_A}{(-r_A)_{exit}}$$

If a plot is drawn between X_A and $1/r_A$ as shown in Fig. (8.3), $\left(\frac{V}{F_{A0}}\right)_{CSTR}$ is the area of the rectangle ACDE. The volume of a PFR for the same conversion is obtained by Eq. (8.35)

$$\left(\frac{V}{F_{A0}}\right)_{PFR} = \int_0^{X_A} \frac{dX_A}{(-r_A)}$$

Thus, $(V/F_{A0})_{PFR}$ is the area of the bottom shaded portion ABDE.

Thus, the difference in the time taken for equal-sized PFR and CSTR for the same amount of conversion X_A is given by:

$$\left(\frac{V}{F_{A0}}\right)_{CSTR} - \left(\frac{V}{F_{A0}}\right)_{PFR} = \left(\frac{V}{\upsilon}\right)_{CSTR}\left(\frac{1}{C_{A0}}\right) - \left(\frac{V}{\upsilon}\right)_{PFR}\frac{1}{C_{A0}}$$

Thus,

$$\frac{\tau_{(CSTR)} - \tau_{(PFR)}}{C_{A0}} = \text{area of BCD}$$

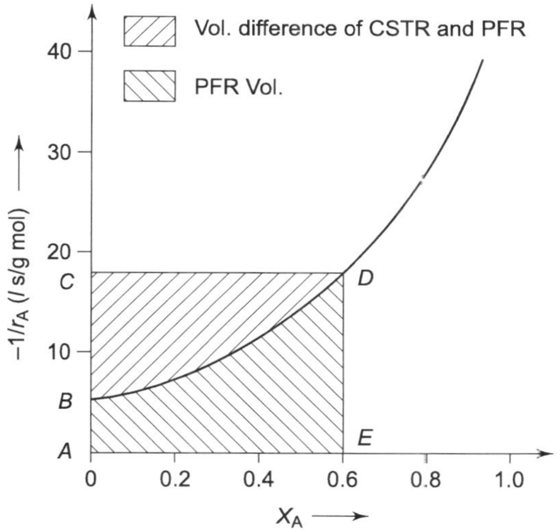

Fig. 8.3 *Comparison of the volumes of CSTR and PFR for a similar conversion (Reprinted from "Elements of Chemical Reaction Engineering", Fogler, H.S., p. 37, Copyright © (1986), with permission from Oldenbourg, Industrieverlag, GmbH, Munchen.)*

Example 8.3

If the data in Table E8.1 were to represent data in a PFR for different space times, find the space time for concentration to reach 12 kg moles/m³. Find the volume of the reactor for the volumetric flow rate of 5.5×10^{-3} m³/h. Compare the space time with those of Example 8.1 and Example 8.2.

Solution It is direct application of Eq (8.34). We evaluate the integral by graphical integration method. For this purpose we need to convert the C_A data into X_A by using the formula

$$X_A = \frac{C_{A0} - C_A}{C_A} = \frac{C_{A0}}{C_A} - 1$$

We calculate X_A by using the above formula, and prepare tabular data in Table E8.3.

Table E8.3 *X_A vs $(1/r_A)$ data for Example 8.3*[♦]

C_A	r_A	X_A	$1/r_A$
68	2.82	0.028	0.355
54.6	4.71	0.22	0.212
33	4.31	0.53	0.232
20	3.74	0.714	0.267
12.2	3.16	0.83	0.316
7.4	2.46	0.89	0.406
4.5	1.91	0.94	0.523
2.7	1.4	0.96	0.71

The data in last two columns are plotted in Fig. E8.2

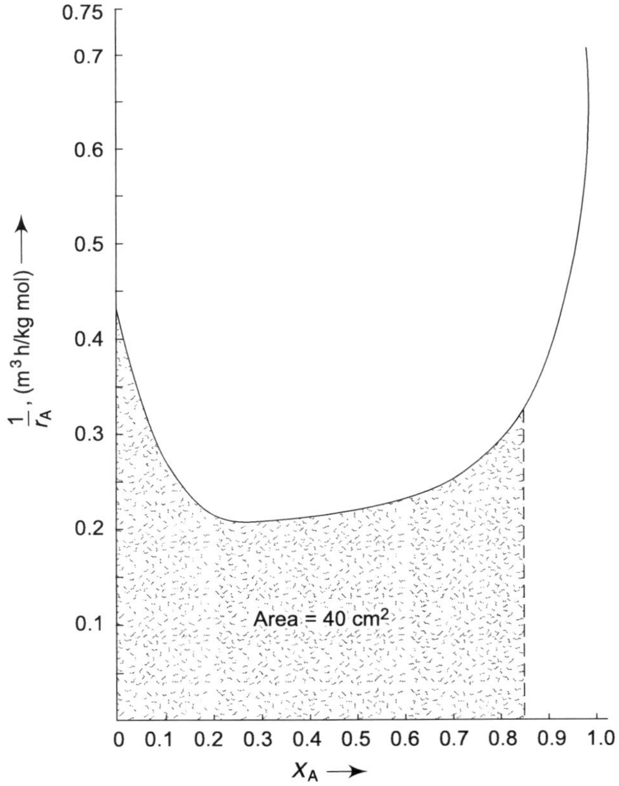

Fig. E8.2 *X_A vs $(1/r_A)$ for Example 8.3*

• $C_{AO} = 70$ kg moles/m^{3}h

For $C_A = 12$ kg-moles/m^3, $X_A = \dfrac{70-12}{70} = 0.829$

Area under the curve $X_A = 0$ and $X_A = 0.83$ is measured from the Fig. E8.2 and is found to be 0.2

τ is given by Eq. (8.34) and is calculated to be

$$= 70 \times 0.2 = 14 \text{ h}$$

$$V = \tau\,\upsilon = 14 \times 5.5 \times 10^{-3} = 0.077 \text{ m}^3$$

τ is same as in Example 8.1 and is less than that of Example 8.2 which is evident from Fig. 8.3.

8.3 MULTIPLE REACTORS

As we have already mentioned earlier, the basic assumptions in CSTR is that the moment the reactants enter the CSTR, all the contents are thoroughly mixed, and the concentration of any component in the CSTR is the same and is equal to the exit concentration (Section 5.7). Thus, under steady state conditions, the concentration of various species in the reactor is constant and is easily calculable on the basis of desired conversion. The reactor volume can be calculated by applications of Eq. (8.28). The volumetric flow rate, υ (in m^3/s), can be obtained from

$$\upsilon = \frac{F_{A0}}{C_{A0}} = \frac{\text{mol/s}}{\text{mol/m}^3} = \text{m}^3/\text{s} \tag{8.36}$$

By incorporating υ, Eq. (8.28) can be written as

$$V = \frac{C_{A0}\, X_A\, \upsilon}{(-r_A)} \tag{8.37}$$

In the above equation, $(-r_A)$ is based on the concentration of the reactant prevailing in the reactor, which is also equal to the exit concentration, and hence will be the lowest. If the concentration of the reactants is brought down from C_{A0} to C_A, r_A is based on C_A and hence will be the lowest. In Eq. (8.37), as $(-r_A)$ decreases, V will increase, which results in the highest volume requirements in case of CSTR for a given conversion.

Resnick (1981) explains this concept with a numerical example which shows that CSTR requires four times the volume of a batch reactor, in which 75% of the component A in one litre of a feed material, with C_{A0} being 3 g mol/l, would be converted to products in ½ hour. The second-order reaction rate constant (k_2) for the reaction

$$2\text{A} \xrightarrow{k_2} \text{R} \tag{8.38}$$

would be 2*l*/(g mol h)

$$(-r_A) = k_2\, C_A^{\,2} = k_2\, [C_{A0}\,(1 - X_A)]^2$$
$$= 2[3(1 - 0.75)]^2 \tag{8.39}$$

By substituting the numerical values into Eq. (8.17), we get for a batch reactor

$$t = 3 \int_0^{0.75} \frac{dX_A}{2 \times 3^2 \times (1 - X_A)^2}$$

or

$$t = \frac{1}{6} \int_0^{0.75} \frac{dX_A}{(1 - X_A)^2} = \frac{1}{6}\left(\frac{1}{1 - X_A}\right)_0^{0.75} = \frac{1}{2}\text{h}$$

If a similar reaction is taken up in a CSTR, the material converted in ½ hour at 75% conversion is $3 \times 0.75 = 2.25$ g mol/l. In 1 hour it would be 4.5 g mol/h and the flow rate of A would be 2 l/h. The CSTR volume for similar operation can be found out from Eq. (8.37) in which $\upsilon = 2$ l/h

$$V = \frac{(3)(0.75)(2)}{2[3(1-0.75)]^2} = 4\text{l} \tag{8.40}$$

Thus, for the above conversion, CSTR requires four times larger volume as compared to a batch reactor. This happens because in a batch reactor, the concentration of A falls gradually from 3 g mol/l to 0.75 g mol/l, and hence the rate of reaction falls gradually from the maximum at the beginning to the minimum at the end, i.e. when C_A falls to 0.75 g mol/l. Whereas, in a CSTR, the concentration (C_A) would fall to 0.75 g mol/l initially itself, and hence a minimum reaction rate is realised throughout, which results in higher reactor volume.

In the above example, it is obvious that if the concentration in the reactor could be brought down gradually in a number of CSTRs, then higher reaction rates could be realised, with the result that the overall volume of all the CSTRs would be less than that of the single CSTR. Let us see this with the above numerical example, that how the CSTR volumes would change if we keep two CSTRs in series. The outlet concentration of the first reactor would be the inlet concentration for the second CSTR. For the sake of simplicity, we take that the two CSTRs are of equal volume.

Figure 8.4 shows the schematic representation of two CSTRs in series.

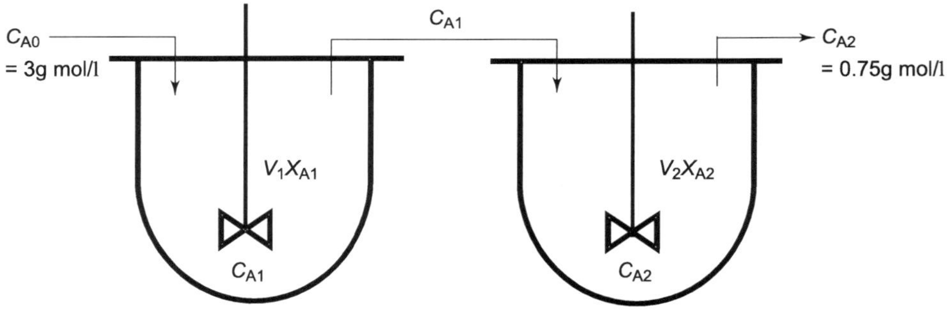

Fig. 8.4 *Schematic representation of two CSTRs in series*

$$C_{A1} = C_{A0}(1 - X_{A1}) \tag{8.41}$$

$$C_{A2} = C_{A1}(1 - X_{A2}) = C_{A0}(1 - X_{Af}) \tag{8.42}$$

$$\frac{C_{A2}}{C_{A0}} = 1 - X_{Af} \tag{8.43}$$

$$\frac{C_{A2}}{C_{A0}} = \frac{C_{A2}}{C_{A1}} \times \frac{C_{A1}}{C_{A0}} = (1 - X_{A1})(1 - X_{A2}) \tag{8.44}$$

Combining Eqs. (8.43) and (8.44), we get

$$(1 - X_{Af}) = (1 - X_{A1})(1 - X_{A2}) \tag{8.45}$$

The volume of each reactor is given by Eq. (8.37)

$$V_1 = \frac{C_{A0} X_{A1} \upsilon}{(r_{A1})} \tag{8.46}$$

$$V_2 = \frac{C_{A1} X_{A2} \upsilon}{(r_{A2})} \tag{8.47}$$

where $$r_{A1} = k_2(C_{A1})^2 \tag{8.48}$$

and $$r_{A2} = k_2(C_{A2})^2 \tag{8.49}$$

Substituting for r_{A1} and r_{A2}, and equating V_1 and V_2, we get

$$\frac{C_{A0} X_{A1} \upsilon}{k_2 (C_{A1})^2} = \frac{C_{A1} X_{A2} \upsilon}{k_2 (C_{A2})^2} \tag{8.50}$$

X_{A1} and X_{A2} in Eq. (8.50) can be evaluated by trial-and-error method to be

$$X_{A1} = 0.564 \quad \text{and} \quad X_{A2} = 0.426.$$

By substituting X_{A1} and X_{A2} in Eqs. (8.46) and (8.47), we get

$$V_1 = V_2 = 0.99 \text{ l} \cong 1.0 \text{ l}$$

Thus, the total volume of two reactors would be equal to only 2 l as compared to 4 l for a single CSTR.

Graphical method of calculating V for CSTR The method of evaluation of V can also be shown graphically more conveniently because it may not always be possible to express the rate equation as a simple analytical expression to be used in Eq. (8.37). In such cases, it is resorted to measure the reaction rate as described in Section 7.2. The rate of reaction $(-r_A)$ is plotted against C_A as shown in Fig. 8.5. The material balance equation, i.e. Eq. (8.37) can be written as

$$\upsilon C_{A0} = V(r_A) + \upsilon C_A \tag{8.51}$$

substituting for $$X_A = 1 - (C_A/C_{A0})$$

Equation (8.51) can also be written as

$$C_A = C_{A0} - \frac{V}{\upsilon} (r_A) \tag{8.52}$$

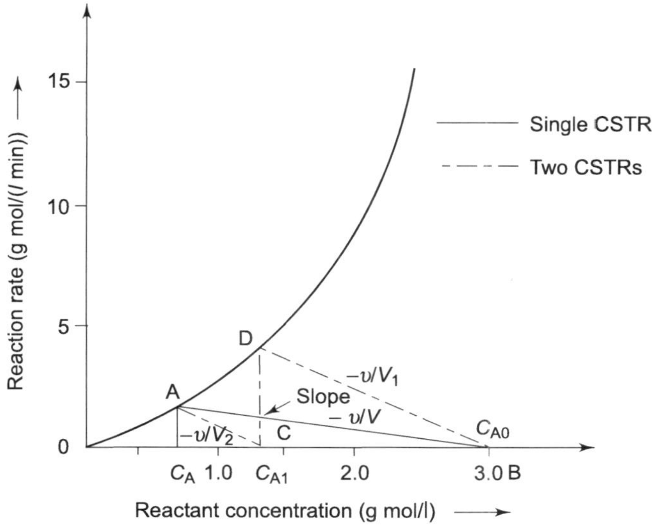

Fig. 8.5 *Graphical construction for CSTR calculations.*
(Reprinted from "Process Analysis and Design for Chemical Engineers", Resnick, W., p. 116, Copyright © (1981), with permission from McGraw-Hill, Inc., New York).

In Fig. 8.5, (C_A, r_A) point is noted on the curve (as point A). This point is joined to C_{A0} (as point B). The slope of the line joining these two points A and B would give the value of $-\,\upsilon/V$. From the knowledge of the volumetric flow rate (υ), V for the CSTR can be found.

The graphical procedure can also be extended to two CSTRs in series. The point C_{A1} is chosen at designer's discretion (Resnick, 1981) and is plotted at C on X-axis. Point D on the curve is plotted corresponding to (C_{A1}, r_{A1}).

$$\text{Slope of BD} = -\,\upsilon/V_1$$
$$\text{Slope of AC} = -\,\upsilon/V_2.$$

Thus, V_1 and V_2 are calculated. By trial-and-error in choosing C_{A1}, V_1 and V_2 can be made equal, by making slopes of BD and AC equal.

By increasing the number of CSTRs in series, we can maintain higher concentration of A in different reactors, and hence higher r_A can be realised. This helps reduce the total volume of CSTRs. This was explained with a numerical example by Lee (1994) for processing a feed solution containing 4.0 kg mol/m^3 of reactant A with the following conditions in CSTR:

$$\upsilon = 1.2 \times 10^{-4}\ \text{m}^3/\text{s}$$
$$X_A = 0.69$$

r_A versus C_A data were obtained in a laboratory experiment. The graphical calculations show the following reactor volumes:

No. of tanks	*Volume of each tank (m³)*	*Total volume (m³)*
1	3.16	3.16
2	1.06	2.12
3	0.6	1.8

Thus, these results illustrate the general conclusion that by increasing the number of CSTRs in series, we can almost reach to the minimum volume of a plug flow reactor for the given duty of operation. This is shown schematically in Fig. 8.6.

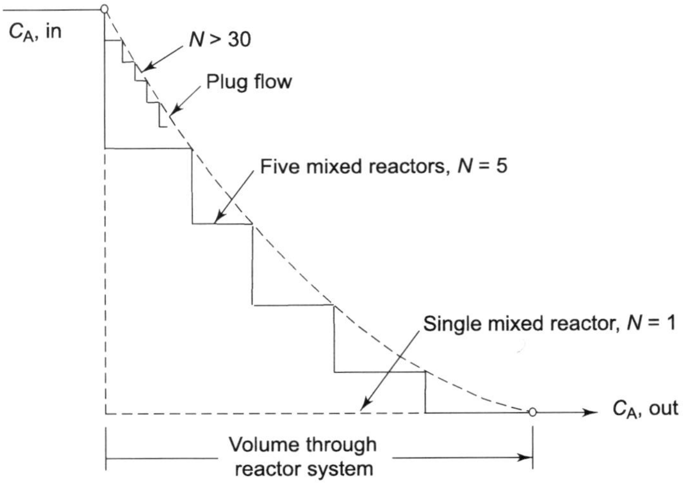

Fig. 8.6 *Concentration profile through an N-stage mixed reactor system compared with single flow reactors. (Reprinted from "Chemical Reaction Engineering", Levenspiel, O., 2nd Edn., p. 135, Copyright © (1974), with permission from John Wiley & Sons, Inc.)*

In a plug flow reactor, the concentration falls progressively along the length of the reactor, whereas in a CSTR, the concentration falls suddenly to the lowest value which is equal to that in the exit flow. That is why a PFR is considered to be more efficient as compared to a CSTR for any order of reaction (other than zero order) in which the rate of reaction is dependent on the concentration. In other words, a PFR can be viewed as a series of micro CSTRs with the concentration in each micro CSTR corresponding to the concentration at that particular point of PFR.

Example 8.4

The fermentation in Ex 8.2 with the concentration versus the rate data (Example 8.1) were carried out in two equal volume CSTFs. Using the graphical construction procedure, find the volume of each reactor. Note how much is the difference in total volume as compared to single CSTF.

Solution The data in Table E8.2 on C_A vs r_A are plotted in Fig. E8.3 corresponding to final concentration $C_A = 12$ kg-moles/m^3, point A is noted on the curve. Point B is noted on abscissa corresponding $C_{AO} = 70$ kg-moles/m^3. AB is joined. The slope of the line corresponds to $(-\upsilon / V)$. Knowing υ, volume V can be found.

$$\text{Slope of AB} = 0.055$$

$$\upsilon = 5.5 \times 10^{-3} \text{ m}^3/\text{h}$$

$$\therefore \quad V = -\frac{\upsilon}{(\text{slope})} = \frac{5.5 \times 10^{-3}}{0.055} = 0.1 \text{ m}^3$$

This is what we obtained in Example 8.2 for a single CSTF.
Now we shall try to find out the volume of two equal CSTFs i.e. $V_1 = V_2$

$$\therefore \quad \frac{\upsilon}{V_1} = \frac{\upsilon}{V_2}$$

We have to locate the point C on abscissa which corresponds to C_{A1} which is the concentration prevailing in the first CSTF. Once we locate C we can locate D on the curve.

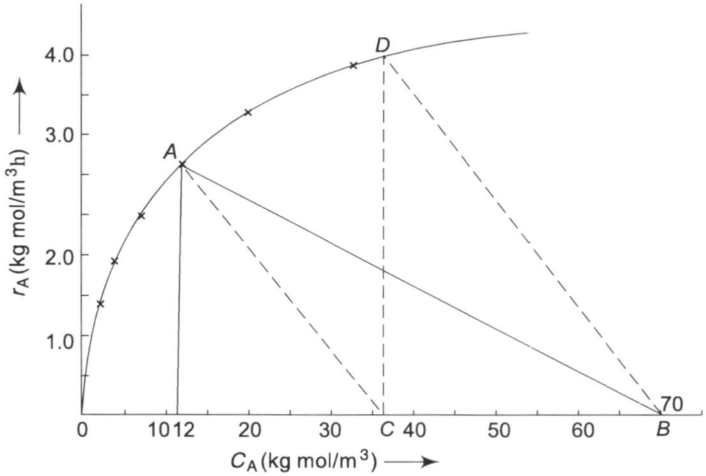

Fig. E8.3 *Graphical construction for CSTF calculations of Example 8.4*

$$\text{Line } BD \text{ will have slope } = \frac{\upsilon}{V_1}$$

$$\text{Line } CA \text{ will have slope } = \frac{\upsilon}{V_2}$$

$$\text{Since } V_1 = V_2$$

$$\frac{\upsilon}{V_1} = \frac{\upsilon}{V_2}$$

i.e, slope of *BD* and *CA* are same, which indicates lines *BD* and *CA* are parallel.

In other words, we draw two parallel lines from point *A* and *B* with varying slopes until the points *C* and *D* fall on the same coordinate.

This exercise is done in Fig. E8.3 to locate points *D* and *C* such that they are on the coordinate.

$$-\text{Slope} = 0.131 = \frac{\upsilon}{V_1} = \frac{\upsilon}{V_2}$$

where $\upsilon = 5.5 \times 10^{-3}\ \text{m}^3/\text{h}$

$$V_1 = V_2 = \frac{0.131}{5.5 \times 10^{-3}} = 0.041\ \text{m}^3$$

Total volume of two reactors = 0.041 + 0.041 = 0.082 m^3

It is less than the single CSTF of 0.1 m^3 volume.

The concentration in the first CSTF is 36.5 kg-moles/m^3 (from Fig. E8.3) as shown by the point *C*.

Equal sized CSTRs in Series Let *N* number of equal sized CSTRs be connected in series as shown in Fig. 8.7.

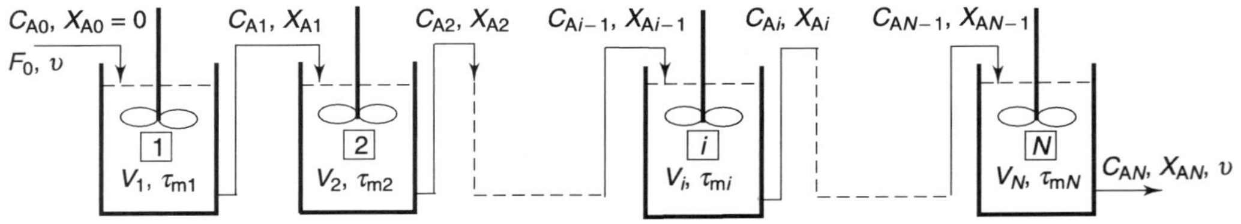

Fig. 8.7 *N equal-sized mixed flow reactors in series*

From Eq. (8.27), we have for a CSTR,

$$\tau = \frac{C_{A0} - C_A}{(-r_A)}$$

And for first-order reaction,

$$-r_A = k_1 C_A$$

∴ The above equation becomes

$$\tau = \frac{C_{A0} - C_A}{k_1 C_A}$$

or

$$k_1 \tau = \frac{C_{A0}}{C_A} - 1$$

$$\frac{C_{A0}}{C_A} = 1 + k_1 \tau \tag{8.53}$$

$$\frac{C_{A0}}{C_{AN}} = \frac{C_{A0}}{C_{A1}} \times \frac{C_{A1}}{C_{A2}} \cdots \frac{C_{AN-1}}{C_{AN}}$$

$$= (1 + k_1\tau_1)(1 + k_1\tau_2) \ldots (1 + k_1\tau_N)$$

For N number of CSTRs in series, assuming that there is no change in density, υ becomes same throughout, and we have

$$\tau_1 = \tau_2 = \tau_i = \tau_N = V_i / \upsilon \tag{8.54}$$

where V_i is the volume of any one reactor and is equal to V/N, and V is the total volume of all N reactors put together.

Hence,

$$\frac{C_{A0}}{C_{AN}} = (1 + k_1 \, t_i)^N \tag{8.55}$$

or

$$1 + k_1 \, \tau_i = \left(\frac{C_{A0}}{C_{AN}}\right)^{1/N} \tag{8.56}$$

or

$$1 + \frac{k_1 \, \tau}{N} = \left(\frac{C_{A0}}{C_{AN}}\right)^{1/N}$$

where $\tau = V/\upsilon = N\upsilon_i$.

Symbols

a	: stoichiometric coefficient for A
A, B, C, D	: reactants [Eqs. (8.4) and (8.5)]
b	: stoichiometric coefficient for B
c	: stoichiometric coefficient for C
C	: concentration (g/l or mol/m^3)
d	: stoichiometric coefficient for D
F	: molar flow rate (mol/s or kg mol/h)
k_2	: second-order reaction rate constant
N	: number of moles of component, number of CSTRs in series in Eq. (8.55)
r	: rate of reaction (mol/(m^3 s) or mol/(m^3 h))
s	: space velocity (1/s)
t	: time (s or h)
t_m	: average residence time in mixed flow reactor (s)
υ	: volumetric flow rate (m^3/s)
V, V_1, V_2	: reactor volume (m^3)
V_i	: volume of ith reactor [Eq. (8.54)]
X	: conversion
$Y_{C/A}$	: yield factor of C based on A

Subscripts

0	: initial conditions
A, B	: components A, B
f	: final or outlet flow
i	: component i
m	: mixed flow reactor
N	: N^{th} reactor or vessel

Greek Letters

τ	: residence time in flow reactor (s)

References

8.1 Erickson, L.E., and Stephanopoulos, G. (1987), "Biological reactors", *Chemical Reaction and Reactor Engineering*, Carberry, J.J., and Varma, A. (Eds.), Marcel Dekker, Inc., New York, p. 816.

8.2 Fogler, H.S. (1986), *Elements of Chemical Reaction Engineering*, Prentice-Hall, New Jersey, pp. 36–37.

8.3 Ghose, T.K. (Ed.) (1990), *Bioprocess Computations in Biotechnology*, Vol. 1, Ellis Horwood Ltd., Chichester, pp. 53–55.

8.4 Lee, J.C. (1994), "Reactor design—General principles", *Coulson & Richardson's Chemical Engineering*. Richardson, J.F., and Peacock, D.G. (Eds.), 3rd Edn, Vol. 3, Pergamon Press, London, pp. 48–49.

8.5 Levenspiel, O. (1974), *Chemical Reaction Engineering,* 2nd Edn., Wiley Eastern Pvt. Ltd., New Delhi, pp. 93–109.

8.6 Resnick, W. (1981), *Process Analysis and Design for Chemical Engineers*, McGraw-Hill, Inc., New York, pp. 113–119.

Review Questions

8.1 What are the ideal flow conditions in a bioreactor?

8.2 What are the salient features that you consider as important in the design of a bioreactor?

8.3 Write the material balance equation with schematic representation for a bioreactor.

8.4 Write the energy balance equation with schematic representation for a bioreactor.

8.5 Define and explain the terms:
  (i) Conversion
  (ii) Yield
  (iii) Selectivity.

8.6 Write the design equation for a batch reactor.

8.7 Define and explain the terms:
  (i) Space-time
  (ii) Space-velocity.

8.8 Write the design equation for: (i) CSTR, (ii) PFR.

8.9 Explain how multiple reactors will help in reducing the reactor volume.

8.10 Explain the graphical method of calculating the volume of a CSTR.

8.11 For N number of CSTRs in series, derive $\dfrac{C_{A0}}{C_A} = \left(1 + \dfrac{k_1\,\tau}{N}\right)^N$

Problems

8.1 Penicillin is produced by the fermentation of glucose by a penicillium mould, using phenyl acetic acid as precursor, as shown below:

$$1.67C_6H_{12}O_6 + 2NH_3 + 1/2O_2 + H_2SO_4 + C_6H_5CH_2COOH \rightarrow C_{16}H_{18}O_4N_2S + 2CO_2 + 9H_2O$$

(Glucose) (Phenyl acetic acid) (Penicillin)

In a batch process, 200 kg of glucose was taken. At the end of the reaction time, unconverted glucose was 50 kg, and 160 kg of penicillin was produced. Find the (i) Conversion of glucose, (ii) Yield of penicillin based on glucose, and (iii) Selectivity of the production of penicillin.

(Ans: (i) 75%
(ii) 107% (kg/kg); 57.5% (mol/mol)
(iii) 96%)

8.2 A certain fermentation process is carried out in a two-stage continuous bioreactor system. The maximum attainable concentration of the microorganism in the second reactor (x_2) is equal to (e – 1) times that in the first reactor (x_1), i.e. $x_2 \leq 1.718\, x_1$ (Ghose, 1990).[1] The following data are available for the first bioreactor of capacity 1×10^{-2} m^3:

$$\upsilon = 1 \times 10^{-3}\ \text{m}^3/\text{h}$$
$$x_0 = 0.5\ \text{kg/m}^3.$$

Growth rate of cells in the first bioreactor = 0.29 kg/(m^3 h)

Specific growth rate, $\mu = 0.2\ \text{h}^{-1}$

Find the maximum attainable cell concentration in the second bioreactor.

(Ans: x_2 = 17.18 kg/m^3)

---

1 It is advised that the reader should practise the derivation. Here e stands for exponential, and is equal to 2.718.

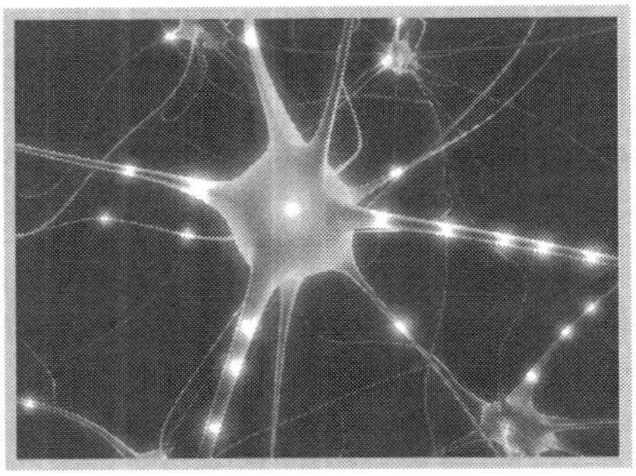

chapter 9

Multiple Reactions

Depending upon the nature of the reactants, reaction conditions and thermodynamic feasibility, the reactions can be classified as simple or multiple. The multiple reactions can be further sub-classified as:

- parallel reactions (also called as competing reactions or simultaneous reactions)
- series reactions (also called as consecutive reactions)
- combination of series–parallel reactions.

All the chemical or biochemical reactions fall into one of the above sub-classes.

Before we discuss in detail the above reaction schemes, it is worth considering the various classes of fermentation type reactions. Deindoerfer (1960) has categorised the fermentation processes based on the fermentation kinetics (Table 9.1).

Table 9.1 *Various types of fermentation reactions*

Type	*Description*
Simple	Nutrients converted to products in a fixed stoichiometry without accumulation of intermediates
Simultaneous	Nutrients converted to products in variable stoichiometric proportions, without accumulation of intermediates
Consecutive	Nutrients converted to products, with accumulation of an intermediate
Stepwise	Nutrients completely converted to intermediate before conversion to product Or Nutrient selectively converted to product in preferential order

(Reprinted from "Fermentation kinetics and model processes", Deindoerfer, F. H., Advances in Applied Microbiology, Umbreit, WW. (Ed.) Vol. 2, p. 327, Copyright © (1960), with permission from Elsevier).

Simple reactions The simple types of reactions are those, which can be both growth- and non-growth-associated reactions. They have been dealt with in detail (Section 7.4) for growth-associated and non-growth-associated kinetics. Figure 9.1 shows the kinetics of growth of *Aerobacter cloacae* which consumes the carbon in the glucose and results in its growth (indicated in the form of cell carbon) and formation of acetic acid as metabolic product.

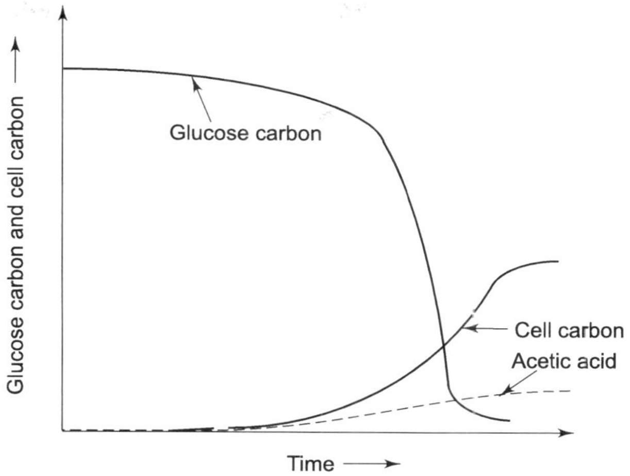

Fig. 9.1 *Simple growth of Aerobacter cloacae during glucose metabolism (Reprinted from "The oxygen requirement of growing cultures of an Aerobacter species determined by means of the continuous culture technique", Pirt, S.J., J. Gen. Microbiol., Vol. 16, p. 71, Copyright © (1957), with permission from Society for General Microbiology)*

Stepwise reactions The stepwise reactions are somewhat typical of microbiological reactions, which we normally do not come across in chemical reactions. One such stepwise reaction is the phenomenon of enzyme selectivity and adaptation. If there are two carbohydrate sources, such as glucose and sorbitol, *E. coli* first utilises the glucose completely and then only starts consuming sorbitol, which Monod (1949) termed as "diauxie" type of reaction (Fig. 9.2).

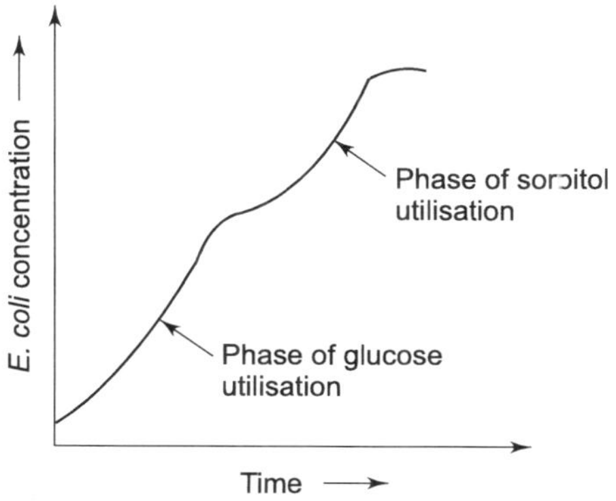

Fig. 9.2 *Diphasic growth of Escherichia coli in a two carbohydrate source*

9.1 PARALLEL REACTIONS

If a reactant A forms both products B and C as follows:

$$A \begin{cases} \xrightarrow{k_1} B \\ \xrightarrow{k_2} C \end{cases} \tag{9.1}$$

the scheme is called a 'parallel reaction', i.e. the reactant A is converting to both products B and C competitively. They are similar to simultaneous reactions, in which the substrate converts to more than one product. For example, glucose is oxidised to acetic acid and gluconic acid by acetic acid bacteria. The relative formation of products depends upon the nutrient concentration and availability. Figure 9.3 shows the conversion of sugar into protein and cell fat during *Rhodotorula gracilis* growth (Enebo et al., 1946). The process consists of two stages, the phases of protein culture. Initially for the first 12.5 hours, only the yeast quantity (cell protein) increases. After about 17 hours, the fat formation process starts. Thus, initially the sugar gets converted only to cell protein. After a build-up of certain amount of cell protein, it (sugar) starts converting to fat.

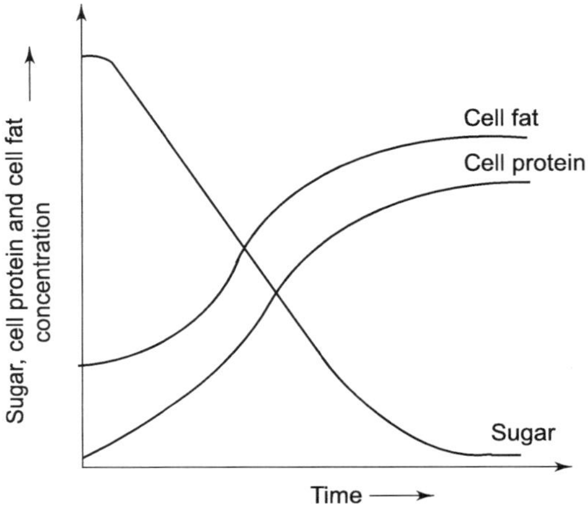

Fig. 9.3 *Simultaneous conversion of sugar into cell protein and cell fat during Rhodotorula gracilis growth (Reprinted from "Microbiological fat synthesis by means of Rhodotorula yeast", Enebo, L., Anderson, L.G., and Lundin, H., Arch. Biochem, vol. 11, p. 389, Copyright © (1946), with permission from Elsevier.)*

Obviously, in the above type of reaction scheme, sometimes only one of the products would be the desired product, and the other the undesired product. We may write Eq. (9.1) as follows:

$$A \xrightarrow{k_D} D \text{ (desired)}$$

$$A \xrightarrow{k_U} U \text{ (undesired)} \quad (9.2)$$

For above reaction, the rates of formation r_D and r_U for the desired and undesired product, respectively, can be written as

$$r_D = k_D\, C_A^{a_1} \quad (9.3)$$

$$r_U = k_U\, C_A^{a_2} \quad (9.4)$$

We always wish that the rate of formation of the desired product D should be maximum, and that of the undesired product U should be minimum, since otherwise the separation costs will be prohibitive (Fogler, 1986).

If we define the selectivity parameter, *S, as the ratio of number of moles of the reactant converted into desired products to the number of moles of reactant converted into undesired products,* then

$$S = \frac{r_D}{r_U} = \frac{k_D}{k_U} C_A^{a_1 - a_2} \quad (9.5)$$

Obviously we desire S to be maximised. Depending upon the reaction kinetics and numerical values of a_1 and a_2, there are two possibilities which are described as follows:

9.1.1 Possibility 1: $a_1 > a_2$

If $a_1 > a_2$, $a_1 - a_2$ is positive, and hence Eq. (9.5) can be maximised if C_A is maintained as large as possible in the reaction system. For maintaining the highest values of C_A, a batch or a plug flow reactor should be used, since in these reactors, the concentration of C_A is high initially and falls gradually. This is in contrast to a CSTR in which the concentration C_A falls to its lowest value the moment the reactants enter the CSTR. Thus, batch or a plug flow reactor system is the obvious choice. Resnick (1981) suggests that if for any operational considerations, a mixed flow reactor (CSTR) is to be used, a number of them should be used in series, which would increase the selectivity. The capacity of the CSTRs should be progressively increasing also.

9.1.2 Possibility 2: $a_1 < a_2$

If $a_1 < a_2$, $a_1 - a_2$ becomes negative, which means that C_A will be in the denominator of Eq. (9.5). Hence, maintaining the lowest C_A will obviously increase the value of S and we can achieve higher selectivity for the desired product. For such a situation, CSTR is the best choice. If due to any operational reasons, other than kinetic considerations, a batch reactor or a plug flow reactor is to be used, then it is suggested that stepwise addition of the reactant to the reactor would maintain lower C_A, which in turn would increase the selectivity.

In the above two cases, we have gone purely by the considerations based on reaction order, and the numerical values of a_1 and a_2. However, the reaction temperature also could play a significant role, which in turn is dependent upon the activation energies of reactions given by Eqs. (9.3) and (9.4). In Eq. (9.5), the ratio k_D/k_U can be expressed as follows, by applying Arrhenius equation (6.6),

$$\frac{k_D}{k_U} = \frac{k_{D0}}{k_{U0}} e^{-(E_D - E_U)/RT} \tag{9.6}$$

where k_{D0} and k_{U0} are the Arrhenius constants, and E_D and E_U are the activation energies for desired product and undesired product, respectively.

9.1.3 Possibility 3: $E_D > E_U$

As shown in Eq. (9.6), if $E_D > E_U$, k_D/k_U will increase as temperature increases. Thus, higher selectivity is possible at higher temperatures. The highest possible temperature can be employed for getting maximum desired product. However, the higher temperature should not lead to any undesired chemical or physical reactions.

9.1.4 Possibility 4: $E_D < E_U$

If $E_D < E_U$, the reaction should be carried out at as low temperature as possible. The only practical limitation imposed on the system is that the temperature should not be so low that the desired reaction (Eq. (9.2)) does not proceed at all.

It is also possible that parallel reactions may take place with two reactants as follows:

$$\begin{aligned} A + B &\longrightarrow D \\ A + B &\longrightarrow U \end{aligned} \tag{9.7}$$

They can also be handled in a similar way and are described in any standard book on chemical reaction engineering [Fogler (1986), Levenspiel (1974(a)), and Resnick (1981)].

9.2 Series Reactions

If the reactant A converts to product R, which in turn converts to product S, as follows

$$A \xrightarrow{k_1} R \xrightarrow{k_2} S \tag{9.8}$$

the scheme is called a 'series reaction'.

They are similar to consecutive reactions in which the reactant converts to an intermediate product. After the intermediate product R builds up to a suitable level of concentration, it further converts to form products. The antibiotics formations mostly fall into this category, in which there is invariably formation of non-active intermediates. Figure 9.4 shows the concentration profiles of the reactant, and the intermediate product. Deindoerfer (1960) cites an example for series reactions as conversion of progesterone to 6β, 11α-dihydroxy-progesterone through the formation of intermediate 11α-hydroxyprogesterone, by the action of *Aspergillus ochraceus*. Acetic acid bacteria of genera *Acetobacter* (viz. *Acetobacter aceti*) overoxidises ethanol to acetic acid and further to CO_2 and water, whereas *Gluconobactor* bacteria restricts the oxidation of ethanol to acetic acid alone.

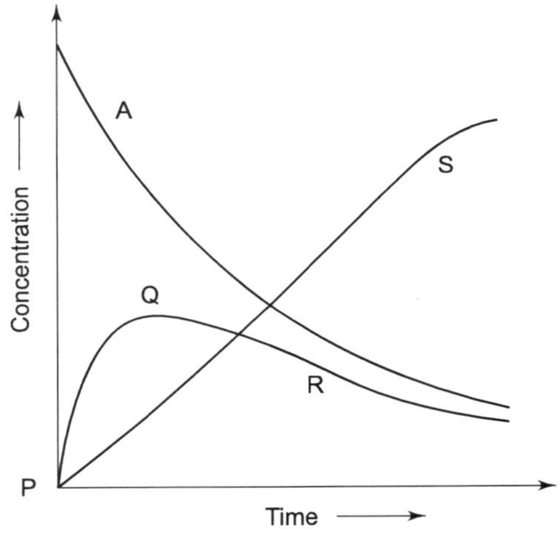

Fig. 9.4 *Concentration profiles of various components in a consecutive reaction scheme* $A \rightarrow R \rightarrow S$

Another type of series reactions is 'stepwise reactions'. An example for this is the bio-oxidation of glucose to 5-ketogluconic acid by *Acetobacter suboxydans*, in which initially all the glucose is converted to gluconic acid. After the complete conversion of glucose, gluconic acid starts converting to 5-ketogluconic acid (Stubbs, 1940). The concentration profiles are shown in Fig. 9.5.

In Section 9.1, for parallel reactions, we have seen that higher selectivity can be achieved towards the desired product by the adjustment of the concentration or by a proper choice of the reactor. However, in case of series reactions, the most important parameter is time. It could be space-time in case of PFR or CSTR, or real-time for batch reactor. In Eq. (9.8), if R is the desired product, which is normally so in most of the cases, the maximum conversion to R depends upon the relative values of k_1 and k_2. If $k_1 > k_2$, getting a higher yield of R is very easy. But if the reaction is allowed to continue too long in a batch reactor or has very high residence times, then the desired product R will be converted to S. On the contrary, if $k_2 > k_1$, it becomes difficult to get higher yields of R. We may have to devise a mechanism for removing R as soon as it is formed by means of some separation technique such as distillation or by selectively solubilising in a solvent (so that the concentration of it can be kept low).

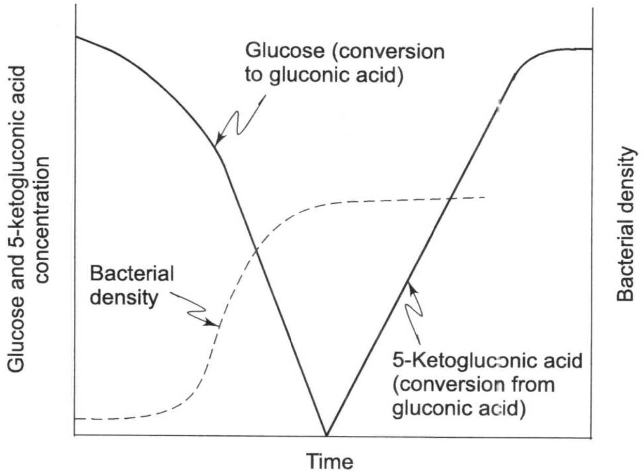

Fig. 9.5 *Diphasic bio-oxidation of glucose to 5-ketogluconic acid by Acetobacter suboxydans (Reprinted from "Ketogluconic acids from glucose: Bacterial production", Stubbs, J.J., Lockwood, L.B., Roe, E.T., Tabenkin, B., and Ward, G.F., I & EC, Vol.32, p 1627, Copyright © (1940), with permission from American Chemical Society)*

9.2.1 Maximizing R in a Batch Reactor/PFR

For a batch or a tubular flow reactor, we can write, on the basis of material balance for Eq. (9.8),

$$-\frac{dC_A}{dt} = r_A = k_1 C_A \tag{9.9}$$

$$\frac{dC_R}{dt} = k_1 C_A k_2 C_R \tag{9.10}$$

$$\frac{dC_S}{dt} = k_2 C_R \tag{9.11}$$

The above three simultaneous equations can be solved for the initial condition that

$$C_A = C_{A0} \quad \text{and} \quad C_{R0} = C_{S0} = 0, \text{ when } t = 0 \tag{9.12}$$

The reactant concentration is given by

$$C_A = C_{A0}\, e^{-k_1 t} \tag{9.13}$$

The desired product concentration, C_R, is given by

$$C_R = C_{A0} + \frac{k_1}{k_2 - k_1}(e^{-k_1 t} - e^{-k_2 t}) \tag{9.14}$$

and the undesired product concentration, C_S, is given by

$$C_S = C_{A0}\left(1 + \frac{k_2 e^{-k_1 t}}{k_1 - k_2} - \frac{k_1 e^{-k_2 t}}{k_1 - k_2}\right) \tag{9.15}$$

At any time,

$$C_{A0} = C_A + C_R + C_S \tag{9.16}$$

The concentrations at any particular time or space-time can be conveniently plotted on a triangular diagram as single points (Fogler, 1986), with the vertices corresponding to pure A, R and S (Fig. 9.6). It can

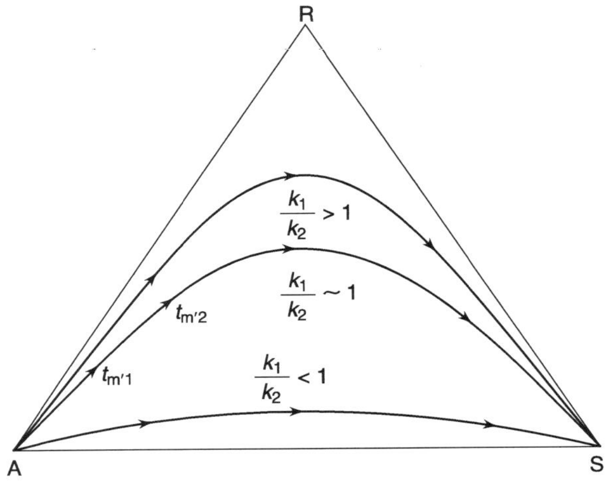

Fig. 9.6 *Concentration profiles of reactant, intermediate product and final product in a consecutive reaction scheme for different ratios of k_1/k_2*

be seen from the figure that the formation of the desired component R is high when $k_1/k_2 >> 1$. Similarly, very little R is formed when $k_1/k_2 << 1$.

The concentration profiles given by Eqs. (9.13)–(9.15) are shown in Fig. 9.4, irrespective of the value of k_1 and k_2. The desired product concentration, C_R, goes through a maximum and then falls. In the region PQ, the concentration of R builds up, even though there will be some conversion of R to S. After the point Q, the concentration of R becomes so much that its decomposition dominates over its formation.

Hence, the optimum time for maximum concentration of R ($C_{R,max}$) can be found by differentiating Eq. (9.14) with respect to t and equating it to zero, which gives

$$t_{opt,\,p} = \frac{\ln(k_2/k_1)}{k_2 - k_1} \tag{9.17}$$

The subscript p for $t_{opt,p}$ stands for PFR.

Obviously, it is at this point of time, the rate of formation of S will also be maximum. $C_{R,\,max}$ can be obtained by replacing t in Eq. (9.14) with $t_{opt,p}$ given by Eq. (9.17):

$$C_{R,\,max} = C_{A0}\left(\frac{k_1}{k_2}\right)^{k_2/(k_2-k_1)}$$

9.2.2 Maximizing R in CSTR

By carrying out material balance, we can derive similar type of equations for a CSTR (Eq. 8.53)

$$\frac{C_A}{C_{A0}} = \frac{1}{1 + k_1 t_m} \tag{9.18}$$

where t_m stands for space-time in CSTR (mixed flow reactor). Material balance on R for a volumetric flow rate υ gives,

$$\upsilon C_{R0} = \upsilon C_R + (-r_R)V$$

$$\therefore \quad \upsilon C_{R0} = \upsilon C_R - k_1 C_A V + k_2 C_R V \tag{9.19}$$

$-r_R$ is the rate of conversion of R in the reactor volume V. The LHS term is the quantity of R entering CSTR, whereas the first term in the RHS is the quantity of R leaving the CSTR, the second term indicates the amount of R formed due to reactions in the CSTR, while the third term in the RHS is the quantity of R consumed in the CSTR because of reaction. Noting that $V/\upsilon = t_m$ and $C_{R0} = 0$, we get from Eqs. (9.18) and (9.19):

$$\frac{C_R}{C_{A0}} = \frac{k_1\, t_m}{(1 + k_1\, t_m)(1 + k_2\, t_m)} \tag{9.20}$$

C_S can be found from Eq. (9.16), knowing C_A and C_R from Eqs. (9.18) and (9.20), respectively. The $t_{opt,m}$ for maximum C_R can be found by

$$\frac{dC_R}{dt} = 0$$

which simplifies to give

$$t_{opt,\,m} = \frac{1}{(k_1\, k_2)^{1/2}} \tag{9.21}$$

Figure 9.7 shows that maximum conversion to R in the CSTR is always less than what we get in PFR/batch reactor (Resnick, 1981). Thus, the undesirable effect of ideal mixing is evident in the figure.

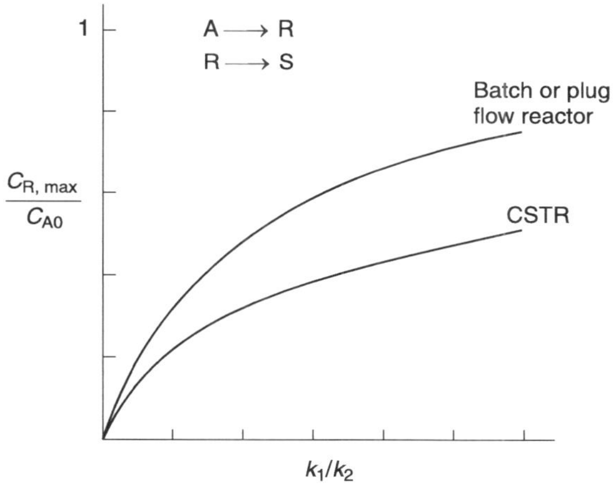

Fig. 9.7 *Function of $C_{R,\,max}$ versus k_1/k_2 for PFR and CSTR*
(Reprinted from "Process Analysis and Design for Chemical Engineers", Resnick, W., p. 126, Copyright © (1981), with permission from McGraw-Hill, Inc., New York)

Example 9.1

Glucose can be catabolised by yeast to produce ethanol and CO_2 or to other metabolic products, depending upon the culture and environmental conditions. Assuming that formation of other biosynthetic products is a consecutive reaction after the formation of alcohol which is the desired reaction, $A \xrightarrow{k_1} R \xrightarrow{k_2} S$. Plot the graphs for C_R/C_{A0} vs k_1t_m where t_m is the average space-time in CSTR, and is in the range of 1 – 8 hours. The following are the other process parameters.

$$k_1 = 0.3\ h^{-1}$$
$$k_2 = 0.1\ h^{-1}$$
$$C_{A0} = 20\ kg/m^3$$

Calculate the maximum possible C_R and compare it with the graphically determined value.

Solution Given

$$k_1 = 0.3\ \text{h}^{-1}$$
$$k_2 = 0.1\ \text{h}^{-1}$$
$$C_{AO} = 20\ \text{kg/m}^3$$

C_A stands for glucose concentration in moles/m^3.

Molecular weight of glucose= 180

$$\therefore \quad C_{AO} = \frac{20}{180} = 0.11\ \text{kgmol/m}^3$$

From Eq. (9.20), we note down

$$\frac{C_R}{C_{A0}} = \frac{k_1\ t_m}{(1 + k_1\ t_m)(1 + k_2\ t_m)}$$

We prepare the tabular data for $(C_R)/(C_{AO})$ and $k_1 t_m$ for different values of t_m ranging from 1–8 hours in Table E9.1. With the data, Fig. E9.1 is drawn.

From the graph, we find that max. C_A/C_{AO} occurred

when $\quad t_{mopt} = 5.8$ h

and $\quad (C_R/C_{AO})_{max} = 0.403$

$$(C_R)_{max} = 0.403 \times 0.11 = 0.044\ \text{kgmol/m}^3$$

Table E9.1 *Data on $k_1 t_m$ vs C_R/C_{AO} to draw Fig. E9.1*

t_1 (h)	$k_1 t_m$	$k_2 t_m$	C_R/C_{AO}
1	0.3	0.1	0.21
2	0.6	0.2	0.3125
3	0.9	0.3	0.364
4	1.2	0.4	0.3896
5	1.5	0.5	0.40
6	1.8	0.6	0.402
7	2.1	0.7	0.398
8	2.4	0.8	0.392

If the reaction follows first—order kinetics,

$(t_m)_{opt}$ is given by Eq. (9.21)

$$(t_m)_{opt} = \frac{1}{(k_1\ k_2)^{1/2}} = \frac{1}{(0.3 \times 0.1)^{0.5}} = 5.77\ \text{h}$$

Using this value of $(t_m)_{opt}$ in Eq. (9.20), we can calculate $(C_R/C_{AO})_{max}$

$$(C_R/C_{AO})_{max} = \frac{0.3 \times 5.77}{((1 + (0.3 \times 5.77))(1 + (0.1 \times 5.77))}$$
$$= 0.402$$

$$\therefore \quad (C_R)_{max} = 0.402 \times 0.11 = 0.044\ \text{kgmol/m}^3$$

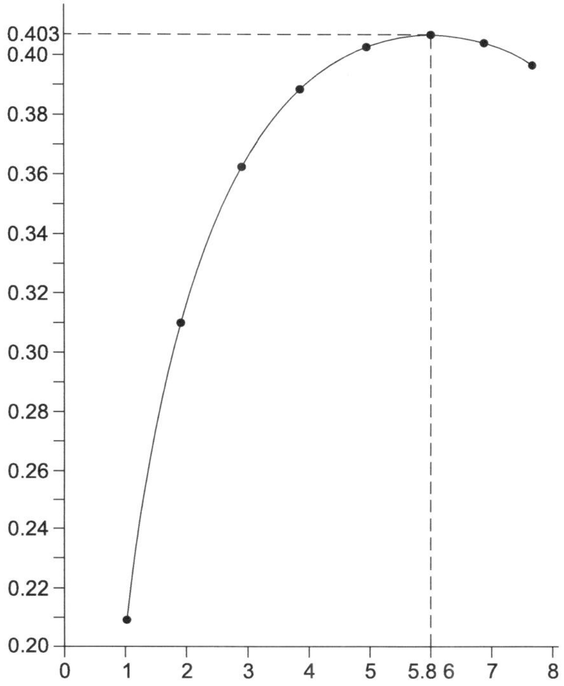

Fig. E9.1 *Plot of t vs (C_R/C_{Ao}) for Example 9.1*

9.2.3 Reactor Choice for Series Reactions

For maximising C_R, it is evident that k_1/k_2 should be high, which is not in our control. One possibility could be to evaluate the effect of temperature on k_1 and k_2 by means of Arrhenius equation. If $E_D > E_U$ in Eq. (9.6), we can resort to higher temperatures, as was mentioned in Section 9.1.3. If $E_D < E_U$, lower temperatures would favour higher conversion to R. As can be seen from Fig. 9.7, for any value of k_1/k_2, a PFR is preferred to a mixed flow rector. However, as has been mentioned in Section 9.1.1, if a CSTR is to be used for any other considerations, several small CSTRs in series should be used to approach the performance of PFR (Fig. 8.6).

If k_1/k_2 is low, naturally the formation of R is less and its conversion to undesired S will be more. In such cases, we have to keep the conversion per pass to be as little as possible, and whatever be the small amount of R formed, should be withdrawn in a separator, and the unreacted A should be recycled to inlet (Fig. 9.8). Large amount of efforts will have to be put in the separator to separate huge quantities of A, because of deliberately keeping low conversions per pass. In such cases, the final choice of fixing the per-pass conversion is based on the economics, i.e. the cost considerations predominate the technical considerations.

9.3 SERIES-PARALLEL REACTIONS

It is also possible that the reactions are as given below:

$$A + B \xrightarrow{k_1} R$$

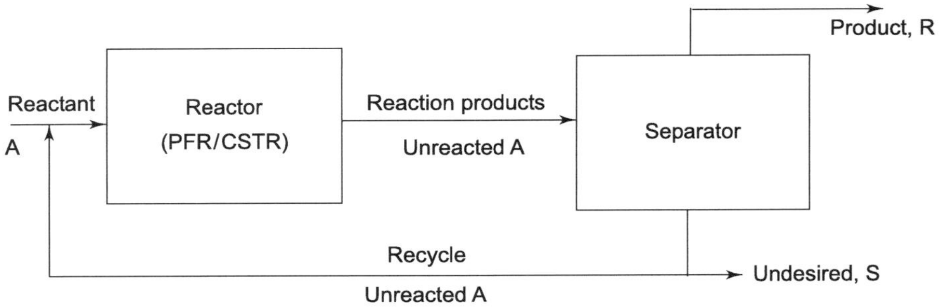

Fig. 9.8 *Reaction–Separation system*

$$B + R \xrightarrow{k_2} S$$

$$A + S \longrightarrow T \qquad (9.22)$$

In such case, we call the reaction scheme as a series–parallel reaction. The conversion of R to S is a series reaction, but the conversion of B to S is a parallel reaction, since B is getting converted to form products R and S concurrently or in parallel.

Thus, the series–parallel reactions are complex in nature, and may involve a combination of the type of reactions shown in Table 9.1, which goes beyond the 'stepwise' reactions. We come across this kind of complex reactions in fermentation processes like in the production of penicillin. The penicillin production curve with synthetic medium, shown in Fig. 9.9, was described as an example of this type of reactions by Deindoerfer (1960). The penicillin production curve exhibits diphasic characteristics, and shows the accumulation of an intermediate product, which could be 6-amino penicillinic acid.

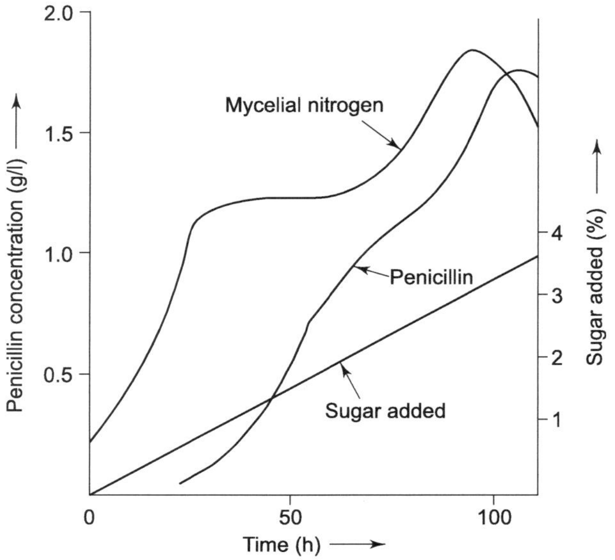

Fig. 9.9 *Penicillin production in synthetic medium with continuous addition of glucose (Reprinted from "Penicillin from chemically defined media", Holser, P., and Johnson, M.J., I&EC, Vol. 45, p. 873, Copyright © (1953), with permission from American Chemical Society)*

We restrict our discussion on the series–parallel complex reactions to the first two reactions given in Eq. (9.22). Assuming constant density, the rate expressions for A, R and S can be written as follows:

$$r_A = -k_1 C_A C_B \tag{9.23}$$

$$r_B = -k_1 C_A C_B - k_2 C_R C_B \tag{9.24}$$

$$r_R = k_1 C_A C_B - k_2 C_R C_B \tag{9.25}$$

$$r_S = k_2 C_R C_B \tag{9.26}$$

The product distribution of R and S depends upon how we mix A and B. Levenspiel (1974 (a)) considers this system in three ways:

(i) adding A slowly into bulk of B and mixing
(ii) adding B slowly into bulk of A and mixing
(iii) mixing A and B together

Case 1: Addition of A slowly into bulk of B The reactant A is added slowly into reactant B in bulk. The contents are thoroughly mixed. Addition of A should be in such a way that all A is consumed before the next batch of A is added. This results in depletion of B continuously, and instantaneous accumulation of R which reacts immediately with B to form product S. This situation continues until all B is consumed. Subsequently, any addition of A will not result in any further reaction. This type of methodology does not help in anyway to get the product R.

Case 2: Addition of B slowly into bulk of A In this case, bulk A is taken, into which B is added slowly. The small amount of B added will react immediately to form product R. There will not be any B left out for further reaction with R. Thus, the reaction system is starved of B, which really has to participate in the second reaction (Eq. (9.22)). Hence, there will be scope for the accumulation of product R. After certain amount of R is accumulated, it starts reacting with B and forms the product S.

Case 3: Mixing A and B together In this type of mixing, the concentration profiles are same as those in the earlier cases. The concentration profiles of A, R and S are similar to those of series reaction schemes (Fig. 9.4). Thus, the reaction scheme may be represented as

$$A \xrightarrow{+B} R \xrightarrow{+B} S \tag{9.27}$$

Another important observation is that the concentration of B does not have any effect on the product distribution for Cases 2 and 3, i.e. whether B is added slowly or all the B and A are mixed together. Only in Case 1, where A is in limited quantity, it does not have any accumulation of product R.

9.3.1 Product Concentration for a PFR or Batch Reactor

Dividing Eq. (9.25) by Eq. (9.23) gives

$$\frac{r_R}{r_A} = \frac{dC_R}{dC_A} = \frac{k_2}{k_1}\frac{C_R}{C_A} - 1 \tag{9.28}$$

The solution of the above differential equation gives (Levenspiel, 1974(a)):

$$\frac{C_R}{C_{A0}} = \frac{1}{1 - k_2/k_1}\left[(C_A/C_{A0})^{k_2/k_1} - (C_A/C_{A0})\right] + \frac{C_{R0}}{C_{A0}}(C_A/C_{A0})^{k_2/k_1} \tag{9.29}$$

if $k_2/k_1 \neq 1$.

If $k_2/k_1 = 1$,

$$\frac{C_R}{C_{A0}} = \frac{C_A}{C_{A0}}\left(\frac{C_{R0}}{C_{A0}} - \ln\frac{C_A}{C_{A0}}\right) \tag{9.30}$$

9.3.2 Product Concentration in CSTR

For a CSTR,

$$t_m = \frac{C_{A0} - C_A}{-r_A} = \frac{C_{R0} - C_R}{-r_R} \tag{9.31}$$

Replacing r_A and r_R from Eqs. (9.23) and (9.25)

$$t_m = \frac{C_{A0} - C_A}{k_1\, C_A\, C_B} = \frac{C_{R0} - C_R}{k_2\, C_R\, C_B - k_1\, C_A\, C_B} \tag{9.32}$$

On rearranging, we get

$$\frac{C_{R0} - C_R}{C_{A0} - C_A} = \frac{k_2 C_R}{k_1 C_A} - 1 \tag{9.33}$$

$$C_R = \frac{C_A\,(C_{A0} - C_A + C_{R0})}{C_A + (k_2 / k_1)(C_{A0} - C_A)} \tag{9.34}$$

9.4 NON-ISOTHERMAL REACTIONS

The reactions in which the temperatures are not constant are known as **non-isothermal reactions**. They are in contrast to **isothermal reactions** in which the temperatures are maintained constant. In an isothermal reaction, since the temperature is constant, rate of reaction is constant for a given space-velocity in flow reactors and varies with concentration in a batch reactor from maximum to a minimum. In a non-isothermal reaction system, the temperature is not maintained constant, and hence the rate of reaction varies. Non-isothermal reactions in which heat is neither supplied nor removed after the reaction sets in, are called **adiabatic reactions**. It is difficult to control non-isothermal reactions, particularly those which are **exothermic** in nature (in which heat is generated during the reaction), causing a temperature rise. In turn, the rate of reaction increases, resulting in higher heat generation, which further raises the temperature. Thus, the temperature rises rapidly until the whole reactant is consumed (to what products, is anybody's guess!). This either disturbs the selectivity of the reaction or results in explosion of the reactor. We call such a situation as **run away of the reaction,** i.e. the reaction runs away from our control.

Fermentation of glucose to alcohol (ethanol) by yeast is an example of exothermic reaction. The fermentation is favourable in the temperature range 23–30 °C, with the heat of reaction being equal to 605 kJ/kg of sugar. If the temperatures, in the absence of control, reach as high as 37 °C or so, the yeast will not be able to withstand the temperature and the cells would die. This would result in cessation of the fermentation. Hence, normally in alcohol fermentation process, the temperatures in the fermenters are maintained by circulating cooling water in the fermenter's jacket, or by passing the fermentation broth through an outside heat exchanger.

9.4.1 Design Procedure for Non-isothermal Reactions

The temperature, composition and reaction rates are interrelated and can be shown graphically. The size of the reactor for achieving the desired conversion can be calculated based on the temperature progression (Levenspiel, 1974 (b)).

Adiabatic operations For a PFR or mixed flow reactor with a reactant feed stream of specific heat $C_{p.f}$ and product stream of specific heat $C_{p.P}$, we can write the energy balance equation over a reference temperature of T_1 as follows:

Input energy = output energy + energy accumulated
+ energy consumed in the reaction scheme (9.35)

Input energy = 0, for an adiabatic process

Output energy = $C_{p,P}\,(T_2 - T_1)\,X_A$

Accumulated energy = $C_{p,f}\,(T_2 - T_1)\,(1 - X_A)$

Energy consumed/disappeared by reaction

$= (\Delta H_{r1})\,X_A$

where ΔH_{r1} is the heat of reaction per mole of reactant A.

Thus,

$$0 = C_{p,P}\,(T_2 - T_1)\,X_A + C_{p,f}\,(T_2 - T_1)\,(1 - X_A) + (\Delta H_{r1})\,X_A \tag{9.36}$$

which on rearrangement gives

$$X_A = \frac{C_{p,f}\;\Delta T}{(-\Delta H_{r1}) - (C_{p,P} - C_{p,f})\Delta T} \tag{9.37}$$

The denominator in the above equation will be equal to ΔH_{r2}, the heat released by reaction at the temperature of T_2.

$$X_A = \frac{C_{p,f}\;\Delta T}{(-\Delta H_{r2})} \tag{9.38}$$

The relationship between X_A and temperature, as given by Eq. (9.38), was shown by Levenspiel (1974) in the form of a graph (Fig. 9.10). The best type of reactor, which has minimum V/F_{A0}, can be found from the above graph.

1. If r_A progressively decreases with conversion, which is normally the case with endothermic reactions, a PFR is the optimum choice.
2. For small $C_p/(\Delta H_r)$ ratios, mixed flow reactor is the best.
3. For high $C_p/(\Delta H_r)$ ratios, PFR is the best.

9.4.2 Thermal Behaviour of a CSTR

The way the temperature effects are realised in CSTR is very interesting for exothermic reactions, since an exothermic reaction evolves heat and increases the temperature in the CSTR (unless the heat is removed). It results in higher reaction rates, and hence higher heats of reaction, ultimately leading to a situation of "run away" reactions which we have mentioned earlier in this section. So, if we draw a graph for temperature versus rate of heat generation, we get a sigmoidal curve, shown in Fig. 9.11. The plateauing of the curve at higher temperature is mainly due to depletion of the reactant. We shall try to derive an expression for the heat generation rate and also that for the heat removal rate, so that these two could be equated to evaluate the thermal stability of CSTR.

Material balance on component A for a first-order reaction kinetics is (from Eq. (8.27))

$$\tau = \frac{V}{\upsilon} = \frac{C_{A0} - C_A}{-r_A}$$

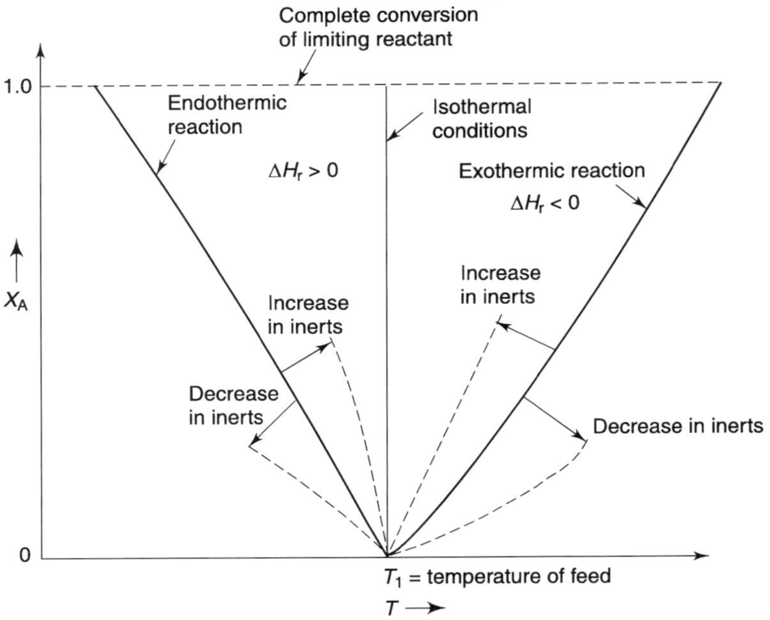

Fig. 9.10 *Graphical representation of energy balance equation for adiabatic operation (Reprinted from "Chemical Reaction Engineering", Levenspiel, O., 2nd Edn, p. 223, Copyright © (1974), with permission from John Wiley & Sons, Inc.)*

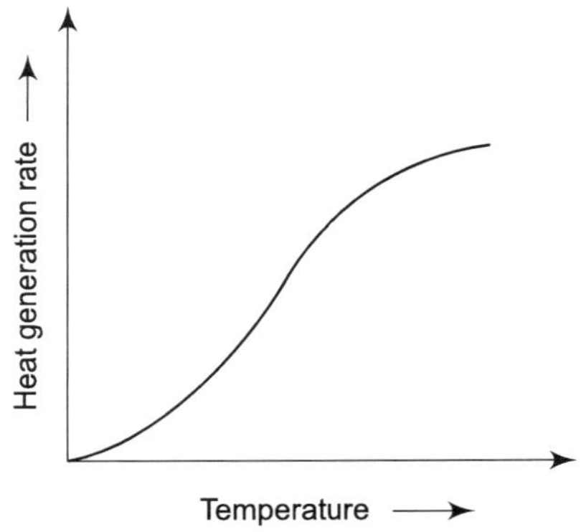

Fig. 9.11 *Heat generation rate versus temperature in a CSTR*

which on rearrangement and replacing $-r_A$ by $k_1 C_A$ for first order kinetics gives

$$\upsilon C_{A0} = \upsilon C_A + V k_1 C_A$$

from which

$$C_A = \frac{C_{A0}}{1 + k_1 V/\upsilon} \tag{9.39}$$

The above equation is similar to Eq. (9.18), except that t_m is replaced by V/υ.

The rate of heat generation due to chemical reaction (Q_R) is given by multiplying the number of moles of A reacted by the heat of reaction.

Thus,
$$Q_R = (-r_A)V(\Delta H_r)$$
$$= -k_1 C_A V(\Delta H_r) \tag{9.40}$$

The negative sign in the above equation only indicates the direction of heat flow. It shows that the heat is transferred from the system, because of which the system is depleted of so much of energy.

By replacing C_A in Eq. (9.40) by Eq. (9.39), we get

$$Q_R = \frac{C_{A0}\, \upsilon\, \Delta H_r}{1 + (\upsilon/V k_0)e^{E/RT}} \tag{9.41}$$

in which k_0 is Arrhenius factor in the expression $k = k_0\, e^{-E/RT}$ and T is expressed in degree Kelvin.

Under the steady state conditions in CSTR, Q_R is equal to heat removal rate Q_r, which is the sum of the heat taken away by the product stream with reference to a standard base temperature of T_0, and the heat removed by heat transfer. The heat removal rate is represented by Q_r.

Thus,
$$Q_r = \upsilon\, \rho\, C_p(T - T_0) + UA(T - T_c) \tag{9.42}$$

where T_c = cold fluid temperature
A = heat transfer area
U = heat transfer coefficient.

Equation (9.42) can be linearised in T and can be written as

$$Q_r = (\upsilon\, \rho\, C_p + UA)T - (\upsilon\, \rho\, C_p\, T_0 + UAT_c) \tag{9.43}$$

Equation (9.43) is shown in Fig. 9.12 along with Eq. (9.41). The straight lines 1, 2, 3 and 4 are the rate of heat removal lines at different temperatures, while the sigmoid curve is for rate of heat generation.

Under steady state conditions,

$$Q_R = Q_r$$

And this occurs at the points of intersections (viz. a, b, c or d) of the heat generation curve and heat removal lines.

From Fig. 9.12, we see that only one steady state point occurs at high and low temperatures of heat removal, represented by lines 1 and 4. Lines 2 and 3 have multiple stable points. In the case of line 2, the operation

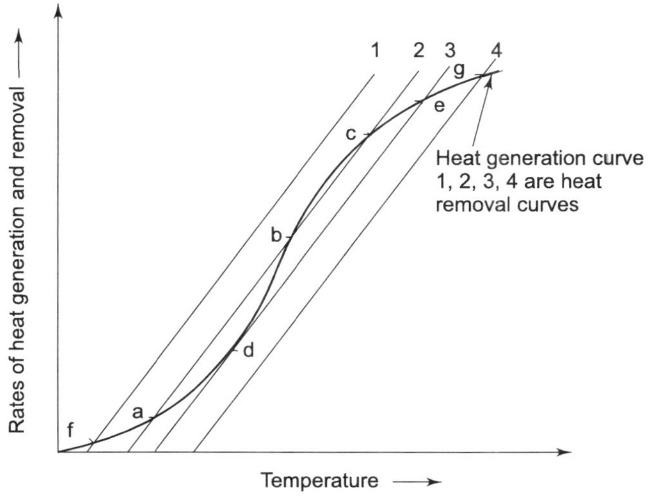

Fig. 9.12 *Heat generation and removal rates with temperature in CSTR*
(Reprinted from "Process Analysis and Design for Chemical Engineers", Resnick, W., p. 134, Copyright © (1981), with permission from McGraw-Hill, Inc., New York)

at point a or c is stable, whereas that at point b is highly unstable. Any small increase in reactor temperature at this point would result in higher heat generation by reaction as compared to heat removal, and would only attain stability at point c. The reverse is also true. Any small decrease in temperature would push the stabilisation point to a.

A similar type of explanation holds good for point d on line 3. Any increase in temperature would push the stabilisation point to e. Unusually the upper stable or stationary state is the desired one (Resnick, 1981).

Example 9.2

Glucose is converted to alcohol in a CSTF (0.15 m^3 capacity) which is having an effective heat transfer area of 1 m^2. The conversion process can be considered to be an overall first order reaction with the rate constant $k_1 = 0.3\ h^{-1}$. The reaction produces heat at the rate 605 kJ/kg of sugar converted. The initial concentration of sugar is 0.11 kg-mol/m^3. The dilution rates of substrate vary between 0.2 and 1.0 h^{-1}. Find the heat released (Q_R) for each dilution rate. Is Q_R varying considerably in each case? If so, how much is the percentage variation. If the feed is entering at the reaction temperature (T), and the fermenter is having a very low heat transfer coefficient of 0.3 kW/(m^2 °C) and the cooling water temperature (Tc) is 25°C, find the reaction temperature, T.

Solution Given

$$\text{Vol. of CSTR} = V = 0.15\ m^3$$
$$A = 1\ m^2$$
$$k_1 = 0.3\ h^{-1} = 8.33 \times 10^{-5}\ s^{-1}$$
$$\Delta H_r = 605\ \text{kJ/kg of sugar} = 605 \times 180 = 1.1 \times 105\ \text{kJ/kg mole}$$
$$C_{AO} = 0.11\ \text{kg mol/m}^3$$
$$D \text{ varies between } 0.2 \text{ to } 1\ h^{-1}$$
$$U = 0.3\ \text{kW/m}^2\ °C$$
$$T_c = 25\ °C$$

Find Q_R variation with D

$$D = \upsilon / V$$

$\therefore$
$$\upsilon = DV = 0.15\ D$$

Q_R can be found by simple application of Eq. (9.40)

$$Q_R = -k_1\, C_A\, V\, (\Delta Hr)$$

C_A is replaced by Eq. (9.39) for first order kinetics

$$C_A = \frac{C_{A0}}{1 + k_1\,(V/\upsilon)} = \frac{C_{A0}}{1 + k_1\,\tau}$$

If D varies between 0.2 to 1 h^{-1}, τ varies between 1 – 5

$$\text{since } \tau = 1/D = V/\upsilon$$

We tabulate data for different Ds and hence different τ in Table E9.2

From Table E9.2, we notice that variation in Q_R is quite significant which is of the order of 48% (= (418.8 – 217.8)/418.8)

But $Q_R = Q_r$ where Q_r is heat removal rate and Q_r is given by Eq. (9.43)

$$Q_r = (\upsilon\,\rho\,C_P + UA)\,T - (\upsilon\,\rho\,C_P\,T_o + UA\,T_c)$$

In the above, we take T_o is equal to T_c which is equal to 25°C.

Table E9.2 *Data for calculating Q of Ex. 9.2*

D (1/h)	τ (h)	C_A (kg mol/m^3)	Q_R (kJ/h)	υ (m^3/h)
0.2	5.0	0.044	217.8	0.03
0.3	3.33	0.055	272.4	0.045
0.4	2.5	0.063	311.8	0.06
0.5	2.0	0.069	340.3	0.075
0.6	1.66	0.073	363.5	0.09
0.7	1.43	0.077	381.0	0.105
0.8	1.25	0.08	396.0	0.12
0.9	1.11	0.082	408.5	0.135
1.0	1.0	0.085	418.8	0.15

ρ and C_P are taken to be those of water

$$\rho = 1000 \text{ kg/m}^3$$

$$C_P = 4.2 \text{ kJ/kg °C}$$

When $Q_r = 217.8$, kJ/h and $\upsilon = 0.03$

$$217.8 = (0.03 \times 1000 \times 4.2 + 0.3 \times 1)\, T - (0.03 \times 1000 \times 4.2 \times 25 + 0.3 \times 1 \times 25)$$

from which $T = 26.7$°C

Similarly, when $Q_r = 418.8$ kJ/h and $\upsilon = 0.15$ m^3/h

$$418.8 = (0.15 \times 1000 \times 4.2 + 0.3 \times 1)\, T - (0.15 \times 1000 \times 4.2 \times 25 + 0.3 \times 1 \times 25)$$

from which $T = 25.7$°C

The reaction temperature could be approx. 26°C

$$T = 26°\text{C}$$

9.5 DESIGN PRINCIPLES

The cost fraction of a fermenter in the overall project cost of any biotechnological process (or for that matter, any chemical process) is very small. Machinery and equipment component includes capital investment in feed preparation line, product purification section, tankage, piping, controls and ancillaries. However, from the technology point of view, the fermenter is most important. The economical operation of the plant depends upon the successful design and operation of the fermenter. Hence, it is imperative that greater care is to be exercised in the design of the fermenter, since maximisation of conversion and yield will lead to maximum profits. Obviously, profit maximisation is the objective function appropriate to most industrial optimisations.

An experienced engineer works within the practical design constraints, using judgment based largely on experience (Nauman, 1987). His main objective is to design the reactor that is technically feasible and economically plausible.

In this section, we deal more with the chemical design aspects rather than the economics.

The design features include:

- type of reactor
- reactor volume
- reaction time

- feeding mechanism
- substrate concentration
- medium design
- nutrient composition.

The last two aspects are beyond the purview of process design.

We have discussed in the earlier sections of this chapter, the type of reactor and reaction time for various types of reaction systems, viz.

- simple reactions
- parallel reactions
- series reactions
- series–parallel reactions, etc.

It is obvious that a thorough and detailed knowledge of the kinetics of the reaction is desired for any rational design of the fermenters. It needs to be combined with the biochemist's knowledge of the chemistry of various reactions. Maintaining an optimum and active cell growth of the microorganisms, which is normally taken care by the microbiologists, is an additional requirement in case of fermentation process.

Highly efficient and effective reactor schemes, which are costly, will generally produce products with a very little or little undesired product, and hence will have least separation costs. On the other hand, if the reactor is inexpensive and produces products with a lot of undesirable or unconverted reactants, the separation costs will be higher. Thus, the economic incentives will sometimes decide the design of the reactors. An efficient reactor system will be at a point where both the costs are optimum, unless the reaction system demands it to be otherwise.

Symbols

a_1, a_2	: order of reactions
A	: reactant A
A	: heat transfer area [in Section 9.4.2]
B	: reactant B
C	: concentration
C_p	: specific heat (kJ/kg K)
D	: desired product
E	: activation energy
ΔH_r	: molar heat of reaction (kJ/mol)
k	: reaction rate constant
k_0, k_{U0}, k_{D0}	: Arrhenius constants
k_1, k_2	: reaction rate constants
Q_r	: heat removal rate (kJ/h)
Q_R	: heat generation rate due to reaction (kJ/h)
r	: rate of reaction
R	: product
R	: universal gas constant = 8.31 J/(mol K)
S	: product
S	: selectivity parameter [Eq. (9.5)]
t	: time (s or h)
t_m	: space-time in mixed flow reactor (s)
T	: temperature (°C or K)
T_c	: cold fluid temperature

U	: undesired product
U	: heat transfer coefficient (kW/(m^2 K))
v	: volumetric flow rate (m^3/s or m^3/ h)
V	: reactor volume (m^3)
X	: conversion

Subscripts

0	: initial conditions
1, 2	: conditions at 1 and 2
A	: component A
B	: component B
D	: desired reaction
f	: feed
m	: mixed flow
max	: maximum
opt	: optimum
p	: plug flow reactor
P	: product
R	: component R
S	: component S
U	: undesired reaction

Greek Letter

ρ	: density (kg/m^3)

References

9.1 Deindoerfer, F.H. (1960). "Fermentation kinetics and model processes", *Advances in Applied Microbiology*, Umbreit, W.W. (Ed.), Vol. 2, Academic Press, New York, pp. 321–333.

9.2 Enebo, L., Anderson, L.G. and Lundin, H. (1946), "Microbiological fat synthesis by means of Rhodotorula yeast", *Arch. Biochem.* 11, 383–395.

9.3 Fogler, H.S. (1986), *Elements of Chemical Reaction Engineering,* Prentice-Hall, Englewood Cliff, New Jersey, pp. 452–464.

9.4 Hosler, P. and Johnson, M.J. (1953), "Penicillin from chemically defined media", *I&EC*, 45, 871–874.

9.5 Levenspiel, O. (1974), *Chemical Reaction Engineering*, Wiley Eastern Pvt. Ltd., New Delhi, (a) pp. 163–199, (b) pp. 218–226.

9.6 Monod, J. (1949), "The growth of bacterial cultures", *Ann. Rev. Microbiol.,* 3, 371–394.

9.7 Nauman, E.B. (1987), *Chemical Reactor Design*, John Wiley & Sons, New York, pp. 153–156.

9.8 Pirt, S.J. (1957), "The oxygen requirement of growing cultures of an *Aerobacter* species determined by means of the continuous culture technique", *J. Gen. Microbiol.*, 16, 59–75.

9.9 Resnick, W. (1981), *Process Analysis and Design for Chemical Engineers,* McGraw-Hill, Inc., New York, pp. 103–135.

9.10 Stubbs, J.J., Lockwood, L.B., Roe, E.T., Tabenkin, B., and Ward, G.F. (1940), "Ketogluconic acids from glucose: Bacterial production", *I&EC*, 32, 1626–1631.

Review Questions

9.1 What are the various types of fermentation reactions? Explain with examples.

9.2 Define the "Selectivity parameter" in parallel reactions. Explain the various possibilities of parallel reactions. Recommend a suitable flow reactor for each.

9.3 In the series reaction scheme in PFR

$$A \xrightarrow{k_1} R \xrightarrow{k_2} S$$

derive expressions for concentrations of R and S. At what space-time the C_R will be maximum?

9.4 Derive an expression for $C_{R,max}$ in a CSTR for the above series reaction scheme.

9.5 Explain which type of reactor do you choose for series reaction scheme $A \xrightarrow{k_1} R \xrightarrow{k_2} S$ for different combinations of k_1 and k_2.

9.6 Write expressions for C_R in a series–parallel reaction scheme of the following type $A \xrightarrow{+B} R \xrightarrow{+B} S$: (i) in a PFR (ii) in a CSTR.

9.7 Write a note on non-isothermal reactions.

9.8 Draw a graph for conversion versus temperature for an adiabatic operation based on the energy balance equation.

9.9 Describe the effect of non-control of temperature for an exothermic reaction in a CSTR.

9.10 Describe the thermal stability criterion for an adiabatic reaction system in a CSTR.

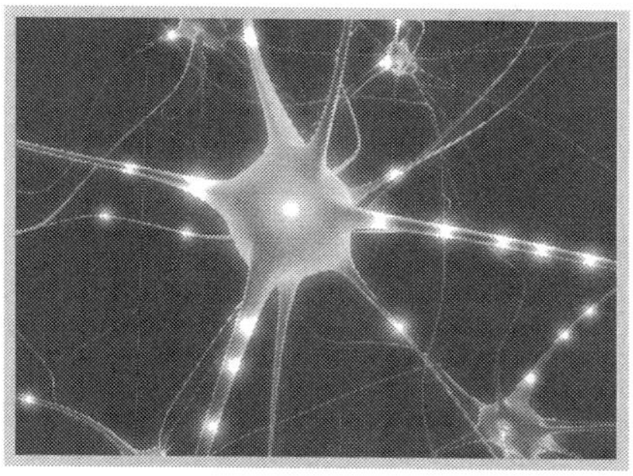

chapter 10

Non-Ideal Flow

So far we have seen two ideal flows in continuous flow reactors, viz.

- plug flow
- completely mixed flow.

A plug flow is characterized by the fact that there will not be any mixing in the axial direction, and hence there will not be any velocity gradients in that direction. There will be only changes in the concentration in the axial direction due to reaction. The reactant concentration will be falling and the product concentration will be building up in the axial direction. There will not be any changes in concentration with time, once the steady state is reached. Absolutely there will not be any intermixing in the plug flow reactor. Exactly opposite to this will be the completely mixed flow reactor, which we call as CSTR. The reactants are fully mixed, the moment they enter the CSTR. The initial concentration of the reactant, C_{A0}, will immediately fall to C_A, the concentration prevailing in the reactor and it will also be the exit concentration of the reactant. The concentration of the reactants or products at any corner of the reactor will be the same. Perfect mixing conditions prevail in the CSTR. Thus, a CSTR and PFR are two extreme conditions of ideality of perfect mixing conditions and completely non-mixing conditions. Hence, they are called as ideal reactors. In reality, it is difficult to attain either of these states. Any deviations from these conditions of ideality is known as **non-ideality,** which forms the subject matter for this chapter, and will be discussed in detail.

10.1 REASONS FOR NON-IDEALITY

Non-ideality prevails in the flow pattern of the fermenters or reactors mainly due to the following reasons:

(i) stagnant zones
(ii) short circuiting
(iii) channelling
(iv) recirculation.

All the above non-ideal features result in reduced or changed residence time, τ, which was defined as (Eq. (8.19)):

$$\tau = \frac{V}{v}$$

In the above equation, V is the volume of the reactor. In view of the non-ideal flow conditions, V will not be the same as measured, and thus it affects the value of τ, which in turn was related to the conversion by the following equations:

For a CSTR (Eq. (8.27)):

$$\tau = t_{Av} = \frac{C_{A0} - C_A}{-r_A}$$

For a PFR (Eq. (8.34):

$$\tau = t_{Av} = C_{A0} \int_0^{X_A} \frac{dX_A}{-r_A}$$

Thus, the flow with non-idealities will give erroneous results.

Let us see the non-ideal flow characteristics. Figure 10.1 shows some typical non-ideal flows in reactors. Figure 10.1(a) shows stagnant zones formed at the corners of the CSTR. They make the reactor volume less than the actual value. The magnitude of reduction in volume depends upon:

- type of agitator
- speed of agitator
- L/D ratio of the reactor tank
- the shape of the tank in the corners
- presence and size of the baffles in the reactor
- the nature of fluid
- the velocity of the fluid, etc.

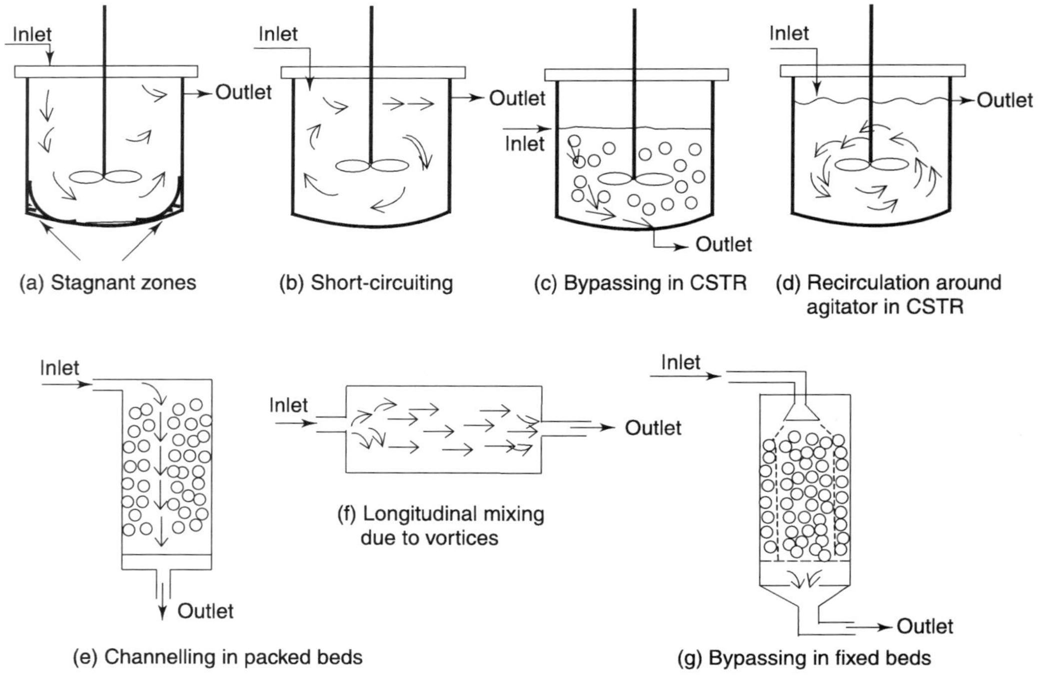

Fig. 10.1 *Non-ideal flow characteristics*

Figure 10.1(b) shows short circuiting. The inlet fluid takes a short-cut route to the outlet, without undergoing conversion in the entire bulk of the reactor. This results in lower conversions. Similar to short-circuiting is bypassing, shown in Fig. 10.1(c) and (g), in which the inlet fluid bypasses the solid (biocatalyst) particles, and reaches the outlet. This results in most of the catalyst particles becoming ineffective, and hence gives less conversions. Channelling is also similar to bypassing. The inlet flow makes a channel through the packed beds of solids (Fig. 10.1(e)), with the result that most of the solid particles do not come in contact with the inlet fluid before it reaches the outlet. Another typical problem which we come across during CSTR operation is that the inlet fluid will be recirculating around the agitator, with the result that this packet of inlet fluid spends more time in the reactor and does not allow another fresh packet of fluid to stay in the reactor for adequate amount of time (Fig. 10.1(d)). In the case of PFR, the entering fluid suffers from longitudinal mixing at the entrance and exit of the reactor (Fig. 10.1(f)), causing longitudinal mixing in the PFR, which leads to non-ideality.

In addition to the above, there could be a number of other fluid behaviours also, which causes non-ideality in the reactors and affects the reactor performance. To find out the extent of non-ideality, we study how the residence time of various packets or particles of incoming fluid in the reactor is varying as compared to that given by Eq. (10.19). These studies are called as **residence time distribution** (RTD) studies (Danckwerts, 1953). The RTD studies have been applied to find out the type and extent of non-ideality prevailing in the reactor. Levenspiel (1974) described the RTD studies for chemical reactions in chemical reactors suffering from non-ideality. However, such studies related to fermentation processes are scarce, and were limited to processes involving yeasts and bacteria (Charles, 1978). Most of the fermentation broths exhibit non-Newtonian behaviour (i.e. the shear stress is not directly proportional to the shear strain), and hence the application of RTD studies to fermenters is complicated due to mixing, mass transport and microbial kinetics, which are once again interrelated. To a great extent, the RTD studies in fermenters have been well-described by Bailey and Ollis (1986).

10.2 RTD STUDIES

Because of non-ideality of flow in the reactors, different packets of the inlet fluid spend different times in the reactor, i.e. their residence times in the reactor are distributed. Some packets may spend less time and come out of the reactor quickly, while some others spend more time and stay for longer periods in the reactor. To know as to what extent the elements are staying in the reactor, we resort to **stimulus–response studies** by introducing a tracer to give stimulus. For this purpose, we choose a proper material as tracer. The tracer should have the following characteristics

- it should not react with the reactants
- it should be easily detectable
- it should not disturb the flow pattern.

Normally, materials like some salts, an acid or a base or a hydrocarbon gas are introduced as a stimulus in any of the following ways (Levenspiel, 1974):

- step input
- pulse input.

The stimulus is given to a reactor and the response is measured on a detector (Fig. 10.2(a)). Now let us see the two types of inputs, and study their responses. In the step input, the tracer is injected in the form of a step, and the input continues for some time (Fig. 10.2(b)). Let us say the time of starting the input as zero. The input continues at the same concentration for some time. On the contrary, in the case of a pulse input, the tracer is injected only once as a pulse (Fig. 10.2(c)). After the injection is over, there will not be any more supply of tracer. Subsequently we can measure the concentration of the tracer in the outlet fluid stream. We draw various types of curves for the response, based on the type of reactor and the flow conditions in the reactor.

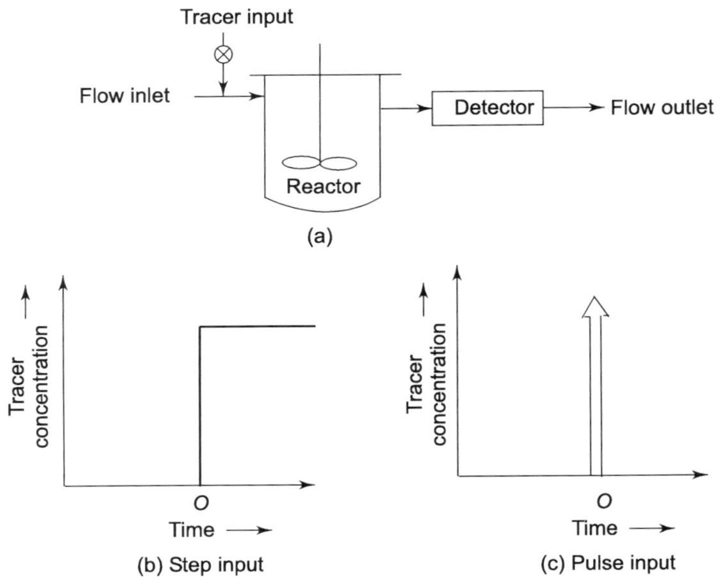

Fig. 10.2 *Stimulus – Response studies on a reactor*

10.2.1 Exit Age Distribution

Different elements of fluid entering the reactor through an inlet will spend different times in the reactor, i.e. they will have different **ages**, which we call as residence times before they leave the reactor. Let Q be the total quantity of the tracer introduced. The tracer spends different time-periods in the vessel (reactor) before it leaves because of non-ideality in flow in the vessel. That means, different fractions of the tracer (hence the fluid) spend different times in the reactor. Let us consider a fluid element fraction which spends a time between t and $t + dt$. If we study all the fractions of the tracer which have spent different times in the reactor, the study is known as 'exit age distribution', which is classically known as function **E**. Now, all the fractions of the fluid which have spent a time between 0 and t is given by

$$\int_0^t \mathbf{E}dt$$

Similarly, all the fractions of the fluid which have spent time between t and ∞ is given by

$$\int_t^\infty \mathbf{E}dt$$

Obviously, the summation of these two fractions will be equal to the total fraction, which is equal to 1.0.

Thus, $$\int_0^t \mathbf{E}dt + \int_t^\infty \mathbf{E}dt = \int_0^\infty \mathbf{E}dt = 1 \tag{10.1}$$

The fraction of material which has an age between t and $t + dt$ will be $\mathbf{E}(dt)$.

10.2.2 F-curve

The **F**-curve is drawn for step input. Initially there will not be any tracer. At a certain point, the tracer is introduced at a concentration of C_0. The concentration of the exit stream is noted, and is divided by the initial concentration, so that the exit concentration is obtained as a fraction of the initial concentration and varies from zero to one. This is drawn against the time, as shown in Fig. 10.3. The *Y*-axis is C/C_0. Since it shows the exit age distribution (fraction of the tracer at time *t* in the exit stream), it is equal to the fraction of the exit stream younger than the age *t*.

$$\int_0^t \mathbf{E}dt \tag{10.2}$$

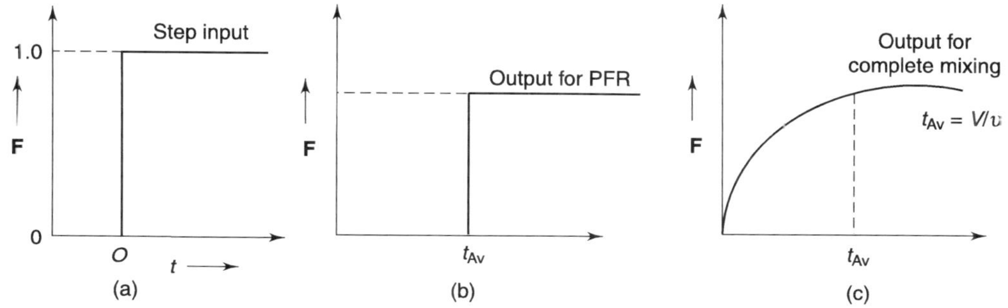

Fig. 10.3 *F-curves: (a) Step input, (b) Output for PFR, (c) Output for CSTR*

The step input (**F**-curve) given to PFR will give out the response after completion of time t_{Av}. Until that time, the material will be in the PFR and the first element of it comes out only after the completion of the space-time (t_{Av}). The response is shown in Fig. 10.3(b). Obviously, in a PFR, the response is similar to the input. Only difference is that the response manifests only after a time of t_{Av} from the time of injection of the stimulus. Similarly in the case of CSTR, all the contents will mix instantaneously, and hence the first element tracer comes out immediately after injection, and the concentration of it slowly builds up. This is shown in Fig. 10.3(c). The response curve reaches the maximum at t_{Av}, and subsequently it continues.

10.2.3 C-curve

The **C**-curve is drawn for pulse input, and is normalised by noting the fractional concentration on the *Y*-axis. Hence,

$$\mathbf{C} = \frac{C}{Q} \tag{10.3}$$

where *C* is the concentration of the tracer in the response at any time, and *Q* is given by Eq. (10.4)

$$\mathbf{Q} = \int_0^\infty C dt \tag{10.4}$$

The response curve for a pulse input is thus equal to the exit age distribution

Thus, $\mathbf{C} = \mathbf{E}$

$$\int_0^\infty \mathbf{C} dt = \int_0^\infty \frac{C}{Q} dt \tag{10.5}$$

The relationship amongst the three types of curves is given as

$$\mathbf{E} = \mathbf{C} = \frac{d\mathbf{F}}{dt} \tag{10.6}$$

When the **C**-curve is drawn for pulse input (Fig. (10.4)) in a PFR, it comes out in the form of a pulse only after a time of t_{Av}. Figure 10.4(b) shows the response curve. The stimulus curve and the response curve are the same, except that the response is manifested only after a time period of t_{Av}; up to that time, there will not be any tracer in the response curve, and all the tracer will come at t_{Av} only. In a CSTR, the concentration of the contents in the reactor will become uniform, the moment the contents enter the reactor. Thus, the maximum concentration of the pulse input is realised in the response curve for CSTR from the time zero onwards (Fig. 10.4(c)), which slowly falls. The situation is contrary to the PFR. The first manifestation of the tracer is realised immediately after the input. **F**– and **C**– curves are drawn for some non-ideal flows in PFR and CSTR (Levenspiel, 1974; Danckwerts, 1953). They are shown in Fig. 10.5. The response curve for an arbitrary flow (non-ideal flow) is intermediate to the flow curves of PFR and CSTR. It will start responding out somewhere in between time 0 and t_{Av}. The curve is somewhat parabolic and is slightly vertical. This shows that the non-ideal flow is in between the flow patterns of PFR and the CSTR (Fig. 10.3(b) and (c)).

As expected, the C-curve for arbitrary flow shows the curve which is intermediate to the response of the PFR and the CSTR (Fig. 10.4(b) and (c)).

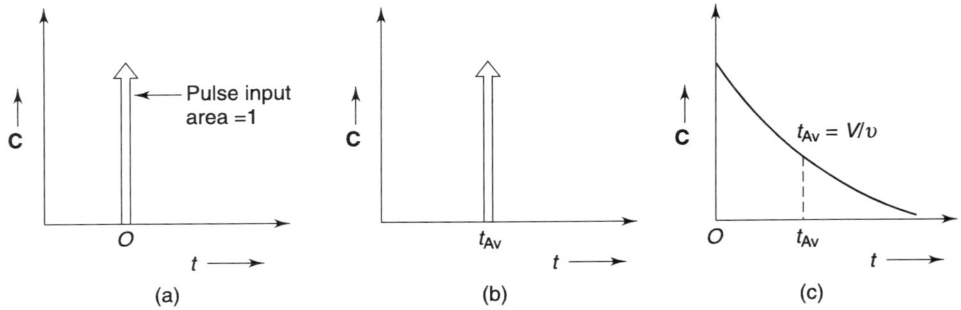

Fig. 10.4 *C – curves (a) Pulse input, (b) Output for PFR, (c) Output for fully mixed reactor*

10.2.4 Mean and Variance of Residence Time

Since different packets of the reactant will stay for different times in the reactor, it is better to know the mean residence time. It is given by the ratio of summation of the fraction of the tracer material multiplied by its stay period in the reactor to the total quantity of the tracer material introduced. It can be expressed mathematically as (Levenspiel, 1974; Lee,1994):

$$t_{Av} = \frac{\int_0^{\infty} tCdt}{\int_0^{\infty} Cdt} = \frac{\sum_{i=1}^{\infty} t_i\, C_i\, \Delta t_i}{\sum_{i=1}^{\infty} C_i\, \Delta t_i} \tag{10.7}$$

The spread of the distribution of residence times over the average residence time (t_{Av}) is defined as the **variance**. Thus,

$$\sigma^2 = \frac{\int_0^{\infty} (t - t_{Av})^2\, Cdt}{\int_0^{\infty} Cdt} = \frac{\sum_{i=1}^{\infty} t_i^2\, C_i\, \Delta t_i}{\sum_{i=1}^{\infty} C_i\, \Delta t_i} \tag{10.8}$$

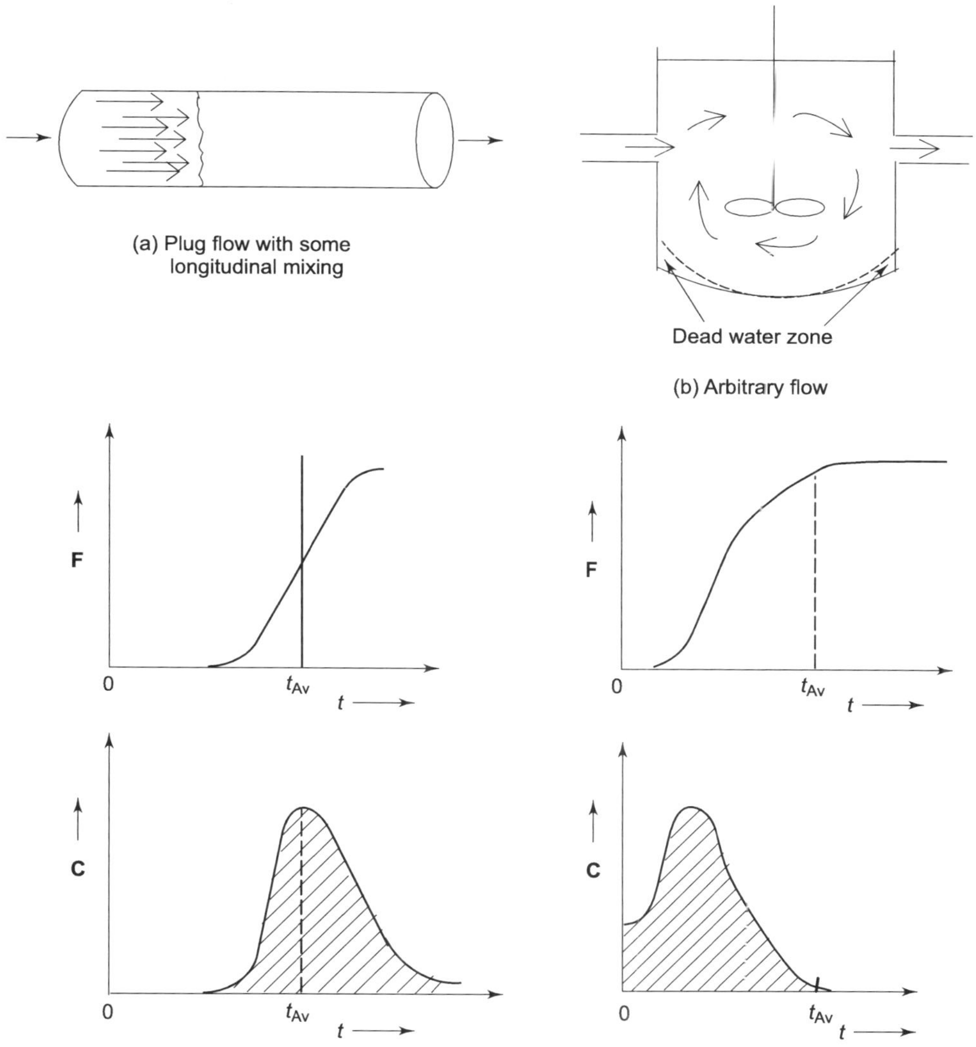

Fig. 10.5 *C and F-curves for different non-ideal types of flow in reactors*
(Reprinted from "Continuous flow systems: Distribution of residence time", Danckwerts, P.V., Chem. Eng. Sci., Vol. 2, pp. 2-3, Copyright © (1953), with permission from Elsevier)

Example 10.1

A particular fermentation is to be carried out in a chemostat. Before carrying out the actual fermentation, it was decided to evaluate the flow characteristics of the chemostat by introducing a tracer in the form of pulse input (delta function). The time versus concentration of the tracer data are presented in Table E10.1. Find the average residence time and plot the exit age distribution E.

Table E10.1 *Time vs concentration data for the tracer.*

Time, min	0	10	20	30	40	50	60	70
Concn. of tracer in the response g/l	0	2	6	7	5	3	1	0

Solution t_{AV} is calculated by application of Eq. (10.7).

$$t_{AV} = \frac{\sum t_i\, C_i\, \Delta t_i}{\sum C_i\, \Delta t_i} = \frac{\sum t_i\, C_i}{\sum C_i}$$

Table E10.2 is prepared using data in Table E10.1.

Table E10.2 *Data for calculating t_{AV}*

t_i (min)	C_i (g/l)	$t_i\, C_i$ (g min/l)	$E = C/Q$
0	0	0	0
10	2	20	8.33×10^{-3}
20	6	120	0.025
30	7	210	0.029
40	5	200	0.021
50	3	150	0.0125
60	1	60	4.16×10^{-3}
70	0	0	
	$\Sigma\, C_i = 24$	$\Sigma t_i\, C_i = 760$	

$$\therefore \quad t_{AV} = \frac{760}{24} = 31.7 \text{ min}$$

The E-curve is drawn using the expression

$$\boldsymbol{E} = \boldsymbol{C} = C/Q$$

in which Q is given by $\Sigma\, C_i\, \Delta t$

$$Q = (\Sigma\, C_i)\, \Delta t$$

and $\Delta t = 10$ min

$\therefore \quad Q = 24 \times 10 = 240$ g-min/l

E data are presented in Table E10.2

From the first and last columns, Fig. E10.1 is drawn to show the E curve.

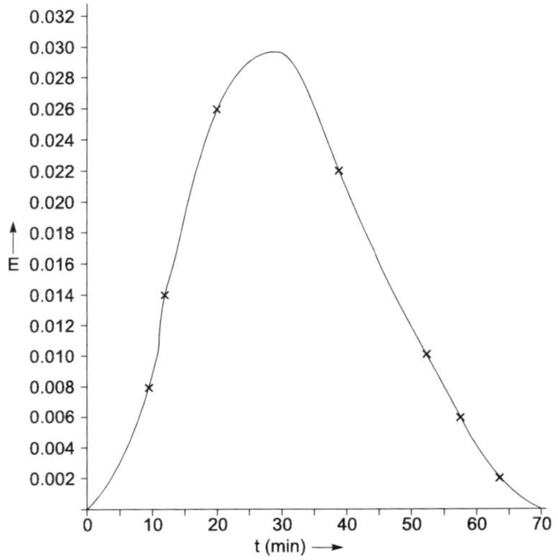

Fig. E10.1 *E-curve for E10.1.*

Example 10.2 *For the data of Table E10.1, find the variance.*

Solution Variance is calculated by Eq. (10.8)
Data required for calculating variance is tabulated in Table E10.3 from Table E10.1

Table E10.3 *Data for calculating σ^2*

t_i	C_i	$t_i^2 C_i$
0	0	0
10	2	200
20	6	2400
30	7	6300
40	5	8000
50	3	7500
60	1	3600
70	0	0
	$\Sigma C_i = 24$	$\Sigma t_i^2 C_i = 28000$

$$\sigma^2 = \frac{\sum t_i^2 C_i}{\sum C_i} - t_{AV}^2 = \frac{28000}{24} - (31.66)^2$$

$$= 1166.7 - 1004.9 = 161.77 \text{ min}^2$$

10.2.5 Conversions Using Tracer Information

By measuring the exit age distribution of the tracer, it is possible to measure the conversion of the reactant, provided the process is linear (Levenspiel, 1974). That is, the exit age distribution of the reactant is assumed to be similar to that of the tracer. Thus, the tracer age distribution (**E**) is same as the outlet concentration distribution of the reactant. The expression for the concentration of the unconverted reactant was given by Levenspiel and Bischoff (1963) as follows:

Average concentration of the reactant in the outlet

$$= \sum \begin{pmatrix} \text{Concentration of the reactant} \\ \text{in the element which has an} \\ \text{age between } t \text{ and } t + dt \end{pmatrix} \begin{pmatrix} \text{Fraction of the exit stream in the} \\ \text{overall stream which has an age} \\ \text{between } t \text{ and } t + dt \end{pmatrix} \tag{10.9}$$

i.e.

$$C_{A.Av} = \int_0^\infty (C_{A.reactant})(\mathbf{E}dt) \tag{10.10}$$

in which $C_{A,reactant}/C_{A0} = e^{-k_1 t}$ for first-order reaction kinetics (Eq. (9.13)).
Thus, the above equation can be written as

$$\frac{C_{A,Av}}{C_{A0}} = \sum e^{-k_1 t_i} (\mathbf{E}_i \, \Delta t) \tag{10.11}$$

The exit age distribution (**E**) of the tracer is used in the above equation.

Example 10.3 *The bioreactor of Example 10.1 is being used for fermentation of molasses which obeys an overall first order reaction kinetics with $k_1 = 0.3\ h^{-1}$. Find the fractional conversion of the reactant.*

Solution

$$k_1 = 0.3\ \text{h}^{-1} = 5 \times 10^{-3}\ \text{min}$$

$$\frac{C_{AO} - C_A}{C_{AO}} = X_A$$

$$\therefore \quad \frac{C_A}{C_{AO}} = 1 - X_A$$

C_A/C_{AO} can be evaluated by Eq. (10.11)

The data required for above is given in Table E10.4

Table E10.4 *Data for calculation of C_A/C_{AO}*

t_i	C_i	E	k_1t_i	$e^{-k_1t_i}$	$e^{-k_1t_i}\,E\,\Delta t$
0	0	–	0	1.0	
10	2	0.0083	0.05	0.9513	0.079
20	6	0.025	0.1	0.905	0.2262
30	7	0.0292	0.15	0.86	0.2513
40	5	0.0208	0.20	0.818	0.1703
50	3	0.0125	0.25	0.779	0.0974
60	1	0.0042	0.30	0.741	0.0311
Total					0.8552

$$\frac{C_A}{C_{AO}} = \sum e^{-k_1t_i} E\,(\Delta t) = 0.8552$$

$$\therefore \quad X_A = 1 - C_A/C_{AO} = 1 - 0.8552 = 0.144$$

$$= 14.4\%$$

Comparison with a plug flow reactor

On integration of Eq. (8.32 (a)), and substituting

$-r_A = k_1 C_A$ for first-order reaction, we get

$$\tau = -\int_{C_{AO}}^{C_A} \frac{dC_A}{(-r_A)} = -\int_{C_{AO}}^{C_A} \frac{dC_A}{(k_1C_A)}$$

or

$$\int_{C_{AO}}^{C_A} \frac{dC_A}{(C_A)} = -k_1\tau$$

$$\ln \frac{C_A}{C_{AO}} = -k_1\tau$$

$$\therefore \quad C_A/C_{AO} = e^{-k_1\tau}$$

$$= e^{-(5 \times 10-3)(31.66)} = 0.854$$

There is not much difference which indicates that the reactor is behaving almost ideally.

10.3 DIAGNOSIS OF ILLS OF FLOW REACTORS

A reactor which is having some inconsistencies in its flow systems will behave non-ideally. It could be by the way of having some stagnant zones, bypassing, and/or internal recirculation, etc. This can be found out by carrying out the pathological tests on the reactors in the form of tracer studies. The stimulus response data will give us some clues to know what is happening in the reactor, and accordingly corrective measures can be initiated. For example, the diagnosis of a PFR can be carried out by noting down its curve by giving a pulse input. The details are shown schematically (Levenspiel, 1974) in Fig. 10.6. Figure 10.6(a) is for a perfect PFR in which we find that the response curve for the pulse input will also come out in the form of a pulse at t_{Av}. Any deviations from it can indicate the prevalence of non-ideal flow conditions in the reactor. For example, Fig. 10.6(b) indicates the pulse response earlier than t_{Av}, which shows that there is some amount of short circuiting or bypassing. Similarly, Fig. 10.6(c) shows multiple pulses indicating the probability of existence of internal recirculation. A delayed response (which is delayed more than t_{Av}) also indicates that the fluid elements are spending more time in the reactor than t_{Av} which results in ineffective utilisation of the reactor volume. Figure 10.6(d) shows the curve. Figure 10.6(e) shows two pulse stimuli. This is an indication of existence of two parallel paths, out of which one is coming earlier than t_{Av}, and the other later than t_{Av}.

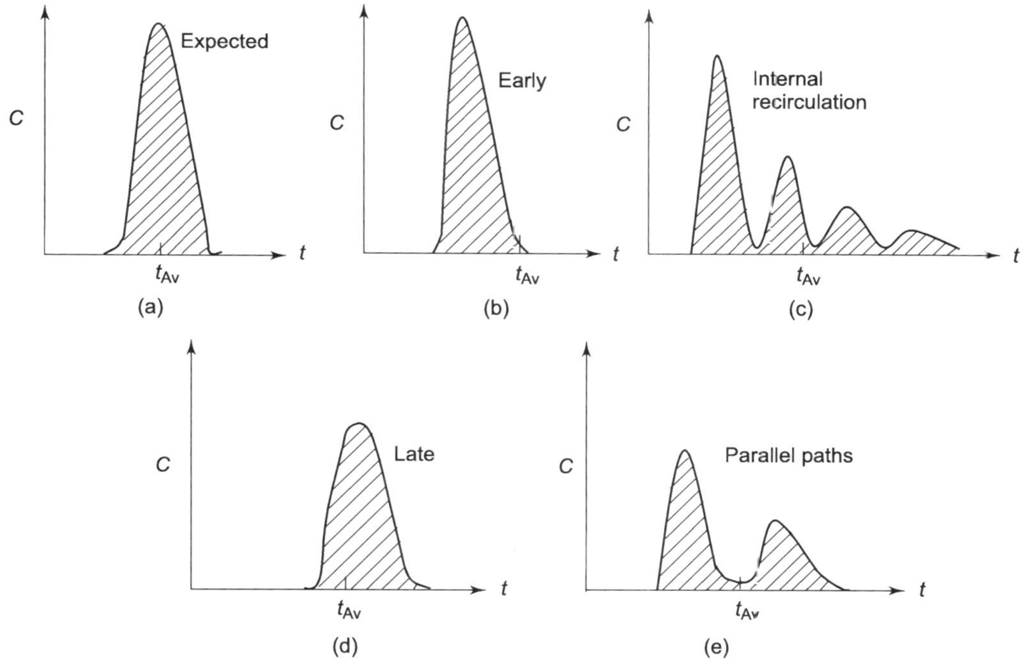

Fig. 10.6 *Pathological tests in a PFR by noting exit concentration of tracer (Reprinted from "Chemical Reaction Engineering", Levenspiel, O., 2nd Edn., Ch. 9, p. 305, Copyright © (1974), with permission from John Wiley & Sons, Inc.)*

Similar diagnostic tests can be carried out on a fully mixed flow reactor (CSTR) with a pulse input. Figure 10.7 shows five curves along with the expected one. The response for a perfect CSTR is shown in Fig. 10.7(a) in which the response curve starts with the maximum value and gradually falls with the mean value arriving at t_{Av}.

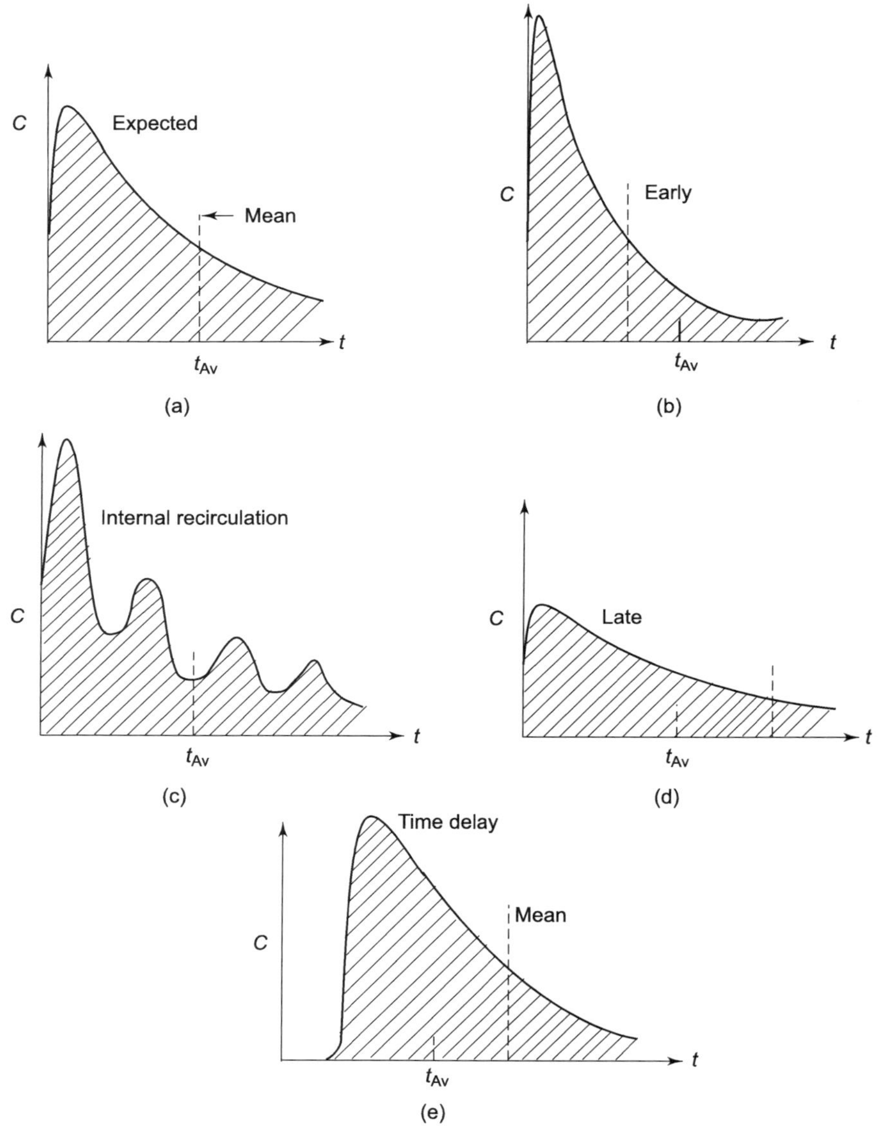

Fig. 10.7 *Pathological tests for a CSTR by noting the exit concentration of tracer (Reprinted from "Chemical Reaction Engineering", Levenspiel, O., 2nd Edn., Ch. 9, p. 306, Copyright © (1974), with permission from John Wiley & Sons, Inc.)*

If any stagnant zones are present, the entire volume of the reactor is not effectively used, with the result that the residence time is reduced; and hence the tracer starts coming out earlier than expected. It is shown in Fig. 10.7(b). Thus, early coming out of the tracer is an indication of stagnant zones, bypassing or channelling (of course, channelling does not take place in CSTR). If there is any internal recirculation, the tracer concentration will have more than one peak, as shown in Fig. 10.7(c). The tracer may also exit later than expected (Fig. 10.7(d)). This happens if any elements of fluid spend time more than t_{Av} because of internal circulation of fluid elements around the agitator, as shown in Fig. 10.1(d). There is also one more possibility of late exit of the tracer, as shown in Fig. 10.7(e). This generally occurs because of shift in time, which probably indicates an error in instrumentation. It does not warrant any corrective action to the reactor, except correcting the measuring instruments. It is similar to Fig. 10.7(a), except that there is some lag, and hence shifts the mean away from t_{Av}.

Thus, the RTD studies give us a clue to identify the ills of the reactor. However, the corrective steps are to be incorporated in the form of changes to hydrodynamic system. The results of the RTD studies were concluded by Nauman (1987) as follows:

Residence time measurements are a diagnostic aid in such cases, but the actual solution will probably require a hydrodynamic analysis.

10.4 MODELLING OF NON-IDEAL BEHAVIOUR

The non-ideal flow behaviour has been conveniently modelled on the basis of the dispersion characteristics. Various models are available in literature (Levenspiel, 1974), out of which the following two models are very prominent and will be discussed here:

(i) Dispersion model
(ii) Tanks-in-series model.

The **dispersion model** is applicable to a plug flow reactor system, whereas the **tanks in-series model** builds a chain or network of ideal mixers to approach the performance of a PFR.

10.4.1 Dispersion Model

This model is based on the assumption that there are

(i) no stagnant zones
(ii) no gross bypassing
(iii) no short circuiting

The above assumptions clearly indicate that the model does not fit for CSTR. The non-ideality is mainly due to certain amount of back mixing or intermixing in the plug flow reactor.

In any transport process, certain amount of mixing takes place due to the disturbance in flow conditions between different layers. This kind of intermixing by molecular movement is called **molecular diffusion** and is governed by Fick's law, which is given by Eq. (10.12) in X-direction:

$$\frac{\partial C}{\partial t} = \text{Đ} \frac{\partial^2 C}{\partial X^2} \tag{10.12}$$

where Đ is the diffusivity coefficient.

In a similar manner, the intermixing or back mixing can also be modelled as follows:

$$\frac{\partial C}{\partial t} = D_e \frac{\partial^2 C}{\partial X^2} \tag{10.13}$$

where D_e is the 'axial dispersion coefficient', and it represents the axial mixing due to velocity distribution (Fig. 10.5(a)). Now we define two dimensionless parameters for length and time as follows:

$$z = \frac{X}{L} \tag{10.14}$$

$$\theta = \frac{t}{t_{Av}} = t\frac{u}{L} \tag{10.15}$$

where z is the dimensionless length, or in other words, it is the fractional length in the PFR; θ is a dimensionless time parameter and is a fraction of the average residence time (which is obtained by dividing the length of the PFR with the linear fluid velocity u).

Substituting Eqs. (10.14) and (10.15) in Eq. (10.13), we get

$$\frac{\partial C}{\partial \theta} = \left(\frac{D_e}{uL}\right)\frac{\partial^2 C}{\partial Z^2} - \frac{\partial C}{\partial Z} \tag{10.16}$$

In the above equation, D_e/uL is a dimensionless number and is called as the 'vessel dispersion number'. It is a measure of the extent of axial dispersion.

If $D_e/uL \rightarrow 0$, the dispersion is negligible, and hence the flow is plug flow.

If $D_e/UL \rightarrow \infty$, the dispersion is very large, and hence the flow is a mixed flow. The above model suitably predicts the performance of packed bed reactors (like immobilized bio-reactors), in which the flow does not deviate too much from plug flow).

$1/(D_e/uL)$ is a form of dimensionless number, called as **Pecklet number** (N_{Pe}) which is defined as

$$N_{Pe} = \frac{\text{Rate of transport by convection}}{\text{Rate of transport by diffusion or dispersion}}$$

$$= \frac{uL}{D_e} \tag{10.17}$$

We also define one more dimensionless number called **Damkohler number** (N_{Da}), which is a ratio of the consumption of reactant A by chemical reaction and the transport of reactant A by convection. Thus,

$$N_{Da} = \frac{\text{Rate of consumption of A by reaction}}{\text{Rate of transport of A by convection}} \tag{10.18}$$

$$= k_1\, t_{Av} \text{ (for first-order reaction)}$$

$$= k_1 \frac{L}{u} \tag{10.19}$$

We also define here a dimensionless variance:

$$\sigma_\theta^2 = \frac{\sigma^2}{t_{Av}^2} \tag{10.20}$$

The relationship of σ_θ^2 with N_{Pe}, which will help us find the conversion using the dispersion model, is as follows (Fogler, 1986):

$$\sigma_\theta^2 = \frac{2}{N_{Pe}} - \frac{2}{N_{Pe}^2}\left(1 - e^{-N_{Pe}}\right) \tag{10.21}$$

By analysing the tracer response information, we will be able to calculate t_{Av} and σ^2. Using Eq. (10.21), N_{Pe} can be calculated by trial-and-error. The procedure is described with a numerical example in the Appendix.

Example 10.4

A tubular bioreactor is used for carrying out a particular fermentation process to evaluate the non ideal characteristics of the PFR. Tracer studies were carried out with pulse input, and the data are presented in Table E10.5. Find the dispersion coefficient of the PFR for a linear flow rate of substrate of 0.75 m/min and the length of the reactor is 10 m.

Table E10.5 *Time vs concentration data for E10.4*

Time	0	12	24	36	48	60	72	84
Concentration g/l	0	2.5	5.2	6.5	4.0	3.1	1.0	0

Solution The dispersion number is calculated by trial and error method using Eq. (10.17) and Eq. (10.21) with its approximations if possible.

The required data are calculated and presented in Table E10.6

Table E10.6 *Data for calculation of dispersion number.*

t_i (min)	C_i (g/l)	$t_i C_i$	$t_i^2 C_i$
0	0	0	0
12	2.5	30	360
24	5.2	124.8	2995.2
36	6.5	234	8424
48	4.0	192	9216
60	3.1	186	11160
72	1.0	72	5184
84	0	0	
Total	22.3	838.8	37339.2

$$t_{AV} = \frac{\sum t_i C_i}{\sum C_i} = \frac{838}{22.3} = 37.6 \text{ min}$$

$$\frac{\sum t_i^2 C_i}{\sum C_i} = \frac{37339.2}{22.3} = 1674 \text{ min}^2$$

$$\sigma^2 = \frac{\sum t_i^2 C_i}{\sum C_i} - t_{AV}^2 = 1674 - (37.6)^2 = 259 \text{ min}^2$$

σ_θ^2 is given by Eq. (10.20)

$$\sigma_\theta^2 = \frac{\sigma^2}{t_{AV}^2} = \frac{259}{(37.6)^2} = 0.183$$

By Eq. (10.21), we have

$$\sigma_\theta^2 = 2\frac{(De)}{uL} - 2\frac{(De)^2}{uL}\left(1 - e^{-uL/De}\right)$$

First approximation,

$$\sigma_\theta^2 = 2\,(De/uL)$$

$$\frac{De}{uL} = 0.0916$$

$$1 - \exp^{(-1/0.0916)} = 1 - 1.8095 \times 10^{-5} = 0.99998$$
$$\approx 1.0$$

The above value may be taken as 1.0

Hence Eq. (10.21) may be written as

$$\sigma_\theta^2 = 2\,(De/uL) - 2\,(De/uL)^2$$

Now we tabulate (De/uL) vs σ_θ^2 for different values of (De/uL) as follows:

De/uL	$\sigma_\theta^2 = 2\,(De/uL) - 2\,(De/uL)^2$
0.0916	0.1664
0.10	0.18
0.11	0.1958
0.102	0.1832

$$\frac{De}{uL} = 0.102 \text{ is matching with } \sigma_\theta^2$$

$$\therefore \quad \frac{De}{uL} = 0.102$$

$$\therefore \quad De = 0.102 \times 75 = 0.765 \text{ m}^2/\text{min} = 0.0127 \text{ m}^2/\text{s}$$

Estimating the conversion using dispersion model The conversion in a tubular reactor with dispersion can be calculated by applying the mass balance equation to a differential length of the reactor, and integrating the resulting differential equation. The mass balance equation should account for the mass flow in the inlet and outlet flows of the reactor, mass flow by dispersion and material disappearing by chemical reaction. However, the material accumulated will be equal to zero under steady state conditions. The details of the derivation are available in Levenspiel (1974) textbook. The final equation in one direction (X-direction) given by Danckwerts (1953) for a first-order reaction kinetics is as follows:

$$\frac{d^2 C_A}{dX^2} - \frac{u}{D_e}\frac{dC_A}{dX} - \frac{k_1 C_A}{D_e} = 0 \tag{10.22}$$

By replacing $z = X/L$, and multiplying with (D_e/u), we get

$$\frac{D_e}{uL}\frac{d^2 C_A}{Ldz^2} - \frac{1}{L}\frac{dC_A}{dz} - \frac{k_1 C_A}{u} = 0$$

Multiplying throughout by L and noting $L/u = t_{Av}$, we get

$$\frac{D_e}{uL}\frac{d^2 C_A}{dz^2} - \frac{dC_A}{dz} - k_1 C_A t_{Av} = 0 \tag{10.23}$$

Equation (10.23) is similar to Eq. (10.16), in which θ is replaced by t/t_{Av}, and dC_A/dt is taken as the rate of conversion and is replaced by k_1C_A for a first-order kinetics. Since only axial movement is considered, the partial derivatives are replaced by ordinary derivatives.

Equation (10.23) can be written as follows, in terms of fractional conversion, X_A, by noting down that $1-X_A = C_A/C_{A0}$,

$$\frac{D_e}{uL}\frac{d^2 X_A}{dz^2} - \frac{dX_A}{dz} + k_1 t_{Av}(1 - X_A) = 0 \tag{10.24}$$

The analytical solution to the above equation was given by Wehner and Wilhelm (1956) as follows

$$\frac{C_A}{C_{A0}} = (1 - X_A)$$

$$= \frac{4q \exp(N_{Pe}/2)}{(1+q)^2 \exp(N_{Pe}q/2) - (1-q)^2 \exp(-N_{Pe}q/2)} \quad (10.25)$$

where

$$q = (1 + 4(N_{Da}/N_{Pe}))^{1/2} \quad (10.26)$$

10.4.2 Tanks-in-Series Model

In this model, we view the reaction system to be consisting of a number (N) of equal-sized ideal stirred tanks (Fig. 10.8). It is also a one-parameter model with N, the number of ideal reactors. The exit age distribution data is related to N as follows (Levenspiel, 1974):

$$t_{i,\text{Av}}\,\mathbf{E} = \left(\frac{t}{t_{i,\text{Av}}}\right)^{N-1} \frac{1}{(N-1)!} e^{-t/t_{i,\text{Av}}} \quad (10.27)$$

in which $t_{i,\text{Av}}$ is the mean residence time of the ith tank. The total average residence time for all N tanks is

$$t_{\text{Av}} = N\, t_{i,\text{Av}}$$

Similarly,

$$\theta_i = t/t_{i,\text{Av}}$$

and

$$\theta = \frac{t}{t_{\text{Av}}} = \frac{t}{N t_{i,\text{Av}}} \quad (10.28)$$

$$\sigma_\theta^2 = \frac{\sigma^2}{t_{\text{Av}}^2} = 1/N \quad (10.29)$$

and

$$\mathbf{E}_\theta = t_{\text{Av}}\,\mathbf{E} \quad (10.30)$$

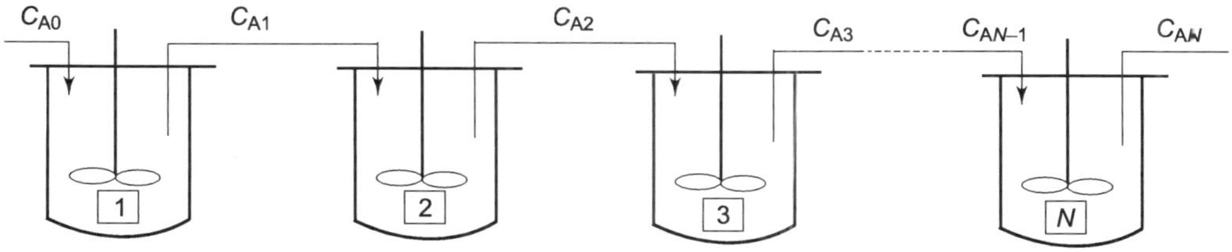

Fig. 10.8 *Tanks-in-series model representation with N number of equal-sized fully mixed tanks*

Estimating the conversion using tanks-in-series model The conversion in one CSTR for a first-order reaction is given by Eq. (8.27):

$$\frac{C_A}{C_{A0}} = \frac{1}{1 + k_1 t_{\text{Av}}}$$

or

$$\frac{C_{A0}}{C_A} = 1 + k_1 t_{\text{Av}} \quad (10.31)$$

If there are N reactors in series, the outlet of the first reactor will be the inlet for the second reactor, and so on (Fig. 10.8).

The conversion in the first reactor is given by

$$\frac{C_{A0}}{C_{A1}} = 1 + k_1\, t_{1,\,Av} \tag{10.32}$$

Similarly, for the second reactor,

$$\frac{C_{A1}}{C_{A2}} = 1 + k_1\, t_{2,\,Av} \tag{10.33}$$

and for the Nth reactor,

$$\frac{C_{AN-1}}{C_N} = 1 + k_1\, t_{N,\,Av} \tag{10.34}$$

$$\frac{C_{A0}}{C_{AN}} = \frac{C_{A0}}{C_{A1}} \times \frac{C_{A1}}{C_{A2}} \times \cdots \times \frac{C_{AN-1}}{C_{AN}} \tag{10.35}$$

$C_{AN} = C_A$ = Final outlet concentration.
Thus, Eq. (10.35) becomes

$$\frac{C_{A0}}{C_A} = \left(1 + k_1 t_{1,Av}\right)\left(1 + k_1 t_{2,Av}\right)\cdots\left(1 + k_1 t_{N,Av}\right) \tag{10.36}$$

Since all the reactors are of equal volumes and the flow rates are same (because it is assumed that there is no change in density), hence

$$t_{1,\,Av} = t_{2,\,Av} = t_{N,\,Av} = \frac{V}{\upsilon} \tag{10.37}$$

where V is the volume of each reactor and υ the volumetric flow rate. The overall space-time is equal to the summation of individual space-times.
Thus,

$$t_{Av} = \sum_{i=1}^{N} t_{i,Av} = N t_{i,Av} \tag{10.38}$$

Combining Eqs. (10.36) and (10.38), we get

$$\frac{C_{A0}}{C_A} = \left\{1 + k_1 \left(t_{Av}/\mathrm{N}\right)\right\}^N \tag{10.39}$$

$$\frac{C_A}{C_{A0}} = 1 - X_A = \frac{1}{\left\{1 + k_1 \left(t_{Av}/N\right)\right\}^N}$$

$$X_A = 1 - \frac{1}{\left\{1 + k_1 \left(t_{AV}/N\right)\right\}^N} \tag{10.40}$$

The application of the tanks-in-series model is shown in the Appendix with a numerical example (Fogler, 1986).

Symbols

C	: concentration (kg mol/m^3 or kg/m^3)
$\mathbf{C}$	: pulse input response curve [Eq. (10.5)]
D_e	: dispersion coefficient (m^2/s)

E	: exit age distribution function
F	: step input response curve [Eq. (10.4)]
k_1	: first-order reaction rate constant (1/s)
L	: length (m)
N	: number of reactors in series
N_{Da}	: Damkohler number [Eq. (10.18)]
N_{Pe}	: Pecklet number [Eq. (10.17)]
q	: parameter [Eq. (10.26)]
Q	: total quantity of tracer [Eq. (10.4)]
t	: time (s or h)
u	: linear velocity (m/s)
υ	: volumetric flow rate (m^3/s or m^3/h)
V	: reactor volume (m^3)
X	: distance in X-direction (m)
X	: conversion
z	: dimensionless length [Eq. (10.14)]

Subscripts

0	: initial conditions
1, 2, ..., N	: 1, 2, …, N^{th} tank in series
A	: component A
Av	: average
i	: component i

Greek Letters

τ	: residence time (s)
σ^2	: variance [Eq. (10.8)]
σ_θ^2	: dimensionless variance [Eq. (10.20)]
θ	: dimensionless time [Eq. (10.15)]
$Đ$	: diffusivity coefficient (m^2/s)

References

10.1 Bailey, J.E. and Ollis, D.F. (1986), *Biochemical Engineering Fundamentals*, 2nd Edn., Mc-Graw-Hill, Inc., New York, pp. 551–586.

10.2 Charles, M. (1978), "Technical aspects of the rheological properties of microbial cultures", *Advances in Biochemical Engineering,* Eds. Ghose, T.K., Fiechter, A., and Blakebrough, N., Springer-Verlag, Berlin, pp. 1–62.

10.3 Danckwerts, P.V. (1953), "Continuous flow systems: Distribution of residence times", *Chem. Eng. Sci.*, 2, 1–13.

10.4 Fogler, H.S. (1986). *Elements of Chemical Reaction Engineering,* Prentice-Hall, Englewood Cliff, New Jersey, Ch. 14, pp. 680–706.

10.5 Levenspiel, O., and Bischoff, K.B. (1963). "Patterns of flow in chemical process vessels", *Advances in Chemical Engineering,* Drew, T.B., Hoopes, Jr., J.W., and Vermeulen, T. (Eds.), Vol. 4. Academic Press, New York, pp. 95–198.

10.6 Levenspiel, O. (1974), *Chemical Reaction Engineering,* Wiley Eastern Pvt. Ltd., New Delhi, Ch.9, pp. 253–308.
10.7 Lee, J.C. (1994), "Flow characteristics of reactors-flow modelling", *Coulson and Richardson's Chemical Engineering*, Richardson, J.F, and Peacock, D.G. (Eds.), Pergamon Press, Oxford, Ch. 2, pp. 71–107.
10.8 Nauman, E.B. (1987), *Chemical Reactor Design*, John Wiley & Sons, New York, Ch. 9, pp. 255–289.
10.9 Wehner, J.F., and Wilhelm, R.H. (1956), "Boundary conditions for flow reactors", *Chem. Eng. Sci.*, 6, 89–93.

Review Questions

10.1 What are the reasons for non-ideality in flow reactors?
10.2 Describe the various non-ideal flow behaviour characteristics in bioreactors with neat diagrams.
10.3 Define residence time in a flow bioreactor.
10.4 What is meant by residence time distribution (RTD)?
10.5 What are the characteristics of a tracer? How can it be used in RTD studies?
10.6 Describe exit age distribution in a non-ideal flow bioreactor.
10.7 Describe **F**-curve and **C**-curve in RTD studies of non-ideal flow bioreactor.
10.8 Define 'mean' and 'variance' for RTD studies.
10.9 Describe systematically how do you calculate the conversion based on the information of tracer studies.
10.10 Describe how the ills of a flow bioreactor can be diagnosed using tracer studies. Substantiate your answer with neat diagrams.
10.11 Describe the two models available for predicting the non-ideal flow behaviour in bioreactors.
10.12 What is meant by a "dispersion number"? What is its importance in non-ideal flow bioreactors?
10.13 Define the following:
  (i) Dispersion coefficient
  (ii) Dispersion number
  (iii) Pecklet number
  (iv) Damkohler number.
10.14 Write the expression for σ_θ^2 (dimensionless variance) in terms of Pecklet number.
10.15 Describe how do you estimate the conversions in a bioreactor with non-ideal flow behaviour using dispersion model.
10.16 Describe how do you estimate the conversions in a bioreactor with non-ideal flow behaviour using tanks-in-series model.

Appendix to Chapter 10

PROCEDURE FOR CALCULATING CONVERSIONS USING DISPERSION MODEL, TANK-IN-SERIES MODEL AND IDEAL FLOWS

A first-order reaction A $\xrightarrow{k_1}$ B was carried out in a tubular reactor (length 6.4 m and diameter 0.1 m), for which $k_1 = 0.25\ \text{min}^{-1}$. The results of the tracer test are given in Table 10A.1.

Table 10A.1 *Tracer output data*

t (min)	0	2	4	6	8	10	12	14
C (mg/l)	0	5	10	6	3	1.5	0.6	0

The data in Table 10A.1 (Fogler, 1986) is used to calculate the conversions by (i) dispersion model and (ii) tanks-in-series model, and compared with a PFR or CSTR.

Solution From the data in Table 10A.1, we draw the **E**-curve

$$\mathbf{E} = \frac{C}{Q}$$

where $$Q = \int_0^\infty C\,dt \qquad \text{[by Eq. (10.4)]}$$

or $$Q = \sum C(\Delta t)$$

$$= (0 + 5 + 10 + 6 + 3 + 1.5 + 0.6 + 0) \times 2 = 52.2 \text{ mg min/l}$$

$$\mathbf{E} = \mathbf{C} = \frac{C}{Q}$$

Average residence time, t_{Av}, is given by Eq. (10.7). The derived data are tabulated in Table 10A.2
Thus,

$$t_{Av} = \frac{\sum t_i C_i (\Delta t_i)}{\sum C_i (\Delta t_i)}$$

$$= \frac{\sum t_i C_i}{\sum C_i} = \frac{132.2}{26.1} = 5.06 \text{ min}$$

$$\sigma^2 = \frac{\sum t_i^2 C_i \Delta t_i}{\sum C_i \Delta t_i} - t_{Av}^2$$

$$= \frac{\sum t_i^2 C_i}{\sum C_i} - t_{Av}^2 = \frac{824.4}{26.1} - (5.06)^2$$

$$= \frac{824.4}{26.1} - 25.06 = 5.93 \text{ min}^2$$

σ_θ^2 is given by Eq. (10.20)

$$\sigma_\theta^2 = \frac{\sigma^2}{t_{\text{Av}}^2} = \frac{5.93}{(5.06)^2} = 0.2316$$

The Pecklet number, N_{Pe}, is calculated by Eq. (10.21) by trial-and-error by initially assuming a value for N_{Pe} in Eq. (10.21)

$$\sigma_\theta^2 = \frac{2}{N_{\text{Pe}}} - \frac{2}{N_{\text{Pe}}^2}\left(1 - e^{-N_{\text{Pe}}}\right)$$

For a first guess of N_{Pe}, we neglect the higher terms on RHS of Eq. (10.21).

$$0.2316 \cong \frac{2}{N_{\text{Pe}}}$$

$$\therefore \quad N_{\text{Pe}} = \frac{2}{0.2316} = 8.63$$

when $N_{\text{Pe}} = 8.63$, $(1 - e^{-N_{\text{Pe}}}) = 0.99982$

when $N_{\text{Pe}} = 7.5$, $(1 - e^{-N_{\text{Pe}}}) = 0.99944$

Hence, this term can be neglected in further calculations and may be taken as being equal to 1.

TAble 10A.2

t_i	C_i	$C = E = C_i/Q$	t_iC_i	$t_i^2C_i$	k_1t_i	$e^{-k_1t_i}$	$Ee^{-k_1t_i}\Delta t$
2	5	0.095	10	20	0.5	0.6065	0.1152
4	10	0.1915	40	160	1.0	0.3678	0.1408
6	6	0.115	36	216	1.5	0.2231	0.0512
8	3	0.0575	24	192	2.0	0.1353	0.01556
10	1.5	0.0287	15	150	2.5	0.082	0.0047
12	0.6	0.0115	7.2	86.4	3.0	0.05	0.01145
14	0	0	0	0	0		
Total	26.1		132.2	824.4	10.5		0.3389

$$\sum E\, e^{-k_1 t}\Delta t = 0.339$$

Table 10A.3 *Trial-and-error calculation of N_{Pe}*

N_{Pe}	σ_θ^2
8.63	0.2048
8.2	0.2141
8.0	
7.8	0.2235
7.6	0.2285
7.5	0.2337
7.4	0.2311
7.48	0.2316

Conversion based on dispersion model From Table 10A.3, $N_{Pe} = 7.48$
Now we apply the dispersion model equation, i.e. Eq. (10.25) to calculate the conversion X_A.

By Eq. (10.26)

$$q = \{1 + 4\,(N_{Da}/N_{Pe})\}^{1/2}$$

where $N_{Da} = k_1 t_{Av}$.

t_{Av} is found by Eq. (10.7).

$N_{Da} = k_1 t_{Av} = 0.25 \times 5.065 = 1.266$

$$q = \{1 + 4(1.266/7.48)\}^{1/2} = (1.677)^{1/2} = 1.295$$

$$\frac{N_{Pe} \times q}{2} = \frac{7.48 \times 1.295}{2} = 4.843$$

From Eq. (10.25)

$$1 - X_A = \frac{4 \times 1.295 \exp(7.48/2)}{(1 + 1.295)^2 \exp(4.843) - (1 - 1.295)^2 \exp(-4.843)}$$

$$= 0.326$$

$$X_A = 0.673 = 67.3\%$$

$X_A = 0.673 = 67.3\%$

Conversion by dispersion model = 67.3%.

Conversion based on tracer information By Eq. (10.11), we have

$$\frac{C_{Av}}{C_{A0}} = \int_0^\infty e^{-k_1 t}\, \mathbf{E} dt = 1 - X_A$$

$$= \sum \left(e^{-k_1 t_i}\, E_i \Delta t\right)$$

From Table 10A.2, we have

$$\sum e^{-k_1 t}\, \mathbf{E}\Delta t = 0.339$$

$\therefore$ $1 - X_A = 0.339$

or $X_A = 1 - 0.339 = 0.661 = 66.1\%$

Conversion based on tracer information is 66.1%.

Conversion based on tanks-in-series model The number of tanks-in-series, N, can be found by Eq. (10.29)

$$\sigma_\theta^2 = 1/N$$

or

$$N = \frac{1}{\sigma_\theta^2} = \frac{1}{0.2316} = 4.32$$

By applying Eq. (10.39)

$$1 - X_A = \frac{1}{(1 + k_1 t_{Av}/N)^N} = \frac{1}{(1 + 1.266/4.32)^{4.32}} = 0.329$$

$\therefore$ $X_A = 1 - 0.329 = 0.6705 = 67.05\%$

Conversion based on tanks-in-series model = 67.05%

Conversion based on PFR model

$$\frac{C_A}{C_{A0}} = 1 - X_A = e^{-k_1 t_{Av}}$$

$$= \exp(-1.266) = 0.2819$$

$$\therefore \quad X_A = 1 - 0.2819 = 0.718 = 71.8\%$$

Conversion based on CSTR

$$\frac{C_{Av}}{C_{A0}} = \frac{1}{1 + k_1 t_{Av}} = \frac{1}{1 + 1.266} = 0.4413$$

$$\therefore \quad X_A = 1 - 0.4413 = 0.559 = 55.9\%$$

i.e., 55.9% conversion would have been achieved in a single CSTR.

In a nut shell, the results may be summarised as given in Table 10A.4.

Table 10A.4 *Results*

Type of model	X_A *(% conversion)*
PFR	71.8
Tracer studies	66.1
Dispersion model	67.3
Tanks-in-series model	67.05
CSTR	55.9

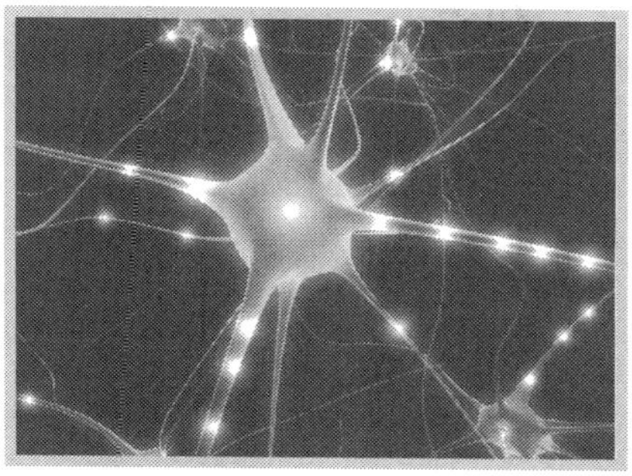

chapter 11

Rheology and Mixing in Fermentation Broths

Transportation or movement of fluids, known as fluid mechanics, plays an important role in chemical engineering and biochemical engineering; and hence abundant literature information is available on it. Before we proceed with the subject matter, let us see what a fluid is, what are its characteristics, and how it is different from the solids.

11.1 NATURE OF FLUIDS AND THEIR CLASSIFICATION

Both gases and liquids are known as fluids. Fluid is a substance that is amenable to distortion, and hence takes the shape of the container; contrary to the characteristic feature of solids, which are rigid and resist distortion. Fluids are flowable in nature. They flow freely in open areas, ducts, conduits or pipes of any cross-section whenever there is a pressure differential. Solids do not flow on their own. They only slide if there is a potential gradience.

At a given temperature and pressure, fluids possess a definite mass per unit volume, which is defined as 'density'. However, a change in temperature or pressure causes the density to change. For some fluids, the change in density is very minimal, and such fluids are called **incompressible fluids**; whereas, for some other fluids, change in density is significant with change in pressure, and such fluids are called **compressible fluids**. Mostly, the gases are compressible, whereas the liquids are incompressible in nature.

11.2 BOUNDARY LAYER CONCEPT

When the fluids are flowing through the pipes or conduits, they suffer some loss of energy (usually known as pressure drop) due to:

(i) friction at the surface of the walls

(ii) friction amongst different layers of the fluids, which is known as **viscosity**.

It is very interesting to note the fluid flow phenomenon in pipes or conduits or on any surface. The fluid layer which is first adhering to the surface will have the velocity of the surface. That means, if the surface is stationary (usually the surface of the pipe walls through which the fluid flows is invariably stationary), the surface layer adhering to the surface of the pipe is also stationary. Layer next to it will have a slightly higher velocity and thus the velocity increases. After certain layers are over, the layers will attain the constant velocity of the fluid. This is schematically represented in Fig. 11.1. The flow profiles on a stagnant flat plate are shown in Fig. 11.2. The layer AA′ will be having zero velocity and is stagnant like the flat plate. AA′ is very thin, and may be of the order of a few microns (the thickness of AA′ is exaggerated in Fig. 11.2 for the purpose of comprehension). The subsequent layers will have increased velocities up to B. The layer beyond BC will have the same velocity as that of the fluid.

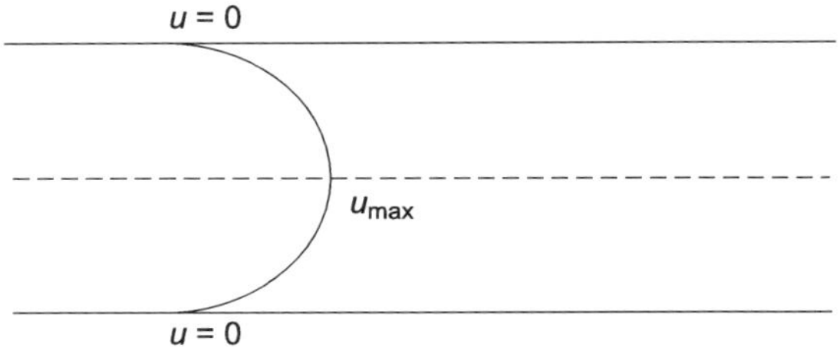

Fig. 11.1 *Velocity profile of a fluid flowing through a pipe*

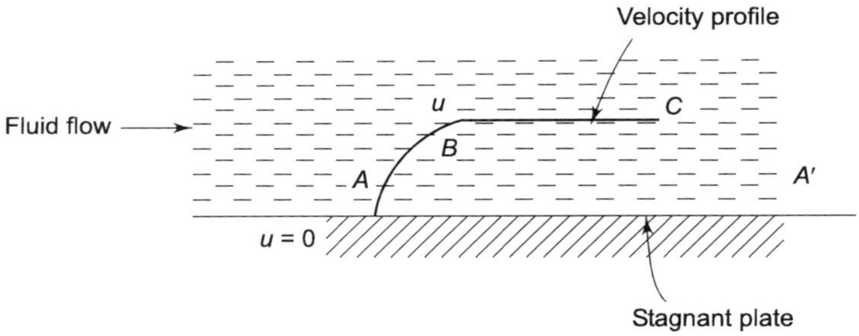

Fig. 11.2 *Velocity profiles on a stagnant flat plate*

We find distinctly a layer of the fluid ABC adjacent to the stagnant flat plate which will be consisting of different sub-layers having velocities ranging from zero to the maximum velocity u. "The layer ABC in which the velocity of the fluid is affected because of the presence of the stagnant flat plate is called as **boundary layer**." Similar type of boundary layer separation is possible in the case of circular pipes, also all along the walls of the pipes. This boundary layer is known as the **hydrodynamic boundary layer** (to contrast it from the thermal boundary layer which we come across in heat transfer having temperature profiles); and it plays a very important role in **transport phenomena**, viz. momentum transfer, heat transfer and mass transfer. The thickness of the boundary layer depends on the fluid velocity u; higher the velocity, thinner is the boundary layer.

If a vertical plate is placed abruptly in the flow of a fluid, we find the boundary layers to be separating out as shown in Fig. 11.3. Immediately after the plate, the layers orient themselves in such a way that up to some distance C, the layers do not meet creating certain amount of vacuum in the region. The vacuum also pulls the layers creating wakes and eddies in the region. The phenomenon of layer separation and creation of vacuum immediately after plate have a lot of utility in fluid mechanics for measurement of fluid velocity and also for making vacuum pumps, etc.

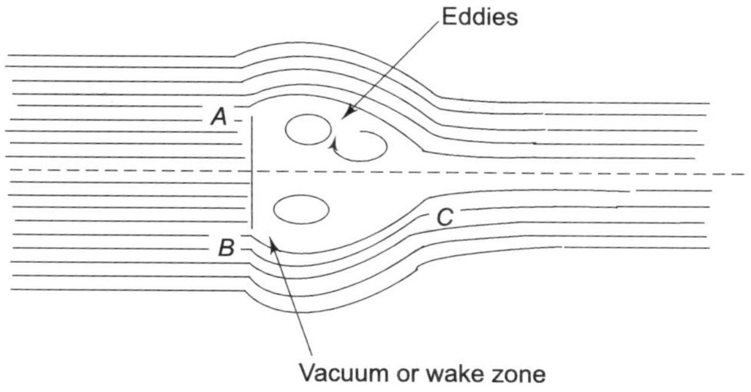

Fig. 11.3 *Formation of vacuum and eddies in laminar layer separation*

Viscosity Let us consider an element of fluid between two plates AB and CD which are initially stagnant, as shown in Fig. 11.4(a). When the steady state is established, let the plate CD move with a velocity of u because of the application of force F. Obviously, as described in the previous section, the layer just touching the plate will also be moving with a velocity of u. As this layer moves, it would also drag EF, the layer adjacent to it, to move. However, the adjacent layer will have inertia to move, but it will be forced by the lower moving layer. Thus, EF also moves, but with a velocity u' which is less than u. Similarly, the layer GH would also be dragged by EF to move, with the result that GH moves, but with velocity u'' which is still less than u'. In this way, as we go up, the layers will be having lesser and lesser velocity until we reach the topmost layer touching the plate AB which is stagnant. Thus, all the layers of the fluid of thickness y between the two plates will be having a velocity profile ranging from 0 to u.

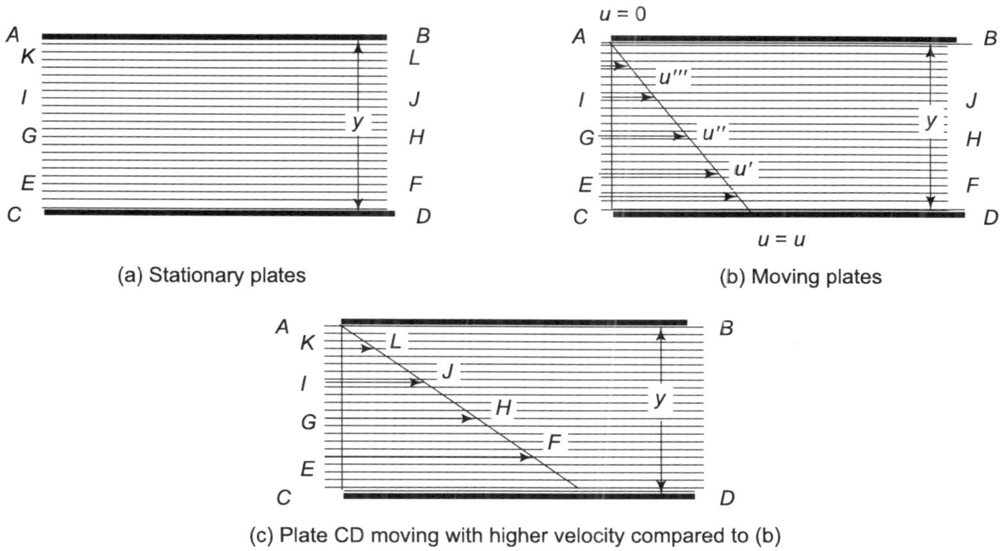

Fig. 11.4 *Concept of viscosity: fluid flow between two parallel plates*

Similarly, if more force is applied to the plate CD, it moves with a higher velocity; and hence the layer adjacent to it will also be moving with a higher velocity. But since AB is still stagnant, the layer adhering to AB will be stagnant. Thus, a different velocity profile forms, as shown in Fig. 11.4(c). Thus, slope of the line joining all the velocity profiles is called velocity gradient, du/dy, and this is found to be proportional to the force applied, i.e. the more is the F, the more is the du/dy. In other words, we can write

$$F \propto \frac{du}{dy} \tag{11.1}$$

Now we define a term called as **shear stress**, which is the force applied per unit area, and is represented by τ. Now,

$$\tau \propto \frac{du}{dy} \tag{11.2}$$

The above proportionality can be represented by an equation, as follows, which is termed as **Newton's law of viscosity**:

$$\tau = -\mu \frac{du}{dy} \tag{11.3}$$

in which μ is the coefficient of viscosity
du/dy is the velocity gradient or shear strain
τ is the shear stress.

The –ve sign in Eq. (11.3) indicates that as y increases, u decreases, the above equation brings an important concept of the proportionality constant μ, which is defined as the **coefficient of viscosity**, and it indicates the resistance to flow. Thus, it leads us to an important concept of 'viscosity' which is defined as the resistance to flow, and this resistance is constant, with the result that as the shear stress (τ) is increased, shear strain also increases proportionately, keeping the coefficient of viscosity μ to be constant. The shear strain or the velocity gradient du/dy is also customarily represented by a term γ. Thus, if we draw a plot of τ versus γ, as shown in Fig. 11.5, we find a straight line passing through the origin, and the slope of the line is equal to the coefficient of viscosity.

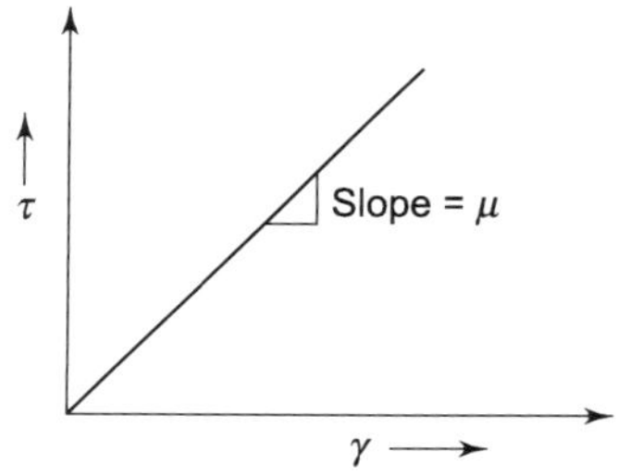

Fig. 11.5 *Shear stress versus shear strain, following Newton's law of viscosity*

11.3 FLOW THROUGH PIPES

When a fluid flows through a pipe or a channel or a conduit of any cross-section with a low velocity (linear velocity), the flow is smooth, or what we call as **streamlined**. All the layers of the fluid will be moving in an orderly manner, one over the other, without having any cross-mixing (Fig. 11.6(a)). If the velocity is slightly increased, disturbances in the flow pattern start cropping up. There will be a small amount of intermixing taking place slowly. If the velocity is further increased, the intermixing also increases, the streamlined flow behaviour gets distorted, and disturbances in the flow pattern become more. The ultimate result is that the flow pattern is chaotic, eddies form inside the flow, and such a flow is called as turbulent flow (Fig. 11.6(b)).

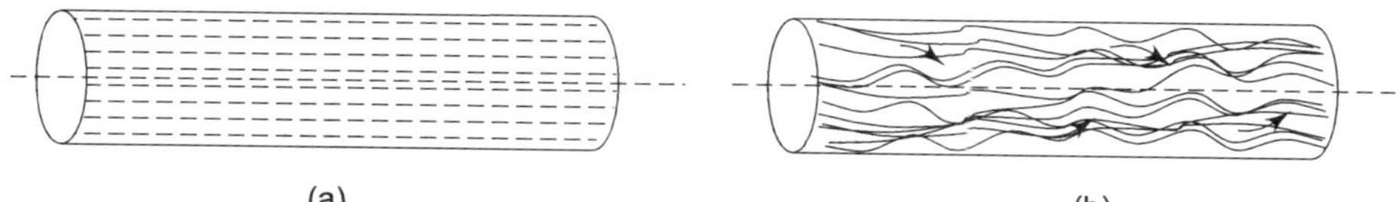

Fig. 11.6 *Flow through pipes: (a) Streamline flow at low fluid velocity, (b) disturbed flow at high fluid velocity*

Thus, the smooth flow is known as **laminar flow,** and the disturbed flow with eddies is known as **turbulent flow**. Both the flow patterns have their own advantages and disadvantages. Mixing, heat transfer and mass transfer are better in turbulent flow, whereas the pressure drop effects are much less in the laminar flow.

Reynolds number In the above discussion, we have only used some qualitative statements like high velocity and low velocity of fluids to characterise whether the flow is turbulent or laminar. We shall try to quantify the velocity values by introducing the concept of **Reynolds number**. It is a dimensionless number which is the *ratio of the kinetic forces to the viscous forces* when the fluid is flowing through a pipe. It is defined as follows:

$$N_{Re} = \frac{\rho d u}{\mu} \tag{11.4}$$

where d is the diameter of the pipe, u the linear velocity of the fluid, and ρ and μ are, respectively, the density and viscosity of the fluid. The number takes care of the nature of the fluid, its flow characteristics, and system geometry. If the number is high, it indicates that the kinetic forces are dominant as compared to the viscous forces, and vice versa. It characterises the flow as follows:

If $N_{Re} < 2100$, the flow is laminar

> 4000, the flow is turbulent

$2100 < N_{Re} < 4000$, the flow is transient

Reynolds number is a very important concept, and we come across this virtually in all transport processes, viz. momentum transfer, heat transfer and mass transfer. The empirical equations in all transport processes incorporate the Reynolds number term.

11.4 NEWTONIAN AND NON-NEWTONIAN FLUIDS

The relationship between the shear stress and shear strain has led to the formulation of Newton's law of viscosity, and gave the concept of viscosity which is a proportionality factor between shear stress and shear strain. This proportionality relationship may not be true for all liquids, which lead us to conclude that there are two types of fluids:

(i) Those which obey Newton's law of viscosity, such liquids are known as **Newtonian fluids**. Water is a very common example. Most of the gases, and aqueous solutions of low molecular weight compounds are some of the examples of Newtonian fluids.

(ii) Those which do not obey Newton's law of viscosity, such liquids are known as **non-Newtonian fluids**.

The relationship between τ and γ is not always a straight line as shown in Fig. 11.5, but the lines are curved as shown in Fig. 11.7. The two kinds of fluids are represented by curves B and C. They start from the origin; one is concave upwards, whereas the other is concave downwards. The fluids represented by curve B are known as **dilatent** fluids, whereas those represented by curve C are known as **pseudo-plastic** fluids. Some of the common dilatent fluids are some solutions of corn flour, sugar and starch, suspensions of iron powder, etc. in low viscosity liquids, whereas, most of the biological fluids, starch suspensions, cellulose

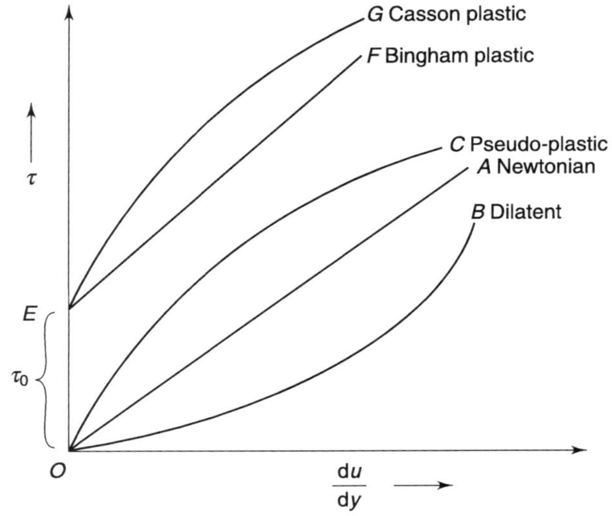

Fig. 11.7 *τ versus du/dy data for Newtonian and non-Newtonian fluids*

acetate slurries are some of the pseudo-plastic fluids. There is another type of fluids represented by straight line EF. They are also like Newtonian fluids represented by line OA, except that they are shifted from the origin by a distance of OE on the *y*-axis. Such type of fluids is known as **Bingham plastics**. Toothpaste is a very common example of Bingham plastics. They do not move or distort unless a certain threshold shear stress is given, which is represented by τ_0. Subsequently, the shear strain is proportional to shear stress like Newtonian fluids. Some plastic melts, cooking fats, and some soap and detergent slurries are some of the Bingham plastics. There is yet another type of non-Newtonian fluids which are known as **Casson plastics**. They also need a threshold shear stress τ_0; subsequently, they move like pseudo-plastics. They are represented by the curve EG (Fig. 11.7). Blood, tomato sauce, and orange juice are some common examples of casson plastic fluids (Atkinson and Mavitona, 1991).

All the above fluids are classified as time-independent non-Newtonian fluids. There is another variety of fluids which are classified as time-dependent non-Newtonian fluids. They are called as time-dependent fluids because their stress–strain relationship has a hysteresis path as shown in Fig. 11.8, i.e. they do not

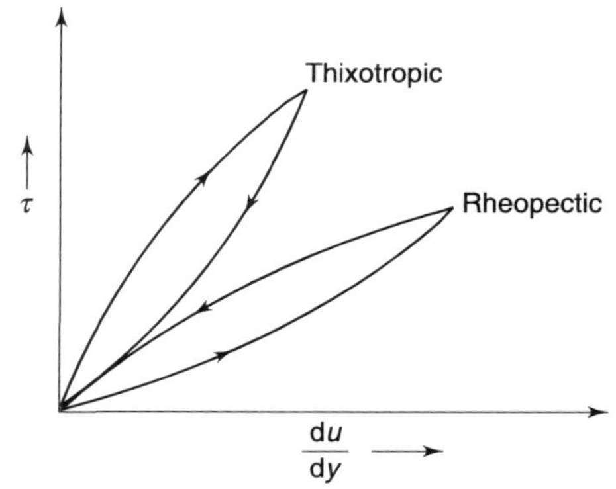

Fig. 11.8 *Stress–strain relationship for time-dependent non-Newtonian fluids*

follow the same path while increasing and decreasing the shear stress. They are further sub-divided into two classes, viz.

(i) thixotropic
(ii) rheopectic.

Some high polymeric melts, greases, many food materials are examples of thixotropic fluids. Bontonite clay solutions and gypsum suspensions are examples of some of the rheopectic fluids (Skelland, 1967). We do not come across much of the time-dependent non-Newtonian fluids in biochemical industries and hence are not discussed much.

Two-Parameter Model Most of the non-Newtonian fluids fall under the category of pseudo-plastic and dilatent fluids. Their rheological behaviour can be conveniently represented by a two-parameter model, known as **Ostwald-de Waele law** or simply known as 'power-law model':

$$\tau = K\gamma^n \tag{11.5}$$

in which K is known as the **consistency index**,

n is known as the **flow behaviour index**.

If $n < 1$, the model represents a pseudo-plastic fluid.

If $n > 1$, the model represents a dilatent fluid.

If $n = 1$, the model represents a Newtonian fluid with $K = \mu$.

For non-Newtonian fluids, as the shear strain changes, the viscosity also changes, and hence we cannot have a single viscosity in the entire range of shear strains applied. However, we can define an apparent viscosity, μ_a, which is strain-specific and may be represented by

$$\mu_a = \frac{\tau}{\gamma} = K\gamma^{n-1} \tag{11.6}$$

For pseudo-plastic fluids, when $n < 1$, as the shear rate increases, the apparent viscosity falls, and hence these fluids are called **shear thinning** fluids; contrary to this behaviour, for dilatent fluids, when $n > 1$, as the shear rate increases, the apparent viscosity also increases, and hence these fluids are known as **shear thickening fluids**.

The Bingham plastics are represented by an equation similar to Eq. (11.3) as follows:

$$\tau - \tau_0 = \mu_B\gamma \tag{11.7}$$

Casson plastics are represented by

$$\tau^{1/2} - \tau_0^{1/2} = K\gamma^{1/2} \tag{11.8}$$

Example 11.1

The following rheological data were obtained with a fermentation broth in the very initial stages of fermentation. Find out whether the fluid is Newtonian or non-Newtonian. If it is a non-Newtonian fluid, find the rheological constants. If it is a Newtonian fluid, find the viscosity of the fluid.

Table Ex11.1 *Rheological data*

γ, s^{-1}	5	10	15	20	32	50
τ, N/m^2	1×10^{-2}	2.1×10^{-2}	2.95×10^{-2}	4.1×10^{-2}	6.2×10^{-2}	0.1

Solution The data in Table E11.1 is used for drawing a graph of ln γ vs ln τ. If the slope is equal to 1.0, the fluid is Newtonian. If the slope is not equal to 1.0, it is a non-Newtonian fluid, and will indicate the value of n. The intercept will give the value of ln K.

Table E11.2 *Data of Example 11.1*

γ, (s^{-1})	$\tau (N/m^2) \times 10^2$	$\ln \gamma$	$\ln \tau$
5	1	1.609	– 4.605
10	2.1	2.302	– 3.863
15	2.95	2.708	– 3.523
20	4.1	3.0	– 3.194
32	6.2	3.466	– 2.78
50	10.0	3.91	– 2.3

The data in the last two columns are plotted in Fig. E11.1

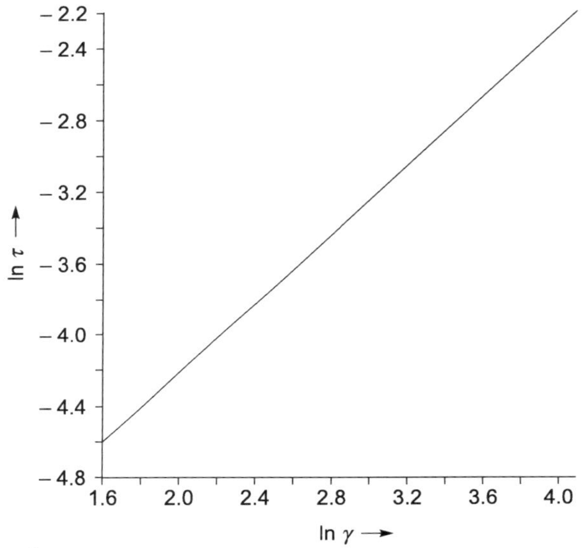

Fig. E11.1 *Solution for Example 11.1*

The line is a straight line with

$$\text{Slope} = \frac{-2.3-(-4.6)}{3.9-1.6} = 1.0$$

Since the slope= 1, the fluid is a Newtonian fluid.

$$\text{Slope} = \frac{-4.6 - y_o}{(1.6-0)} = 1.0$$

$$\therefore \qquad -y_o = (1.0 \times 1.6) + 4.6 = -6.2$$
$$= \ln \mu$$

where μ is viscosity of the fluid.

$$\therefore \qquad \mu = \text{anti} \ln (-6.2) = 2.3 \times 10^{-3} \text{ Pas}$$

11.5 EVALUATION OF RHEOLOGICAL DATA

In a way, the evaluation of rheological data means measuring viscosity. In case of non-Newtonian fluids, it means evaluation of K and n of Eq. (11.5). Essentially, the evaluation of rheological data means finding out the deformity in a fluid (which means shear strain) on application of different amounts of shear stress. There are many types of equipment available in practice, but we restrict ourselves to the following three varieties:

(i) extrusion rheometer
(ii) cone and plate viscometer
(iii) coaxial cylinder viscometer

They are discussed in detail here.

11.5.1 Extrusion Rheometer

It is a capillary tube viscometer consisting of a capillary tube attached to the bottom of a cylindrical container into which the fluid, whose rheological properties are to be measured, is filled (Fig. 11.9). The cylindrical container is pressurised from outside by applying a gas (usually inert gas) at high pressure from a cylinder. Because of the pressure applied on the top, fluid flows through the capillary tube. The flow rate is proportional to the pressure applied for a constant length of the capillary tube at a constant temperature. To obtain data at different temperatures, thermic fluid is passed outside the jacket of cylindrical container (not shown in figures). If the temperatures are above ambient, there may be a slight fall in temperature when the fluid is flowing through the capillary tube. To compensate for the temperature effect, the capillary tube is sometimes kept inside the container as shown in Fig. 11.9(b).

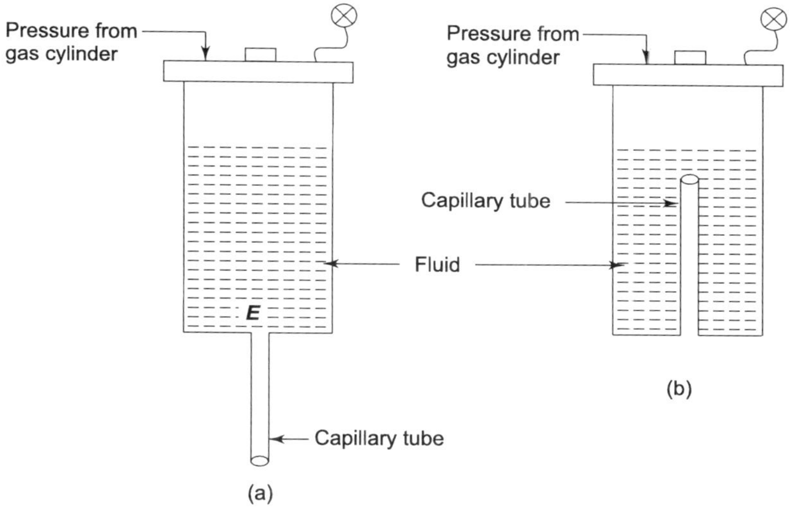

Fig. 11.9 *Extrusion rheometer or a capillary tube viscometer*

The rheological data can be collected by the application of Rabinowitsch–Mooney equation (Skelland, 1967) which gives a relationship for the shear stress and shear rate in terms of measurable parameters.

Thus,

$$\tau = \frac{d(\Delta p)}{4L} \tag{11.9}$$

$$\frac{du}{dy} = \gamma = \frac{8u}{d} \tag{11.10}$$

in which Δp is the pressure differential applied on the top of the fluid surface,
L is the length of the capillary tube, and
d is the diameter of the capillary tube.

The Rabinowitsch–Mooney equation will make the power-law fluid model (Eq.(11.5)) as follows:

$$\tau_w = \frac{d(\Delta p)}{4L} = K\left(\frac{8u}{d}\right)^n \tag{11.11}$$

Using the above equation, we can find K and n for the non-Newtonian fluids by plotting $\ln(d(\Delta p)/4L)$ versus $\ln(8u/d)$, as shown in Fig. 11.10. The slope will give the value of n and the intercept is equal to $\ln K$. Thus,

$$\text{Slope} = n$$
$$\text{Intercept} = \ln K.$$

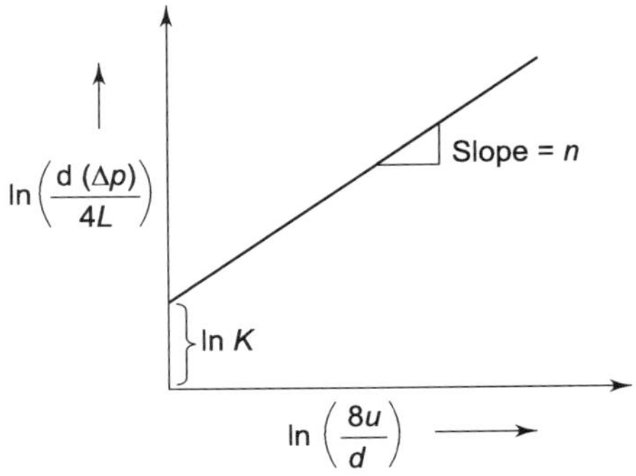

Fig. 11.10 *Experimental method to evaluate the rheological constants*

Extrusion rheometer is one of the most convenient and oldest methods available. In fact, extrusion rheometers are not manufactured, and hence are not generally available in the market. They are made by individual users to suit to their requirement. The capillary sizes vary; generally they are in the range of 0.5–3 mm, depending upon the solution viscosity. The length is usually of the order of 500 mm or less. The longer the tube, the less is the entrance effects. The flow rates through the capillary should be small enough so that the flow can be approximated to be a streamline flow. The arrangement shown in Fig. 11.9(b) is helpful to take care of the temperature variations during flowing, and the entrance effects at **E** (in Fig. 11.9(a)) can be avoided to a great extent.

Measurement of rheological parameters is easy and simple with extrusion rheometer, as it does not involve any sophisticated instrumentation; except that it takes slightly longer times for measurements and too much of calculations are involved. The extrusion rheometer is not suitable for suspension with the solids having tendency to settle. They may even block the capillary tube and clog the flow. It is particularly so with fermentation broths containing filamentous fungi. Thus, the capillary tube viscometers are not suitable for measuring the rheological properties of culture broths.

Example 11.2 *The rheological properties of a fermentation culture broth containing Xanthan campestris culture of 0.5% concentration were studied in laboratory using a capillary tube viscometer by application of various gauge pressures. The diameter of the capillary tube is 2mm and its length is 25 cm. Table E11.3 shows the pressure applied and the quantity of fluid collected at room temperature*

Table E11.3

Δp, MN/m²	0.0431	0.08	0.1065	0.13
Qty. of fluid, ml	10	20	20	50
Time of collection, min	7	5.5	3.5	6.67

Draw the shear stress-shear strain diagram. Find out the consistency index and flow behavior index.

Solution Diameter of the capillary tube = 2 mm

i.e., $d = 2 \times 10^{-3}$ m

Length of the capillary = 25 cm

$L = 0.25$ m

Cross-sectional area of the capillary tube

$$A = \frac{\Pi}{4} \times \left(2 \times 10^{-3}\right)^2 = 3.14 \times 10^{-6}\ \text{m}^2$$

If Q is volumetric flow rate,

$$u = Q/A$$

The rheological data γ vs τ can be evaluated by Eq. (11.9) and Eq. (11.10)

$$\gamma = \frac{8u}{d}$$

and
$$\tau = \frac{\text{d}(\Delta p)}{4L}$$

The data are presented in Table E11.4

Table E11.4 *Data for Example 11.2*

$\Delta p \times 10^2$ (MN/m^2)	Qty of fluid $\times 10^6$ (m^3)	Time (min)	$Q \times 10^6$ (m^3/s)	u (m/s)	γ (s^{-1})	τ (N/m^2)	ln γ	ln τ
4.31	10	7	0.0238	7.58×10^{-3}	30	86.2	3.401	4.456
8.0	20	5.5	0.06	0.019	77.2	160	4.35	5.075
10.65	20	3.5	0.095	0.03	121.3	213	4.8	5.36
13.0	50	6.67	0.125	0.04	159.2	260	5.07	5.56

Data in last two columns of Table E11.4 are used to draw Fig. E11.2

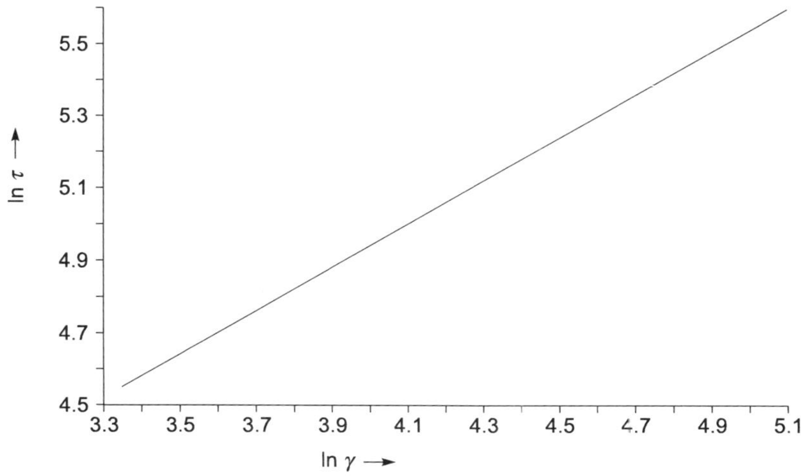

Fig. E11.2 *Solution for Example 11.2*

$$\text{Slope of line in Fig. E11.2} = \frac{5.56 - 4.46}{5.1 - 3.4} = \frac{1.1}{1.7} = 0.65$$

$\therefore \quad n = 0.65$

The intercept is calculated from the last point.

$$y = 5.56$$
$$x = 5.1$$
$$\text{Intercept} = 5.56 - 0.65\,(5.1) = 2.26$$

$\therefore \quad K = e^{2.26} = 9.58\ (\text{N/m}^2)\text{s}^n$

11.5.2 Cone-and-Plate Viscometer

It is one of the simplest viscometers. It consists of a horizontal plate, and an inverted cone standing on it. In fact, the upper plate does not look like a cone; it looks more like a horizontal plate only because the angle between the cone and the bottom horizontal plate is of the order of 3° or less (Fig. 11.11). The liquid whose viscosity is to be measured is loaded in the small clearance between the plate and the cone. The cone rotates at a very low speed (of the order of a few rpm). This rotation (the angular velocity of the cone) develops a shear strain in the liquid. The strain in the liquid offers resistance for the movement of the cone which is measured. The torque applied to the cone will give a measure of the shear stress. Thus, the shear stress and strain relationship is established as follows (Skelland, 1967):

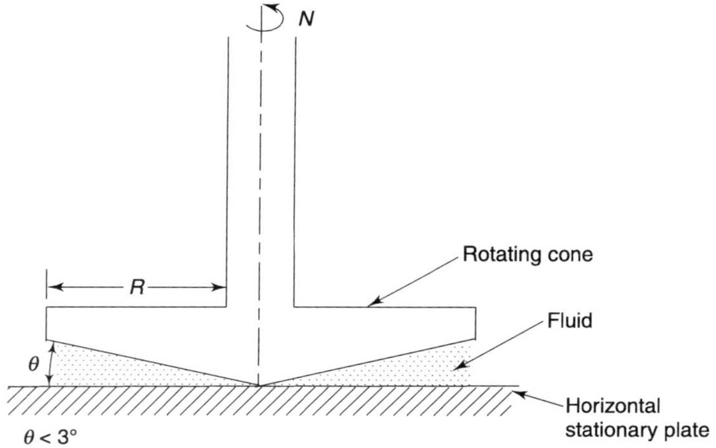

Fig. 11.11 *Cone-and-plate viscometer*

The shear rate is given by

$$\frac{du}{dy} = \frac{2\pi N}{\tan\theta} \approx \frac{2\pi N}{\theta} \tag{11.12}$$

for small values of θ. Similarly,

$$\tau = \frac{3t}{2\pi R^3} \tag{11.13}$$

where t is the torque applied and R the radius of the cone.

The temperature of the liquid is maintained constant at a particular value by circulating a thermic fluid through the bottom plate from a constant temperature bath. This also helps for collecting the data at different temperatures. Since the angular velocity of the cone is very small, the fluid motion can be reasonably assumed to be streamlined.

11.5.3 Coaxial Cylinder Viscometer

It is a popular viscometer for measuring rheological data of non-Newtonian fluids. Brooksfield viscometer is a coaxial cylinder viscometer. It consists of two coaxial cylinders, one rotating inside the other (Fig. 11.12). The outer cylinder is stationary and the inner cylinder is hung into the outer cylinder through a torsion wire. It is connected to a gear motor on the top for rotating the coaxial inner cylinder. The coaxial space between the two cylinders is filled with the fluid whose viscosity is to be measured.

In some other designs, the outer cylinder rotates slowly and it develops certain amount of torque in the inside cylinder (Fig. 11.12) because of the viscosity of the fluid. The more is the viscosity of the fluid, the more is the shear strain developed for a given shear stress which is measured by the torque on a scale. The τ versus du/dy data is generated as follows (Skelland, 1967):

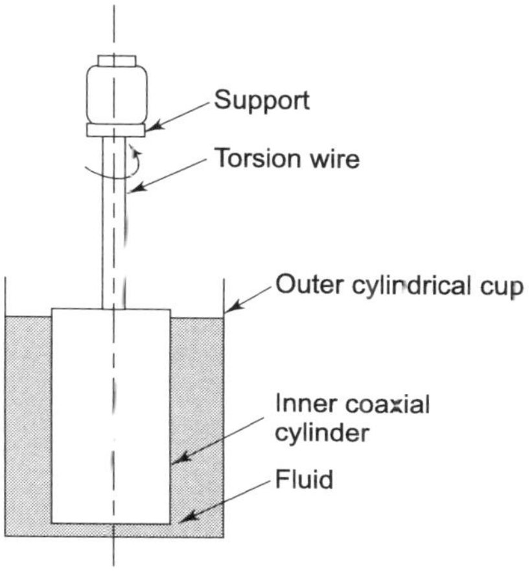

Fig. 11.12 *Coaxial cylinder viscometer*

$$\tau = \frac{t}{2\pi R_b^2 L} \quad (11.14)$$

and

$$\frac{du}{dy} = \frac{4\pi N}{1 - S^{-2}} C_R \quad (11.15)$$

where S is the ratio of radii of outer cylinder and inner cylinder:

$$S = R_a/R_b \quad (11.16)$$

and C_R is a constant to be evaluated.

Skelland (1967) tabulated certain values of C_R for different S and other parameters.

Thus, τ versus du/dy data are generated, from which we can evaluate the rheological parameters K and n. However, in most of the commercial instruments (like Brooksfield viscometer), all the data are incorporated in the form of some coefficients from which we can directly get shear stress–shear strain data or even directly get K and n values.

11.6 MIXING IN FERMENTATION BROTHS

Mixing is one of the most important unit operations in fermentation processes. Most of the fermentation broths contain

- a medium
- a substrate
- cells in suspension
- nutrients
- gas bubbles (mostly in case of aerobic fermentations for air dispersion).

They will increase the viscosity of the broth considerably and make mostly the broths non-Newtonian in character, which results in complicated rheology of the broth. In many circumstances, the broth is initially Newtonian in character, which slowly becomes non-Newtonian with the growth and build-up of cells, and changes the rheological characteristics.

Mixing is also used to dissolve solids in liquids or for dispersing air in the form of small bubbles in the fermentation broth. It will also help in keeping cells in suspension, which otherwise would settle down, and may not be able to participate in biochemical reactions. In addition to the above applications, mixing is also used to improve the heat transfer to or from the fermentation broth (we are considering heat transfer in next chapter).

Thus, mixing is helpful for providing a conducive environment for the cell suspensions to interact with the substrate in the viscous liquid medium and to consume oxygen which is made available to them through the liquid medium by increased interfacial area. The cells also will be able to utilise better the nutrients in the medium; and mixing will help equilibrate the concentration of the nutrients and the substrate in the entire fermentation broth.

Mixing Equipment The above envisaged objectives of mixing can be achieved by having a mixing equipment which can maintain the same composition of various ingredients in any corner of the equipment. In fact, it is too ambitious to expect ideal mixing to take place. Even if it occurs, it occurs only after an infinite mixing time. Hence, we satisfy ourselves with achieving a reasonable amount of mixing, which is neither too far, nor too close to ideality. The problems of mixing become more complicated in case of

- solid–solid mixing
- one component being in very low concentration.

In bioprocessing, our interest is restricted to

- solid–liquid mixing
- mixing of suspensions
- dispersion of gas (usually air in the form of small bubbles in the liquid media).

Mixing is carried out in a stirred tank fermenter, similar to the one shown in Fig. 5.5, provided with an agitator and a gas sparger. Various types of gas spargers used for effective mixing and mass transfer will be discussed in Section 14.5.4. Different types of agitators and impellers used for effective mixing are shown in Fig. 11.13. A typical mixing pattern for an agitator is shown in Fig. 11.14. The agitator helps for effective mixing by intermixing different packets, and pushing and pulling the fluid from different corners.

The specific choice of a mixing equipment for a particular fermentation process is a matter of technical viability, convenience and availability. The technical viability is based on:

- gas–liquid interfacial area
- liquid–liquid interfacial area
- interphase mass transfer limitations
- power requirement in kilowatts to achieve a desired stirrer speed
- cell morphology
- susceptibility of cells to shearing
- susceptibility of cells to sediment
- viscosity, density and other physico-chemical parameters of the broth.

However, on many occasions, the ultimate decision is made on economic considerations.

We discuss, in the following section, the power requirement characteristic for achieving the desired mixing, which in turn is quantified in terms of the stirrer speed, etc.

11.7 POWER REQUIRED FOR MIXING

The electrical power required for driving the motor connected to the agitator/impeller is known as the power required for mixing. The power is required by the impeller blades to overcome the viscous forces (forces opposing the movement) of the fermentation broth, and is a function of

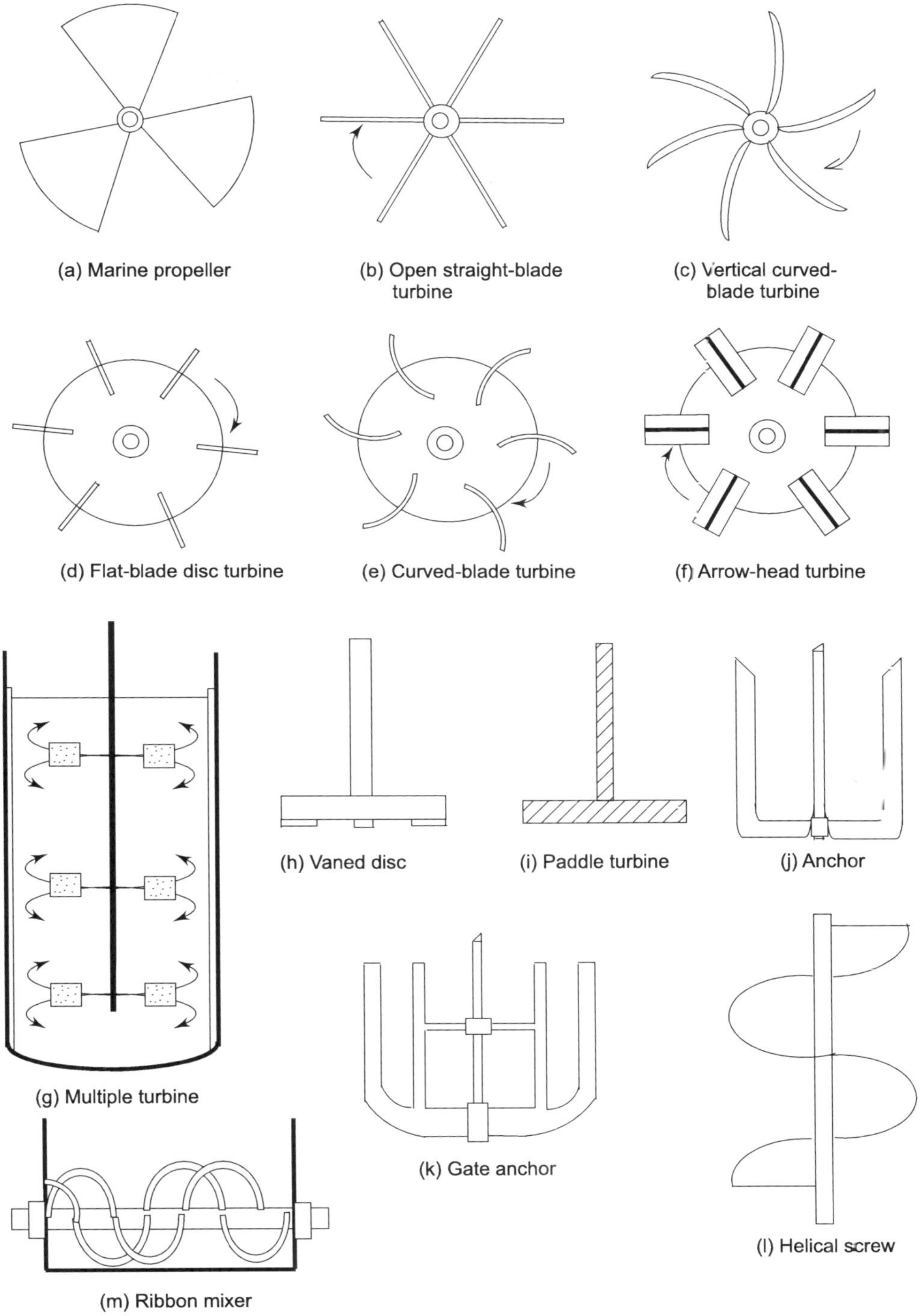

Fig. 11.13 *Various types of agitators*

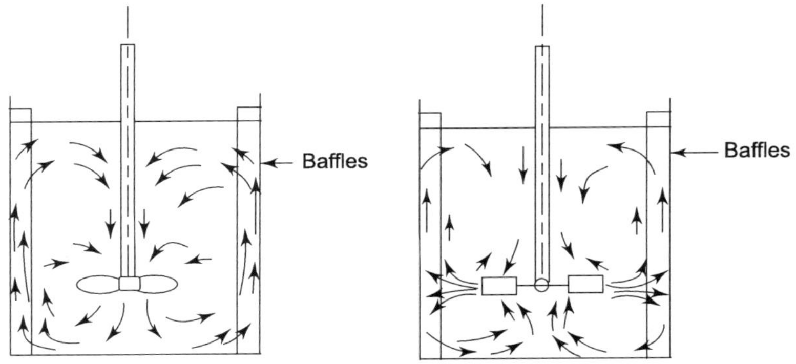

(a) Fluid flow pattern for propeller when baffles are used

(b) Fluid flow pattern of turbine with baffles

Fig. 11.14 *Mixing pattern for some of the agitators*

- the viscosity of the fluid
- the density of the fluid
- the presence of suspended solids, etc.
- the speed of the agitator
- the type and nature of the impeller
- orientation of vanes/blades of the impeller.

The power requirement calculated on the basis of the above obviously does not include the power consumed by

- gear box assembly, etc.
- stuffing box, O-rings or mechanical seals, etc., assembled to the agitator shaft
- friction in the motor
- friction in the assembly
- any other mechanical losses.

Hence, the actual electrical power requirement of the motor connected to the agitator is more than the one calculated by us in the following sections.

When the fermentation broth is sparged with some kind of a gas (usually we sparge air in the case of aerobic fermentations), obviously the zone around the sparger will be less dense than the other zones. Usually, the sparger is located under the impeller. When the system is sparged with gas, the fermentation broth around the impeller is less dense, and requires less power than a system which is not sparged with any gas at all. Hence, we discuss the power requirement for an ungassed system and a gassed system separately.

11.7.1 Power Requirement for Ungassed Newtonian Fluids

As has been mentioned, the power requirement is a function of the fluid properties and the agitator speed. The relationship can be expressed by means of a number called **power number**, which is defined as follows:

$$N_{P0} = \frac{P_0}{\rho N_i^3 d_i^5} \tag{11.17}$$

where N_{P0} is the power number for an ungassed system

P_0 is the power requirement in watts or kilowatts for an ungassed system

N_i is the impeller speed

d_i is the diameter of the impeller.

For a variety of impellers, the relationship between the power number and impeller Reynolds number (N_{Rei}) was established and the data were presented in the form of graphs by Rushton et al. (1950). Subsequently, Bates et al. (1963) explained various geometric parameters of turbine impellers, which are presented here in Fig. 11.15.

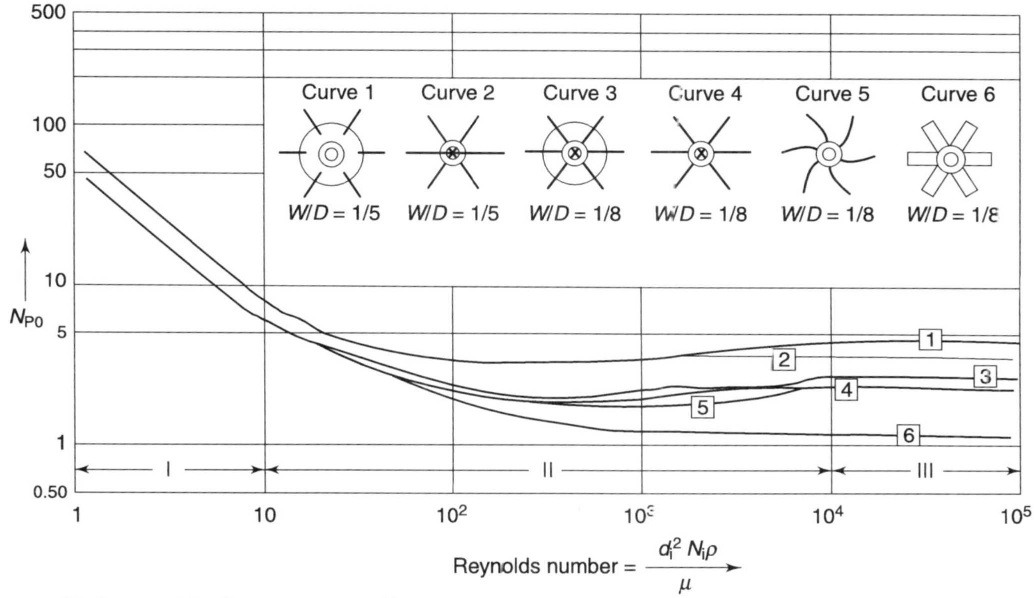

Fig. 11.15 *N_{Rei} versus N_{Po} for various impellers*
(Reprinted from "An examination of some geometric parameters of impeller power", Bates, R.L., Fondy, P.L., and Corpstein, R.R., I&EC, Process Des. Dev., Vol. 2(4), p. 311, Copyright © (1963), with permission from American Chemical Society)

The geometry of the systems considered is as follows:

- generally 4 to 6 baffles
- baffle width is one-tenth of the diameter of the tank (i.e. $t_b = d_t/10$)
- impeller diameter is one-third of the diameter of the tank (i.e. $d_i = d_t/3$)
- the liquid height in the tank is equal to the diameter of the tank (i.e. $H_L = d_t$)
- the height of the liquid in the tank is two-thirds of the height of the tank ($H_L = 2/3H_t$).

The impellers considered are:

(i) Rushton turbine
(ii) Paddle
(iii) Marine propeller
(iv) Anchor
(v) Helical ribbon.

The impeller Reynolds number is defined as

$$N_{Rei} = \frac{\rho N_i d_i^2}{\mu} \tag{11.18}$$

where N_{Rei} is impeller Reynolds number.

From the graphs in Fig. 11.15, we can find out the power number corresponding to the impeller Reynolds number. Once we know the power number, we can calculate the power requirement from Eq. (11.19):

$$P_0 = N_{P0}\,(\rho N_i^3 d_i^5) \tag{11.19}$$

Equation (11.19) holds good for a baffled fermenter. If the fermenter does not have baffles, vortex will form on the top, and Eq. (11.19) is added with an additional dimensionless number called **Froude number** (N_{Fr}), which takes care of the gravitational effect, and is defined as:

$$N_{Fr} = \frac{g}{d_i N_i^2} \tag{11.20}$$

As long as $N_{Rei} < 300$, even for an unbaffled fermenter also, the contribution of Froude number term becomes insignificant; and Eq. (11.19) holds good (Skelland, 1967). For $N_{Rei} > 300$, Eq. (11.19) modifies to

$$N_{P0} = \left(\frac{P_0}{\rho N_i^3 d_i^5}\right) \times \left(\frac{g}{d_i N_i^2}\right)^{\left(\frac{a - \log N_{Rei}}{b}\right)} \tag{11.21}$$

where a and b are coefficients, and are reported for common impellers with $d_t/d_i = 3.0$ as follows (Skelland, 1967):

for propeller, $a = 2.1$ and $b = 18.0$

for a six-blade turbine, $a = 1.0$ and $b = 40.0$

Since we use mostly a baffled fermenter, we use Eq. (11.19) for calculation of power number and hence power requirement.

The plots in Fig. 11.15 are typical and can be divided into three zones, viz.

- zone I, in which N_{P0} is falling as N_{Rei} is increasing, which we call as **laminar zone**
- zone III, in which N_{P0} is constant as N_{Rei} is increasing, which we call as **turbulent zone**
- zone II, in which the plots are curved, and we call this zone as **transition zone**

Laminar zone In the laminar zone, $N_{Rei} \le 10$, in which case the power number is inversely proportional to impeller Reynolds number. For some impellers, the laminar zone extends up to $N_{Rei} \le 100$ also. In this zone, we find

$$N_{P0} \propto \frac{1}{N_{Rei}} \tag{11.22}$$

where

$$N_{P0} = \frac{P_0}{\rho N_i^3 d_i^5} \propto \frac{\mu}{\rho N_i d_i^2}$$

Therefore,

$$P_0 = C_1\,\mu N_i^2\, d_i^3 \tag{11.23}$$

in which C_1 is the proportionality constant, and is reported in the literature for different impellers, when the N_{Rei} is in the range of 1.0, as follows (Doran, 1995):

Impeller	C_1
Turbine (Rushton type)	70
Paddle	35
Propeller	40
Anchor	420
Helical ribbon	1000

Turbulent zone The turbulent zone starts when $N_{Rei} \geq 10^4$, and in this zone, virtually N_{P0} is constant irrespective of the value of N_{Rei}. Here we can write

$$P_0 = N_{P0}' \, \rho \, N_i^3 \, d_i^5 \tag{11.24}$$

where N_{P0}' is the power number in the turbulent zone. Its approximate values are (Rushton et al 1950):
For Rushton turbine, $N_{P0}' = 5$
For paddle impeller, $N_{P0}' = 2$
For other impellers, viz. marine propellers, anchor and helical ribbon, $N'_{P0} = 0.35$

Transition zone The zone in between the laminar and turbulent zones is the transition zone. In this zone, both the viscosity and density of the fluid have effects on the power requirement. The value of N_{P0} can be measured on the graph (Fig. 11.15).

Example 11.3 *A fermentation broth containing a dilute Pullularia pullulans culture is agitated in a 5 lit reactor containing Rushton turbine impeller. The vessel geometry containing four baffles is as per the one described by Rushton et al (1950). The impeller diameter is 16 cm and is rotating at 300 rpm. The apparent viscosity of the fluid is 8 cp, and its density is 1010 kg/m³. Find the power required to run the agitator using the plots of Bates et al (1963)*

Solution Given data:

$$d_i = 16 \text{ cm} = 0.16 \text{ m}$$
$$N_i = 300 \text{ rpm} = 5 \text{ rps}$$
$$\mu = 8 \text{ cp} = 8 \times 10^{-3} \text{ Pas}$$
$$\rho = 1010 \text{ kg/m}^3$$

For using the data of Bates et al (1963) in Fig. 11.15, we have to calculate N_{Rei} given by Eq. (11.18).

$$N_{Rei} = \frac{\rho N_i d_i^2}{\mu} = \frac{1010 \times 5 \times (0.16)^2}{8 \times 10^{-3}} = 16160$$

From Fig. 11.15, N_{Po} corresponding $N_{Rei} = 16160$ for a Rushton turbine impeller with the vessel containing 4 baffles is

$$N_{Po} = 4.5$$

The power P_o is given by Eq. (11.19)

$$N_{Po} = P_o/(\rho N_i^3 d_i^5)$$

from which we can calculate Po

$$\therefore \quad P_o = 4.5 \times 5^3 (0.16)^5 \times 1010 = 59.6 \text{ watts}$$

The power required for running the agitator = 59.6 W

11.7.2 Power Requirement for Ungassed Non-Newtonian Fluids

It is difficult to estimate the power requirements in case of non-Newtonian fluids, as it is not possible to quantify the term 'viscosity' of the broth. It changes with the shear rate. Since most of the non-Newtonian fluids are highly viscous, they seldom reach the turbulent zone in which the power number is constant. One of the envisaged methodologies for it is to redefine the impeller Reynolds number by replacing the viscosity term with an apparent viscosity term. Thus,

$$\mu_a = K \gamma^{n-1}$$

where μ_a is the apparent viscosity of the broth, and it replaces μ in the Reynolds number as follows:

$$(N_{Rei})_{non\text{-}Newtonian} = \left(\frac{\rho N_i d_i^2}{K\gamma^{n-1}}\right) \tag{11.25}$$

Thus, the impeller Reynolds number for the non-Newtonian broth is found by Eq. (11.25), and the same is used in Fig. 11.15 to evaluate the power number, and hence the power requirement by the application of Eq. (11.19) or (11.24).

Metzner et al. (1961) reported the power number–Reynolds number correlation for a six-blade Rushton turbine impeller for a pseudo-plastic fluid in a baffled tank. The data are shown in Fig. 11.16, and the data are compared with that of a Newtonian fluid. In the laminar zone, where $N_{Rei} < 10$, and the turbulent zone, where $N_{Rei} > 100$, the two curves are the same, provided N_{Rei} is calculated by Eq. (11.25). In the transition zone, where $10 < N_{Rei} < 100$, the N_{P0} values are slightly lower than those for the Newtonian fluid.

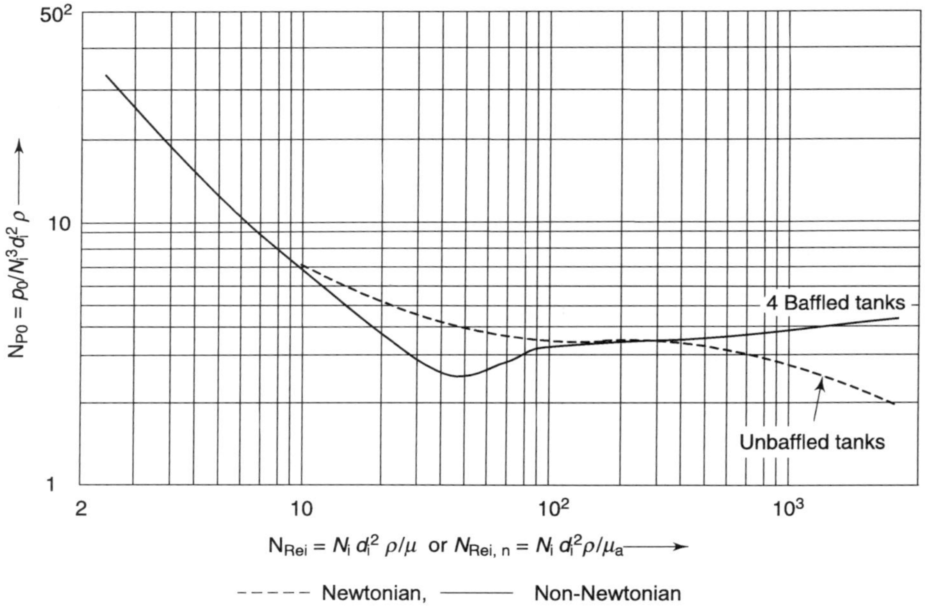

Fig. 11.16 *Agitation and mixing of liquids in unaerated mixing tank with Rushton turbine (Reprinted from "Agitation of viscous Newtonian and non-Newtonian fluids," Metzner, A.B., Feehs, R.H., Lopez Ramos, H., Otto, R.E., and Tuthill, J.D., AIChEJ, Vol. 7(1), p. 4, Copyright © (1961), with permission from John Wiley & Sons, Inc.)*

Application of data in Fig. 11.16 is not straightforward for design purposes in fermentation processes. Errors creep in since the flow pattern in non-Newtonian fluids near the impeller is much different from the rest of the fluid. Because of high viscosity of the fermentation broths, the flow regime in the rest of the fluid bulk is more or less streamlined and consumes less power as compared to the zone near the impeller blades.

11.7.3 Power Requirement for Gassed System

In the case of gassed fluids, the hydrodynamics in the vicinity of the gas sparger changes. It changes the density of the fluid around the sparger as well as near the impeller blades. Thus, the power requirement for agitating a gassed fluid is always less than that for an ungassed system. Hughmark (1980) reported an expression for a turbine impeller to relate the power requirement for ungassed and gassed systems as follows:

$$\frac{P_g}{P_0} = 0.1\left(\frac{v_g}{N_i V}\right)^{-0.25}\left(\frac{N_i^2 d_i^4}{g w_i V^{2/3}}\right)^{-0.2} \tag{11.26}$$

where P_g is the power requirement for a gassed system,
v_g is the volumetric gas flow rate,
V is the liquid volume, and
w_i is the width of the impeller.

P_g/P_0 can be found by application of Eq. (11.26) with a knowledge of the process parameters and impeller dimensions. P_0 can be found by Eq. (11.19) or (11.24). Thus, P_g can be calculated.

The experimental data of P_g/P_0 was shown in Fig. 11.17 as a function of a dimensionless aeration rate N_a (Ohyama and Endoh, 1955) which is defined as:

$$N_a = \frac{v_g}{N_i d_i^3} \tag{11.27}$$

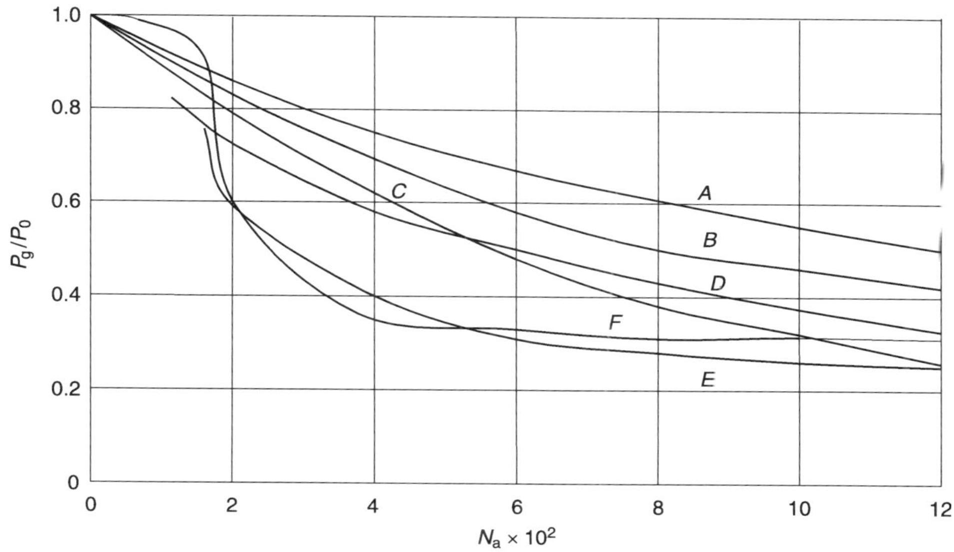

A: Flat-blade turbine (8 blades), *B*: Vaned disc (8 vanes), *C*: Vaned disc (6 vanes), *D*: Vaned disc (16 vanes), *E*: Vaned disc (4 vanes), *F*: Paddle

Fig. 11.17 *P_g/P_o as a function of aeration rate*
(Reprinted from "Power characteristics of gas liquid contacting mixers", Ohyama, Y. and Endoh, K., Chemical Engineering Japan, Vol. 19, p.2, copyright © (1955), with permission from the Society of Chemical Engineering, Japan)

The data are shown for most commonly used impellers, viz.

- flat blade turbine
- vaned discs
- paddles.

N_a can be calculated using system parameters by Eq. (11.27). With the help of Fig. 11.17, we can calculate P_g/P_0, which is less than 1.0. As usual, P_0 can be calculated by Eq. (11.19) or (11.24), and hence P_g can be calculated. Application of Fig. 11.17 can differentiate power requirement for different configurations of impellers, unlike application of Eq. (11.26) which is meant only for a turbine impellers.

As has already been mentioned, the above equations give power requirement of the impeller to cut against the viscous forces of the fermentation broth or the given system geometry. Obviously, the actual power requirement of the motor agitating the impeller will be much more than just what is calculated.

Symbols

a	: coefficient [Eq. (11.21)]
b	: coefficient [Eq. (11.21)]
C_R	: constant [Eq. (11.15)]
C_1	: constant [Eq. (11.23)]
d	: diameter (m)
d_i	: diameter of the impeller (m)
d_t	: diameter of the tank or fermenter (m)
F	: force applied (N)
g	: acceleration due to gravity (m/s^2)
H_L	: height of liquid in the tank (m)
H_t	: height of the tank (m)
K	: consistency index [Eq. (11.5)]
L	: length (m)
N	: rotations per second (s^{-1})
N_a	: dimensionless aeration rate [Eq. (11.27)]
N_{Fr}	: Froude number [Eq. (11.20)]
N_i	: impeller speed in RPS (s^{-1})
N_{P0}, N'_{P0}	: power number for an ungassed system (W or kW)
N_{Re}	: Reynolds number [Eq. (11.5)]
N_{Rei}	: impeller Reynolds number [Eq. (11.18)]
$N_{Rei, n}$	: impeller Reynolds number for non-Newtonian fluid [Eq. (11.25)]
n	: flow behaviour index [Eq. (11.5)]
P_g	: power requirement for a gassed system
P_0	: power required for an ungassed system (W or kW)
Δp	: pressure differential
R	: radius of the cone [Eq. (11.13)]
R_a	: radius of the outer cylinder
R_b	: radius of the inner cylinder
S	: ratio of radii [Eq. (11.16)]
t	: torque applied [Eq. (11.13)]
t_b	: thickness of the baffle (m)
u	: linear velocity (m/s)
V	: liquid volume (m^3)
v_g	: volumetric gas flow rate (m^3/s)
w_i	: impeller blade width (m)
y	: distance between the plates, or thickness of the fluid element (m)

Subscript

w	: wall

Greek Letters

τ	: shear stress (N/m^2)
μ	: viscosity (Pa s or N s/m^2 or kg/(m s))
γ	: shear strain (s^{-1})
τ_0	: threshold shear stress
μ_B	: viscosity index of Bingham plastic fluids
ρ	: density (kg/m^3)
μ_a	: apparent viscosity (Pa s or kg/(m s))
θ	: angle of the cone-and-plate viscometer

References

11.1 Atkinson, B., and Mavituna, F. (1991), *Biochemical Engineering and Biotechnology Handbook*, Macmillan, Basingstoke.

11.2 Bates, R.L. Fondy, P.L. and Corpstein, R.R. (1963). "An examination of some geometric parameters of impeller power", Ind. Eng. Chem., Process Des. Develop, 2(4), 310–314.

11.3 Doran, P.M. (1995), *Bioprocess Engineering Priciples,* Academic Press, London, 129–163.

11.4 Hughmark, G.A. (1980), "Power requirements and interfacial area in gas–liquid turbine agitated systems", *Ind. Eng. Chem. Process Des. Develop.*, 19, 638–641

11.5 Metzner, A.B., Feehs, R.H., Lopez Ramos, H., Otto, R.E., and Tuthill, J.D. (1961), "Agitation of viscous Newtonian and Non-Newtonian fluids", *AIChEJ*, 7(1), 3–11.

11.6 Ohyama, Y., and Endoh, K. (1955), "Power characteristics of gas–liquid contacting mixers", *Chem. Eng. Japan*, 19, 2.

11.7 Rushton, J.H., Costich, E.W., and Everett, H.J. (1950), "Power characteristics of mixing impellers: Part I and Part II", *Chem. Eng. Prog.*, 46, 395–404 and 467–476.

11.8 Skelland, A.H.P. (1967), *Non-Newtonian Flow and Heat Transfer,* John Wiley and Sons, Inc., New York, 27–49.

References for Further Reading

11.9 Charles, M. (1978), "Technical aspects of the rheological properties of microbial cultures", in *Advances in Biochemical Engineering*, Ghose, T.K., Fiechter, A., and Blakebrough, N. (Eds), Springer-Verlag, Berlin, pp. 1–62.

11.10 Deindoerfer, F.H., and West, J.M. (1960), "Rheological properties of fermentation broths", *Advances in Applied Microbiology*. Umbreit, W.W. (Ed.), Vol. 2 Academic Press, New York, 265–273.

11.11 McCabe, W.L., Smith, J.C., and Harriott, P. (1993), *Unit Operations of Chemical Engineering*, 5th Edn., McGraw-Hill, Inc., New York, 235–283.

11.12 Metzner, A.B., and Reed, J.C. (1955), "Flow of non-Newtonian fluids—Correlation of the laminar, transition, and turbulent-flow regions", *AIChE J*, 1, 434–440.

11.13 Oldshue, J.Y. (1966), "Fermentation mixing scale-up techniques", *Biotechnol. Bioeng.*, 8, 3–24.

11.14 Taguchi, H., Miyamoto, S. (1966). "Power requirement in non-Newtonian fermentation broth", *Biotechnol. Bioeng.*, 8, 43–54.

Review Questions

11.1 Define a fluid and differentiate it from solids.
11.2 What are compressible and incompressible fluids? Describe them with examples.
11.3 Describe the fluid flow phenomenon in pipes or conduits.
11.4 What is a boundary layer? Explain the concept of boundary layer with a neat diagram.
11.5 Describe the concept of velocity profiles of a fluid in between two plates, one is stationary and the other moving with velocity u.
11.6 Explain the concept of viscosity with a neat diagram.
11.7 What is meant by shear stress and shear strain? Draw a relationship between them.
11.8 What is Newton's law of viscosity? Define coefficient of viscosity.
11.9 What is Reynolds number? What is its significance?
11.10 What is meant by laminar flow and turbulent flow? How do you quantify them?
11.11 What are Newtonian fluids and non-Newtonian fluids? Write some examples.
11.12 Explain the rheology of non-Newtonian fluids.
11.13 Describe the stress–strain relationship of non-Newtonian fluids.
11.14 What are time-dependent non-Newtonian fluids? Explain with a diagram and examples.
11.15 What is a power law fluid model? Explain its significance.
11.16 What is meant by shear thinning and shear thickening fluids?
11.17 Explain the construction and working of an extrusion rheometer.
11.18 Explain a cone-and-plate viscometer for determining the rheological data of a fermentation broth.
11.19 Describe the working of a coaxial cylindrical viscometer.
11.20 Describe the importance of mixing in bioprocessing.
11.21 What are the different types of agitators used in bioprocessing?
11.22 How do you estimate the impeller power requirement for an ungassed bioreactor?
11.23 What is meant by:
  (i) impeller Reynolds number
  (ii) power number
  (iii) aeration rate
  (iv) Froude number
11.24 How do you estimate the impeller power requirement for a gassed bioreactor?

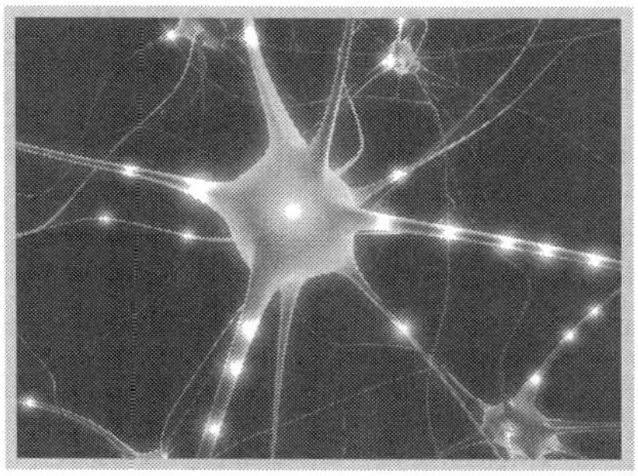

chapter 12

Heat Transfer in Bioprocessing

Heat-transfer plays an important role in bioprocessing, both for fermentation and for enzymatic reactions. The temperature effect on biochemical reaction rates is conveniently represented by the Arrhenius equation (Eq. 6.6). The fermentation broths are either heated to the desired temperatures or maintained at a desired lower temperature by heating or cooling the fermenters, depending upon whether the biochemical reaction is endothermic (absorbing heat) or exothermic (generating heat). Here, we use the term *heat transfer* which means it is the process of transfer of heat; it could be for the purpose of heating or vaporizing or cooling or freezing. The heat-transfer mechanism is the same in both, heating or cooling. The direction of heat-transfer could be; either from the heat source to the body or from the body to the cooling medium.

In addition to heating of fermenters or enzyme reactors, heat-transfer is also used for sterilisation of media or fermenters. Of the various integrated bioprocessing steps (Fig. 1.1), in the step 2, the strain is preserved by freezing or lyophilisation. In the downstream processing side, if the product is recovered by solvent extraction (Sec 17.6), the product is recovered by desolventisation in which the solvent is removed by distillation. Various applications of heat-transfer in bioprocessing are presented in Table 12.1.

Table 12.1 *Heat-transfer applications in biotechnology*

S. No	*HT Operation*	*Remarks*
(i)	Heating of fermentation broth/enzyme reactors	There are many examples; hydrolysis of starch to glucose, etc. The reactor is maintained at 95°C.
(ii)	Cooling of the fermentation broths	There are many examples. Fermentation of molasses to alcohol. The fermenter is maintained at 35 – 37°C.
(iii)	Sterilisation of media	—
(iv)	Lyophilisation	Preservation of the isolated strains by freezing.
(v)	Desolventisation	Desolventisation is done to recover the extracted product.
(vi)	Simultaneous heat and mass transfer operations	Drying, crystallisation, distillation, solvent extraction, etc.
(vii)	Concentrating the substrates/recovered product	Evaporation, crystallisation as in the case of production of citric acid, etc.

First, let us look into the principles of heat-transfer. Heat moves from a body at a higher temperature (usually called *source*) to a body at a lower temperature (usually called *sink*). Any transfer process is always proportional to the

- driving force, and
- contact area

In heat-transfer, the driving force is the temperature difference between the source and sink. Thus,

$$\text{the rate of heat transfer} \quad \alpha\,(\Delta T)$$
$$\alpha\,(A) \tag{12.1*}$$

where ΔT is the temperature difference, and
A is the contact area.

The above equation can also be written as

$$\text{rate process } \alpha \text{ driving force, i.e., } (\Delta T)$$
$$\alpha\ 1/\text{resistance} \tag{12.2}$$

The constant of proportionality is known as the **heat-transfer coefficient,** which we shall discuss later. The heat-transfer process also depends upon whether the materials are in direct contact or otherwise.

Depending upon whether the material is in direct contact with the heating source or otherwise, heat is conceived to transfer by any of the following three fundamental heat-transfer methods:

- Conduction
- Convection
- Radiation

We shall discuss them briefly here.

12.1 HEAT TRANSFER BY CONDUCTION

Conduction is a method of heat-transfer in which the heat is transferred from one body to another with/without direct contact and without the movement of either the source or the sink. This kind of heat-transfer takes place in solids. For example, heating of a metallic rod by putting one end in a hot flame and holding the other end. Initially, the other end is cold. Slowly it gets heated; i.e., the heat travels in the rod from one end to the other without the movement of the rod. How fast the heat transfers depends upon the nature of the material. Some materials conduct heat very fast, and are known as *conductors*; whereas other materials conduct very slowly or virtually do not conduct any heat, and are known as *insulators* or *non-conductors*.

Let us consider a slab of thickness B, one end of which is maintained at a higher temperature T_1, and the other end is maintained at a lower temperature T_2 as shown in Fig. 12.1.

$T_1 > T_2$. Initially, the temperature at one end is T_1 and the other end is at T_2 as shown in Fig. 12.1(a). As time progresses, slowly the temperature profile changes as shown in Fig. 12.1(b) and Fig. 12.1(c). Finally, the steady state is reached when the temperature profile becomes an inclined straight line as shown in Fig. 12.1(d).

12.1.1 Fourier's Law of Heat Conduction

The rate of heat-transfer per unit area is given by Eq. (12.2) as follows

$$\frac{q}{A} \quad \alpha\,(\Delta T)$$
$$\alpha\ 1/B$$

* note the similarity between Eq. (12.1) and Eq. (14.8)

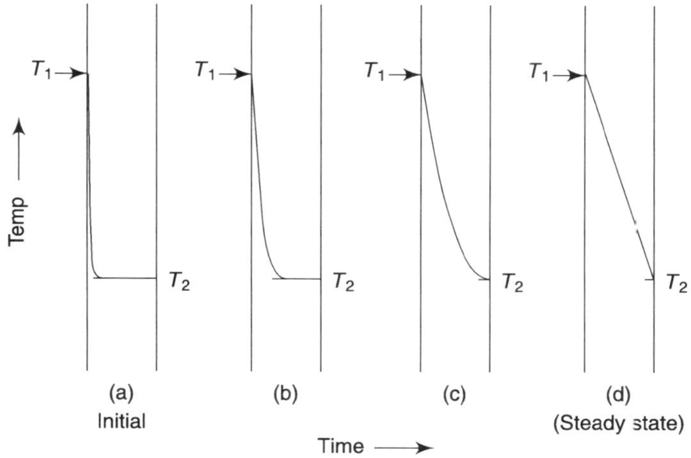

Fig. 12.1 *Temperature profiles of slab as time progresses.*

or
$$\frac{q}{A} = \frac{k}{B}(\Delta T) = \frac{(\Delta T)}{(B/k)} = \frac{\Delta T}{R} \tag{12.3}$$

in which q is the rate of heat-transfer in kJ/s or kW,
k is the thermal conductivity (W/m °C)
ΔT is the temperature difference (°C)
B is the total thickness of the slab (m)

R is known as the resistance for heat-transfer and is equal to (B/k).

$$R = B/k \tag{12.4}$$

Equation (12.3) gives the definition for thermal conductivity (k) of the material. If k is low, the heat-transfer rate is low. Materials like asbestos or cotton have low thermal conductivities of the order of 0.17 W/m°C, and 0.6 W/m°C, and are used to insulate or to prevent escape of heat. Metals like aluminum or copper will have higher thermal conductivities of the order of 2300 and 2100 W/m°C, and are used for conducting heat. In the differential form, Eq. (12.3) can be written as

$$\frac{dq}{dA} = -k\frac{dT}{dx} \tag{12.5}$$

where dx is the thickness of an elemental slab and is measured in the direction of heat-transfer, and dq/dA is known as the heat flux, and is the heat-transfer rate per unit area.

The –ve sign in Eq. (12.5) indicates that as x increases, T decreases since x is measured in the direction of heat-transfer.

Equation (12.3) or (12.5) is known as the Fourier law of conduction, and it is similar to momentum transfer given by Newton's law of viscosity (Eq. (11.3).

12.1.2 Conduction in Slabs in Series

Heat-insulating walls are usually made by having the insulting materials in series. Usually, the furnace walls are made up of a series of

- metallic wall inside
- silo-cel bricks or asbestos
- glass-wool or asbestos

It becomes necessary to find out the heat loss or heat-transfer through such series of conductors/insulators. The conductors in series are shown schematically in Fig. 12.2. The conductors in series are *A*, *B* and *C* with thickness B_A, B_B and B_C, and with thermal conductivities k_A, k_B and k_C. The temperature difference between each slab is ΔT_A, ΔT_B and ΔT_C as shown in Fig. 12.2. If R_A, R_B and R_C are the resistance for heat-transfer in slabs *A*, *B* and *C* respectively, we can write from Eq. (12.3),

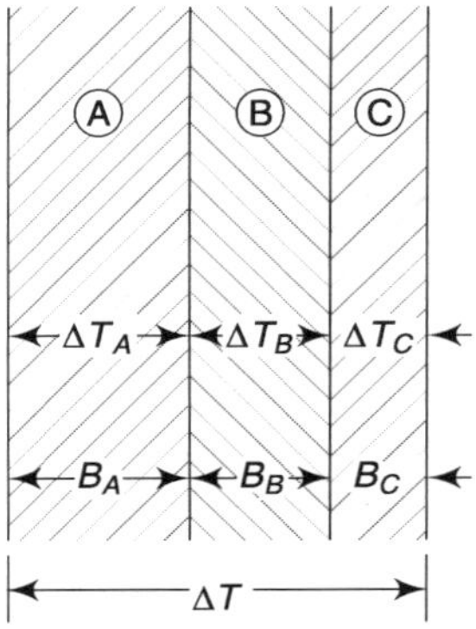

Fig. 12.2 *Conduction in series*

$$\Delta T_A = \frac{q_A}{A}\left(\frac{B_A}{k_A}\right) = \frac{q_A}{A} R_A \tag{12.6}$$

$$\Delta T_B = \frac{q_B}{A}\left(\frac{B_B}{k_B}\right) = \frac{q_B}{A} R_B \tag{12.7}$$

$$\Delta T_C = \frac{q_C}{A}\left(\frac{B_C}{k_C}\right) = \frac{q_C}{A} R_C \tag{12.8}$$

$$\Delta T = \frac{qR}{A} \tag{12.9}$$

The overall temperature difference ΔT is equal to the sum of the individual temperature differences. Thus,

$$\Delta T = \Delta T_A + \Delta T_B + \Delta T_C \tag{12.10}$$

Since the same amount of heat has to transfer through all the slabs,

$$q_A = q_B = q_C = q \tag{12.11}$$

Combining Eq. (12.6) to Eq. (12.11), we get

$$\frac{(q)}{A}(R) = \frac{(q)}{A}(R_A) + \frac{(q)}{A}(R_B) + \frac{(q)}{A}(R_C) = \frac{(q)}{A}(R_A + R_B + R_C) \tag{12.12}$$

Cancelling q/A throughout, we have

$$R = R_A + R_B + R_C \tag{12.13}$$

From the above equation, we can say that the overall resistance is equal to the sum of the individual resistances.

or
$$1/R = 1/(R_A + R_B + R_C) \tag{12.14}$$

Example 12.1

A cold-storage room is constructed using an SS wall followed by a cotton wool wall followed by a thick brick wall to maintain the temperature inside the room at – 20°C. The outside temperature is 28°C. Calculate the rate of heat loss per unit area of the wall and the temperature in between each wall. The properties and thicknesses of the walls are as follows.

Material	*Represented by*	*Thickness (mm)*	*Thermal conductivity W/m °C*
SS	*A*	6	21
Cotton wool	*B*	40	0.04
Brick	*C*	20	0.7

Solution Cross-sectional area of each wall is taken as 1 m^2.

$$R_A = B_A/k_A = \frac{6 \times 10^{-3}}{21} = 2.86 \times 10^{-4}\ m^2\ °C/W$$

$$R_B = 40/0.04 = 1000\ m^2\ °C/W$$

$$R_C = 20/0.7 = 28.5\ m^2\ °C/W$$

$$\therefore \quad R = 1028.57\ m^2\ °C/W$$

$$\Delta T = 28 - (-20) = 48°C$$

$$\frac{q}{A} = \frac{\Delta T}{R} = \frac{48}{1028.57} = 0.047\ W/m^2$$

The rate of heat loss per unit area of wall = 0.047 W

To find temperatures at the junction of each wall

$$\frac{q}{A} = \frac{\Delta T_A}{R_A}$$

i.e., $$0.047 = \frac{\Delta T_A}{2.86 \times 10^{-4}}$$

$$\therefore \quad \Delta T_A = 3.146 \times 10^{-3}$$

The temperature at the junction of the steel wall and cotton wool is almost same as that inside, i.e., – 20°C.

Similarly, $$\frac{q}{A} = \frac{\Delta T_B}{R_B}$$

i.e., $$0.047 = \frac{\Delta T_B}{1000}$$

$$\therefore \quad \Delta T_B = 47°C$$

$$T_{BC} - (-20) = \Delta T_B = 47°C$$

$$\therefore \quad T_{BC} = 47 - 20 = 27°C$$

The temperatures at the junctions of the walls are shown in Fig. E12.1.

From the above calculations, it is clear that most of the resistance for heat-transfer is offered by cotton wool. However, the other two walls (SS and brick) are used to provide adequate mechanical strength for the cold-storage facility.

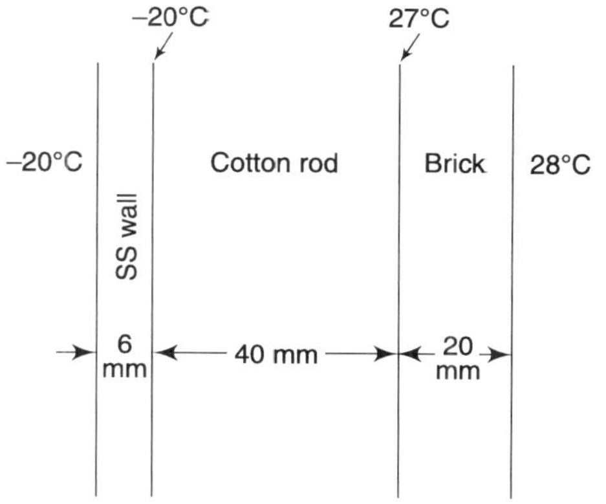

Fig. E12.1 *Solution for Example 12.1*

12.1.3 Heat Conduction through Cylindrical Pipes

Up to now we have seen heat-transfer through slabs which have a definite cross-sectional area like a square or a rectangle. But mostly we find heat-transfer applications to be through cylindrical pipes in which case the heat transfers from inside the pipes to the outside or vice versa. In view of the thickness of the pipe, the contact area on either side is not the same. If r_o and r_i are outside and inside radii, thickness of the pipe is B and the length of the pipe is L,

$$A_o = \text{outside contact area} = 2\Pi\, r_o\, L \tag{12.15}$$

$$A_i = \text{inside contact area} = 2\Pi\, r_i\, L \tag{12.16}$$

and $$B = r_o - r_i$$

Since the two areas are different, confusion arises as to which area is to be taken. One of the approaches could be to take the average area as follows:

$$A_{Av} = \frac{2\Pi L (r_o + r_i)}{2} \quad (12.17)$$

But a rigorous approach could be to use the logarithmic mean radius which is obtained as follows.

Consider a cylindrical pipe of length L with an elemental cylinder at a radius of r and having a thickness of Δr (as shown in Fig. 12.3). We can apply Fourier law for heat conduction by replacing dx with dr in Eq. (12.5).

$$\frac{q}{2\Pi rL} = \frac{-k\,dT}{dr} \quad (12.18)$$

On rearrangement,

$$\frac{2\Pi kL}{q}\int_{T_o}^{T_i} dT = \int_{r_i}^{r_o} \frac{r}{dr}$$

which on integration between the limits gives

$$\frac{2\Pi kL}{q}(T_i - T_o) = \ln(r_o/r_i) \quad (12.19)$$

The negative sign in Eq. (12.18) is taken care of by change of limits.

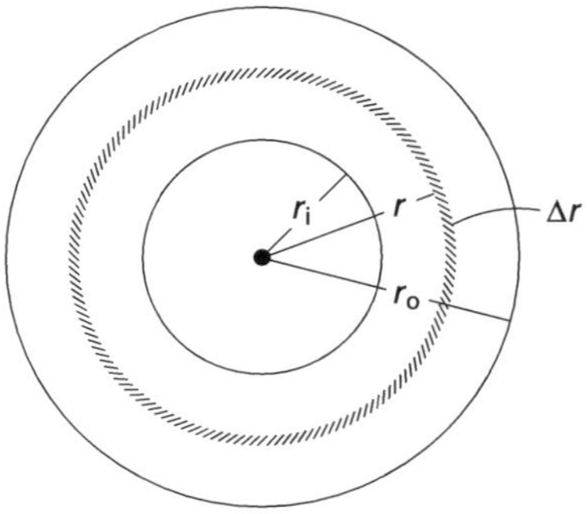

Fig. 12.3 *Conductive heat transfer through a cylindrical pipe of radii r_i and r_o*

or
$$q = \frac{2\Pi kL(T_i - T_o)}{\ln(r_o/r_i)}$$

Multiplying and dividing with $(r_o - r_i)$

$$q = 2\Pi kL \frac{(r_o - r_i)}{\ln(r_o/r_i)} \frac{\Delta T}{(r_o - r_i)}$$

$$= 2\Pi r_L k \frac{\Delta T}{B} \quad (12.20)$$

where
$$r_L = \frac{(r_o - r_i)}{\ln(r_o/r_i)} \quad (12.21)$$

and is known as the *log mean radius*.

or
$$\frac{q}{A_L} = k\frac{\Delta T}{B} \quad (12.22)$$

where
$$A_L = 2\Pi r_L L$$

and is known as log mean area.

Example 12.2 *Steam at 100°C is passing through an SS tube of 5-cm diameter with a wall thickness of 1 mm. The outside temperature is 30°C. Find the rate of heat-transfer through the pipe wall of unit length if the k for steel is 21 W/m°C. Find the % error in q in using average area concept.*

Solution Given L = 1 m

k = 21 W/m°C

$T_i = 100°C$
$T_o = 30°C$
$B = 1 \text{ mm} = 1 \times 10^{-3} \text{ m}$
$r_i = 2.5 \text{ cm} = 2.5 \times 10^{-2} \text{ m}$
$r_o = 2.5 + 0.1 = 2.6 \text{ cm} = 2.6 \times 10^{-2} \text{ m}$

$$r_L = \frac{(r_o - r_i)}{\ln(r_o/r_i)} = \frac{1 \times 10^{-3}}{\ln(2.6/2.5)} = 0.0255$$

$$\therefore \quad A_L = 2\Pi r_L L = 2 \times \Pi \times 0.0255 \times 1 = 0.1602 \text{ m}^2$$

$$\therefore \quad q = A_L k \frac{\Delta T}{B} = 0.1602 \times 21 \times \frac{(100 - 30)}{1 \times 10^{-3}} = 235494 \text{ W}$$

$$= 235.494 \text{ kW}$$

$$\text{Average area} = A_{Av} = 2\Pi L \frac{(r_o - r_i)}{2}$$

$$A_{Av} = \Pi \times 1 \times (2.6 + 2.5) \times 10^{-2} = 0.1602 \text{ m}^2$$

Since there is no difference between A_L and A_{Av}, there is no difference in heat-transfer rates. *We shall find the difference to be significant if the diameter of the pipe is large or the thickness is large.*

12.1.4 Unsteady-State Heat-transfer by Conduction

So far, we have considered the steady-state heat-transfer by conduction; i.e., the temperatures on either sides of the wall are constant, and sufficient time was allowed to reach steady-state conditions*. It was shown in Fig. 12.1(d). Before the steady-state conditions are reached, the heat-transfer is an unsteady state process, in which the temperature along the body of the conductor changes not only with distance (x) but also with time (t), as shown in Fig. 12.1 (b) and Fig. 12.1 (c).

Now we consider the unsteady-state heat-transfer process in a slab which is being heated on both sides at a constant surface temperature of T_s like a slice of bread in a toaster (Fig. 12.4). The situation can be analysed by considering the energy balance equation (Eq. (8.3)) over a small elemental strip of thickness *dx* in Fig. 12.4.

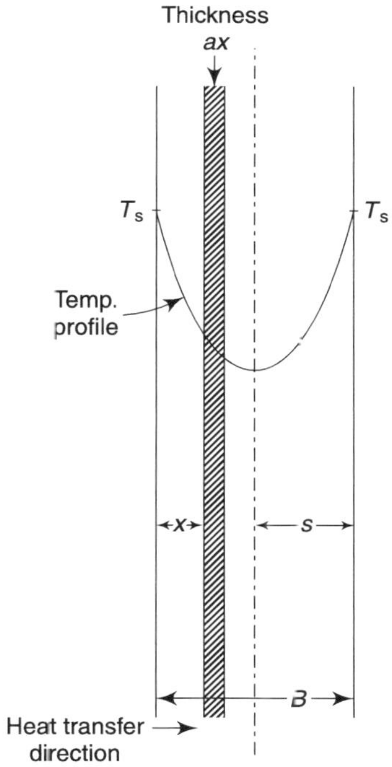

Fig. 12.4 *Unsteady state heat conduction in a slab*

$$\text{Heat entering the strip in time d}t = -(kA)\frac{\partial T}{\partial x}\text{d}t \qquad (12.23)$$

Partial derivatives are used to indicate that T varies both with x and t.

Temperature profile within the strip of thickness dx

$$= \frac{\partial T}{\partial x}\left(\frac{\partial}{\partial x}\right)\frac{\partial T}{\partial x}\text{d}x \qquad (12.24)$$

Heat leaving the strip in time dt

$$= -kA\left(\frac{\partial T}{\partial x} + \frac{\partial}{\partial x}\left(\frac{\partial T}{\partial x}\right)\text{d}x\right)\text{d}t \qquad (12.25)$$

* The process parameters do not change with time.

Heat accumulation in time dt

$$= \text{Heat in by Eq. (12.24)} - \text{Heat out by Eq. (12.26)}$$

$$= -kA\frac{\partial T}{\partial x}\,dt + kA\left(\frac{\partial T}{\partial x} + \frac{\partial}{\partial x}\left(\frac{\partial T}{\partial x}\right)dx\right)dt$$

$$= kA\left(\frac{\partial^2 T}{\partial x^2}\right)dx\,(dt) \tag{12.26}$$

Heat accumulation in the strip is given by

(mass of strip) (specific heat) × (temp variation with time) × (incremental time)

$$= \rho\,(A\,\mathrm{d}x)\,C_\mathrm{p}\left(\frac{\partial T}{\partial t}\right)(dt) \tag{12.27}$$

where ρ is the density of the solid, and
C_p is the specific heat of the solid.

i.e.,

$$kA\,\frac{\partial^2 T}{\partial x^2}\,(dx)\,(dt) = \rho\,(A\,dx)\,C_\mathrm{p}\,\frac{\partial T}{\partial t}\,(dt) \tag{12.28}$$

Dividing throughout by $\rho A\, C_\mathrm{p}\,(dt)\,(dx)$, we get

$$\frac{k}{\rho C_\mathrm{p}}\left(\frac{\partial^2 T}{\partial x^2}\right) = \frac{\partial T}{\partial t} \tag{12.29}$$

Now we define a term known as **thermal diffusivity (α)** where

$$\alpha = \frac{k}{\rho C_\mathrm{p}} \tag{12.30}$$

Combining Eq. (12.29) and Eq. (12.30), we get

$$\frac{\partial T}{\partial t} = \alpha\,\frac{\partial^2 T}{\partial x^2} \tag{12.31}$$

Equation (12.31) is the fundamental equation for unsteady state conduction of heating/cooling.

Integration of Eq. (12.31) gives an expression for unsteady-state heat conduction in slab with half thickness of the slab being equal to s. The solution for Eq. (12.31) was available (Mc Cabe et al, 1993 (a)) relating the surface temperature T_s, initial temperature T_o and average temperature ($\overline{T}$) the slab reaches in total time t_T.

$$\frac{T_\mathrm{s} - \overline{T}}{T_\mathrm{s} - T_\mathrm{o}} = \frac{8}{\Pi^2}\left(\exp\left\{-(\Pi/2)^2\,N_\mathrm{Fo}\right\} + \frac{1}{3^2}\exp\left\{-(3\Pi/2)^2\,N_\mathrm{Fo}\right\} + \frac{1}{5^2}\exp\left\{-(5\Pi/2)^2\,N_\mathrm{Fo}\right\} + \cdots\right) \tag{12.32}$$

Where N_Fo is a dimensionless number and is known as **Fourier number;**

$$N_\mathrm{Fo} = \alpha\,t_\mathrm{T}/s^2 \tag{12.33}$$

in which s is half thickness of the slab

t_T is the total time in which the slab reaches an average temperature $\overline{T}$

Since the heating is from both sides, we take only half the thickness of the slab. In case if the slab is heated from only one side, s is replaced by $2s$ which is equal to B.

Similar kind of expressions can also be obtained by integrating Eq. (12.31) for other shapes of bodies, viz., infinitely long cylinders# and spheres. They are tabulated in Table 12.2. These expressions are useful to find total time required for a given final average temperature ($\bar{T}$) to be reached by calculating the Fourier number. The procedure is complicated and involves trial and error calculations of N_{Fo}. To simplify this procedure, various terms in the expressions can be neglected if $N_{Fo} > 0.1$. Only the first term alone becomes significant. Expressions for total time for various shapes of bodies are given in Table 12.3 for the condition that $N_{Fo} >$ 0.1

Table 12.2 *Unsteady-state heat-transfer solutions for different shapes of material*

S. No	*Material shape*	N_{Fo}	*Integrated solution for* $\frac{T_S - \bar{T}}{T_S - T_O}$	*Reference*	*Eq. no*	*Remarks*
1.	Infinite slab of half thickness, *s*	$\frac{\alpha t_T}{s^2}$	$8/\Pi^2$ (exp $\{-(\Pi/2)^2 N_{Fo}\}$ + $1/3^2$ {exp. $\{-(3\Pi/2)^2 N_{Fo}\}$ + $1/5^2$ exp $\{-(5\Pi/2)^2 N_{Fo}\}$ + ...	Mc-Cabe et al 1993 (a)	12.32	Slab is heated from both sides
2.	Infinitely long cylinder of radius, *r*	$\frac{\alpha t_T}{r^2}$	0.692 exp ($-$ 5.78 N_{Fo}) + 0.131 exp ($-$30.5 N_{Fo}) + 0.00534 exp ($-$74.9 N_{Fo}) + ...	Mc Adams (1954)	12.34	Length of the cylinder *L* is much longer compared to *r*
3.	Sphere of radius, *r*	$\frac{\alpha t_T}{r^2}$	0.608 exp ($-$ 9.87 N_{Fo}) + 0.152 exp ($-$39.5 N_{Fo}) + 0.0676 exp ($-$ 88.8 N_{Fo}) + ...	Carslaw and Jaegar (1959)	12.35	–

Table 12.3 *Expressions for total time taken by different shaped bodies during unsteady state heat conduction to reach final temperature of* $\bar{T}$, *if* $N_{FO} > 0.1$

S. No	*Material shape*	*Expression for* t_T	*Eq. No*
1	Long slab of half thickness, *s*	$\frac{1}{\alpha}\frac{(2s)^2}{\Pi^2}\ln\left(\frac{8(T_s - T_o)}{\Pi^2(T_s - \bar{T})}\right)$	12.36
2	Long cylinder of radius, *r*	$\frac{r^2}{5.78(\alpha)}\ln\left(\frac{0.692(T_s - T_o)}{(T_s - \bar{T})}\right)$	12.37
3	Sphere of radius, r	$\frac{r^2}{9.87(\alpha)}\ln\left(\frac{0.608(T_s - T_o)}{(T_s - \bar{T})}\right)$	12.38

A cylinder with its length much more as compared to its diameter is known as infinitely long cylinder.

12.2 HEAT-TRANSFER BY CONVECTION

Heat-transfer in fluids takes place by convection. In convective heat-transfer, there will be virtually physical movements of layers of the fluid. If we heat water in a vessel or in a beaker, the temperature at the bottom layers of the fluid will rise resulting in decreasing its density. Hence the layer moves up and the high dense upper layers will move downwards and get heated up subsequently. This kind of heat-transfer is known as **convective heat-transfer**. If the temperature of the surface (heating the fluid) is low, only the water vapour bubbles form at the bottom, and they rise. This kind of boiling is known as **nucleate boiling**. If the temperature of the surface is high, the vapour bubbles at the bottom coalesce and form a film. The heat-transfer takes place through the film. This kind of heating is known as **film boiling**.

If the fluid layers move by means of temperature differences alone (and hence density differences), this kind of convection is known as **natural convection or free convection**. On the contrary, if the movement of layers is facilitated by means of an external pump, convection is known as **forced convection**.

12.2.1 Heat-Transfer Between a Metallic Wall and the Fluid

Normally, the fluids are heated in metallic/glass containers. Whenever heat is transferred from a metallic wall to the fluid, there always exists a constant thin film of fluid, which actually offers resistance for the heat-transfer within the film. The temperature varies between the temperature of the surface and the average temperature of the bulk fluid (Fig. 12.5(a)). This concept was originally proposed by Prandtl. The thickness of the film may vary depending upon the movement of the fluid or the degree of agitation given to the fluid.

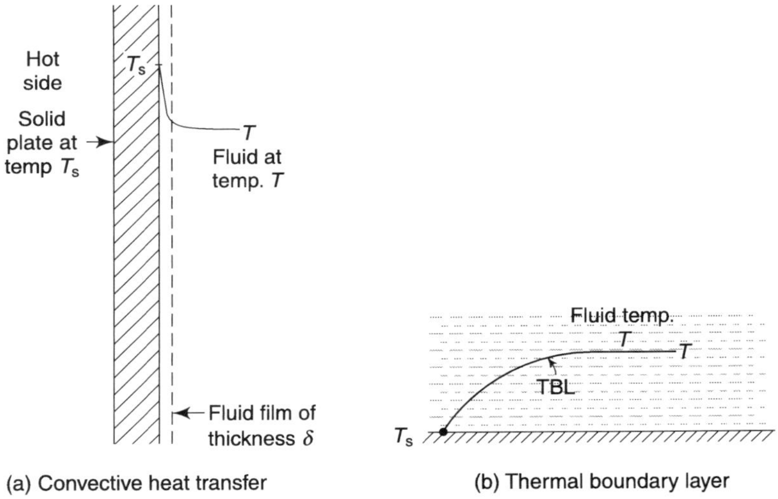

Fig. 12.5 *Convective heat transfer and thermal boundary layer concept*

The film within which the temperature varies is known as the **thermal boundary layer (TBL)**. This is similar to the hydrodynamic boundary layer concept proposed in Chapter 11 and is shown in Fig. 11.2. The thermal boundary layer concept is shown in Fig. 12.6(b).

12.2.2 Heat-Transfer Coefficient

The convective heat-transfer rate is also proportional to the temperature difference and contact area. Thus,

$$q \; \alpha \; \Delta T$$
$$\alpha \; A$$

Combing the above two expressions, we can write

$$q = hA\,(\Delta T) \tag{12.39}$$

in which h is known as the convective **heat-transfer coefficient** and has the units of kJ/(s m^2 °C) or kW/m^2 °C, $1/h$ is known as resistance for heat-transfer

Thus,
$$R = 1/h \tag{12.40}$$

In a differential form, Eq. (12.39) can be written as

$$\frac{dq}{dA} = h\,(T_s - T) \tag{12.41}$$

where T_s is the temperature of the hot surface, and
T is the bulk temperature of the fluid.

h is dependent on the
- fluid properties
- system parameters

and hence needs to be evaluated experimentally

Example 12.3 *A dilute sugar solution is heated continuously from an initial temperature of 30°C to a final temperature of 80°C in a double pipe heat exchanger shown in Fig. E12.2. The flow rate of the sugar solution is 40 lpm. The heat exchanger wall is maintained at 120°C by passing steam through the outer shell. The inside tube diameter is 5 cm and is 3.5 m long. Calculate the heat-transfer rate assuming that the sugar solution has the properties similar to that of water.*

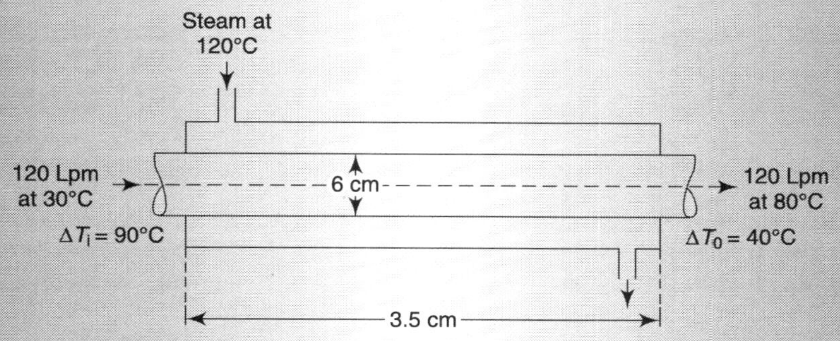

Fig. E12.2 *A double pipe heat exchanger of Example 12.3*

Solution The fluid properties are

$$\rho = 1000 \text{ kg/m}^3$$
$$C_p = 4.2 \text{ kJ/kg°C}$$

Diameter of pipe (d) = 5 cm = 0.05 m

Length of the pipe (L) = 3.5 m

We calculate the heat-transfer rate to the fluid by calculating the heat taken up by the fluid during its travel through the heat exchanger.

$$q = \dot{m}\,C_p\,(\Delta T)$$

where $\dot{m}$ is the mass flow rate
C_p is the specific heat
ΔT is temperature rise = 80 – 30 = 50°C

$$\dot{m} = \text{volumetric flow rate} \times \text{density}$$
$$= 40 \text{ lpm} \times 10^{-3} \times 1000$$
$$= \frac{40}{60} \times 10^{-3} \times 1000 = 0.67 \text{ kg/s}$$

$\therefore$ $$q = 0.67 \times 4.2 \times 50 = 140.7 \text{ kJ/s}$$

By Eq. (12.39), $$q = hA\,(\Delta T)$$

where A is contact area

and ΔT is the temperature difference

Since here the temperature is varying from 30°C to 80°C in the pipe, we find an average temperature difference*, i.e., $\Delta T = \dfrac{\Delta T_i + \Delta T_o}{2}$

where ΔT_i is temperature difference at the inlet and is equal to 120 – 30 = 90°C

ΔT_o is temperature difference at the outlet and is equal to 120 – 80 = 40°C

Hence $\Delta T = \dfrac{90 + 40}{2} = 65°C$

We calculate the contact area based on the inside diameter of the pipe. Hence the heat-transfer coefficient is known as *inside heat-transfer coefficient*, and is represented by h_i

$$A = \Pi\, d\, L = \Pi \times 0.05 \times 3.5$$
$$= 0.55 \text{ m}^2$$

$\therefore$ $$140.7 = h_i \times 0.55 \times 65$$

$\therefore$ $$h_i = \frac{140.7}{0.55 \times 65} = 3.94 \text{ kJ/s m}^2\ °\text{C}$$

12.2.3 Overall Heat-Transfer Coefficient

We come across in many biotechnological processes that heat-transfer between the fluids through a metallic/glass wall when one is hot while the other one is cold. It can be schematically represented by Fig. 12.6(a) in which the hot fluid is at temperature T_h and the cold fluid is at temperature T_c while the surface temperatures of the metallic wall on either side is T_{sh} and T_{sc}. It can better be represented in the form of a shell-and-tube heat exchanger similar to Fig. E12.2. It is shown in Fig. 12.6(b) where the cold fluid, flowing through the tube, is being continuously heated from a temperature at T_{ca} to T_{cb} by exchanging heat with hot fluid flowing through the outside shell, and is getting cooled in the process from an initial temperature of T_{ha} to a final temperature of T_{hb}. The heat-transfer process can be represented as

(i) heat-transfer between the hot fluid and metallic wall by convection in the shell side; heat-transfer coefficient is h_o

(ii) heat-transfer through the metallic wall from the hot side to the cold side by conduction; thermal conductivity of wall is k and the wall thickness is x_w

(iii) heat-transfer between the cold fluid and the hot metallic wall by convection in the inside tube; heat-transfer coefficient is h_i

* In case of shell-and-tube heat exchangers, the temperature difference is calculated by taking the log mean temperature difference (LMTD) which will be described in Ch 13.

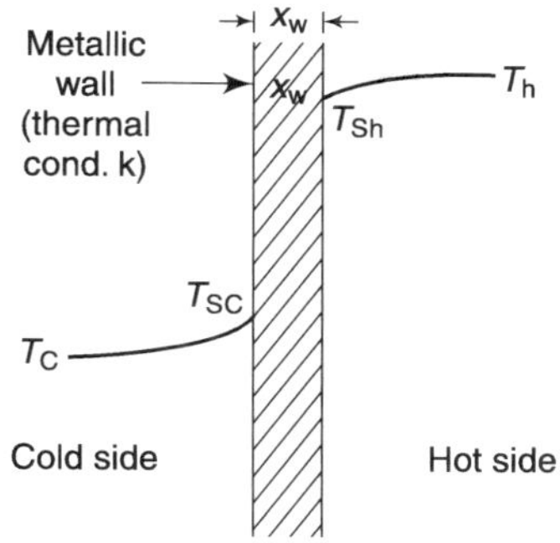

Fig. 12.6 *(a) Schematic representation of convective heat transfer*

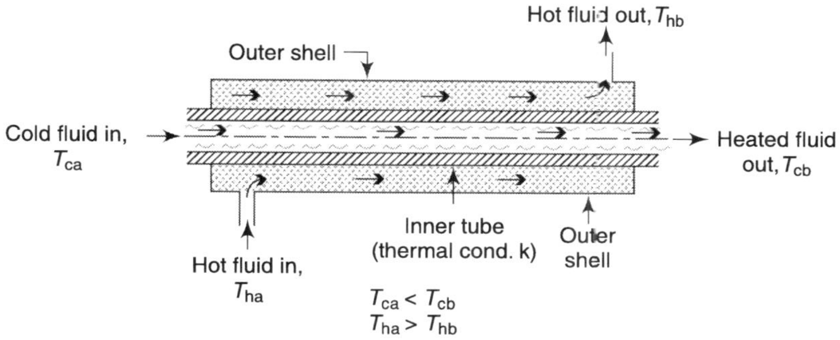

Fig. 12.6 *(b) Convective heat transfer between two fluids*

The contact area inside the tube is A_i and that on the outside of the tube is A_o
Heat flux in the shell side is

$$\frac{dq}{dA_o} = h_o (T_h - T_{sh}) \tag{12.42}$$

Heat flux in the inside of the tube is

$$\frac{dq}{dA_i} = h_i (T_{sc} - T_c) \tag{12.43}$$

Heat flux through the metallic wall is (by Eq. (12.5))

$$\frac{dq}{dA_w} = \frac{k}{x_w} (T_{sh} - T_{sc}) \tag{12.44}$$

The overall temperature difference (ΔT) between the hot fluid and cold fluid is $T_h - T_c$ which can be written as

$$(T_h - T_{sh}) + (T_{sh} - T_{sc}) + (T_{sc} - T_c) = T_h - T_c = \Delta T \tag{12.45}$$

By Eq. (12.42) to Eq. (12.44), we can write

$$\Delta T = dq\left(\frac{1}{dA_i h_i} + \frac{x_w}{k}\frac{1}{dA_w} + \frac{1}{dA_o h_o}\right) \tag{12.46}$$

If we divide and multiply Eq. (12.46) by dA_o, we get

$$\Delta T = \frac{dq}{dA_o}\left(\frac{dA_o}{dA_i}\left(\frac{1}{h_i}\right) + \frac{x_w}{k}\left(\frac{dA_o}{dA_w}\right) + \frac{1}{h_o}\right) \tag{12.47}$$

It is difficult to measure the wall temperatures (T_{sh} and T_{sc}) when the heat-transfer is going on. Hence, we define a term known as the **overall heat–transfer coefficient** based on T_h and T_c (which are easily measurable) and is represented by U.

We now write

$$\frac{dq}{dA} = U(\Delta T) \tag{12.48}$$

The overall heat-transfer coefficients are also represented by inside overall heat-transfer coefficient represented by U_i and outside overall heat-transfer coefficient represented by U_o. Thus,

$$\left.\begin{aligned} \frac{dq}{dA_i} &= U_i(\Delta T) \\ \text{and} \quad \frac{dq}{dA_o} &= U_o(\Delta T) \end{aligned}\right\} \tag{12.49}$$

in all the above equations, the areas are represented by $\Pi(d)L$, and hence the ratios are represented by ratio of the diameters. Thus

$$\frac{dA_o}{dA_i} = \frac{d_o}{d_i} \tag{12.50}$$

and

$$\frac{dA_o}{dA_w} = \frac{dA_o}{dA_L} = \frac{d_o}{d_L} \tag{12.51}$$

where $d_L = 2r_L$

By Eq. (12.47),

$$\frac{dq}{dA_o} = \frac{\Delta T}{\frac{dA_o}{dA_i}\left(\frac{1}{h_i}\right) + \frac{x_w}{k}\left(\frac{dA_o}{dA_L}\right) + \frac{1}{h_o}}$$

$$= \frac{\Delta T}{\frac{d_o}{d_i}\left(\frac{1}{h_i}\right) + \frac{x_w}{k}\left(\frac{d_o}{d_L}\right) + \frac{1}{h_o}} \tag{12.52}$$

Combining this with Eq. (12.49), we write

$$U_o = \frac{1}{\frac{d_o}{d_i}\left(\frac{1}{h_i}\right) + \frac{x_w}{k}\left(\frac{d_o}{d_L}\right) + \frac{1}{h_o}} \tag{12.52}$$

or

$$\frac{1}{U_o} = \frac{d_o}{d_i}\left(\frac{1}{h_i}\right) + \frac{x_w}{k}\left(\frac{d_o}{d_L}\right) + \frac{1}{h_o} \tag{12.53}$$

Similarly, we can write for inside heat-transfer coefficient as

$$\frac{1}{U_i} = \frac{1}{h_i} + \frac{x_w}{k}\left(\frac{d_i}{d_L}\right) + \frac{d_i}{d_o}\left(\frac{1}{h_o}\right) \tag{12.54}$$

For a slab, since the areas are same on either side,

$$\frac{1}{U} = \frac{1}{h_i} + \frac{x_w}{k} + \frac{1}{h_o} \tag{12.55}$$

$1/U$ represents the overall resistance while $1/h_i$, x_w/k and $1/h_o$ represent individual resistance in series

Equation (12.55) is similar to Eq. (12.13) which indicates that the overall resistance is equal to the sum of the individual resistances in series.

12.2.4 Unsteady-state Convective Heat-Transfer

The convective heat-transfer process so far we have seen is restricted to steady-state heat-transfer process, which we rarely come across. Imagine a fermenter containing a fermentation broth at room temperature which needs to be heated to a higher temperature for fermentation to start with; or let us say we want to heat a medium from room temperature to sterilisation temperature at 120°C in a batch process. The heating process is shown in Fig. 12.9(a). In all these cases, what we encounter is an unsteady-state process where the heat-transfer rate varies from the beginning when the broth (fluid) temperature is say T_o till the end when the temperature of the broth will be attaining the desired temperature. The ΔT will be continuously varying maximum to minimum and the q will also be varying from maximum to minimum (Fig. 12.9(b)).

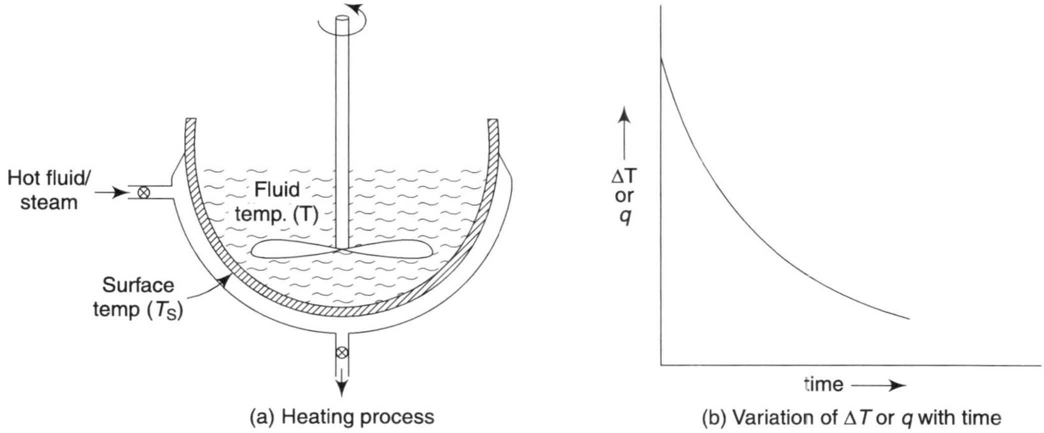

Fig. 12.7 *Unsteady state heat transfer by convection*

We shall consider the situation at any time t when the temperature of the fluid is T. The surface temperature is constant at T_s.

Heat required to raise the temperature of the fluid by dT degrees in a small elemental time of dt is given by

$$Q = \rho V C_p dT$$

Heat-transfer rate,

$$q = \rho V C_p \frac{dT}{dt}$$

which is also equal to $h_i A(T_s - T)$

$$\therefore \quad \rho V C_p \frac{dT}{dt} = h_i A (T_s - T) \tag{12.56}$$

in which h_i is inside heat-transfer coefficient.

The above equation is written with the following assumption:

(i) There is no/negligible internal resistance to heat-transfer.

(ii) The fluid is adequately stirred to ensure that there are no temperature gradients within the fluid; i.e., the fluid temperature is uniform at T.

Equation (12.56) can be written as

$$\frac{dT}{(T_s - T)} = \frac{h_i A}{\rho V C_p} dt \tag{12.57}$$

which on integration between the limits

$$T = T_o \quad \text{when} \quad t = 0$$
$$T = T \quad \text{when} \quad t = t$$

gives

$$\ln\left(\frac{T_s - T_o}{T_s - T}\right) = \frac{h_i A}{\rho V C_p}(t) \tag{12.58}$$

The above equation is applicable for heating/cooling of fluids with negligible internal resistance to heat-transfer, and can be used for evaluating the heat-transfer coefficient by collecting data on T vs t.

Example 12.4

A 55-litre of fermentation medium is heated from an initial temperature of 30°C in a 60 cm dia hemispherical vessel. The jacketed vessel is heated by passing a high-pressure stream at 2 atm pressure. The temperature versus time data are as follows:

Time, min	0	1	2	3	4	5
Temp, °C	30	44	58	70	75	84

(a) Calculate the heat-transfer coefficient assuming that the fluid has properties like those of water.

(b) Calculate the time taken to heat the medium to 100°C.

(c) Calculate the quantity of steam required to heat the medium to 100°C.

Solution From the steam tables, we find that the steam temperature corresponding to a pressure of 2 atm.

$T_s = 120°C$

$\lambda_s = 2202.6$ kJ/kg

Given data:

$T_o = 30°C$

$V = 55$ litre $= 55 \times 10^{-3}$ m^3

Tank dia = 60 cm = 0.6 m

Hemispherical tank radius (r_t) = 0.3 m

Area of hemispherical tank = $2\Pi\, r_t^2$

$A = 2\Pi\,(0.3)^2 = 0.565$ m^2

Properties of medium = properties of water

∴ $\rho = 1000$ kg/m^3

$C_p = 4.2$ kJ/kg

(a) From the given data on t vs T, we tabulate $\ln\left(\frac{T_s - T_o}{T_s - T}\right)$ in Table 12.4, and draw the plot (Fig. E12.3) between t and $\ln((T_s - T_o)/(T_s - T))$.

Table E12.1 *Data of Example 12.4*

t min	*T°C*	*t, s*	$\frac{T_s - T_o}{T_s - T}$	$\ln \frac{T_s - T_o}{T_s - T}$
0	30	0	1	0
1	44	60	1.184	0.169
2	58	120	1.45	0.373
3	70	180	1.8	0.588
4	75	240	2.0	0.693
5	84	300	2.5	0.916

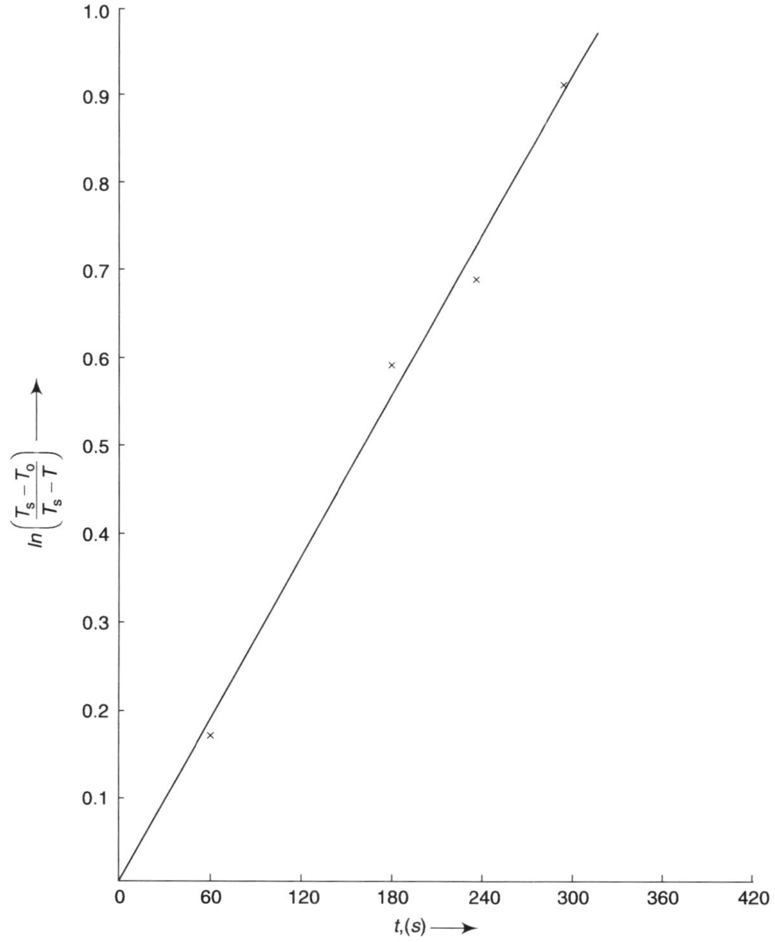

Fig. E12.3 *Data for Example 12.4*

The slope of line in Fig. E12.3 is measured.

$$\text{Slope} = \frac{0.92}{300} = 3.07 \times 10^{-3}\ \text{s}^{-1}$$

By Eq. (12.58), the slope $= \dfrac{h_i\, A}{\rho V C_p}$

i.e., $$3.07 \times 10^{-3} = \frac{h_i \times 0.565}{1000 \times 55 \times 10^{-3} \times 4.2} = 2.446 \times 10^{-3} \times h_i$$

$$\therefore \quad h_i = \frac{3.07 \times 10^{-3}}{2.446 \times 10^{-3}} = 1.255\ \text{kJ/m}^2\text{-s-°C}$$

(b) Time required to heat the medium to 100°C, i.e., $T = 100$°C

$$\ln \frac{T_s - T_o}{T_s - T} = \ln\left(\frac{120 - 30}{120 - 100}\right) = 1.5$$

$$= \frac{h_i\, A}{\rho V C_p}\, t = (3.07 \times 10^{-3})\,(t)$$

$$\therefore \quad t = \frac{1.5}{3.07 \times 10^{-3}} = 490\ \text{s}$$

$$\cong 8.2\ \text{min}$$

(c) Quantity of steam required to heat up to 100°C.

Quantity of heat required to raise the temperature of the medium from 30°C to 100°C.

$$= m\, C_p\,(\Delta T) = (\rho\, V)\, C_p\,(\Delta T)$$

$$= (1000 \times 55 \times 10^{-3}) \times 4.2 \times (100 - 30)$$

$$= 16170\ \text{kJ}$$

Latent heat of steam at a pressure of 2.0 atm is

$$\lambda_s = 2202.6\ \text{kJ/kg}$$

$$\therefore \quad \text{quantity of steam required} = \frac{Q}{\lambda_s} = \frac{16170}{2202.6} = 7.34\ \text{kg}$$

12.2.5 Calculation of Heat-transfer Coefficient

Heat-transfer coefficients are to be generally determined by experimental methods only. However in literature, there are some available correlations for predicting the heat-transfer coefficients for the systems based on

(i) fluid properties
(ii) flow conditions
(iii) system geometry
(iv) heat-transfer conditions
- by natural convection
- by forced convection

(v) type of equipment used

All the correlations are based on certain dimensionless numbers which are important, and are not affected by the units of the system (viz., CGS, FPS or SI units) nor by the process parameters. These dimensionless numbers used in heat-transfer are tabulated in Table 12.4.

Table 12.4 *Dimensionless numbers in heat-transfer*

Dimensionless number	*Notation number*	*Expression*	*Eq. No*
Reynolds no.	N_{Re}	$\rho du/\mu$	11.4
Impeller Reynolds no.	$N_{Re,\,i}$	$\rho d_i^2 N_i/\mu$	11.8
Nusselt no.	N_{Nu}	hd/k	12.59
Prandtl no.	N_{Pr}	$C_p\mu/k$	12.60
Pecklet no.	N_{Pe}	$(N_{Re})(N_{Pr}) = \rho duC_p/k$	12.61
Grashof no.♣	N_{Gr}	$d^3 g\beta\rho^2(\Delta T)/\mu^2$	12.62
Rayleigh no.	N_{Ra}	$(N_{Gr})\,(N_{Pr})$	12.63
Stanton no.	N_{St}	$N_{Nu}/(N_{Pr}\cdot N_{Re})$	12.64
Graetz no.	N_{Gz}	$\dot{m}\ C_p/k\,L = \dfrac{\Pi}{4}\dfrac{d}{L}\,(N_{Fe})$	12.65
Fourier no.♠	N_{Fo}	$\alpha t_T/r^2$	12.66

Calculation of h under laminar flow conditions for natural convection There are some predictive equations available in literature for air and water under natural convection conditions (Mc-Adams, 1954).

For horizontal cylindrical pipes for both heating and cooling,

$$h = 1.3196 \times \left(\frac{\Delta T}{d}\right)^{0.25} \quad \text{for air} \tag{12.67}$$

$$h = 291.1 \times \left(\frac{\Delta T}{d}\right)^{0.25} \quad \text{for water} \tag{12.68}$$

The above equations are also applicable for heat-transfer in vertical pipes with a small change in the numerical coefficient.

$$h = 1.3683 \times \left(\frac{\Delta T}{d}\right)^{0.25} \quad \text{for air} \tag{12.69}$$

$$h = 127.1 \times \left(\frac{\Delta T}{d}\right)^{0.25} \quad \text{for water} \tag{12.70}$$

They are also applicable for vertical flat plates# with the d replaced by L (the length of the flat plate) in the above equations.

Calculation of h under laminar flow conditions for forced convection It is difficult to maintain laminar flow conditions under forced convection. The Graetz numbers approach the limiting value of 3.66 which make the calculation of h difficult (Mc Cabe et al, 1993(b)). However, with reasonable amount of accuracy, we can write for $N_{Gz} > 20$ as follows:

$$N_{Nu} \cong 2.0\,(N_{Gz})^{1/3} \tag{12.71}$$

♣ Ratio of gravitational forces to the buoyancy forces resulting due to ΔT.

♠ in convective heat transfer, $t_T = L/u$ where L is the length of the pipe.

We come across this kind of heat transfer in plate heat exchangers.

If the wall temperatures are very much different from the bulk temperatures of the fluids, we need to account for effect of change of viscosity of the fluid layers near to the wall and viscosity of the bulk fluid at its average temperature. The effect is usually represented by ϕ_v where

$$\phi_v = \left(\frac{\mu}{\mu_w}\right)^{0.14} \tag{12.72}$$

where μ_w represents the viscosity of the fluid at the wall temperature.

Considering the change in viscosity effect, Eq. (12.71) modifies to

$$N_{Nu} = 2.0\,(N_{Gz})^{1/3}\,(\phi_v) \tag{12.73}$$

In some places, the above equation is reported (Toledo, 1997) as follows:

$$N_{Nu} = 1.75\,(N_{Gz})^{1/3}\,(\phi_v) \tag{12.74}$$

For non-Newtonian fluids, Eq. (12.73) modifies to

$$N_{Nu} = 2.0\left(\frac{3n+1}{4n}\right)^{1/3}(N_{Gz})^{1/3}\left(\frac{\mu_a}{\mu_{a,w}}\right)^{0.14} \tag{12.75}$$

where μ_a is the apparent viscosity and is given by

$$\mu_a = K\,\gamma^{n-1} \tag{11.6}$$

in which K is the flow behavior index
n is the consistency index
γ is the shear strain of the non-Newtonian fluid

Calculation of** h **under turbulent conditions Heat-transfer is better under turbulent conditions, and is usually represented by a relationship of N_{Re} and N_{Pr}. For flow through pipes, *Dittus–Boelter equation* adequately represents a relationship for heat-transfer in the form of dimensionless numbers as follows:

$$N_{Nu} = 0.023\,(N_{Re})^{0.8}\,(N_{Pr})^{1/3} \tag{12.76}$$

Sieder–Tate equation modifies above equation by taking into consideration the variations of viscosity as follows:

$$N_{Nu} = 0.023\,(N_{Re})^{0.8}\,(N_{Pr})^{1/3}\,\phi_v \tag{12.77}$$

It is to be noted that the Dittus–Boelter equation or Sieder–Tate equation is applicable for flow through pipes without phase change of the fluid and with the following conditions:

$$10^4 \le N_{Re} \le 1.2 \times 10^5$$

$$0.7 \le N_{Pr} \le 120 \tag{12.78}$$

And $L/D \ge 60$

In case of **stirred vessels,** the heat-transfer coefficient is very much influenced by the degree of agitation which is indicated by the impeller Reynolds number (N_{Rei}) where N_{Rei} was defined as

$$N_{Rei} = \frac{\rho d_i^2 N_i}{\mu} \tag{11.18}$$

For heat-transfer in stirred jacketed vessels (Chilton et al, 1944),

$$N_{Nu} = 0.36\,(N_{Rei})^{2/3}\,(N_{Pr})^{1/3}\,\phi_v \tag{12.79}$$

In case of jacketed vessels, the Nusselt no modifies as follows:

$$N_{Nu} = \frac{h_j\, d_t}{k} \tag{12.80}$$

where h_j stands for heat-transfer coefficient of the jacketed vessel.

If the stirred vessels are heated by a helical coil, the heat-transfer improves, and is given by

$$N_{Nu} = 0.87\,(N_{Rei})^{0.62}\,(N_{Pr})^{1/3}\,\phi_v \tag{12.81}$$

12.2.6 Design of Heat Exchanger Equipment

Design of heat exchanger equipment mainly involves in finding out the heat-transfer area for affecting a desired amount of heat to be transferred or for achieving a desired amount of heat-transfer rates. For this purpose, we need to know what is the total amount of heat to be transferred which we can evaluate based on the

- quantity of fluid / fluid flow rate (m or $\dot{m}$)
- fluid properties such as heat capacity (C_p)
- temperature rise to be achieved (ΔT)

$$Q = \mathrm{m}\, C_p\,(\Delta T) \tag{12.82}$$

or

$$q = \dot{m}\, C_p\,(\Delta T) \tag{12.83}$$

Or in case of microbial fermentation, the heat generated by consumption of oxygen by microbial cells is to be removed. The heat generated by microbial cells is dependent upon

- specific oxygen demand by cells (Q_{SOD})
- cell density (x)
- heat evolved for each gram of oxygen consumed by the cells*.

Thus,

$$q = 14.4 \times 10^3\, Q_{SOD}\, xV \tag{12.84}$$

The heat-transfer rate calculated by the above equations is equated to

$$q = hA\,(\Delta T) \tag{12.39}$$

To evaluate A in the above equation, we need to know the value of h which is calculated by any of the predictive equations (12.71 to 12.81) depending upon the type of heat-transfer equipment used and upon the type of flow conditions.

Example 12.5

A freshly prepared nutrient medium for a fermentation process is heated in a stirred jacketed vessel of 1m dia. The diameter of the stirrer is 0.35 m, and is rotating at a speed of 60 rpm. The physico-chemical and thermal properties of the fluid are as follows:

$$\mu = 3 \times 10^{-3}\ Pas$$
$$k = 0.65\ W/m°C$$
$$\rho = 1010\ kg/m^3$$
$$C_p = 4.4\ kJ/kg°C$$

Neglecting the effect of viscosity changes, calculate the heat-transfer coefficient.

---

* This value is reported to be 460 kJ for every gram mole of oxygen absorbed which amounts to 14.4×10^3 kJ/kg of oxygen absorbed; further details on it are available in Sec. 14.4.3.

Solution *Given data*

$d_t = 1$ m

$d_i = 0.35$ m

$N_i = 1$ rps

Approach

We need to calculate N_{Rei} and N_{Pr}, and then apply Eq. (12.79) to calculate N_{Nu} from which we can evaluate h by Eq. (12.80)

$$N_{Rei} = \frac{\rho d_i^2 N_i}{\mu} = \frac{1010 \times (0.35)^2 \times 1}{3 \times 10^{-3}} = 41{,}242$$

$$N_{Pr} = \frac{C_p \mu}{k} = \frac{4.4 \times 3 \times 10^{-3}}{0.65 \times 10^{-3}} = 20.3$$

$$N_{Nu} = 0.36\,(N_{Re})^{2/3}\,(N_{Pr})^{1/3}$$

$$= 0.36 \times 1193.7 \times 2.728 = 1172.3$$

$$\therefore \quad h = \frac{N_{Nu} \times k}{d_t} = \frac{1172.3 \times 0.65 \times 10^{-3}}{1} = 0.762 \text{ kJ/m}^2\text{s°C}$$

Example 12.6 *The above fermenter vessel is used to heat continuously the nutrient medium flowing at the rate of 1 kg/s from an initial temperature of 30°C to a final desired temperature of 100°C. The fermenter is heated by passing steam through the outer jacket. The steam pressure is 0.2 MN/m². If the liquid height in the fermenter is kept equal to its diameter, is it possible to achieve the final temperature of 100°C? If not, what temperature can be achieved?*

Solution ***Approach***

Heat flow rate requirement is calculated by Eq. (12.83). We try to find out the heat-transfer coefficient by Eq. (12.39) and compare it with the one predicted in Example 12.5, where

$$h_{calculated} = 0.762 \text{ kJ/m}^2\text{s°C}$$

Given data:

$\dot{m} = 1$ kg/s

$T_1 = 30$°C

$T_2 = 100$°C

Steam temperature corresponding to a steam pressure of 0.2 MN/m^2 = 120°C

i.e., $T_{steam} = 120$°C

$$q = \dot{m}\, C_p\, \Delta T = 1 \times 4.4\,(100 - 30) = 308 \text{ kJ/s}$$

This is the heat-transfer rate required to raise the temperature of medium flowing at a flow rate of 1 kg/s and entering at a temperature of 30°C and leaving at 100°C. This heat-transfer rate should be achievable by the vessel to heat the fluid to the desired temperature.

Heat-transfer rate in the vessel is given by

$$q_{vessel} = h\,A\,(\Delta T)_{heating}$$

To evaluate A

The liquid height in the tank = dia of the tank

Contact area= $A = \Pi\, d_t\, L = \Pi\, d_t^2 = \Pi \times 1^2 = 3.14$ m^2

To evaluate $(\Delta T)_{heating}$

The surface temperature = T_s = 120°C

Inlet temp of the fluid = 30°C

Outlet temp of the fluid = 100°C

$(\Delta T)_{heating}$ is evaluated by averaging the temperature difference at inlet and outlet temperatures

$$\therefore \quad (\Delta T)_{heating} = \frac{(120-30)+(120-100)}{2} = 55 \text{ C}$$

Now $\quad q_{vessel} = h\,A\,(\Delta T)_{heating}$

$$= 0.762 \times 3.14 \times 55 = 131.7 \text{ kJ/s}$$

Since q_{vessel} is less than q, it is not possible to achieve desired temperature for the flowing fluid. Let us say under the heat-transfer rate available, the achievable final temperature of the fluid is T_f.

$$q = \dot{m}\,C_p\,(\Delta T) = 1 \times 4.4\,(T_f - 30)$$

$$(\Delta T)_{heating} = \frac{(120-30)+(120-T_f)}{2} = 105 - (T_f/2)$$

$$q_{vessel} = hA\,(\Delta T)_{heating} = 0.762 \times 3.14 \times (105 - T_f/2)$$

$$q = q_{vessel}$$

$$4.4 \times (T_f - 30) = 0.762 \times 3.14 \times (105 - T_f/2)$$

$$4.4T_f - 132 = 251.23 - 1.2T_f$$

$$5.6\,T_f = 383.23$$

$$\therefore \quad T_f = 68.4°\text{C}$$

The above calculations show that it is not at all possible to achieve the final temperature of the medium. Either we have to increase the T_s by increasing steam pressure or reduce the flow rate of the fluid.

12.3 HEAT-TRANSFER BY RADIATION

Radiation is one of the methods of heat-transfer and finds good applications in process operations. Radiative heat-transfer is applicable both for cooling and heating. Any of the materials sterilised and/dried in oven receive radiative heat from the oven walls or furnace walls. Materials kept in cold-storage rooms lose heat by radiation. Freezing of materials or chilling of materials is done by radiation heat-transfer when the materials lose heat by radiation to the walls of the chamber.

Unlike the other two methods of heat-transfer (viz., conduction and convection), radiation heat-transfer takes place without any medium to transfer heat. The transfer of heat is affected by electromagnetic waves emitted by a thermal body since no medium is involved in heat-transfer, the heat absorbed by a body by radiation depends upon the nature of the receiving body, and the orientation of the body with respect to the heat source. For example, the transfer of solar heat from sun to earth is by radiation. The black bodies receive more heat compared to white and shining bodies.

12.3.1 Principles of Radiation

When a radiant heat wave impinges on a flat surface, some part of it would be reflected, and some part of it would be absorbed while the rest of it would be transmitted (as shown in Fig. 12.8). The extent of these transfer processes depends upon the nature of the material. If the body is white in colour and shining, most of

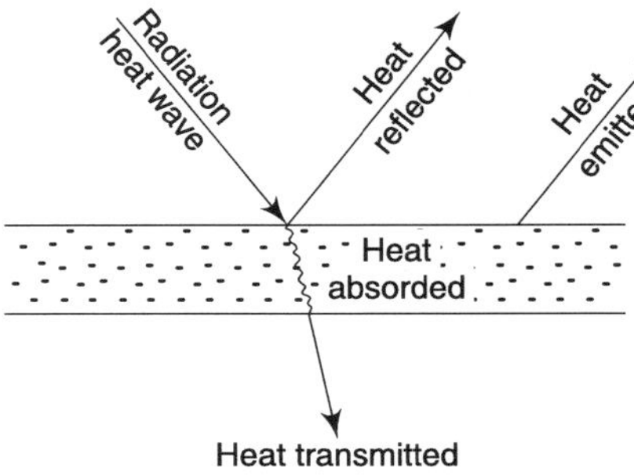

Fig. 12.8 *Radiation heat transfer through and in a body*

the heat would reflect back. In moderate bodies, radiant heat would be partly absorbed and rest of it would be transmitted. In case of black bodies, most of the radiation heat would be absorbed by the body. Based on the above discussion, a perfect black body is defined as one which would absorb all the radiant heat impinging on it, while reflecting none and transmitting none.

The fraction of heat that reflects is known as reflectivity (ν), while the fraction of heat that is transmitted is known as transmissivity (τ). The fraction of heat that is absorbed is known as absorptivity (α).

$$\therefore \quad \alpha + \nu + \tau = 1.0 \tag{12.85}$$

For a perfect black body, $\alpha = 1$. Materials like carbon soot and some grades of carbon black almost approach a black body characteristics.

Emissivity Any body at any temperature will be emitting radiations of heat which may in turn be absorbed by some other bodies. Other bodies may also emit radiations. The emission radiations of one body are independent of the radiations by other bodies. Thus, if the bodies at two different temperatures are in close proximity, the hotter body loses heat at a rate faster by emission than the rate at which it receives radiations from the colder body, with the result that it gets cooled. Similar is the case with the colder body also and it gets hotter in the process. Ultimately, both the bodies reach thermal equilibrium.

A black body will have maximum emissivity as compared to any other bodies which are known as **gray bodies**. Thus, all bodies are classified, on the basis of emissive radiations, as

- black bodies
- gray bodies

Now we define the emission power of other bodies (gray bodies) as the ratio of their emissive power (W) to that of the black body (W_b), i.e.,

$$\varepsilon = \frac{W}{W_b} \tag{12.86}$$

The emission power will have units of W/m^2 and is the heat flux by radiation. The emissivity varies slightly with temperature, and is more dependent upon the state of surface. For example, for brass surface,

for oxidized brass surface at 611°K,	$\varepsilon = 0.22$
polished brass surface at 573°K,	$\varepsilon = 0.032$
tarnished brass surface at 329°K,	$\varepsilon = 0.202$
Similarly for gold, not polished,	$\varepsilon = 0.47$
For gold, polished,	$\varepsilon = 0.025$

Different values for normal emissivity for various bodies are available in literature, and can be found in Kreith and Bohn (2001) textbook.

Kirchhoff's Law If two bodies at different temperatures are brought together, they will attain temperature equilibrium. Kirchhoff's law is applicable when the bodies reach temperature equilibrium. It states that the ratios of their emissive power to absorbivity are equal; and the ratios depend upon the temperature of the bodies. We can write Kirchhoff's law as follows:

$$\frac{W_1}{\alpha_1} = \frac{W_2}{\alpha_2} \tag{12.87}$$

Where W_1 and W_2 are emissive powers of the bodies in Watts/m^2

In the above situation, if one of the bodies is a black body (say body 2) for which the absorbivity is equal to one, we can write

$$\frac{W_1}{\alpha_1} = W_b \tag{12.88}$$

i.e.,

$$\frac{W_1}{W_b} = \alpha_1 \tag{12.89}$$

Comparing Eq. (12.89) and Eq. (12.86), we can write

$$\alpha_1 = \varepsilon_1 \tag{12.90}$$

Thus, *if a body is in temperature equilibrium with its surroundings, its absorbivity is equal to its emissivity*. This forms an alternate statement for **Kirchhoff's law**.

Kirchhoff's law has many practical applications in day-to-day life with reference to radiations. A white paint coated on roof tops will have maximum reflectivity, and hence will have least absorptivity and least emissivity. This will keep the room relatively cooler during hot solar radiations as compared to a room coated with black paint.

Stefan–Boltzman law This law helps us estimate the heat radiations from a radiating body. It states that the *radiating heat flux from a black body is proportional to the fourth power of its temperature measured in degrees Kelvin.*

i.e.,

$$\frac{q_r}{A} \; \alpha \; T^4 \tag{12.91}$$

or

$$\frac{q_r}{A} = \sigma T^4 \tag{12.92}$$

where T is the temperature of the black body in degrees K,
σ is known as Stefan–Boltzman constant and
q_r is radiation heat-transfer rate in W

If A is in m^2, $\sigma = 5.67 \times 10^{-8}$ W/m^{2}K^4

Equation (12.92) clearly shows that any black body, whose surface temperature > 0 K, radiates heat. Even though q_r is independent of surroundings, a net heat-transfer rate can emerge only if the surface temperature of the surroundings or another black body in its vicinity has a different temperature. Let us imagine a black body 1 at the temperature T_1 is enclosed in black surroundings at a temperature T_2; then the black body 1 will be radiating heat to the surroundings, and the surroundings will also be radiating heat to black body 1. Thus the net heat-transfer rate from a black body 1 is

$$q_r = A_1 \, \sigma \, (T_1^4 - T_2^4) \tag{12.93}$$

where A_1 is the surface area of the black body 1.

If the body is other than the black body (which is known as *grey body*), it also will be emitting radiations, but its radiating emission as compared to a black body will be governed by Eq. (12.88). Thus, for any grey body, Eq. (12.93) modifies to

$$q_r = A\,\varepsilon\,\sigma\,(T_1^4 - T_2^4) \tag{12.94}$$

where ε is the emissivity of the body.

12.3.2 Radiative Heat-Transfer Between Two Bodies

If two bodies at two different temperatures, say T_1 and T_2, are in the vicinity, the body at the higher temperature will be losing heat to the body at the lower temperature by radiation and vice versa. The radiating heat-transfer is dependent upon the

- distance between the bodies
- emissivities of both bodies
- nature of the bodies whether they are black bodies or grey bodies
- type of their surfaces (whether they are smooth, or rough, or polished, or dull)
- geometries of the surfaces of the bodies (whether they are convex–shaped or concave-shaped
- orientation of the surfaces of the bodies (whether they are parallel to each other, or perpendicular to each other, or inclined to some angle in between)

For parallel and smooth surfaces, the heat-transfer rate is given by

$$q_r = A\,\sigma\,F\,(T_1^4 - T_2^4) \tag{12.95}$$

where F is a dimensionless geometric factor to account for the above factors.

Heat-Transfer from/to Surroundings Rather than the radiative heat-transfer between two bodies, we realise better applications in biotechnological processes by the radiative heat-transfer between a body and its surroundings. It could be for heating a body in an oven or cooling of a body in a cold room or a cooling chamber. During heat-transfer between a body and its surroundings, usually the body is small in size compared to the surrounding chamber. For example, it could be freeze drying of blood samples or oven drying of baker's yeast etc. Hence we refer the surface area A to that of the body. If we consider the emissivity of the body to be ε, we can write for radiative heat-transfer rate by Eq. (12.94):

$$q_r = A\,\varepsilon\,\sigma\,(T_1^4 - T_2^4)$$

where T_1 is the temperature of the hot surface and T_2 is the temperature of the cold source.

Now we shall try to write an expression for heat-transfer similar to convective heat-transfer by Eq. (12.39):

$$q_c = h_c\,A\,(T_1 - T_2) \tag{12.96}$$

where h_c is convective heat-transfer coefficient.

Combining Eq. (12.94) and Eq. (12.96), we can write

$$h_c\,(T_1 - T_2) = \varepsilon\,\sigma\,(T_1^4 - T_2^4) \tag{12.97}$$

$$T_1^4 - T_2^4 = (T_1^2 - T_2^2)(T_1^2 + T_2^2) = (T_1 - T_2)(T_1 + T_2)(T_1^2 + T_2^2)$$

i.e.,

$$h_c = \varepsilon\,\sigma\,(T_1 + T_2)(T_1^2 + T_2^2) \tag{12.98}$$

Example 12.7

A block of baker's yeast at the room temperature of 30°C is in the form of a cuboid (20 × 10 × 10 cm). It is dried in an oven maintained at 100°C. The emissivity of the yeast block may be taken as 0.55. In addition to radiation heat, the block receives convective heat also from the oven by natural circulation of air. Calculate the total heat-transfer rate.

Solution ***Approach***
Since the block is receiving both convective heat and radiation heat, we useEq. (12.67), Eq. (12.39) and Eq. (12.94)
Given data:

$$T_1 = 100°\text{C} = 373 \text{ K}$$
$$T_2 = 30°\text{C} = 303 \text{ K}$$
$$\Delta T = 70°\text{C}$$
$$A = 3(20 \times 10) + 2(10 \times 10)^*$$
$$= 800 \text{ cm}^2 = 0.08 \text{ m}^2$$
$$\varepsilon = 0.55$$
$$\sigma = 5.669 \times 10^{-8} \text{ W/m}^2 \text{ K}^4$$
$$T_1^4 = (373)^4 = 1.94 \times 10^{10}$$
$$T_2^4 = (303)^4 = 0.84 \times 10^{10}$$

∴ heat-transfer by radiation (Eq. (12.94))

$$q_r = 0.08 \times 5.669 \times 10^{-8} \times 0.55\ (1.94 \times 10^{10} - 0.84 \times 10^{10})$$
$$= 27.26 \text{ W}$$

Heat-transfer by convection We calculate the convective heat-transfer coefficient (h_c) for natural convection by Eq. 12.67.

$$h_c = 1.3196\ (\Delta T/x)^{0.25}$$

where $x = 20 \text{ cm} = 0.2 \text{ m}$, and

∴ $h_c = 1.3196\ (70/0.2)^{0.25} = 1.3196 \times 4.325 = 5.7 \text{ W/m}^2\text{K}$

∴ $q_c = h_c A\ (\Delta T) = 5.7 \times 0.08 \times 70 = 31.96 \cong 32 \text{ W/m}^2\text{K}$

∴ total heat-transfer rate $= q_r + q_c$

$$= 27.26 + 32 = 59.26 \text{ W/m}^2\text{K}$$

Symbols

A	: area, (m^2)
B	: thickness of the slab (m)
C_p	: specific heat (kJ/kg °C)
d	: diameter
d_i	: impeller diameter (m)
d_L	: log mean diameter
d_t	: tank diameter
F	: parameter in Eq. (12.88)
g	: acceleration due to gravity (m/s^2)
h	: heat-transfer coefficient (kJ/(m^2 s °C))
h_j	: heat-transfer coefficient of jacketed vessel
h_r	: radiative heat-transfer coefficient
k	: thermal conductivity (kW/m °C)
K	: flow behaviour index in power law fluid model

*One 10 × 20 cm face is at the bottom and hence is not exposed to heat transfer.

L : length (m)
$\dot{m}$: mass flow rate (kg/s)
n : consistency index in power law fluid model
N_{Fo} : Fourier number (Eq. 12.66 or Eq. 12.33)
N_{Gr} : Grashof number (Eq. 12.62)
N_{GZ} : Graetz number (Eq. 12.65)
N_i : rps of the impeller (s^{-1})
N_{Nu} : Nusselt number (Eq. 12.59)
N_{Pe} : Pecklet number (Eq. 12.61)
N_{Pr} : Prandtl number (Eq. 12.60)
N_{Ra} : Rayleigh number (Eq. 12.63)
N_{Re} : Reynolds number
N_{Rei} : impeller Reynolds number
N_{St} : Stanton number (Eq. 12.64)
q : heat-transfer rate (kW)
q_r : radiation heat-transfer rate (W)
Q : quantity of heat-transferred (kJ)
Q_{SOD} : specific oxygen demand by cells
R : resistance for heat-transfer (m^2/KW)
r : radius (m)
r_t : tank radius (m)
r_L : log mean radius (m)
s : half thickness of the slab (m)
t : time(s or h)
t_T : total time (s)
$\bar{T}$: average temperature (°C)
T_s : surface temp (°C)
ΔT : temperature difference (°C)
u : fluid velocity (m /s)
V : volume (m^3)
W : emissive power of a body (W/m^2)
W_b : emissive power of a black body (W/m^2)
x : distance (m); cell density (kg/m^3)

Subscripts

a : side a
b : side b
A : slab A
A_v : average
B : slab B
C : slab C
c : cold side
h : hot side
i : inside
o : outside
L : log mean values

s	: surface, steam
w	: wall

Greek Symbols

α	: absorptivity in radiation; thermal diffusivity (m^2/s)
γ	: reflectivity
τ	: transmissivity
ε	: emissivity
σ	: Stefan–Boltzman constant
ρ	: density (kg/m^3)
λ_s	: latent heat of steam (kJ/kg)
β	: coefficient of thermal expansion (1/°C)
ϕ_v	: coefficient defined by Eq. (12.57)
μ	: viscosity (Pa s)
μ_a	: apparent viscosity
γ	: shear strain (s^{-1})

References

11.1 Carslaw, HS and Jaeger, JC (1959), *Conduction of Heat in Solids*, Oxford University press, Fair Lawn, NJ, p. 234.

11.2 Chilton, TH, Drew, TB and Jabans, RH (1944), *Heat-Transfer Coefficient in Agitated Vessels*, Ind. Eng. Chem., 36, 510 – 516.

11.3 Kreith, F and Bohn, MS (2001), *Principles of Heat-transfer*, 6th Ed., Brooks/Cole, Pacific Grove (USA), p. A-7

11.4 Mc-Adams, WH (1954), *Heat Transmission*, 3rd Ed., Mc Graw-Hill Pub., New York

11.5 Mc-Cabe, WL, Smith, JC and Harriott, P (1993), *Unit Operations of Chemical Eng.*, 5th Ed., Mc Graw-Hill Pub., New York, (a) p.301, (b) p. 336–342

11.6 Toledo, RT (1997), *Fundamentals of Food Process Eng.*, 2nd Ed., CBS Publishers & Distributors, New Delhi. pp 268–273

Review Questions

12.1 What are various applications of heat-transfer in bioprocessing?
12.2 Describe various methods of heat-transfer.
12.3 Write a note on heat-transfer by conduction.
12.4 What is Fourier's law of conduction?
12.5 Define thermal conductivity.
12.6 Derive an expression for overall resistance for individual resistances in series.
12.7 Derive an expression for heat conduction through cylindrical pipes.
12.8 What is meant by log mean radius of a cylindrical pipe?
12.9 Describe the process of unsteady state-heat-transfer by conduction.
12.10 Write the expression for unsteady state-heat-transfer for a slab.
12.11 What is meant by Fourier number?
12.12 Define thermal diffusivity.

12.13 Write a note on heat-transfer by convection.
12.14 Describe the process of heat-transfer between a metallic wall and a fluid.
12.15 Describe the concept of thermal boundary layer.
12.16 What is meant by convective heat-transfer coefficient?
12.17 Describe the concept of overall heat-transfer coefficient.
12.18 Derive an expression for overall heat-transfer coefficient.
12.19 Describe the process of unsteady-state convective heat-transfer.
12.20 Derive an expression for unsteady-state convective heat-transfer coefficient.
12.21 What are various dimensionless numbers you come across in convective heat-transfer?
12.22 What are various predictive equations for convective heat-transfer coefficient?
12.23 What is Dittus–Boelter equation?
12.24 What is Sieder–Tate equation?
12.25 Describe the design procedure for heat-transfer equipment.
12.26 Describe the process of heat-transfer by radiation.
12.27 Define the terms:
    (i) Reflectivity
    (ii) Absorptivity
    (iii) Transmissivity
    (iv) Emissivity
12.28 Describe what is meant by a black body and grey body.
12.29 What is Kirchoff's law?
12.30 What is Stefan–Boltzman law?
12.31 Describe the process of radiative heat-transfer between two bodies.
12.32 Describe the process of radiative heat-transfer between a body and its surroundings.

Problems

12.1 A furnace wall is composed of a metallic wall and a brick wall. The temperature inside the furnace is 500°C and that outside is 40°C. The thermal conductivity of the brick wall is 0.21 W/ m K. If the tolerable heat loss from the furnace wall is 1.1 kW/m^2, what should be the thickness of the brick wall?
(Ans : 8.8 cm)

12.2 A double-walled jacketed fermenter with agitation is used to heat a dilute medium solution. The wall of the fermenter is maintained at 100°C. The tank diameter is 2 m, and the agitator diameter is 1/3 of the tank diameter. The agitator rotates at 90 rpm. Find the heat-transfer coefficient assuming that the physico–chemical properties of the dilute medium solution may be taken as those of water.
(Ans : h_j = 1.76 kJ/m^2 °C)

12.3 In the above fermenter, 1400 litres of the feed solution at 30°C is heated to a final temperature of 85°C. Find the time taken by the feed solution.
(Ans : t = 19.5 min)

12.4 The outside surface of a copper vessel is at a temperature of 80°C. Calculate the radiative heat emitted per unit area of the vessel if the emissivity of copper is 0.03 W/m^2 K^4.
(Ans : 26.36 W)

12.5 A block of baker's yeast is kept in a deep freezer at –10°C. The initial temperature of the block is 30°C. The average exposed area of block is 0.08 m^2 and its emissivity is 0.55. Calculate the radiative heat-transfer rate.
(Ans : 9W/m^2)

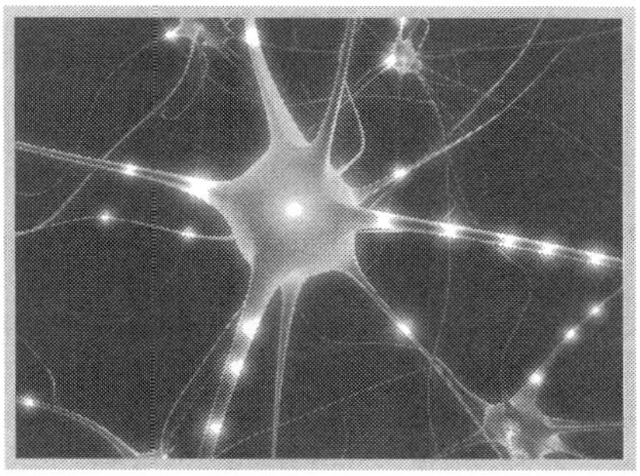

chapter 13

Heat Transfer Equipment

We have seen so far in the last chapter how important is heat transfer in bioprocessing operations, and what are the principles of heat transfer. We have also noted that heat transfer takes place by

- conduction in which the material is in contact with the hot surface, and hence occurs in metallic walls/surfaces
- convection in which the fluid material is in direct contact with the hot surface or cold surface
- radiation in which the material and the heating (or cooling) source are not in direct contact and heat transfer takes place by electromagnetic waves
- a combination of all the above three methods
- infrared heating or
- microwave heating/dielectric heating

For realising all the above methods of heat transfer, we need equipment for the heat to transfer. Heating may be achieved by any of the following methods:

- By passing hot water or thermic fluid in a shell and a tube exchanger
- By directly condensing steam in a steam jacketed kettle

- By passing hot flue gases/exhaust gases
- By means of electrical heating in which the electricity is passed through a resistance which generates heat; generally electric heating is not used in industries in view of its high cost
- By passing steam or hot flue gases through a tube or pipe which is wound outside the heating chamber

We shall discuss some of them in the following sections. A detailed description of them is beyond the scope of this book, and is available in any standard textbook on heat transfer.

13.1 PROCESS EQUIPMENT FOR HEAT TRANSFER

The most frequently used and most versatile process equipment for heat transfer is a jacketed kettle which consists of a vessel with an outside jacket in which the hot or cold fluid passes. It may have stirring / agitating arrangement or all other associated gadgets as shown in Fig. 5.5. In its simple form it is shown in Fig. 13.1. The outside jacket can be used for passing the hot fluid for heating the contents or a cold fluid for cooling the contents or steam which condenses and gives heat. The movement of the stirrer does not hinder the passage of outside fluid since the fluid passes in the outside jacket. All other gadgets like thermowell or pressure gauge or air inlet sparger (which are all shown in Fig. 5.5, but not shown in Fig. 13.1) are all connected to the upper flange, and do not interfere with the operation of the vessel. Cleaning of the vessel inside is easy after removing the upper flange. Aseptic conditions can be better maintained. A number of baffles provided in the tank (usually 2 or 4) helps in breaking the vortex-formation which in turn results in good heat and mass transfer.

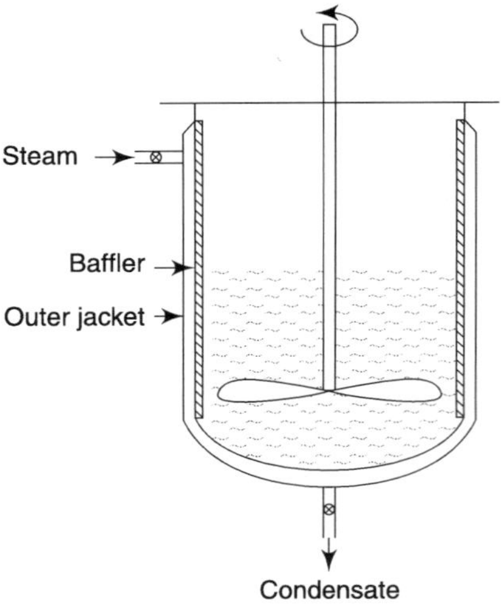

Fig. 13.1 *Jacketted heat exchanger*

The heat transfer area is restricted to the jacketed portion. Hence, realising higher transfer rates are not possible. The only possible alternative may be by increasing the temperature of the outside fluid. This may not be always possible because increasing the temperature of the outside fluid would result in higher thermal gradients which may be detrimental for the survival of the cells in the thermal boundary layer.

In the jacketed vessel described above, the heat transfer is by convection. Instead of having an outside jacket, there are some vessels as shown in Fig. 13.2. The thermic fluid passes through the outside coil. Heat transfer is by means of conduction. This kind of equipment is frequently used for the deep freezer box of refrigerators. The coils can also be used for passing steam as well for heating. The upper flange can have all the gadgets as described for a jacketed vessel. All other advantages and disadvantages are the same as those of a jacketed vessel. The outside coil is used especially when the thermic fluid is to be collected for processing and reusing like a refrigerant. If the outside fluid is not needed to be collected as the condensate of the steam, we may go for jacketed vessels.

Heating vessels with **immersion coils** is another type of arrangement in which the heating coil is directly immersed into the heating vessel (Fig. 13.3). The thermic fluid passes through the coil and can be collected for reprocessing and reuse. Since the coil is directly in contact with the fluid, higher heat transfer rates can be realised. Heat transfer takes place by convection in this arrangement. The immersion coil itself serves the purpose of baffles. The immersion coil is usually coupled to the upper flange by means of union joints or by some other type of flexible coupling arrangement which makes it easy to change the coil to increase the

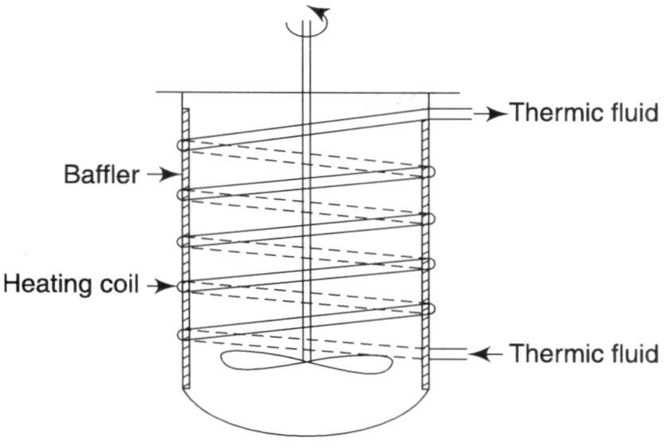

Fig. 13.2 *Outside coil heat exchanger*

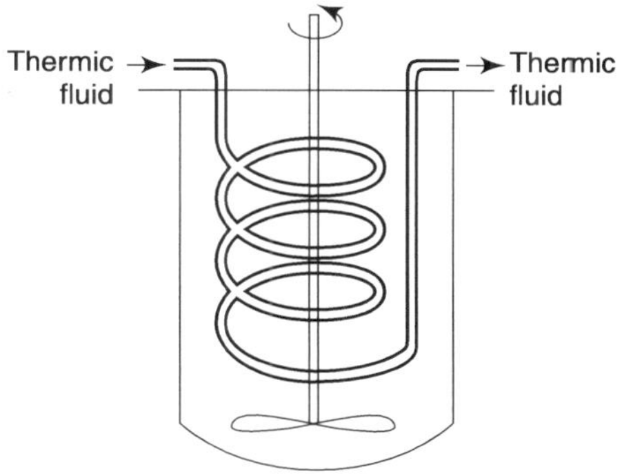

Fig. 13.3 *Immersion coil heating arrangement*

heat transfer area if so desired. The only disadvantage with this arrangement is that the immersion coil may obstruct the rotation of the stirrer. Cleaning of the coil is difficult. Improper clearing may result in non-aseptic conditions in the vessel which are detrimental for the fermentation process.

An **external heat exchanger** is another type of heat transfer equipment we usually use in fermentations. The arrangement is shown in Fig. 13.4 in which a heat exchanger is coupled outside to the process vessel. The fluid in the vessel passes to the heat exchanger either by natural circulation or by forced circulation (shown in Fig. 13.4). This is an excellent arrangement with a lot of flexibility as far as the heat transfer area is concerned. It does not require any separate stirring arrangement since the mixing in the fluid is achieved by means of recirculation which would be adequate enough to provide the desired agitation to the fluid.

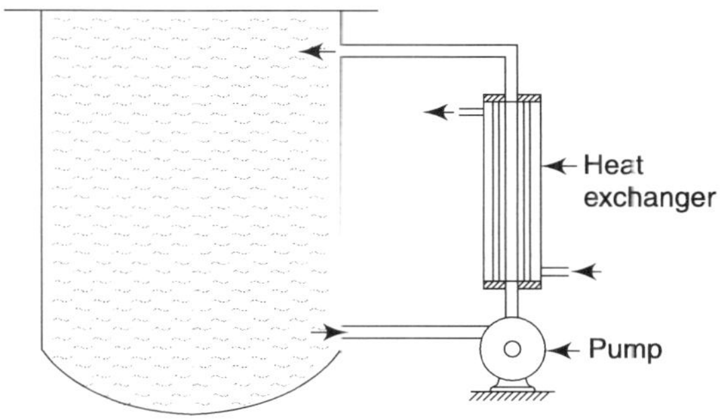

Fig. 13.4 *External heat exchanger arrangment*

The forced circulation systems may have some disadvantage that the shear sensitive cells may collapse during passage through the external pump. In the absence of an external pump, the natural circulation may not be sufficient enough to give adequate mixing since the viscosity of most of the fermentation broths is considerably high.

13.2 DOUBLE PIPE HEAT EXCHANGER

Heat exchanger is an equipment in which heat exchanges or transfers between two fluids which are usually separated by a metallic wall (shown in Fig. E12.2 and Fig. 12.6(b)). One fluid flows inside the pipe and the other fluid flows outside the pipe which is usually known as shell-side. Since the equipment consists of two coaxial pipes, one inside the other, it is known as a double pipe heat exchanger. It is simple in construction and easy in operation. It is also very versatile in the sense that the ΔT can be considerably varied.

13.2.1 Cocurrent and Counter-current Flow Heat Exchangers

The flow pattern of the two fluids in the tube-side and shell-side can be cocurrent or counter current.

In the **cocurrent** arrangement, both the fluids flow cocurrently; i.e., in the same direction. The tube-side fluid flows from left to right, and the shell-side fluid also flows from left to right as shown in Fig. 12.6(b). Let us see how the temperature profile would be all along the length of the heat exchanger. ΔT_a is the temperature difference at the inlet position a, and ΔT_b is the temperature difference at the exit *b*:

$$\Delta T_a = T_{ha} - T_{ca}$$

$$\Delta T_b = T_{hb} - T_{cb}$$

The temperature profiles are shown in Fig. 13.5(a). The temperature difference varies considerably throughout the length of the heat exchanger. Initially ΔT_a is very high which gradually falls to ΔT_b. This is advantageous if the fluid (fermentation broth) is highly viscous, and requires higher thermal gradience. But mostly in fermentations, high thermal gradients are detrimental as the cells are subjected to sudden thermal shock which may result in their mortality. ΔT_b is usually very low making very little heat transfer as we approach to the end of the heat exchanger. Thus, some part of the heat exchanger is ineffective. T_{cb} can never be more than T_{hb}; i.e., the outlet temperature of the tube-side fluid can never be more than that of the shell-side fluid.

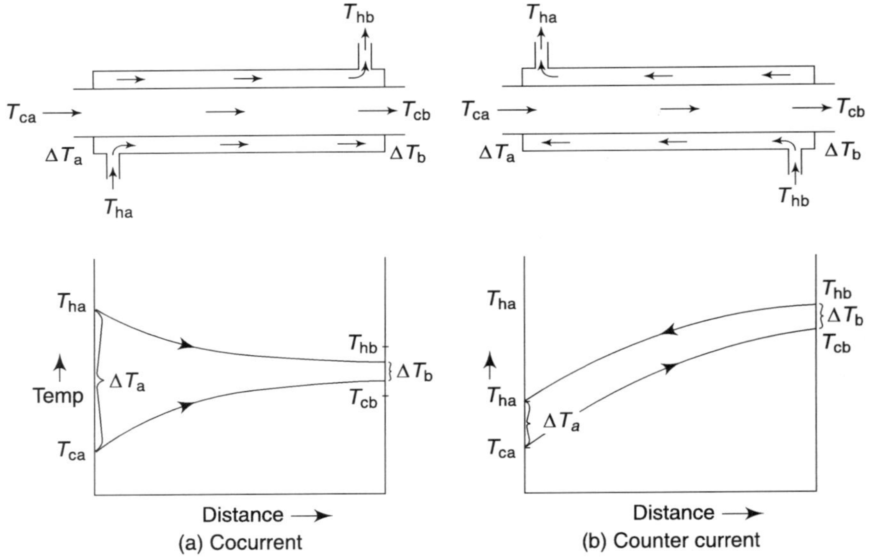

Fig. 13.5 *Temperature profiles*

In the **counter-current flow** arrangement, the tube-side fluid enters in one direction while the shell-side fluid enters in the other direction; i.e., the cold fluid enters at the position *a* and leaves at the position *b*, while the hot fluid enters at the position *b* and leaves at the position *a*. The arrangement is shown in Fig. 13.5(b) along with the temperature profiles. We notice that ΔT does not vary considerably as we move along the length of the heat exchanger. This arrangement helps in maintaining constant thermal gradience throughout. Hence the cells in the fermentation broth are not subjected to high thermal shock. All the area of the heat exchanger is effectively utilised. Another interesting feature is that T_{cb} can be greater than T_{ha}; i.e., the outlet temperature of the tube-side fluid can be more than the outlet temperature of the shell-side fluid depending upon the flow conditions.

Comparison in a nut shell of both the flow patterns is given in Table 13.1.

Table 13.1 *Comparison of cocurrent and counter current flow patterns.*

Feature	*Cocurrent flow*	*Counter current flow*
Temperature gradience	Varies considerably from one end to other	Does not vary much throughout
Effect on thermally sensitive cells	Thermally sensitive cells are subjected to higher thermal shock	Good for thermally sensitive cells
Utilisation for high viscosity fluids	Higher temperature gradience at the inlet is helpful for highly viscous fluids	This advantage cannot be realised.
Effective utilisation of entire surface area	ΔT becomes less as we approach the outlet. Hence the entire surface area is not effectively utilised.	Since ΔT is almost constant throughout, the entire surface area is effectively utilised.
Outlet temperatures of both the fluids	Outlet temperature of cold fluid can never be more than the outlet temperature of the hot fluid	Outlet temperature of the cold fluid can be more than the outlet temperature of the hot fluid

13.2.2 Log Mean Temperature Difference (LMTD)

As can be seen from Fig. 13.5, we notice that ΔT varies as we go along the length of the heat exchanger. Hence it becomes difficult to note down ΔT for using in heat-transfer rate expression (Eq. (12.39)). One approach which we followed earlier was to use average ΔT. A more rational approach is described here, which is based on integration over total surface.

The heat transfer rate is proportional to the temperature difference at any time and at any length of the heat exchanger. Assuming that

(i) the overall heat transfer coefficient (U) is constant throughout the heat exchanger
(ii) the fluid properties do not change considerably in the temperature range
(iii) heat losses from the system are negligible
(iv) the system is under steady state condition so that ΔT does not vary with time, but only varies with distance

we can draw a plot of ΔT vs q in Fig. 13.6 for counterflow conditions. A similar plot can be drawn for cocurrent flow pattern also.

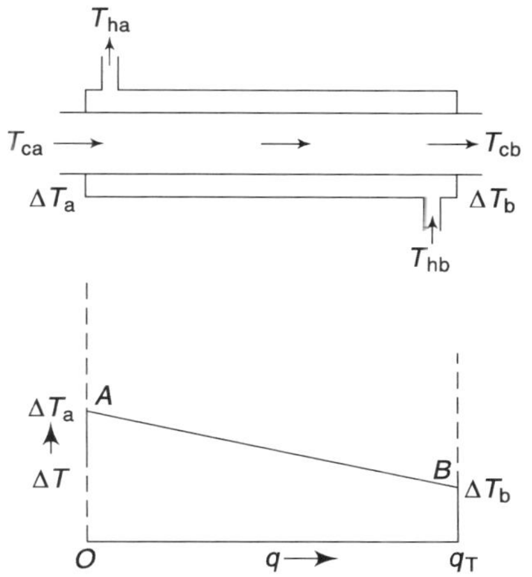

Fig. 13.6 *ΔT vs q for a counter current heat exchanger*

The slope of line AB

$$= \frac{d(\Delta T)}{dq} = \frac{\Delta T_b - \Delta T_a}{q_T} \tag{13.1}$$

By Eq. 12.48, we have

$$\frac{dq}{dA} = U(\Delta T)$$

Replacing for dq by Eq. (13.1), we have

$$\therefore \quad \frac{d(\Delta T)}{U(\Delta T)dA} = \frac{\Delta T_b - \Delta T_a}{q_T} \tag{13.2}$$

Equation (13.2) can be written as

$$\frac{d(\Delta T)}{(\Delta T)} = \frac{U(\Delta T_b - \Delta T_a)}{q_T}(dA)$$

or

$$\int_{\Delta T_a}^{\Delta T_b} \frac{d(\Delta T)}{(\Delta T)} = \frac{U}{q_T}(\Delta T_b - \Delta T_a) \int_0^A (dA)$$

On integration, we get

$$\ln (\Delta T_b/\Delta T_a) = \frac{U}{q_T} (\Delta T_b - \Delta T_a)\, A \tag{13.3}$$

$$q_T = \frac{U\,A\,(\Delta T_b - \Delta T_a)}{\ln (\Delta T_b / \Delta T_a)} = U\,A\,\Delta T_L \tag{13.4}$$

where

$$\Delta T_L = \Delta T_b - \Delta T_a/(\ln (\Delta T_b/\Delta T_a)) \tag{13.5}$$

ΔT_L is known as the logarithmic mean temperature difference which is popularly known as LMTD.

Conventionally, the inlet is represented by the position 1 and the outlet is represented by the position 2, and hence LMTD is written as

$$\Delta T_L = \frac{\Delta T_2 - \Delta T_1}{\ln (\Delta T_2 / \Delta T_1)} \tag{13.6}$$

$$\Delta \bar{T} = \frac{\Delta T_1 + \Delta T_2}{2} \tag{13.7}$$

There will be some difference between these two ΔTs, particularly if both the ΔTs are far apart.

Example 13.1

A particular dilute nutrient medium is heated in a double pipe counter-current-heat exchanger from an initial temperature of 28°C to a final temperature of 60°C. The hot fluid passing through the shell-side is cooled from 95°C to 70°C. Under steady-state conditions, calculate the LMTD. Also, calculate average ΔT and find the difference between the two.

Solution The heat exchanger configuration is shown in Fig. E13.1.

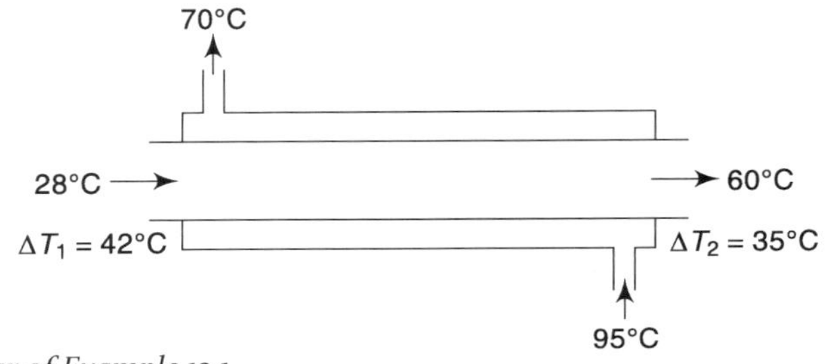

Fig. E13.1 *Heat exchanger of Example 13.1*

$$\Delta T_1 = 70 - 28 = 42°\text{C}$$

$$\Delta T_2 = 95 - 60 = 35°\text{C}$$

$$\Delta T_L = \frac{\Delta T_2 - \Delta T_1}{\ln (\Delta T_2 / \Delta T_1)} = \frac{35 - 42}{\ln (35/42)} = \frac{-7}{-0.1823} = 38.4°\text{C}$$

$$\Delta \bar{T} = \frac{\Delta T_1 + \Delta T_2}{2} = \frac{35 + 42}{2} = 38.5°\text{C}$$

Since ΔT_1 and ΔT_2 are close by, ΔT_L and $\Delta \bar{T}$ are almost equal.

13.3 SHELL AND TUBE HEAT EXCHANGERS

In a double pipe heat exchanger, the inner pipe is referred as tube, and the fluid passing through the tube is known as tube-side fluid. The outer pipe is known as shell, and the fluid passing through the shell is known as shell-side fluid. It is an excellent arrangement for heat transfer between two fluids. The outer surface area of the tube accounts for the heat transfer or heat transfer rate. Obviously, the more is the contact area, the more would be the heat transfer rate. But there is a limitation for the length of the tube also. Usually, the tubes are 20' (6.1 m) long. If the whole length of the tube is used, it will sag at the centre. Hence an alternate arrangement would be to provide a bend after every 10' length or so. The arrangement is shown in Fig. 13.7. The bends at the ends are standard return bends connected to the tube using standard flanged joints. Thus, any number of sets of shells and tubes can be arranged; but the whole set-up becomes cumbersome and occupies a lot of floor space. To circumvent this, an envisaged methodology could be that all the sets of tubes be arranged in between two circular plates and all the tubes are enclosed in a shell. The arrangement is known as tube-and-shell heat exchangers, and will be discussed in the following sections.

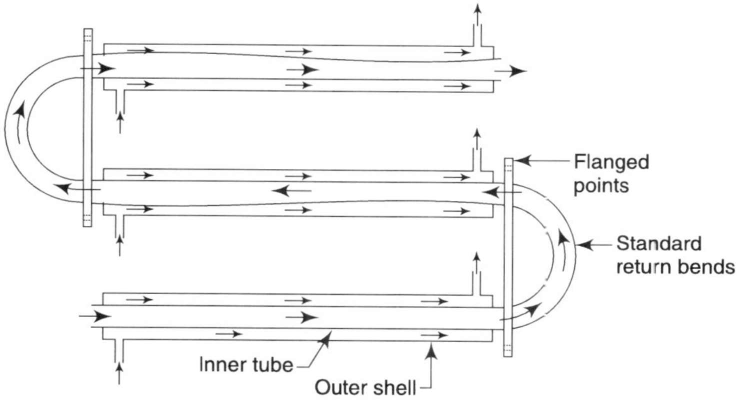

Fig. 13.7 *Long shell and tube heat exchanger with standard return bends.*

13.3.1 1,1 Shell and Tube Heat Exchanger

A set of tubes are enclosed inside a shell as shown in Fig. 13.8. The tubes are held at either end by circular plates. The arrangement is shown in Fig. 13.9 (a). The pitch of the shells could be either in the form of a triangle [Fig. 13.9 (b)] or a rectangle [Fig. 13.9 (c)]. The triangular pitch helps in better exchange of heat with the shell-side fluid, as the tube's position is zig–zag, and will interrupt the flow of the shell–side fluid. This is an advantage. But the arrangement also posses a difficulty while cleaning the tubes for removing scales etc. This is the disadvantage with a triangular pitch.

The tube-side fluid enters into the compartment at one end, say on the left-hand side, and passes through all tubes. After exchanging heat with the fluid on the shell-side during travel, the tube-side fluid enters at the other end (say right-hand side) and is discharged. The shell-side fluid also enters the shell at the left-hand side end and leaves the shell from the right-hand side end. During the process of travel, the shell-side liquid is made to undertake a tortious path by putting baffles in the shell-side. Various arrangements of baffles are shown in Fig. 13.10 (Kreith and Bohn, 2001). These baffles help deflect the flow of the shell-side fluid by avoiding short circuiting or channeling. Usually, the segmented baffles are used. They not only channelise

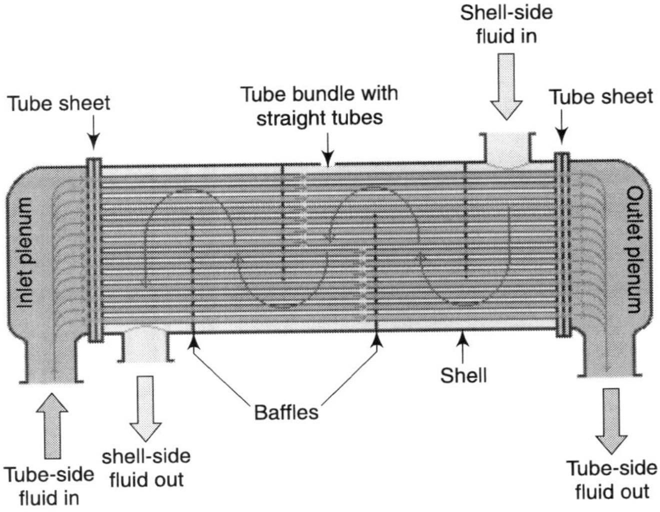

Fig. 13.8 *1,1 – Shell and tube heat exchanger*

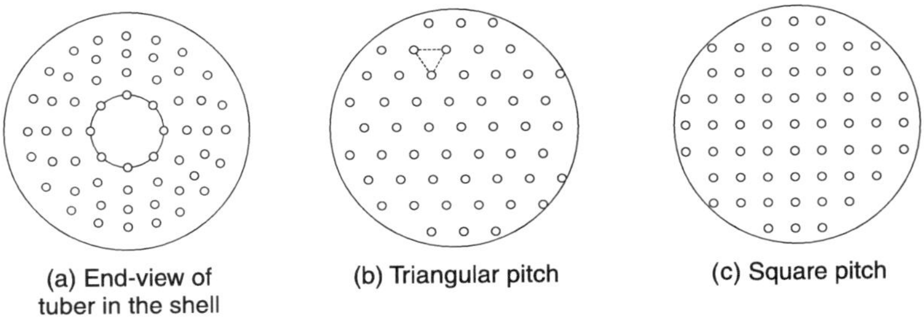

Fig. 13.9 *Tubes configuration in the shell and tube heat exchanger*

the flow of the shell-side fluid, but also provide mechanical support to the tubes which otherwise would tend to sag.

This kind of heat exchanger is known as 1,1 – heat exchanger because the flow of fluids in the shell-side is only one–pass and that in the tube-side is also one–pass; i.e., both the flows are either cocurrent or counter-current. Since the pass is a single path, these devices are known as 1,1 – shell and tube heat exchangers.

13.3.2 1,2 Shell and Tube Heat Exchanger

This heat exchanger is also similar to the 1,1 heat exchanger except that the tube-side fluid has two passes as shown in Fig. 13.11. The tube-side fluid enters from the right-hand side of the heat exchanger, and also leaves from the same side. During the process of flow, the fluid passes through half the number of tubes in one direction, and in the reverse direction through the rest of the tubes. However, the shell-side fluid flows in one direction with the result that we can realise both the co–current flow and the counter-current flow conditions. Initially, when the tube–side fluid is cold, we realise higher ΔT because of co–current conditions. When sufficient flow conditions have been achieved because of higher ΔTs, the later fluids will be in counter-current flow conditions when the ΔT is fairly constant. Thus, we realise the advantages of both the flow conditions.

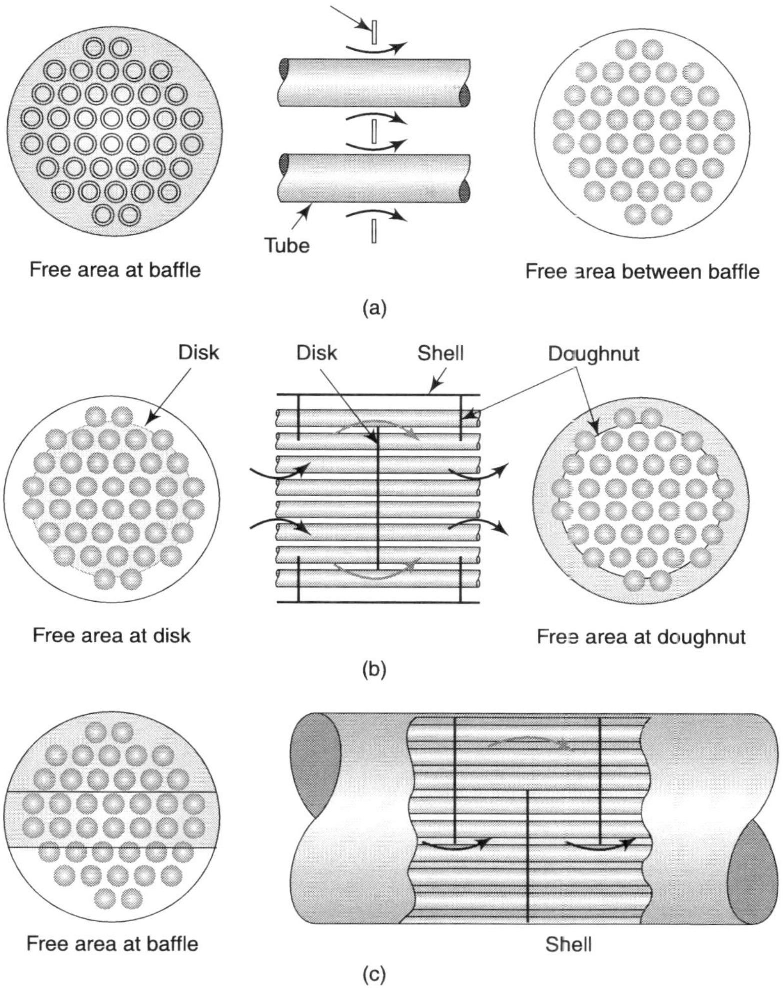

Fig. 13.10 *Three type of baffles used in shell-and-tube heat exchanters: (a) orifica baffle (b) disc-and-doughnut baffle (c) segmental baffle.*
(From Kveith/Bhon, Principles of heat transfer, p. 489, 6th Edn© 2001. Cengage Learning, a part of Cengage Learning Inc. Reproduced by permission. www.cengage.com/permissions)

Since the flow is in two directions in the tube-side and in one direction in the shell-side, this kind of heat exchanger is known as 1,2 – shell and tube heat exchanger, which indicates a single pass in the shell-side and two passes in the tube-side.

13.3.3 2,4 – Shell and Tube Heat Exchanger

This is also same as above heat exchangers except that the shell side fluid is also made to change its direction and will have two-way flow whereas the tube-side fluid has four passes as shown in Fig. 13.12. Hence, this heat exchanger is known as 2,4 – shell and tube heat exchanger. In this arrangement, we realise two times co-current conditions, and two times counter-current conditions. Hence, the heat exchanging is better; but since the fluids are made to change their flow directions many times, it would result in higher pressure drops.

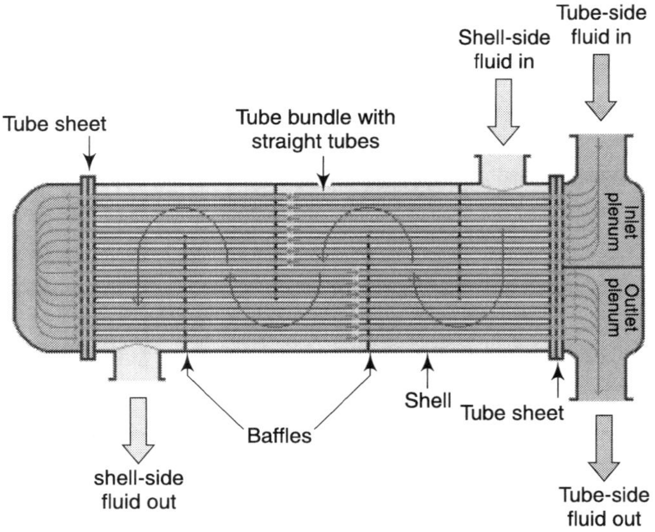

Fig. 13.11 *1,2 – Shell and tube heat exchanger*

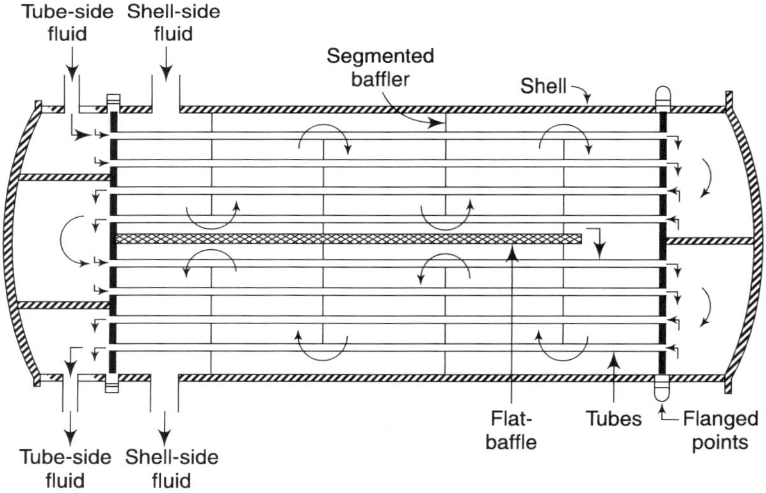

Fig. 13.12 *2,4 – Shell and tube heat exchanger*

13.3.4 Scraped Surface Heat Exchanger

Heating of certain high viscous liquids like corn syrups, etc., is not easy in view of their poor flowability. Neither can they be heated in a shell and tube heat exchanger nor in a kettle heater (Fig. 12.7(a)). In such cases we need to have a heat exchanger which has an arrangement for scrapping the surface continuously so that the liquid/paste sticking to the surface is wiped off. The scraped surface heat exchangers can also be used for concentrating liquid syrups for crystallisation (Sec 17.9) containing up to 75% solids. The scrapping would also improve the heat transfer coefficients as much as by 500% (Uhl and Gray, 1966) with non-Newtonian fluids. One type of batch arrangement is shown in Fig. 13.13 which is similar to Fig. 12.7(a) except

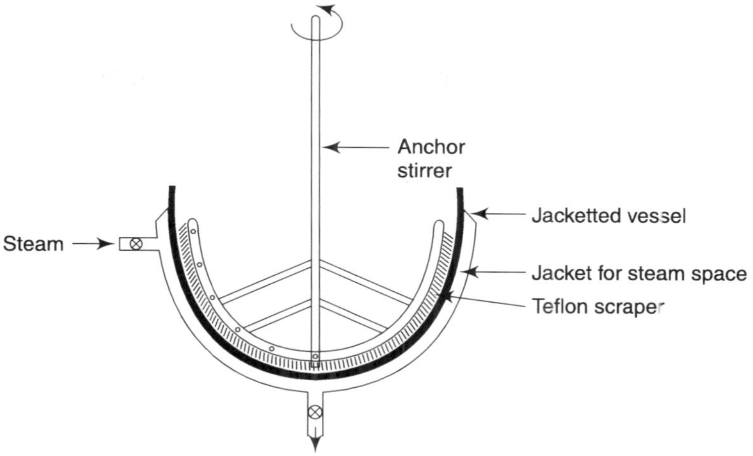

Fig. 13.13 *Batch scraped surface heat exchanger*

that the stirrer is fixed in the form of a scraper. The stirrer is in the form of an anchor shown in Fig. 11.13(k). The edge of the stirrer blade is fixed with a teflon strip or some other type of non-toxic flexible ribbons so that the surface is continuously scraped. Usually, the speed of the stirrer is very low which is of the order of few revolutions per minute. In fact, the stirrer is not meant for stirring, but only to scrape the surface.

Continuous scraped surface heat exchangers are used at industrial scale for heavy duty operations. They are double pipe heat exchangers with large inner pipe diameters which are of the order of 10–30 cm. Steam or cooling water flows through the outer shell, also known as the jacket. The material to be heated or cooled is passed through the inner tube which has a provision for scraping (Fig. 13.14). The tube-side fluid is continu-

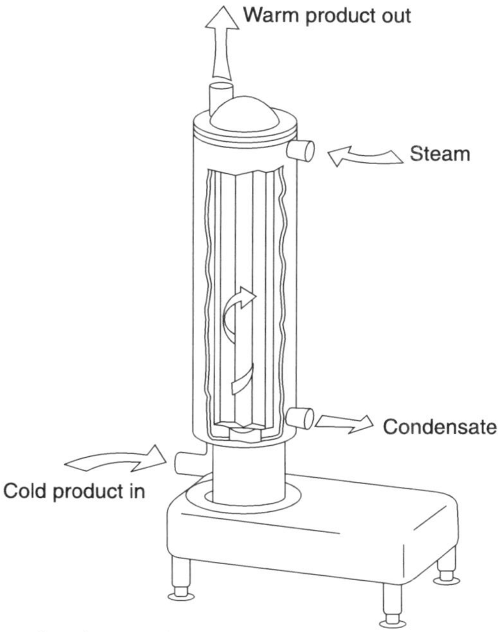

Fig. 13.14 *Continuous scraped surface heat exchanger*

ously agitated and scraped by a rotor operated by a motor. There will be a provision for lifting or lowering of the rotor depending upon the need. The equipment can be operated both horizontally and vertically; and is generally meant for heavy-duty operations.

The heat-transfer coefficient in a scraped surface heat exchange (h_j) is reported (McCabe et al, 1993) as follows:

$$h_j = [2\, k\, \rho\, C_p\, n\, B/\Pi]^{1/2} \tag{13.8}$$

where B is no. of blades carried by shaft
n is rps of shaft
k, ρ and C_p are thermal conductivity, density and specific heat of the fluid

Interestingly, the equation does not contain the viscosity term whereas the scraped surface heat exchangers are virtually meant for highly viscous fluids. Nor does it contain the fluid velocity term. In that way, Eq. (13.8) seems to be over simplification of the heat transfer process.

Another empirical expression based on Nusselt number, Reynolds number and Prandtl number is as follows (Heldman and Lund, 1992):

$$N_{Nu} = (2/\Pi)\, (N_{Re}\, N_{Pr}\, n)^{0.5} \tag{13.9}$$

13.3.5 Plate Heat Exchanger

As the name indicates, the heat exchanger is in the form of a series of SS plates fitted into a frame, so that some amount of space is there in between each plate. The arrangement is shown in Fig. 13.15. The product to be heated or cooled is passed in between the alternate spacing between the plates while the steam or coolant passes through in the remaining plate spacings. Thus each plate will separate the product and the heating medium. The spacing in between the plates is very small which is of the order of 1" / 32 or so. When the fluid is passing through such small clearances, it will be virtually in the form of a thin film. Hence, the heat transfer is better. The plates have corrugations on the surface to induce turbulence when the fluid is flowing

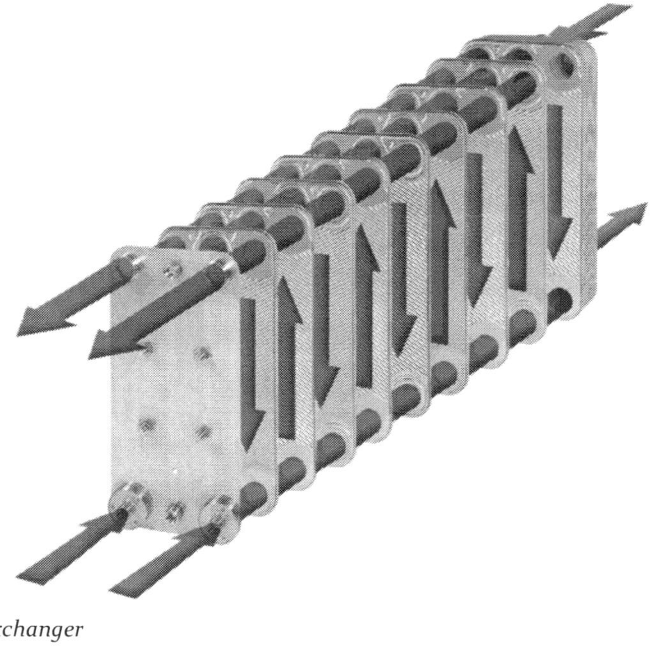

Fig. 13.15 *Plate heat exchanger*

through, which in turn would help mixing and hence better heat-transfer rates. The spacing in between the plates is made leakproof by providing gaskets made out of synthetic rubber. The gaskets should withstand the high temperature operations.

These heat exchangers are used for quick heating or quick cooling of fluids. They are mainly used for continuous sterilisation of medium (Sec. 6.5). It was shown in Fig. 6.9 (a) . These heat exchangers are best suited for quickly raising the temperature or for quickly lowering the temperature. The quick rising of temperatures is used for 'High Temperature Short Time (HTST)' treatment of media. The Sterilisation can be done by this method, but the proteins, etc. do not get denatured. Pasteurisation of milk for preservation over a longer period of time is done by plate heat exchangers alone.

They are also used for 'Ultra High Temperature (UHT)' operations since the temperatures in plate heat exchangers can go up to 200°C. The only drawback with them is that they cannot be used for viscous fluids where the flowability is low. Fluid viscosities should be less than 5cp for using in plate heat exchangers.

13.4 CONDENSERS

Condensers are also simple heat exchangers, but essentially one of the fluids is a vapour, and hence the condensation of it involves change of phase. Phase change can occur when the latent heat of vapourisation of the fluid is removed. During desolventisation of an extracted product by liquid–liquid extraction, the solvent vapours need to be condensed for recovery and reuse. The product also becomes free of solvent. For this kind of downstream processing step, we use condensers. There could be many such applications in bioprocessing. The condensers are usually of two types:

(i) Shell and tube (heat exchanger) condensers
(ii) Direct contact condensers

13.4.1 Shell and Tube Condensers

They contain a bundle of tubes enclosed in a shell like 1,1 – tube and shell heat exchanger, shown in Fig. 13.8. The only envisaged problem in condensers is higher thermal stresses developed in the tubes due to high ΔTs. The cooling liquid enters the tube side, and receives heat from the vapours. The vapours entering will be usually at their boiling point, whereas the cooling liquid will be at either room temperature or below the room temperature. This results in high ΔTs, and hence the tubes are exposed to high thermal stresses. This results in sagging of the tubes due to high thermal expansions. Sometimes, the high thermal expansions are so severe that they may buckle the tubes or they may pull the tubes loose from the tube sheets. To overcome this problem, one end of the tubes along with the holding sheet is left free; i.e., the bundle of tubes is left floating with only one end fixed (Fig. 13.16). The floating side will absorb any expansion of the tubes.

The inlet of the vapours is not in the form of a loose pipe inlet or a tubular inlet. The vapours are admitted at the middle of the shell-side with a perforated plate over the vapour inlet. This arrangement would help prevent cutting of the tubes by drops of liquid which are likely to be carried by the vapours.

Usually the vapours may also contain some non–condensable gases (like air, etc.) and enter the shell-side. Hence the condensers contain one outlet vent in the shell-side for escape of any condensables.

13.4.2 Direct Vapour Contact Condensers

In this type of condensers, the vapours are directly mixed (or brought in contact) with the cooling fluid known as coolant. These condensers are usually used for condensing water vapours. Here, the condensing action is direct. The vapours mix with the coolant, instantly release their latent heat of vaporization to the coolant, and they both leave together from the single outlet, as shown in Fig. 13.17. The coolant water is sprinkled in the form of a spray where the vapours are entering the condenser. The spray will instantaneously condense

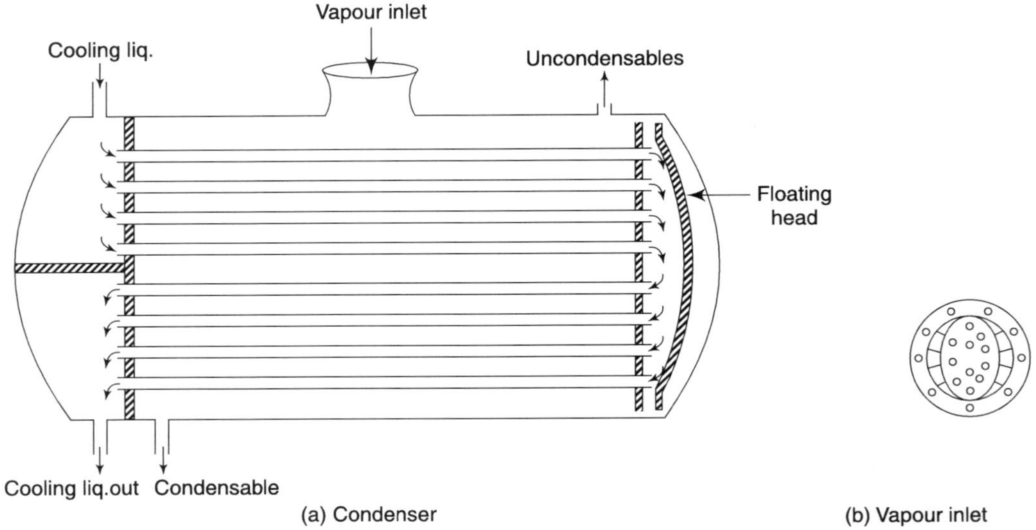

Fig. 13.16 *Two-pass shell and tube condenser with floating head*

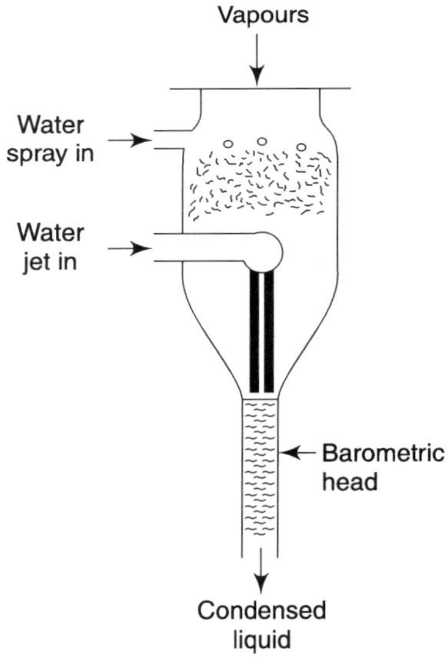

Fig. 13.17 *Direct vapour contact condenser*

the vapours. Some more water is also added through the inlet. Because of condensation of vapours, it creates vacuum in the condenser to drag the vapours into the condenser. The bottom outflow is usually achieved by means of a barometric fluid leg maintained between the condenser and condensate–receiving tank. The barometric leg is a long tube of 10 m sealed at the bottom by condensate receiving tank.

This kind of condensers are usually small in size. High heat transfer rates are realised as compared to shell and tube condenser. They are economical also. The only drawback with these condensers is that they are not suitable if the condensate is the desired product.

Symbols

A	: heat transfer area (m^2)
A_L	: log mean area
B	: no. of blades on the scraper
C_p	: specific heat of liquid (kJ/kg°C)
d_i	: inside diameter of the tube (m)
h_j	: jacketed vessel heat transfer coefficient (kW/m^2 °C)
k	: thermal conductivity of fluid (W/m °C)
N_{Re}	: Reynolds number
N_{Pr}	: Prandtl number
N_{Nu}	: Nusselt number
n	: agitator speed (s^{-1})
q	: heat transfer rate (kJ/s or kW)
q_T	: total heat transfer rate (kJ/s or kW)
ΔT	: temperature difference
U	: overall heat transfer coefficient (kW/m^2 °C)

Subscripts

a	: inlet position
b	: outlet position
1	: position 1
2	: position 2

Greek Symbols

μ	: viscosity (Pa s)
ρ	: density (kg/m^3)

References

13.1 Heldman, D R and Lund, D B (1992), "*Handbook of Food Engineering*", Marcel Dekker, Inc., New York, pp. 266–269.

13.2 Krecth, F. and Bohn, MS., (2001), "*Principles of heat transfer*", 6th Edn., Books/Cole, Pacific grove, (USA), p.489.

13.3 Mc Cabe, W L, Smith, J C and Harriott, P (1993), "*Unit Operations of Chemical Engineering*", 5th Ed., Mc Graw-Hill, Inc., New York, Ch-15.

13.4 Uhl, V W and Gray, J B (1966), "*Mixing*", vol.1, Academic press, New York, pp. 298-303.

Review Questions

13.1 What are different ways of heating?

13.2 What are the various kinds of heat-transfer equipment you usually come across in bioprocessing?

13.3 Describe the advantages and disadvantages of various heat-transfer equipment.

13.4 Describe what is meant by cocurrent flow and counter-current flow conditions.

13.5 Draw the temperature profile for cocurrent and counter-current flow.

13.6 Tabulate various merits and demerits of cocurrent flow and counter-current flow.
13.7 What is meant by LMTD?
13.8 Derive an expression for LMTD.
13.9 Describe various shell and tube heat exchangers.
13.10 Describe the working of 1,1-shell and tube heat exchanger with a neat diagram.
13.11 Describe the working of 1,2-shell and tube heat exchanger with a neat diagram.
13.12 Describe the working of 2,4-shell and tube heat exchanger with a neat diagram.
13.13 Describe the working of batch-scraped surface heat exchanger with a neat diagram.
13.14 Describe the working of a continuous scraped surface heat exchanger with a neat diagram.
13.15 Describe the working of a plate heat exchanger with a neat diagram.
13.16 What are condensers and where are they used in bioprocessing?
13.17 Describe the working of a shell and tube condenser with a neat diagram.
13.18 Describe the working of a direct vapour contact condenser with a neat diagram.

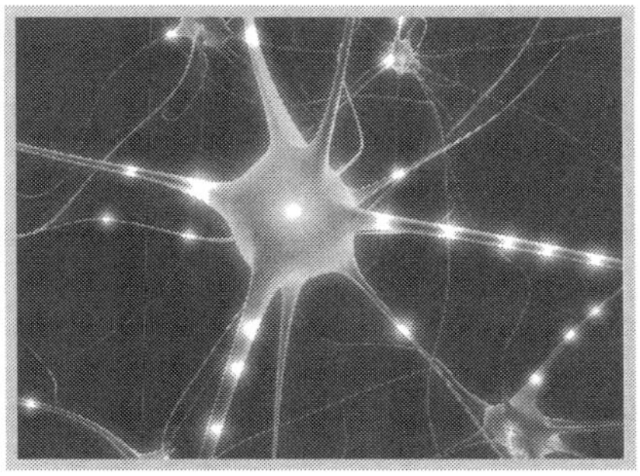

chapter 14

Mass Transfer in Bioprocessing Operations

Mass transfer operations in bioprocessing is one of the most widely studied subjects. In fact, the wealth of literature on physical gas–liquid mass transfer operations (as contrast to mass transfer with chemical reaction) and correlations for interfacial area in chemical reaction engineering studies were mostly related to fermentation broths. Still, most of the correlations were based on aqueous systems that do not adequately represent the true picture of a fermentation broth, which is indeed complicated because of

- complex rheology of fermentation broths
- presence of living organisms which are active
- the oxygen demand varies with the nature and morphology of the cells
- presence of other solids and nutrients in the medium
- presence of foam-causing agents like proteins, etc.

In view of the above, there is perennial dearth of information which prompts further studies, and adds to the bulk of literature information with ever-opening opportunities for further probing.

Mass transfer operations in bioprocessing may be studied under the following main categories:

(i) diffusional/convective mass transfer
(ii) liquid–liquid extraction
(iii) mass transfer with phase change.

These transfer processes are known as transport operations, as contrast to material handling. The transfer of material in bulk from one place to another by means of trolleys or screw conveyors, etc., is a part of material handling operations, and is not considered as a mass transfer process in the true sense. We call a mass transfer or transport process as a mass transfer operation, only if the transfer of mass takes place by concentration difference in two phases. It could be a

Gas–liquid system
Liquid–liquid system (immiscible)
Liquid–solid system
Gas–solid system
Gas–liquid–solid system

The following are some of the examples of mass transfer applications in bioprocessing:

(i) In aerobic fermentations, oxygen mass transfer from the air bubble to the bulk liquid in the soluble form to be available for microorganisms.
(ii) Liquid–liquid extraction technique for the extraction of penicillin from aqueous phase into organic phase, using an organic solvent like butyl acetate.
(iii) Drying of solids as in dehydration of baker's yeast in the dry powder form.
(iv) Distillation of ethanol from the fermented molasses broth (containing approx. 8% alcohol) for industrial or potable use.

Of the above, liquid–liquid extraction and mass transfer with phase change in drying of the wet product to get a pure dry product (free of solvent), will be taken up separately in Chapter 17. These operations are known as downstream processing steps.

In this chapter, we concentrate on mass transfer by diffusion and by convection, for which the concentration difference is the driving force.

14.1 MASS TRANSFER BY DIFFUSION

In molecular diffusion, mass transfer takes place purely by concentration gradients only. That is, the molecules from the higher concentration region will move to the lower concentration region, and try to equilibrate the concentrations in both the regions. If the two regions are in different phases and are separated by a boundary layer, diffusion takes place through the boundary layer. If the solute (active material) in the lower concentration region is constantly depleted by consumption by microorganisms or by chemical reactions, the concentration difference exists, and diffusion continues to take place. This is known as **continuous molecular diffusion** process, and it takes place only because of the concentration differences. Any amount of agitation given to the system by means of stirring, or gas sparging, will only help equilibrate the concentration differences within the bulk of the single phase; but it does not affect the movement of the solute within the boundary layer or the thin film which separates the two phases. Vigorous agitation will only help make the boundary layer thin, but it does not totally eliminate the boundary layer. The interphase mass transfer has to take place by diffusion through the interface. The situation can be represented schematically as given in Fig. 14.1.

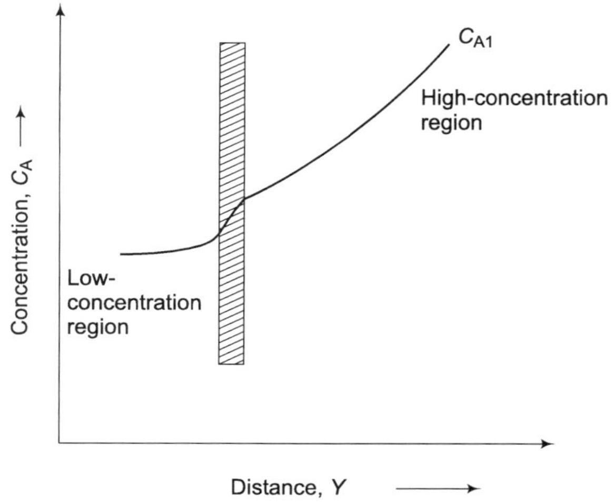

Fig. 14.1 *Schematic representation of molecular diffusion*

The mass flux J_A (mol or kg/(m^2 s)) is given by dividing the mass flow rate by cross-sectional area A. Thus,

$$J_A = \frac{N'_A}{A} = -D_{AB}\frac{dC_A}{dY} \qquad (14.1)$$

where D_{AB} is the diffusion coefficient of component A into the bulk of B and is given in m^2/s; Y is the distance measured from lower concentration zone to higher concentration zone, i.e. the direction of Y is opposite to the direction of the mass flow rate. The negative sign in Eq. (14.1) is mainly due to the fact that mass transfer direction (which is from higher concentration to lower concentration) is opposite to the direction of the concentration gradient. That is to say that the slope of C_A vs Y will be increasing as Y increases (i.e., the slope is positive) while the direction of mass transfer is in the negative direction of Y. Equation (14.1) is known as **Fick's law of diffusion.**

14.2 THEORIES OF DIFFUSIONAL MASS TRANSFER

Various theories have been proposed from time to time, to predict the performance of mass transfer by diffusion, even though the basic purpose of all the theories is to evaluate an expression for mass transfer coefficient.

14.2.1 Film Theory

The most classical theory amongst them is the **Film theory,** first proposed by Whitman (1923), based on the "diffusional-layer" idea of Nernst (originally proposed in the year 1904), which proposes the concept that a stagnant film of constant thickness (δ) exists between the phases at the interphase boundary (Fig. 14.2). Generally there will not exist any concentration gradient within the phase, since it can be equilibrated by agitation. But, however vigorous the agitation may be, the diffusional film exists, which in fact offers resistance for the mass transfer to take place. In this film, the mass transfer takes place solely by diffusion alone.

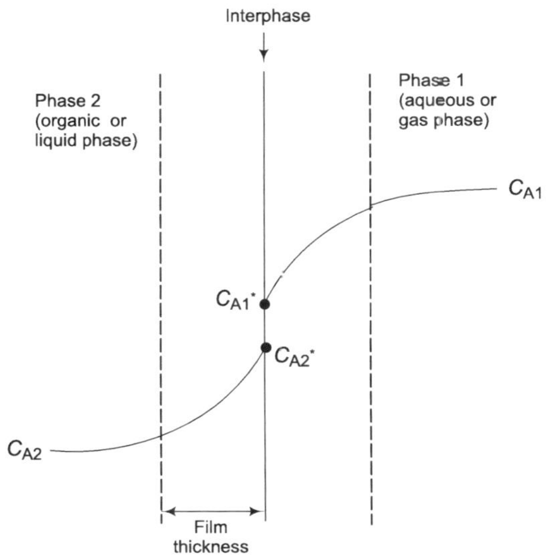

Fig. 14.2 *Schematic representation of film theory*

In Fig. 14.2, the two immiscible phases could be gas and liquid, or an organic and aqueous phase (viz. chloroform and water, or hexane and water, etc.). Any amount of turbulence by agitation in the mixing system can only avoid the concentration gradients within the bulk of either phase. The C_{A1} in phase 1 will fall down to C_{A1}^* within the film and at the interphase boundary. The concentration C_{A1}^* is represented with a * to indicate that it is the equilibrium solubility of the solute in that phase.

Similarly, the concentration in phase 2 will be C_{A2}, which falls from C_{A2}^* at interphase. C_{A1}^* and C_{A2}^* are the equilibrium solubilities of the solute in the two phases. The resistance to mass transfer lies mostly in the liquid side. The boundary generally does not offer any resistance for mass transfer.

In case of gas–liquid systems (for example, air bubbles in a penicillin broth), generally we conceive only a liquid film at the interphase towards the liquid bulk, and the diffusional resistance for oxygen to diffuse from the air bubble into the viscous fermentation broth lies in the thin liquid film that forms at the gas–liquid interphase. Film theory proposes that the mass transfer coefficient, k_L, is given by (Danckwerts, 1970):

$$k_L = \frac{D_{AB}}{\delta} \tag{14.2}$$

where δ is the film thickness, and represents the hydrodynamic characteristics of the system. δ depends upon the system geometry, physical properties of the liquid and agitation.

Even though film theory is the simplest, it has the serious drawback in contending that there exists a real discontinuity between the phases; and more so that the film has a constant thickness. The credit which film theory carries with it is that the gas can get into the liquid by dissolution and molecular diffusion before it is really transported into the bulk liquid by agitation/convection.

Example 14.1

The diffusivity coefficient of oxygen into a fermentation broth was noted to be 7.5×10^{-10} m²/s. The thickness of the stagnant liquid film was calculated to be 5.2 microns. Calculate the mass transfer coefficient using film theory.

Solution It is a direct application of Eq. (14.2)

Given: $D_{O2} = 7.5 \times 10^{-10}$ m²/s

$\delta = 5.2$ microns

$= 5.2 \times 10^{-6}$ m

Now $k_L = \dfrac{D_{O2}}{\delta} = \dfrac{7.5 \times 10^{-10}}{5.2 \times 10^{-6}} = 1.442 \times 10^{-4}$ m/s

14.2.2 Penetration Theory

The Penetration theory, originally proposed by Higbie (1935), is based on the transient mass transfer. The model proposes that every element of the liquid surface is exposed to the gas for the same time period, θ, before it is replaced by the bulk liquid. During the time of exposure, the liquid element absorbs the gas as if it was a stagnant film and infinitely deep. The mass transfer is an unsteady state process. The concentration profiles of the solute changes not only with distance, but also with time. The rate of change of solute, $\partial C_A/\partial t$, in the liquid (which is the rate of gas absorption) will also be changing along the X-axis (one-dimensional). It can be represented by Fick's second law of diffusion:

$$\frac{\partial C_A}{\partial \theta} = D_{AB} \frac{\partial^2 C_A}{\partial X^2} \tag{14.3}$$

with the boundary condition that

when $\theta = 0, X > 0, C_A = C_{A0}$

$\theta > 0, X = 0, C_A = C_A^*$

$\theta > 0, X = \infty, C_A = C_A$

Solving Eq. (14.3) yields an expression for the rate of mass flux (Aiba et al., 1965):

$$N_A' = 2\sqrt{\frac{D_{AB}}{\pi\theta}}\Delta C \tag{14.4}$$

Comparing Eq. (14.4) with the mass transfer equation,

$$N_A = k_L a(C_{AL}^* - C_{AL})$$

$$k_L = 2\sqrt{\frac{D_{AB}}{\pi\theta}} \quad \text{or} \quad \sqrt{\frac{4\,D_{AB}}{\pi\theta}} \tag{14.5}$$

in which θ is defined as the time taken by a single bubble to travel a distance equal to its own diameter in the bulk fluid, and is given as:

$$\theta = \frac{d_B}{u_B} \tag{14.6}$$

Thus, the penetration model predicts that k_L varies as ½ power of D_{AB} (i.e. $k_L \propto \sqrt{D_{AB}}$) unlike that predicted by **film theory** ($k_L \propto D_{AB}$). Some scanty evidence shows (Danckwerts, 1970) that k_L varies as $\sqrt{D_{AB}}$ in packed beds and stirred vessels. Thus, Higbie's model seems to be a more rationale model. However, **film theory** is always credited for its concept of conceiving the presence of a stagnant film at the phase boundary, i.e. in between the interphase and the bulk fluid.

The serious drawback of the Higbie model is that it is unrealistic in conceiving the same time of exposure for all elements of surface.

Example 14.2

In the aerobic fermentation process of Example 14.1, air is bubbled from bottom through a sparger. The air bubble has a diameter of 2.8 mm and has a rise velocity of 16 cm/s. Find the mass transfer coefficient based on penetration theory.

Solution It is also direct application of Eq. (14.6) and Eq. (14.5). We calculate the penetration time period, θ, by Eq (14.6) and then calculate k_L by Eq. (14.5)

Given:

Diffusivity, $D_{O2} = 7.5 \times 10^{-10}\ m^2/s$

Bubble dia, $d_b = 2.8\ mm = 0.28\ cm$

Bubble rise vel., $u_b = 16\ cm/s$

$$\theta = \frac{d_b}{u_b} = \frac{0.28}{16} = 0.0175\ s$$

$$k_L = (4\,D_{O2}/(\Pi\,\theta))^{1/2} = (4 \times 7.5 \times 10^{-10}/(\Pi \times 0.0175))^{1/2} = 2.336 \times 10^{-4}\ m/s$$

14.2.3 Surface Renewal Theory

This model, proposed by Danckwerts (1951), is also based on the transient diffusion (unsteady state diffusion) like the Higbie's model. It is further refined by introducing the concept that the surface renews at every fixed interval of time. Thus, according to the model,

$$k_L = \sqrt{D_{AB}\, s} \tag{14.7}$$

where s is the fraction of the area of surface which is replaced with fresh liquid in a unit time and will have the units of s^{-1}. The surface renewal theory also predicts the dependence of k_L on the 1/2 power of D_{AB} (i.e. $k_L \propto \sqrt{D_{AB}}$), similar to that of the penetration model.

The surface renewal model seems to be more realistic in explaining the situation of pseudo-fast reactions.[1] The hydrodynamic properties of the system are taken care by the sample parameter s.

14.3 MASS TRANSFER BY CONVECTION

The mass transfer in any fluid in motion essentially takes place by the means of turbulence or eddies from the region of higher concentration to the region of lower concentration. This is in addition to the diffusional mass transfer, and is comparatively faster. However, the convective mass transfer process is complicated because of the influence of extent of mixing on mass transfer. Thus, the following parameters sway the mass transfer:

(i) extent of mixing or agitation
(ii) type of agitator
(iii) method of dispersion of one fluid into other
(iv) flow rate of the fluid.

To characterise the influence of all the above parameters on the mass transfer process, let us try to understand the basic mass transfer process.

The mass transfer operation can be considered as a transfer process; any transfer rate is always proportional to the driving force and interfacial area, i.e.

$$\text{rate} \propto (\text{driving force})\ (\text{interfacial area}) \tag{14.8}$$

Here, the driving force is the concentration difference. That is, the mass flows/transfers from the region of higher concentration to that of lower concentration. Thus,

$$N_A \propto (\Delta C)(a) \tag{14.9}$$

where N_A is the rate of mass transfer per unit volume and is given in kg/(m^3 s), ΔC is the concentration difference in kg/m^3, and a is the interfacial area, which is the area of contact available between the two phases per unit volume, and will have the units of m^2/m^3 or m^{-1}. The proportionality constant in the above equation is known as **mass transfer coefficient**

$$\therefore \qquad N_A = k_L\,(a)\,(\Delta C)$$

where k_L is the liquid side mass transfer coefficient with units of m/s. Or in other words,

$$N_A = \frac{\Delta C}{R} \tag{14.10}$$

where $R = 1/(k_L a)$ and is known as the **mass transfer resistance.**

---

[1] The concept of surface renewal model was well-explained by Rao (1990) during oxidation of hydrocarbon, in which it was proposed that whatever be the reaction that took place, would be taking place in the film during its short stay of existence. When the surface is renewed, the contents in the film would mix with the bulk fluid, and accordingly bring down the concentration of reactant (hydrocarbon). Next time when the surface renews, the film will have the new concentration of the reactant (hydrocarbon), which is obviously less than what it was in the earlier film. Thus, the reaction progresses depleting the concentration of the reactant.

14.3.1 Gas–Liquid Mass Transfer

As has been mentioned in the earlier sections, mass transfer will take place between any two phases:

gas–liquid phases
liquid–liquid phases
solid–liquid phases.

However, the discussion in the present chapter is mostly restricted to gas–liquid systems only, since we come across, generally, the distribution or dissolution of oxygen in liquid fermentation broths (as in the production of penicillin, or treatment of effluents). Mass transfer in biochemical reactors was reviewed by Moo-Young and Blanch (1981), and Blanch and Clark (1996).

Mass transfer of A through the liquid film is given by

$$N_{AL} = k_L\, a(C_{A2}^* - C_{A2}) \tag{14.11}$$

Similarly, mass transfer through the gas film is given by

$$N_{AG} = k_G\, a(C_{A1} - C_{A1}^*) \tag{14.12}$$

Here, C_{A1}^* and C_{A2}^* are the equilibrium concentrations of the solute in the gas phase and liquid phase, respectively. They are not equal, but they hold a certain relationship, viz.

$$\frac{C_{A1}^*}{C_{A2}^*} = m \tag{14.13}$$

where m is known as the **distribution coefficient.**

The gas–liquid mass transfer in most of the bioreactors is considered from the standpoint of oxygen dissolution in the fermentation broth, or the transfer of the carbon dioxide from the fermentation broth. It is important to note that most of the fermentation broths are highly viscous and pose more problems for gas–liquid mass transfer. Customarily, Eq. (14.11) is written as

$$N_A = k_L\, a(C_{AL}^* - C_{AL}) \tag{14.14}$$

In the above equation, measurement or evaluation of $k_L a$ is important.

It is interesting to note that the unit of $k_L a$ is s^{-1}, which speaks about the mass transfer rate. Similarly, the unit of the reaction rate constant k_1 (first-order reaction rate constant in Eq. (7.6)) is also s^{-1}, which speaks about the chemical reaction rate. Thus, these two values are compared whenever we want to identify which of the rate process (mass transfer or chemical kinetics) is really the rate controlling step, so that we can put in efforts to improve that value. Whichever is less, is the rate controlling step, i.e. if $k_L a < k_1$, it indicates that the mass transfer is the rate limiting step, and hence efforts should be made to improve $k_L a$ by:

- improving the rate of agitation (rpm)
- increasing the gas (air) flow rate
- providing a suitable impeller
- providing a suitable gas distribution system, etc.

On the contrary if $k_L a > k_1$, it indicates that the chemical reaction rate is the limiting step, and hence efforts need to be put up to increase k_1 by:

- increasing the temperature of reaction (provided there are no side effects)
- choosing a suitable reactor
- choosing a suitable catalyst/biocatalyst
- choosing a suitable combination of reactor configuration, etc.

Thus, measurement/calculation of $k_L a$ by Eq. (14.14) is important to decipher the relative influence of mass transfer on chemical reaction rate.

14.3.2 Correlation for Mass Transfer Coefficient

Various efforts have been made for the last half a century to find some meaningful correlations for evaluating the mass transfer coefficient k_L, or interfacial area a. The efforts have been partly successful.

Mass transfer coefficient The liquid side mass transfer coefficient can be correlated with Sherwood number (N_{Sh}), Grashof number (N_{Gr}) and Schmidt number (N_{Sc}), as follows, for bubbles less than 2.5 mm in diameter (Calderbank and Moo-Young, 1961):

$$N_{Sh} = 2.0 + 0.31\,(N_{Gr})^{1/3}\,(N_{Sc})^{1/3} \qquad (14.15)$$
$$= 2.0 + 0.31\,(N_{Ra})^{1/3}$$

and for bubbles larger than 2.5 mm diameter,

$$N_{Sh} = 0.42\,(N_{Gr})^{1/3}\,(N_{Sc})^{1/2} \qquad (14.16)$$

where

$$N_{Sh} = \frac{k_L\, d_B}{D_{AB}} \qquad (14.17)$$

$$N_{Gr} = \frac{d_B^3\, \rho_L\, g\left(\rho_L - \rho_g\right)}{\mu_L^2} \qquad (14.18)$$

$$N_{Sc} = \frac{\mu_L}{\rho\, D_{AB}} \qquad (14.19)$$

$$N_{Ra} = \frac{d_B^3\, g\Delta\rho}{\mu_L\, D_{AB}} = N_{Sc}\, N_{Gr} \qquad (14.20)$$

For mould pellets and other suspensions, Calderbank (1967) reported that

$$k_L = \frac{0.13\left\{(P/V)\mu_L/\rho_L^2\right\}^{1/4}}{\left(N_{Sc}\right)^{2/3}} \qquad (14.21)$$

where P/V is the power input per unit volume supplied by the agitator.

Interfacial area The interfacial area in an agitated bubble absorber is given by (Calderbank, 1958):

$$a = 1.44\left[\frac{(P/V)^{0.4}\,\rho_L^{0.2}}{\sigma_L^{0.6}}\right]\left[\frac{u}{u_t}\right]^{1/2} \qquad (14.22)$$

where σ_L is the surface tension of the liquid, u the superficial gas velocity, and u_t the terminal velocity of the bubble. Typically, the terminal velocity of the bubbles is of the order of 26.5 cm/s (Danckwerts, 1970). All the terms in the square brackets in Eq. (14.22) should be in CGS units to account for non-dimensionality of constant 1.44.

Empirical correlations are available even for calculating $k_L a$. The one that is widely applicable for stirred vessels in highly non-viscous media is (Van't Riet, 1979):

$$k_L a = 2 \times 10^{-3}\,(P/V)^{0.7}\,u^{0.2} \qquad (14.23)$$

in which P, V and u are in units of W, m^3 and m/s, respectively. The above equation shows the dependence of $k_L a$ on gas flow rate (u) and the power input per unit volume (P/V) which takes care of the agitator speed and turbulence prevailing in the reactor. The above expression predicts the values of k_La with an accuracy of 20–40%, which is more or less tolerable for design purposes. Within this range, most of the predictions in the literature agree. Obviously, these expressions do not speak about the type of agitator, sparger, etc. Van't Reit (1979) reported from literature observations that there is no influence of

(i) type of stirrer and number of stirrers
(ii) type of sparger
(iii) fluid height
(iv) stirrer position

and many other operating factors and vessel geometry characteristics.

There is a number of correlations available in the literature with suitable limiting conditions (Danckwerts, 1970; Erickson and Stephenopoulos, 1987). Whatever is described above is not exhaustive; it is only representative for agitated gas–liquid reactors. The original references may have to be consulted for the exhaustive list of correlations, with limitations of their applicability. In general, we can conclude that as long as the vessel is stirred properly (i.e. absence of flooding and air entrainment, and a suitable sparger to distribute the gas), and proper care is taken to account for the ionic strength of the solution, the k_La values are dependent upon P/V and u.

Example 14.3

Yeast is being grown in a 48 liter capacity standard aerobic fermenter in a pilot plant experimentation. The fermentation broth is agitated with a turbine impeller. The dimensions of the bioreactor and the liquid height in the vessel are as per standard dimensions. Air is being blown into the fermenter @ 12 lps. The gas hold-up is estimated to be 18%. The turbine impeller is operated with 0.1 hp motor, out of which only 10% of the power is being used for agitating the impeller. Estimate the mass transfer rate (k_La).

Solution It is direct application of Eq. (14.23).

$$k_La = 2 \times 10^{-3} (P_g/v_L)^{0.7} u_g^{0.2}$$

Given:

Vol. of fermenter vessel = 48 l = 48×10^{-3} m^3

hp of motor = 0.1 hp

10% of the motor power is used for stirring

P_g = 0.01 hp = 0.01 × 745 = 7.45 watts

Gas hold up (ε) = 18%

Air vel = 12 lps = 12×10^{-3} m^3/s

We need to calculate V_L and u_g

Since it is a standard stirred fermenter, the liquid is filled up to 2/3 vol. of the vessel

$$V_L = 48 \times 2/3 = 32 \text{ l} = 32 \times 10^{-3} \text{ m}^3$$

The liquid height in the tank (h_L) = diameter of the tank (d_t).

$$\therefore \quad \text{liq. vol.} = \frac{\Pi}{4} d_t^2 \times h_L = \frac{\Pi}{4} d_t^3$$

$$\therefore \quad 32 \times 10^{-3} = \frac{\Pi}{4} d_t^3$$

from which d_t = 0.34 m = 34 cm

Hence the tank dia is taken as 35 cm,

i.e. $d_t = 35 \text{ cm} = 0.35 \text{ m}$

Vol. of liquid $= v_L = \dfrac{\Pi}{4} \times (0.35)^2 \times (0.35) = 0.0337 \text{ m}^3$

Gas velocity (u_g) is calculated using gas hold-up and the tank diameter

$$u_g = 12 \times 10^{-3} \times \frac{0.18}{\frac{\Pi}{4} d_t^2} = \frac{2.16 \times 10^{-3}}{0.785 \times (0.35)^2}$$

$$= 0.0225 \text{ m/s}$$

$$u_g^{0.2} = 0.468$$

$\therefore$ $$k_L a = 2 \times 10^{-3} \times (7.45 / 0.0337)^{0.7} \times 0.468 = 0.041 \text{ s}^{-1}$$

Example 14.4

For the fermenter in Example 14.3, the diffusivity coefficient of oxygen is 1.4×10^{-9} m²/s. The bubble diameter is approximately 2 mm. The physical properties of the fermentation broth are like that of water [$\mu_L = 1 \times 10^{-3}$ Pa s, $\rho_L = 1000$ kg/m³, $\sigma_L = 70.4$ mN/m]. The density of air may be found using ideal gas law at 35 °C. The terminal bubble velocity may be taken as 26.5 cm/s. Calculate k_L and a.

Solution Equation (14.15) and (14.22) may be used to calculate k_L and a separately.

Given:

Bubble dia, $d_B = 2$ mm
Diffusivity, $D_{O2} = 1.4 \times 10^{-9}$ m²/s
Density of liquid broth, $\rho_L = 1000$ kg/m³
Viscosity of liq., $\mu_L = 1 \times 10^{-3}$ Pa s
Surface tension, $\sigma_L = 70.4$ mN/m
Terminal velocity, $u_t = 26.5$ cm/s $= 0.265$ m/s

To find k_L we use Eq. (14.15) to calculate the Sherwood number (N_{Sh}). First we have to find the Rayleigh number (N_{Ra}). For this we need to calculate the gas density by using the ideal gas law*.

$$pV = n\,RT$$

in which p is the pressure of gas
V is the volume of gas
T is the temperature of gas in degree Kelvin
R is the universal gas constant (Doran, 1995, p. 20)

Volume occupied by 1 mole of air (28 g) at atmospheric pressure and 35°C (35 + 273 = 308 K) is given by

$$V = 0.00082057 \times (273 + 35)$$
$$= 0.0253 \text{ m}^3/\text{g mole}$$
$$= 0.0253 \text{ m}^3/28 \text{ g of air}$$

$\therefore$ $$\rho_g = \frac{28 \times 10^{-3}}{0.0253} = 1.108 \text{ kg/m}^3$$

By Eq. (14.20), $$N_{Ra} = \frac{d_B^3\, g\, \Delta\rho}{\mu_L\, D_{O2}}$$

$$= \frac{\left(2 \times 10^{-3}\right)^3 \times 9.81 \times (1000 - 1.108)}{1 \times 10^{-3} \times \left(1.4 \times 10^{-9}\right)} = 5.6 \times 10^7$$

---

* in which the universal gas constant, R, is noted to be 0.000082057 atm.m³/g mol.K (Doran, 1995, p. 20)

By Eq. (14.15)

$$N_{Sh} = 2.0 + 0.31\,(N_{Ra})^{1/3}$$
$$= 2.0 + 0.31\,(5.6 \times 10^7)^{1/3}$$
$$= 2.0 + 0.31\,(382.6) = 120.6$$

By Eq. (14.17)

$$N_{Sh} = \frac{k_L\, d_B}{D_{O2}} = 120.6$$

$$\therefore \qquad k_L = \frac{120.6 \times D_{O2}}{d_B} = \frac{120.6 \times 1.49 \times 10^{-9}}{2 \times 10^{-3}} = 9 \times 10^{-5} \text{ m/s}$$

We calculate a by Eq. (14.22):

$$a = 1.44 \left\{ \frac{(P/V)^{0.4}\, \rho_L^{0.2}}{(\sigma_L)^{0.6}} \right\} \left[\frac{u}{u_t} \right]^{1/2} \qquad (14.22)$$

The units are to be in CGS units or in SI units.

$$(u/u_t)^{0.5} = (0.0225/(26.4 \times 10^{-2}))^{0.5} = (0.085)^{0.5} = 0.291$$

$$\frac{P}{V} = \frac{7.45}{32 \times 10^{-3}} = 232.8 \text{ W/m}^3$$

$$(P/V)^{0.4} = 8.85$$

$$(\rho_L)^{0.2} = (1000)^{0.2} = 3.98$$

$$\sigma_L = 70.4 \text{ mN/m} = 70.4 \times 10^{-3} \text{ N/m}$$

$$(\sigma_L)^{0.6} = (70.4 \times 10^{-3})^{0.6} = 0.203$$

$$a = 1.44 \times \frac{(8.85 \times 3.98)}{0.203} \times 0.291$$
$$= 1.44 \times 173.6 \times 0.291 = 72.73 \text{ m}^{-1} = 0.727 \text{ cm}^{-1}$$

14.3.3 Measurement of k_La

k_La measurement techniques have been described considerably in chemical engineering literature (Danckwerts, 1970) as well as in biochemical engineering literature (Blanch and Clark, 1996; Doran, 1995). Basically there are three methods which are frequently described:

(i) Static method
(ii) Dynamic method
(iii) Sulphite oxidation method

They are briefly described here.

Static method This is based on the oxygen balancing in the gas–liquid contactor and making use of the basic mass transfer equation (Eq. (14.14)):

$$N_A = k_La\,(C_{AL}^* - C_{AL})$$

in which N_A is the total quantity of gas (oxygen) absorbed by the system per unit time and per unit volume. Thus,

$$N_A = \frac{(\upsilon_G\, C_{AG})_{in} - (\upsilon_G\, C_{AG})_{out}}{\text{volume of the liquid}} \qquad (14.24)$$

where C_{AG} is the concentration of the gas in the gas stream with flow rate of υ_G. Thus, $(\upsilon_G C_{AG})_{in}$ gives the quantity of the gas entering per unit time, while $(\upsilon_G C_{AG})_{out}$ indicates that in the outlet stream per unit time. The numerator indicates the quantity of gas absorbed (consumed) in unit time. Combining Eqs. (14.14) and (14.24) gives

$$k_L a = \frac{\text{Rate of absorption of gas } \left(\text{kg}/\left(\text{m}^3\ \text{s}\right)\right)}{\left(C_{AL}^* - C_{AL}\right)} \tag{14.25}$$

In Eq. (14.24), C_{AG} can be replaced by

$$C_{AG} = \frac{p_{AG}}{RT} \tag{14.26}$$

assuming ideal gas law; p_{AG} is the partial pressure of the gas in the gas stream, R the universal gas constant, and T the temperature in degrees Kelvin.

Now, Eq. (14.24) may be written as

$$N_A = \frac{1}{R}\left[\frac{\upsilon_G\, p_{AG}}{T}\right]_{in} - \left[\frac{\upsilon_G\, p_{AG}}{T}\right]_{out} \tag{14.27}$$

For applying Eq. (14.25), we need to have a methodology for accurate measurement of C_{AL} in the liquid phase, and p_{AG} or C_{AG} in the inlet and outlet streams. The oxygen electrode used for measuring C_{AL} should give accurate results. Otherwise, measurement of $k_L a$ may go wrong and give erroneous results. Except for this drawback, this method is simple and most reliable, and can also be used when the fermenter is in operation.

Dynamic method It is an unsteady state process. Initially the fermentation broth is made bereft of oxygen by stopping the supply of oxygen. Whatever oxygen is available in the system is consumed by the microorganisms. (In case of non-fermentation systems, the oxygen is stripped by sparging nitrogen.) Thus, the concentration of oxygen reaches the lowest value at point P (Fig. 14.3); later the oxygen supply is restarted to the system at a constant flow rate of air. Now the oxygen concentration starts building up and the concentration profile takes the shape of PQ. If the air (oxygen) supply is more than the consumption by the cells, the oxygen concentration builds up and reaches a constant stagnant value of $\bar{C}_{AL}$.

In the PQ zone, we choose two points R and S at times t_1 and t_2, respectively, having the oxygen concentrations at C_{AL1} and C_{AL2}. The rate of change of concentration of oxygen in the PQ region is given by

$$\frac{dC_{AL}}{dt} = k_L\, a\left(\bar{C}_{AL} - C_{AL}\right) \tag{14.28}$$

which on integration, assuming $k_L a$ to be independent of time, yields

$$k_L a = \frac{\ln\left[\dfrac{\bar{C}_{AL} - C_{AL1}}{\bar{C}_{AL} - C_{AL2}}\right]}{t_2 - t_1} \tag{14.29}$$

Thus, $k_L a$ can be measured experimentally by measuring C_{AL1}, C_{AL2}, t_1 and t_2, by the application of Eq. (14.29). This is based on two points. For achieving better accuracy, we can collect a set of date points of C_{AL} and t. They can be plotted on a graph (Fig. 14.4) taking $\ln\left[\dfrac{\bar{C}_{AL} - C_{AL1}}{\bar{C}_{AL} - C_{AL2}}\right]$ on Y-axis and t_2–t_1 on X-axis. The slope of the straight line passing through the origin gives an average value of $k_L a$.

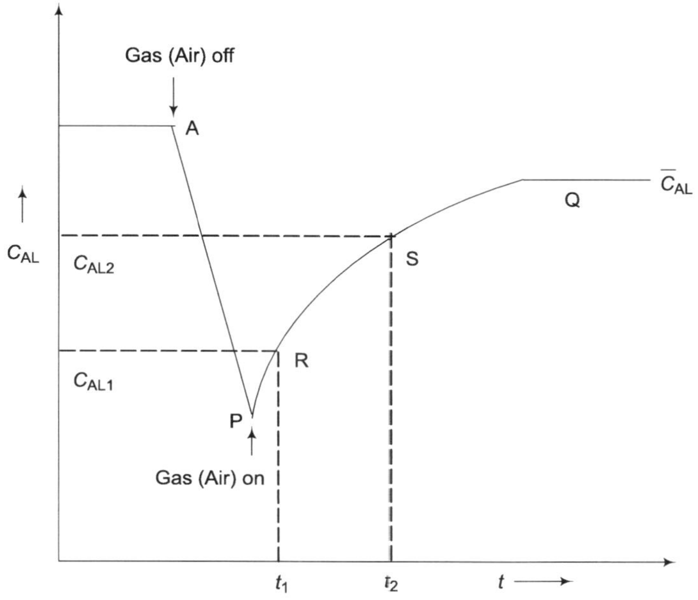

Fig. 14.3 *Oxygen stripping and absorption method for measurement of k_La by Dynamic gassing out method*

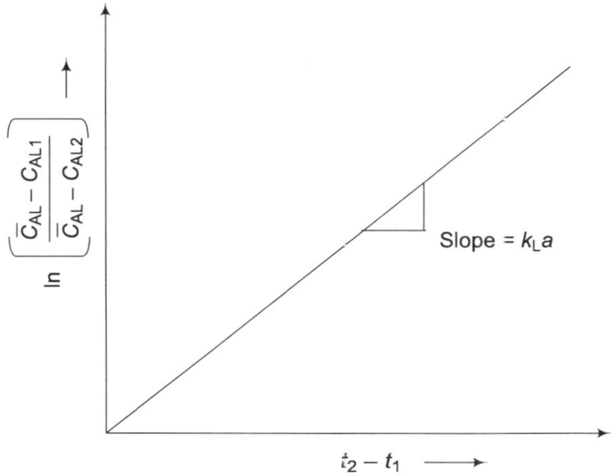

Fig. 14.4 *Plotting of Eq. (14.29) to measure $k_L a$ from a set of experimental data points*

This method is an ideal one for measuring the k_La in the experimental fermentor, provided the oxygen electrode measuring the oxygen concentration can respond quickly to the changes in oxygen concentration. Normally, the time required to reach the saturation oxygen concentration $\bar{C}_{AL}$ is only of the order of a few seconds. The oxygen electrode has to cope up with the quick response. Some of the other limitations associated with this method were dealt by Van't Reit (1979).

Example 14.5

For calculating $k_L a$ in 45-l stirred fermenter, dynamic gassing out method is used. The data are presented in Table E14.1. The steady-state gas (oxygen) concentration is 6 ppm. Calculate $k_L a$.

Table E14.1 *Time vs DO data*

Time, (s)	0	5	10	15	20	25	30
DO conc. (ppm)	0.5	1.5	2.5	3.1	3.8	4.1	4.5

Solution $k_L a$ can be calculated by applying Eq. (14.29) and drawing a plot similar to Fig. 14.4. The data for drawing the plot are given in Table E14.2.

Table E14.2 *Data for drawing Fig. E14.1*

Time (s)	C_{AL} *(ppm)*	$(t_2 - t_1)$ s	$ln\left(\dfrac{\bar{C}_{AL} - C_{AL1}}{\bar{C}_{AL} - C_{AL2}}\right)$
0	0.5	–	0
5	1.5	5	0.2
10	2.5	10	0.45
15	3.1	15	0.64
20	3.7	20	0.87
25	4.1	25	1.06
30	4.5	30	1.3

$$C_{ALO} = 0.5 \text{ ppm}$$
$$C_{AL} = 6 \text{ ppm}$$

With the data in the last two columns in Table E14.2, Fig. E14.1 is drawn.

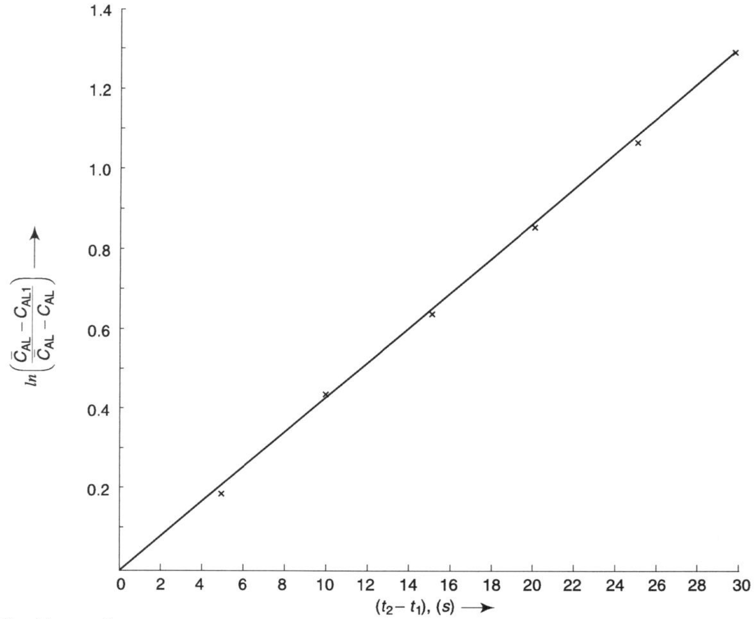

Fig. E14.1 *Solution for Example 14.5*

The slope is measured.

$$\text{slope} = \frac{1.3}{30} = 0.043$$

$$k_L a = 0.043 \text{ s}^{-1}$$

Sulphite oxidation method This is one of the oldest methods and referred to as chemical method (as contrast to the physical method of measuring the oxygen bubbles with an optical probe by Calderbank (1958)). The method is based on the chemical oxidation of sodium sulphite to sulphate by the dissolved oxygen in the presence of a Cu^{2+} or Co^{2+} catalyst.

$$SO_3^{2-} + \frac{1}{2} O_2 \xrightarrow[Co^{2+}]{\text{Catalyst}} SO_4^{2-} \tag{14.30}$$

In this method, the sodium sulphite is oxidised by sparging oxygen for a known amount of time with agitation. The concentration of the sodium sulphite before the start of the reaction and at the end of the reaction is measured to find out the amount of sulphite oxidised. From this, we can know the amount of oxygen absorbed and consumed in the reaction; usually, the oxygen is sparged for about 4–20 min (Cooper et al., 1944). In this system, the chemical reaction is so fast that it is almost instantaneous The concentration of oxygen in the liquid bulk is zero. Thus, $C_{AL} \approx 0$, and the system is a mass transfer controlled one. Hence, the measured rate of reaction is actually the mass transfer rate, from which $k_L a$ can be calculated as follows:

$$r_A = k_L\, a\, C_{AL}^{*} \tag{14.31}$$

The kinetic reaction (Eq. (14.30)) is complex. The rate of absorption is reported to be independent of the sulphite ion concentration in the range 1.0–0.015 molar (Finn, 1967). In a typical example, Danckwerts (1970) reported for the reaction catalysed by Co^{2+}

$$k_r = 0.85 \times 10^8 \text{ cm}^3/(\text{g mol s})$$

and

$$C_{AL}^{*} = 1.35 \times 10^{-7} \text{ g mol/cm}^3$$

for cobalt concentration of 10^{-5} molar and sulphite concentration of 0.5 molar.

The sulphite oxidation method is mainly credited for its simplicity and ease of applicability. Its greatest utility lies in the fact that this method allows for characterising the performance of the vessel under specified and reproducible conditions. Whatever be the purpose for which the vessel (fermenter) is used, $k_L a$ values can be measured, and they give a guess of the values which are adequate for design purposes.

This method has some of the inherent difficulties. It depends very much on the reaction kinetics which changes depending upon the concentration and type of the catalyst and the concentration of sulphite. The conversion is dependent on the reaction rate constant and the order of the reaction. Viscosity of the liquid medium also plays a vital role as observed by some workers earlier (Van't Riet, 1979), and hence cannot simply be applied to any viscous media. Incidentally, most of the fermentation broths are highly viscous, and hence its application for measuring $k_L a$ in fermenters is discouraged.

14.4 OXYGEN TRANSFER METHODOLOGY IN FERMENTERS

The transfer of gas from a bubble to a liquid medium in which the gas is dispersed in the form of bubbles is a typical case of gas–liquid mass transfer, and has been dealt extensively in chemical engineering literature (Danckwerts, 1970). In fermenters, the subject is more or less restricted to oxygen transfer into fermentation broths. In typical chemical reacting systems, the transferred oxygen is consumed in the chemical reaction (as in oxidation of hydrocarbons), and hence the mass transfer rate is enhanced due to chemical reaction, which we call as a classical case of **mass transfer with chemical reaction** (Astarita, 1967). Similarly, in aerobic fermentations, the oxygen is consumed by the microorganisms, and thus the oxygen transfer rate is enhanced.

In this section, we consider the oxygen transfer methodology from air bubbles in a fermentation liquid broth. It is shown schematically in Fig. 14.5. There are about eight steps involved. The gaseous (oxygen) component has to overcome these barriers before it reaches the active site in the cell or in the cell cluster (or cell flock) for biochemical reaction to take place. Some of them are limiting and some are not (Bailey and Ollis, 1986; Doran, 1995). They are shown in Table 14.1.

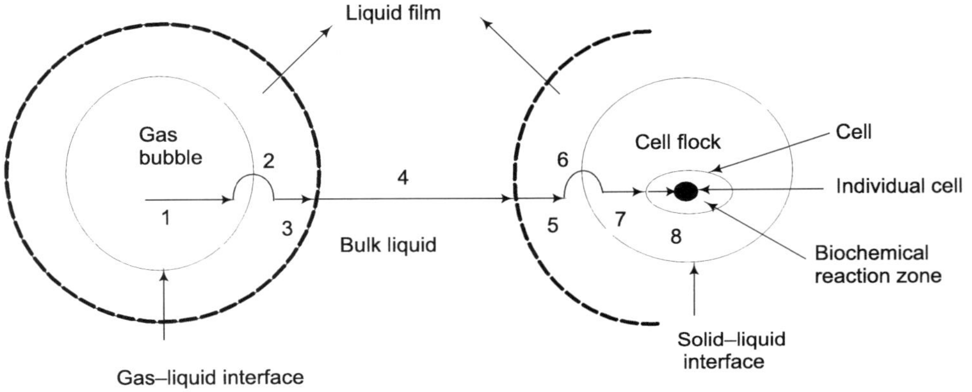

Fig. 14.5 *Schematic representation of oxygen transfer methodology from gas to liquid bulk to active site in the cell cluster*

Table 14. 1 *Various steps involved in the transport of gas from gas bubble to the bulk liquid and to the active site in the cell cluster*

Step	*Transfer operation*	*Remarks*
1	Transport of the gaseous component (oxygen) from the bulk gas (air) to the gas–liquid interface.	This step is relatively fast, and no resistance is involved in it.
2	Crossing the gas–liquid interface.	Resistance is negligible.
3	Diffusion of the gaseous component through the relatively stagnant liquid film into the well-mixed bulk liquid.	The liquid film offers resistance for gas diffusion, and hence may be **rate limiting**.
4	Transport of the gaseous component through the relatively well-mixed bulk liquid.	Resistance is negligible.
5	Diffusion of the gaseous component through the relatively stagnant liquid film surrounding the liquid– solid interface.	The resistance is negligible if the liquid–solid interface surrounds a single cell. In case of lumps of cells, the resistance is considerable.
6	Transport of gaseous component through the liquid–solid interface.	Generally, the resistance is negligible
7	Diffusion of gaseous component within the internals of the cell cluster.	Internal mass transfer resistance is significant, and hence may be **rate limiting**.
8	Transport within the cell to reach the active site (cytoplasm) for the biochemical reaction to take place.	Since the cell is very small in diameter, distances travelled are less, and hence the resistance is negligible.

14.4.1 Specific Oxygen Uptake Rate

The oxygen dissolution in fermentation broths is a tricky job. The dissolution/absorption rates are dependent upon

- various physical parameters
- various chemical parameters
- nature of the substrate
- nature of carbon source in the medium
- morphology and growth phases of the culture.

We have the expression for oxygen transfer as:

$$N_A = k_L a\,(C_{AL}^* - C_{AL})$$

For N_A to be high, either $k_L a$ should be high or $(C_{AL}^* - C_{AL})$ should be high.

As a rule of thumb, k_L values vary between 3–4 × 10^{-4} m/s and 1 × 10^{-4} m/s for bubbles of diameter around 2–3 mm and for bubbles of diameter ≤ 1 mm, respectively. This makes typically the $k_L a$ to vary between 0.02 and 0.25 s^{-1}. The other term $(C_{AL}^* - C_{AL})$ can be higher either if C_{AL}^* is high or C_{AL} is low. The least value for C_{AL} can be zero. C_{AL}^* cannot be high because it is approximately of the order of 10 ppm in pure water. Finn (1967) summarised some of the typical values of oxygen concentration as 0.018–0.049 mg mol/l at 30 °C in the presence of *Azotobacter vinelandi*, and 0.009 mg mol/l for *Penicillin chrysogenum*.

Thus, realising high values of N_A is very difficult in fermentation broths. The transfer rate needs to be monitored based on the oxygen demand by the cells. The oxygen demand varies with time in a batch fermenter. It is typically shown in Fig. 14.6, which was reported earlier by Darby and Goddard (1950) during the respiration of the mycelium of fungus. The specific oxygen demand by the cells (Q_{SOD}) is the oxygen requirement in grams per gram of the cells per second, and is related to the total oxygen uptake rate (Q_{OUR}) as follows:

$$Q_{OUR} = x\,Q_{SOD} \tag{14.32}$$

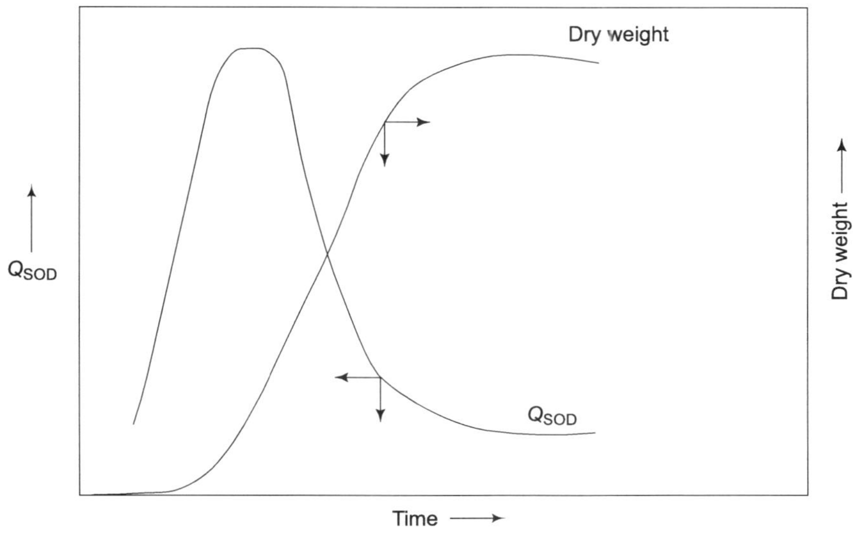

Fig. 14.6 *Variation of specific oxygen uptake rate in a batch culture*

where x is the concentration of the cells in grams of cells in one litre or kg/m^3. In the initial stage of the batch culture, Q_{SOD} reaches a maximum and slowly it falls down and finally flattens. The dry weight of the cells (x) goes on increasing and becomes larger at the later stage.

14.4.2 Critical Oxygen Concentration

The cells' respiratory capacity is high in the initial stage and it falls down gradually as is shown in Fig. 14.6. This shows that the Q_{SOD} depends upon the biochemical nature of the cells and the environment. The specific oxygen demand also depends upon the oxygen concentration in the fermentation broth. As the C_{AL} increases, Q_{SOD} also increases, and after a certain concentration of the oxygen, the Q_{SOD} levels off. Thus, the **critical oxygen concentration**, $C_{AL,crit}$, is defined as that concentration at which the Q_{SOD} is maximum. So we need to maintain C_{AL} in such a way that it is always greater than the critical value $C_{AL,crit}$. Generally, the $C_{AL,crit}$ is approximately 5–10% of the air saturation concentration (Doran, 1995).

14.4.3 Maximum Cell Concentration

The oxygen transfer into the broth is given by

$$N_A = k_L a\,(C_{AL}^* - C_{AL})$$

and all the transferred oxygen is taken up by the cells. Thus,

$$Q_{OUR} = N_A = k_L a\,(C_{AL}^* - C_{AL}) \tag{14.33}$$

or

$$k_L a\,(C_{AL}^* - C_{AL}) = Q_{SOD} \times x \tag{14.34}$$

If all the oxygen that is transferred is being taken up by the cells, then there will not be any oxygen left out in the broth. Thus, $C_{AL} = 0$. So, Eq. (14.34) becomes

$$Q_{SOD} \times x_{max} = k_L a\, C_{AL}^*$$

or

$$x_{max} = \frac{k_L\, a C_{AL}^*}{Q_{SOD}} \tag{14.35}$$

x_{max}, calculated by Eq. (14.35), should be always more than the cell density required by the fermentation system. If not, it is an indication that efforts should be put in to increase $k_L a$ either by increasing the impeller speed, or by increasing the air flow rate.

Doran (1995) also mentions the criterion of heat transfer to realise the value of x_{max} based on the fact that approximately 460 kJ of heat is evolved for each gram mole of oxygen consumed by the cells. Thus,

$$\text{heat generated} = (460/32) \times Q_{SOD} \times x_{max}\, V \tag{14.36}$$

where V is the volume of the fermentation broth and 460/32 is the conversion factor to convert the heat generated in kJ/(g mol) to kJ/g of oxygen. This is typically equal to $UA(\Delta T)$,

where U is the heat transfer coefficient (kW/(m^2 °C)),
A is the heat transfer area (m^2), and
ΔT is the temperature difference between the temperature of the fermentation broth and that of the cooling water.

$\therefore$

$$UA\Delta T = 14.4 \times Q_{SOD} \times x_{max}\, V \tag{14.37}$$

or

$$x_{max} = \frac{UA\Delta T}{14.4 \times Q_{SOD} \times V} \tag{14.38}$$

If the x_{max} calculated by Eq. (14.38) is less than the actual requirement, it indicates that efforts are required to improve the heat transfer mechanism, which can be done relatively easily by manipulating with ΔT, by decreasing the cooling water temperature if possible.

Example 14.6 *The oxygen requirement for cell growth in glucose can be represented by the following equation (Mateles, 1971).*

$$r_{O2} = \frac{32\,N_c + 8(N_{H_2}) + 16\,(N_{O_2})}{Y_{x/s}\,M} + y_{O2} - 2.67\,y_c + 1.714\,y_{N_2} - 8y_{H_2}$$

in which r_{O2} is the oxygen required for each gram of cells produced, N stands for no. of atoms present in each molecule of substrate, y stands for mass fractions, and M is the molecular weight of the substrate. The yeast cell may be considered to be $CH_{1.8}O_{0.5}N_{0.2}$ Calculate r_{O2} if the yield factor ($Y_{x/s}$) is 0.45 gram of cells produced for each gram of substrate consumed.

Solution From the given data, we can write for glucose ($C_6H_{12}O_6$)

$$N_C = 6$$
$$N_{H2} = 12$$
$$N_{02} = 6$$
$$Y_{x/s} = 0.45$$

Mol. wt. of glucose = 180 = M

Yeast cell is $CH_{1.8}\,O_{0.5}\,N_{0.2}$

$$\therefore \quad y_C = \frac{12}{12 + (1.8) + (16 \times 0.5) + (14 \times 0.2)} = \frac{12}{24.6} = 0.49$$

Similarly,

$$y_{H2} = \frac{1.8}{24.6} = 0.073$$

$$y_{O2} = \frac{16 \times 0.5}{24.6} = 0.325$$

$$y_{N2} = \frac{14 \times 0.2}{24.6} = 0.114$$

$$\therefore \quad r_{O2} = \frac{(32 \times 6) + (8 \times 12) - (16 \times 6)}{0.45 \times 180} + (0.325) - (2.67 \times 0.49) + (1.714 \times 0.114) - (8 \times 0.073)$$

$$= 0.998 \approx 1.0 \text{ g of oxygen/g of substrate}$$

14.5 FACTORS AFFECTING OXYGEN TRANSFER RATE

Various operating parameters and the physico-chemical properties of the broth affect the mass transfer rate which we quantify in terms of $k_L a$. As has been mentioned, the mass transfer rate can be increased by increasing the mass transfer coefficient (k_L) or interfacial area (a) or the driving force ($C^*_{AL} - C_{AL}$). The possibility to increase $C^*_{AL} - C_{AL}$ is very less, because C^*_{AL} itself is very small (of the order of 10 ppm or less). Hence, the option left out is to increase the k_L or a. Interestingly, the factors which increase a have a negative effect on k_L, which is caused by decreasing the bubble size or bubble diameter. The effect of the following on mass transfer rate will be discussed in this section:

(i) bubble size
(ii) gas hold-up
(iii) gas velocity
(iv) type of gas sparger
(v) type of agitation
(vi) power input to agitator
(vii) temperature
(viii) pressure
(ix) antifoaming agents
(x) presence of cells
(xi) surface-active solutes.

14.5.1 Bubble Size

The bubble size has a significant effect on the mass transfer rate. As the gas is sparged discretely in the fermentation broth, in the form of small bubbles, the interfacial area will increase and it is possible to increase the mass transfer rate.

For example, a gas bubble of 2 mm diameter will have a volume of 4.19 mm^3, and a surface area 12.57 mm^2. If the same is divided into two bubbles, each will be of 2.1 mm^3 volume, and will have a radius of 0.795 mm and surface area of 7.91 mm^2. Thus, fragmentation of the bubble into smaller sizes will be increasing the interfacial area. However, the smaller bubbles will have a negative effect on the mass transfer coefficient. They contain a smaller quantity of oxygen, which would be transferred quickly. Subsequently, what is left out is only nitrogen, which does not contribute for mass transfer. The smaller bubbles will have a rigid surface. The bubble behaves like a rigid sphere with less internal recirculation, so that the fresh batch of oxygen gas within the bubble does not reach the surface. If the bubble diameter is less, the bubble rise velocity will also be less. Thus, the swirling of the bubble will also be less, with the result that the chances for new surface area to come into contact with the bulk liquid will also be less. This results in low mass transfer coefficients.

Number of bubbles	*Bubble vol. (mm^3)*	*Bubble dia. (mm)*	*Surface area (mm^2)*	*Increase in area (%)*
1	4.19	2	12.57	—
2	2 × 2.1	1.59	2 × 7.91	26.0
3	3 × 1.4	1.39	3 × 6.0	44.2

Similarly, bigger bubbles will have higher rise velocity, and hence their stay in the broth is less; with the result that adequate time is not given for the entire oxygen to transfer to the liquid. Table 14.2 shows these effects, even though the results are not quantified. Bubbles are normally of the size 2–3 mm in diameter. Bubbles with diameter greater than 3 mm are called as large bubbles, and those with diameters less than 2 mm are classified as small bubbles (Calderbank, 1959).

Table 14. 2 *Effect (qualitative) of bubble size on interfacial area and various other parameters*

Bubble diameter	*a*	*k_L*	*Bubble rise velocity*	*Gas Hold-up*	*Internal recirculation*	*Bubble rigidity*
Big	Small	High	High	Less	Yes	No
Small	High	Small	Low	High	No	Yes

14.5.2 Gas Hold-up

Gas hold-up is the volume fraction of the gas held up in the total volume comprising the liquid and the held-up gas together, i.e.

$$\varepsilon = \frac{V_G}{V_G + V_L} = \frac{V_G}{V} \tag{14.39}$$

Thus, higher values of ε indicates higher amount of gas held up in the system. Even though ε does not speak of the bubble size, generally smaller bubbles stay for longer times in the liquid broth and contribute for higher ε. Thus, high ε results in higher interfacial area, and hence higher mass transfer rates. Normally, ε values range anywhere up to 0.2. However, smaller bubbles (diameter << 1 mm) can be a nuisance in bioreactors. For example, very small air bubbles will transfer the oxygen content in them to the liquid broth quickly, and what is remaining is only nitrogen which does not contribute anything for the performance of the fermenter, but simply will show higher gas hold-up. Thus, gas hold-up may not be directly proportional to the mass transfer rates. One of the main reasons for it is that the k_L values decrease with decreasing bubble diameters below 2 mm.

14.5.3 Gas Velocity

The superficial gas velocity (u_G) is the linear velocity of the gas obtained by dividing the volumetric flow rate of the gas with the cross-sectional area of the vessel. The mass transfer rates increase with u_G, as is given by Eq. (14.23). If the gas flow rate is very high, it may not allow the solute to dissolve in the liquid, and hence may escape in the outlet. In agitated vessels, the effect is compounded by the agitator speed also. Different gas flow regimes develop. If the gas flow rate is high, and the stirrer speed is low, the gas does not get dispersed properly. The gas bubbles rise up just like in a bubble column. The contacting pattern is poor, resulting interfacial is less, and most of the gas escapes unutilised. The situation is called **impeller flooding**, i.e. the gas is being supplied to the system at a rate which the impeller is not able to disperse. To overcome this, either the gas flow rate should be reduced or the agitator speed is to be increased. If the flow is too low compared to the agitator speed, then the gas bubbles will be simply recirculating; and hence the gas hold-up will increase, but mass transfer coefficient will reduce, with the result that the mass transfer rate will come down.

14.5.4 Type of Gas Sparger

The effect of spargers on mass transfer has not been studied extensively. Hassan and Robinson (1977) reported that sparger design did not have much effect on gas–liquid dispersion in aerated aqueous phases. However, they felt that it may have effect in large-scale industrial tanks in which a ring-type sparger would yield more uniform distribution of the gas as compared to other types of spargers. The spargers are designed based on experience, and are used based on convenience and availability. The most commonly used variety is a porous sparger (Fig. 14.7(a)). It is made out of a metal or glass or ceramic material with a small amount of projection. The diameter of the sparger is almost equal to that of the impeller, and as a rule of thumb, will be one-third the diameter of the tank, and is positioned below the agitator. The number of holes and their matrix is a matter of convenience and is seldom disclosed as a design secret. The cells may grow along the holes, particularly during off-operation period. They may block the holes and may cause nuisance. Another variety is a ring sparger (Fig. 14.7(b)). It is also known as **orifice sparger**. A number of holes is made to a pipe, and it is made into a ring. The ring diameter is again comparable to that of the agitator and is kept below it. This kind of sparger is used in single-cell protein production and in waste-water treatments. Figure 14.7(c) shows nozzle sparger with the holes along the length of the sparger. Sometimes the holes are made in the bottom of the sparger pipe, particularly when the flow rates are low, to effect better mixing and also to avoid that the gas bubbles directly come under the impeller. Pipe sparger (Fig. 14.7(d)) is simply a pipe through which gas is sparged. It also can be used when the gas flow rates are very low.

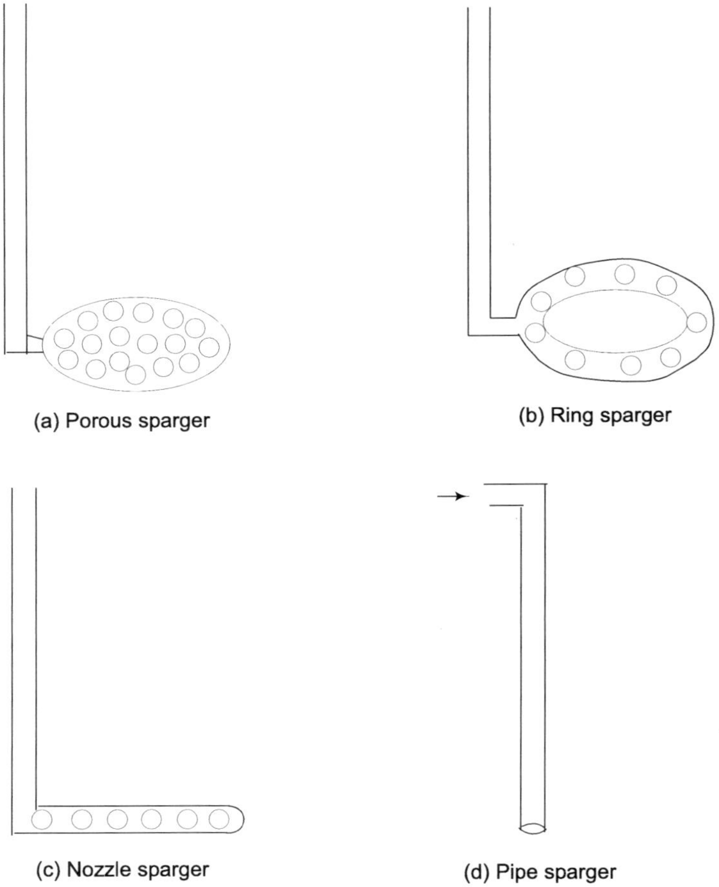

Fig. 14.7 *Different types of gas spargers*

14.5.5 Type of Agitator

Type of agitator and agitator design for effective mixing have been a subject of a lot of discussion in the chemical engineering and biochemical engineering literature. Various types of agitators have been designed for mixing of

gas–liquid
liquid–liquid
liquid–solid
gas–liquid–solid systems.

The subject received a lot of attention from the standpoint of gas–liquid reaction systems, different types of agitators used and the power input requirements, correlation for power input for gassed and ungassed systems, etc. Here we restrict out discussion to various types of agitators used, as shown in Fig. 11.13. They can be broadly classified as:

(i) propeller
(ii) turbine

(iii) paddle
(iv) vaned discs
(v) anchor
(vi) helical screw etc.

Mixing patterns for some of the agitators are shown in Fig. 11.14.

Vaned disc or turbine impellers are used for better gas–liquid contacting. Anchor-type agitators are used for highly viscous liquids and slurries. Helical and ribbon-type agitators are used for highly viscous semi-solid masses.

14.5.6 Power Input to Agitators

The effect of power input on a was shown by Eq. (14.22). The interfacial area varies as 0.4 power of the power input per unit volume (P/V). The power input is also related to agitator speed through the power number (N_P), where

$$N_P = \frac{P}{\rho N_i^3 d_i^5} \tag{11.17}$$

or

$$P = N_P \rho N_i^3 d_i^5$$

The dependence of power number on the agitator speed varies based on the type of flow. For example,

$$\left.\begin{aligned} P &\propto N_i^2, \text{ for laminar regime} \\ P &\propto N_i^3, \text{ for turbulent regime.} \end{aligned}\right\} \tag{14.40}$$

The dependence of $k_L a$ on N_i was demonstrated by Yagi and Yoshida (1975) for gas absorption in agitated vessels by Newtonian and non-Newtonian fluids with a superficial gas velocity of 3.81×10^{-3} m/s (Fig. 14.8).

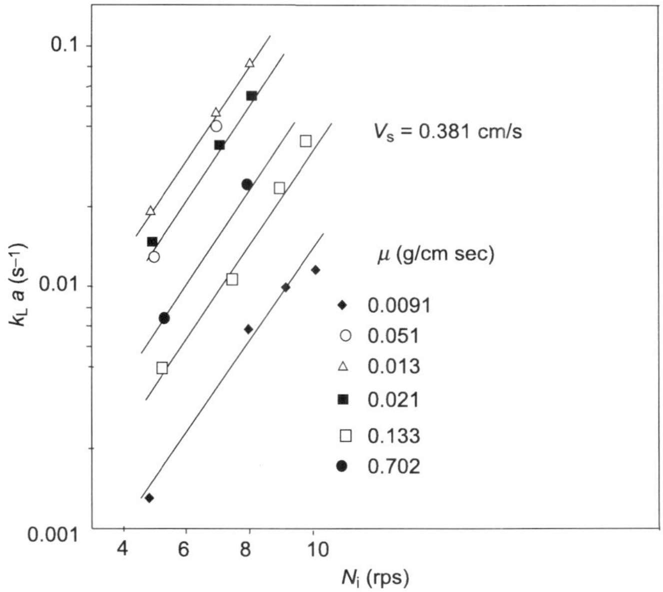

Fig. 14.8 *Effect of N_i on $k_L a$*
(Reprinted from Ind. Eng. Chem., Process Des. Dev., Yagi, H., and Yoshida, F., Vol. 14(4), p.490 Copyright © (1975), with permission from American Chemical Society)

14.5.7 Temperature

The temperature has both the effects on mass transfer. It increases the diffusivity of the gas into the liquid and hence increases the value of k_L. However, increase in temperature decreases the solubility of the gas, and hence reduces C^*_{AL}, which in turn reduces the driving force $(C^*_{AL} - C_{AL})$ for the mass transfer. Hence, it is reported that in the temperature range 10–40 °C, rise in temperature increases the oxygen mass transfer, whereas when the temperature is above 40 °C, the mass transfer rate decreases (Doran, 1995).

14.5.8 Pressure

Pressure affects the mass transfer by increasing the solubility of the gas in the liquid phase, which is given by Henry's law:

$$p_{AG} = H \times C^*_{AL} \tag{14.41}$$

The partial pressure and total pressure of the system are related by

$$p_{AG} = p_T \times y_A \tag{14.42}$$

Thus, as total pressure p_T increases, p_{AG} increases, and hence C^*_{AL} increases, which in turn increases the driving force $(C^*_{AL} - C_{AL})$. Generally, we do not come across high-pressure systems in fermentation processes.

14.5.9 Antifoaming Agents

Most of the fermentation broths contain proteins which cause foaming. Thus, foaming is an uninevitable nuisance in fermentation broths, and should be avoided. Otherwise,

- it may choke the pipelines
- it may choke the measuring instrument lines
- it may harbour unnecessary microorganisms to thrive and thrash the fermentation subsequently
- the choked pipe lines are difficult to be cleaned and may be a potential source for development of toxins.

Hence, ways and means have been devised to avoid foaming. They are

- mechanical means
- adding chemical antifoaming agents.

Of them, the latter is a very common method. Some of the silicon-based antifoam agents are added, which affect the surface chemistry of the bubbles by reducing the surface tension. One positive effect of it is that they reduce coalescence of the smaller bubbles into bigger bubbles. Thus they increase the interfacial area. However, these surface active agents reduce the mobility of the gas–liquid interface and reduce the mass transfer coefficient. Thus, the increase in a is countered by reduction of k_L, which is larger, with the result that the overall k_La is reduced (Kawase and Moo-Young (1990)).

14.5.10 Cells

Oxygen transfer (or mass transfer) in fermentation broths is greatly influenced by the presence of cells. Their influence on the oxygen transfer can be described by the following two ways.

(i) Physical influence They interfere in the break-up and make-up of the gas bubbles by influencing the surface properties. The cells sometimes get absorbed at the gas–liquid interface, and cause surface-blanketing which hinders the mass transfer. They also do not allow smaller bubbles to coalesce into bigger bubbles. This effect increases the interfacial area. However, surface-blanketing reduces the mass transfer coefficient.

(ii) Quantitative influence The cells absorb oxygen during the process which increases the driving force term ($C_{AL}^* - C_{AL}$). As the oxygen is absorbed, C_{AL} will be becoming smaller, and the term $C_{AL}^* - C_{AL}$ will be higher. This results in higher N_A. The effect is similar to enhancement of mass transfer due to chemical reactions, which is frequently described in chemical reaction engineering literature (van de Vusse, 1961; Danckwerts, 1970).

The influence of cells in enhancing the mass transfer rate depends upon

- the type of cells
- morphology of cells
- concentration of cells

However, the information on the above parameters is only qualitative. The quantitative influence is system-specific.

14.5.11 Surface-Active Solutes

The surface-active solutes, which are hydrophilic in nature, will alter the surface characteristics of the gas–liquid interface, and do not allow the gas bubbles to coalesce. This results in increased interfacial area. The concentration of the solute could be very low (as little as 0.05%), but its effect in increasing the surface area could be large (Calderbank, 1958). Since the concentrations of the solutes are very low, they do not affect the interfacial tension to any appreciable or measurable extent. Calderbank (1958) explained this with a solute like alcohol in water did not allow the air bubbles to coalesce in the aqueous media. The bubbles form stable rafts of bubbles which maintain their identity without coalescing until they bursted.

Symbols

a	: interfacial area (m^2/m^3 or m^{-1})
A	: cross-sectional area for mass transfer or heat transfer (m^2)
ΔC	: concentration difference (kg/m^3)
C_A	: concentration of A (kg/m^3 or mol/m^3)
C_A^*	: equilibrium solubility concentration at the interface (kg/m^3 or mol/m^3)
$\bar{C}_{AL}$	: average maximum oxygen concentration [Eq. (14.29)]
d_b	: diameter of bubble (m or mm)
d_i	: diameter of the impeller (m)
d_t	: tank diameter (m)
D_{AB}	: diffusivity coefficient (m^2/s)
g	: acceleration due to gravity (m/s^2)
h_L	: height of liquid (m)
H	: Henry's law constant
J_A	: mass flux (mol or kg/(m^2 s))
k_1	: first-order reaction rate constant (s^{-1})
k_G	: gas-side mass transfer coefficient (m/s)
k_L	: liquid-side mass transfer coefficient (m/s)
k_r	: rate constant for sulphite oxidation
m	: distribution coefficient
N'_A	: mass flow rate (kg/s)
N_A	: mass transfer rate (kg/(m^3. s))

N_{Gr}	: Grashof number [Eq. (14.18)]
N_i	: impeller speed (s^{-1})
N_P	: power number
N_{Ra}	: Rayleigh number [Eq. (14.20)]
N_{Sc}	: Schmidt number [Eq. (14.19)]
N_{Sh}	: Sherwood number [Eq. (14.17)]
p_{AG}	: partial pressure of gas A
p_T	: total pressure
P	: power input (W)
Q_{OUR}	: oxygen uptake rate by the fermentation broth (g/(l s))
Q_{SOD}	: specific oxygen demand by cells (g/(g s))
r_A	: rate of absorption/consumption of A (kg or kg mol/(m^3 s))
R	: universal gas constant,
	: resistance to mass transfer (s) [Eq. (14.10)]
s	: number of times surface renews (s^{-1}) [Eq. (14.7)]
t	: time (s or h)
T	: temperature (K)
ΔT	: temperature difference (°C)
u	: linear velocity (m/s)
u_B	: raising velocity of bubble in the liquid [Eq. (14.6)]
u_t	: terminal velocity of the gas bubble (m/s)
U	: heat transfer coefficient (W/(m^2 °C))
V	: volume of the fermenter broth (m^3)
x	: cell concentration (g/l)
X	: distance in X-direction (m)
y	: mass fractions in Example 14.6
y_A	: fractional concentration of A in the gas phase
Y	: distance (m)
$Y_{x/s}$	: yield factor

Subscripts

0	: initial
1, 2	: phase 1 and phase 2
A	: component A, usually oxygen
B	: bubble
crit	: critical value
G	: gas phase
L	: liquid phase
max	: maximum value

Greek Letters

σ	: surface tension
ρ	: density

ε	: gas hold-up (dimensionless)
θ	: time [Eq. (14.3) or Eq. (14.6)]
δ	: film thickness (m)
μ_L	: viscosity of liquid phase (kg/(ms))
υ_G	: volumetric flow vate of gas (m^3/s)

References

14.1 Aiba, S., Humphrey, A.E., and Millis, N.F. (1965). *Biochemical Engineering*, Academic Press, New York, p. 136.

14.2 Astarita, G. (1967), *Mass Transfer with Chemical Reaction*, Elsevier, Amsterdam.

14.3 Bailey, J.E., and Ollis, D.F. (1986), *Biochemical Engineering Fundamentals*, 2nd Edn, McGraw-Hill, Inc., New York.

14.4 Blanch, H.W., and Clark, D.S. (1996), *Biochemical Engineering*, Marcel Dekkar, Inc., New York, 353–415.

14.5 Calderbank, P.H. (1958), "Physical rate processes in industrial fermentations, Part I", *Trans. Inst. Chem. Eng.,* 36, 443–463.

14.6 Calderbank, P.H. (1959), "Physical rate processes in industrial fermentations, Part-II", *Trans. Inst. Chem. Eng.,* 37, 173–185.

14.7 Calderbank, P.H. (1967), "Mass transfer in fermentation equipment", *Biochemical and Biological Engineering Science*, Blakebrough, N. (Ed.), Vol. 1, Academic Press, London, 101–180.

14.8 Calderbank, P.H., and Moo-Young, M.B. (1961), "The continuous phase heat and mass transfer properties of dispersions", *Chem. Eng. Sci.*, 16, 39–54.

14.9 Cooper, C.M., Fernstram, G.A., and Miller, S.A. (1944), "Performance of agitated gas–liquid contactors", *Ind. Eng. Chem.*, 36(6), 504–509.

14.10 Danckwerts, P.V. (1951), "Significance of liquid flow co-efficient in gas absorption", *I&EC*, 43, 1460–1467.

14.11 Danckwerts, P.V. (1970), *Gas–Liquid Reactions*, McGraw-Hill, Inc., New York.

14.12 Darby, R.T., and Goddard, D.R. (1950), "Studies on the respiration of the mycelium of the fungus *Myrothecium verrucaria*", Am. J. Bot., 37, 379–387.

14.13 Doran, P.M. (1995). *Bioprocess Engineering Principles*, Academic Press, London, 196–213.

14.14 Erickson, L.E., and Stephanopoulos, G. (1987), "Biological reactors", *Chemical Reaction and Reactor Engineering*, Carberry, J.J., and Varma, A. (Eds.), Marcel Dekkar, Inc., New York, 779–838.

14.15 Finn, R.K. (1967). "Agitation and aeration", *Biochemical and Biological Engineering Science*, Blakebrough, N. (Ed.), Vol. I, Academic Press, London, 69–99.

14.16 Hassan, I.T.M., and Robinson, T.W. (1977), "Stirred tank mechanical power requirement and gas hold up in aerated aqueous phases, *AIChEJ*, 23(1), 48–56.

14.17 Higbie, R. (1935), "The rate of absorption of pure gas into a still liquid during short periods of exposure", *Trans*. AIChE, 31, 365.

14.18 Kawase, Y., and Moo-Young, M. (1990), "The effect of antifoam agents on mass transfer in bioreactors", *Bioprocess Eng.*, 5, 169–173.

14.19 Mateles, R.I. (1971), "Calculation of the oxygen required for cell production", *Biotechnol. Bioeng.*, 13, 581–582.

14.20 Moo-Young, M., and Blanch, H.W. (1981), "Design of biochemical reactor: Mass transfer criteria for simple and complex systems", *Advances in Biochemical Engineering*, Fiechter, A. (Ed.), Vol. 19, Springer-Verlag, Berlin, pp. 1–69.

14.21 Rao, D.G. (1990), "A semi-empirical model for the oxidation of cyclohexane", *I&EC, Process Des. Dev.*, 29, 696–699.
14.22 Van't Reit, K. (1979), "Review of measuring methods and results in nonviscous gas–liquid mass transfer in stirred vessels", *Ind. Eng. Chem.*, Process Des. Dev., 18(3), 357–364.
14.23 van de Vusse, J.G. (1961), "Mass transfer with chemical reaction", *Chem. Eng. Sci.*, 16, 21–30.
14.24 Whitman, W.G. (1923), "Preliminary experimental confirmation of two-film theory of gas absorption." *Chem. Met. Eng.*, 29, 146–148.
14.25 Yagi, H., and Yoshida, F. (1975), "Gas absorption by Newtonian and non-Newtonian fluids in sparged agitated vessels", *Ind. Eng. Chem.*, Process Des. Dev., 14, 488–493.

Review Questions

14.1 Write some applications of mass transfer in bioprocessing.
14.2 Explain the process of diffusion in bioprocessing.
14.3 What are various theories of diffusional mass transfer? Explain them.
14.4 Write a short note on the following theories on diffusional mass transfer:
  (i) film theory
  (ii) penetration theory
  (iii) surface renewal theory.
14.5 Explain the process of mass transfer by convection.
14.6 Write a note on gas–liquid mass transfer.
14.7 What are the various correlations for evaluating mass transfer coefficient?
14.8 Write the correlation for evaluating interfacial area.
14.9 Describe the various methods for measuring $k_L a$.
14.10 Compare and contrast the static method and dynamic method for measurement of $k_L a$.
14.11 What are the various drawbacks associated with sulphite oxidation method for the measurement of $k_L a$?
14.12 Describe the process of oxygen transfer methodology from the air bubble to the cell or cluster of cells in fermentation broths.
14.13 What is meant by "specific oxygen uptake rate"?
14.14 Describe what is meant by "critical oxygen concentration" in fermentation broths.
14.15 Derive an expression for the maximum cell concentration.
14.16 What are the various factors affecting oxygen mass transfer in fermentation broths?
14.17 How do you assess the effect of bubble size on mass transfer rate in fermentation broths?
14.18 Describe the effect of gas velocity on the mass transfer rate in fermentation broths.
14.19 Describe the various types of gas spargers you come across in bioprocessing.
14.20 Describe the various types of agitators (impellers) you come across in bioprocessing.
14.21 Why foaming should be controlled in fermentation broths? What are the various methods available for controlling foam?

Problems

14.1 In a typical aerobic fermentation process, the diffusivity coefficient for oxygen into the fermentation broth is 8×10^{-10} m^2/s. The stagnant liquid film thickness was calculated to be 6 microns. Find the mass transfer coefficient for oxygen based on the film theory.
(Ans: $k_L = 1.33 \times 10^{-4}$ m/s)

14.2 In an aerobic fermentation process, the typical average bubble diameter is 3 mm, with an average raise velocity of 18 cm/s. If the diffusivity coefficient is 8×10^{-10} m^2/s, find the mass transfer coefficient on the basis of the penetration theory.
(Ans: $k_L = 2.47 \times 10^{-4}$ m/s)

14.3 A 50 *l* stirred fermenter is used to culture *Torula utilis* at 28 °C. The dissolved oxygen concentration was measured (Table 14.3) after initially shutting off the oxygen supply and then restarting, for calculating k_La using dynamic gassing out method. The steady state oxygen concentration is 7.05 ppm.

Table 14.3

Time (s)	0	5	10	15	20	25	30	37
DO concentration (ppm)	0.5	2.5	3.7	4.4	5.1	5.5	6.0	6.3

Calculate k_La by using dynamic method.
(Ans: $k_La = 0.056$ s^{-1})

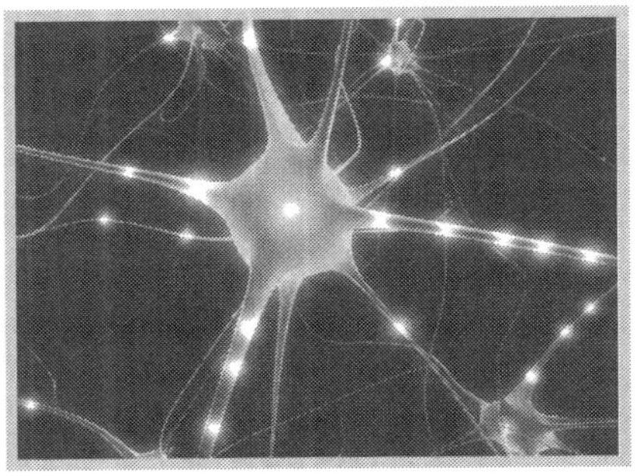

chapter 15

Heterogeneous Reaction Systems

As we have seen in Sec. 4.5, reaction systems may be of two types:

(i) homogeneous
(ii) heterogeneous.

In heterogeneous reaction systems, we encounter two phases; mostly the biocatalyst is in one phase and the substrate in another phase. Invariably, the biocatalyst is an immobilized enzyme system or immobilized cells in the solid phase, and the substrate is usually in the liquid phase. In a way, we can say that the heterogeneous reaction systems are mostly encountered in immobilized enzyme systems or in immobilized bioreactors. Hence, the discussion in this chapter is mostly restricted to immobilized enzyme systems in the solid phase (Section 4.6), and may be considered as a continuation to Chapter 4.

The reaction systems involving solid biocatalysts are important in bioprocessing operations, and they are complex for analysis in view of the fact that they require high level of mathematical treatment. The solid biocatalysts could be

- immobilized cells
- immobilized enzymes in the form of pellets
- flocks of cells
- clumps formed naturally by bacteria or fungi.

15.1 MASS TRANSFER CONSIDERATIONS

These heterogeneous systems operate under high concentration gradients, which are caused by intra-particle diffusion. The situation in a solid–fluid system can be represented by Fig. 15.1. The solid catalyst particle is composed of a large number of micro- and macro-pores. The path inside the pores is tortuous, and exerts resistance to mass transfer to take place inside the pores. The solid catalyst particle is also surrounded by a thin fluid film of the substrate (as represented by film theory). Initially, the substrate has to diffuse through the fluid film, which results in some concentration gradience. Subsequently, the fluid has to diffuse into the pores before it actually reaches the active site in the pellet for the biochemical reaction to take place. The concentration profiles can be represented by Fig. 15.2. The travel of the substrate from the bulk fluid into the particle and up to the active site is picturised in Fig. 15.3, and various process steps are shown in Table 15.1.

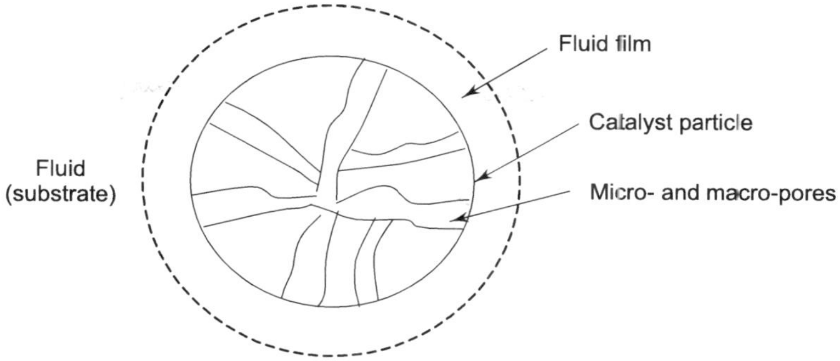

Fig. 15.1 *A fluid–solid catalyst system*

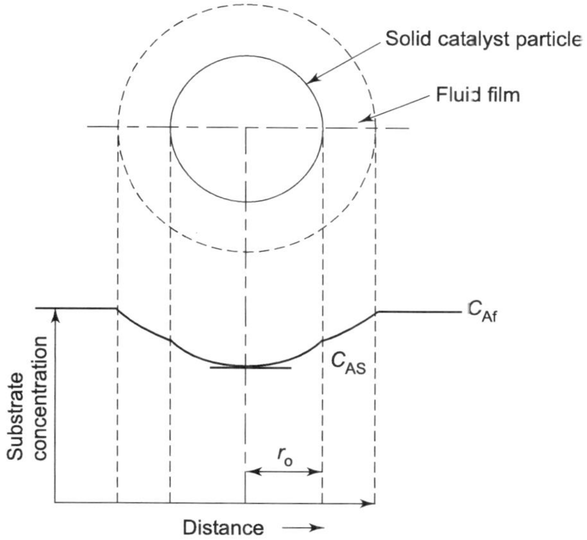

Fig. 15.2 *Concentration profiles in the solid pellet*

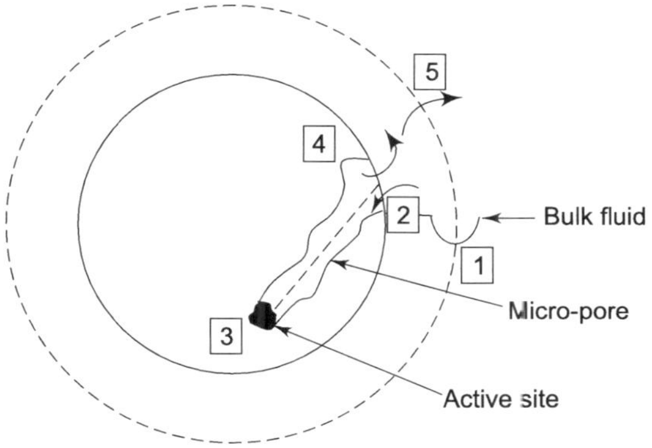

Fig. 15.3 *External and internal diffusion steps*

Table 15.1 *Various steps involved in the transport of substrate to active site and desorption of products*

Step	*Transport operation/reaction*	*Remarks*
1	Diffusion of the substrate into the external fluid film from the bulk.	Represents external mass transfer resistance. Can be overcome by increasing fluid velocity.
2	Diffusion of the substrate into the pores from the fluid film to reach to the active site.	Represents internal mass transfer resistance. Can be overcome by changing the particle size of base material.
3	Biochemical reaction of the substrate at the active site.	Represents the biochemical reaction step. Can be increased only by increasing the reaction temperature.
4	Desorption/Diffusion out of the metabolic product from the active site through the pore to the surface of the solid.	—
5	Diffusion out of the product from the solid surface through the fluid film into the fluid bulk.	—

As tabulated in Table 15.1, the substrate has to encounter two mass transfer resistances before it reaches the active site for conversion to metabolic products. They are:

(i) external mass transfer resistance
(ii) internal mass transfer resistance.

(i) External mass transfer resistance The fluid surrounding the solid particle forms an inter-phase between the solid phase and the fluid phase. This causes resistance for the mass transfer to take place, which we call as the **external mass transfer resistance.** It results in a fall in concentration in the substrate from initial C_{Af} to C_{AS}, which is the concentration at the surface (Fig. 15.2). This resistance can be reduced considerably by increasing the flow rate of the fluid (substrate), until there is no change in the overall reaction rate.

(ii) Internal mass transfer resistance Once the substrate reaches the surface, it has to diffuse into the pores to reach the active site. The pores could be micro-pores, or macro-pores, depending upon the pore diameter vis-à-vis the molecular diameter of the substrate. The pores could be straight or tortuous. The pores may end in between or may join with another pore. All these situations cause resistance known as **internal mass transfer resistance** for the flow (diffusion) of the substrate into the pores to reach the active site for the biochemical reaction to take place. To overcome this resistance, we prepare biocatalyst pellets with different particle sizes, so that the pellet will have more of macro-pores, and hence the internal resistance becomes negligible. To ensure the absence of this, we go on making pellets with different particle sizes, until there is no change in the overall reaction rate. This may not always be possible.

After the product is formed, the situation is similar for the product to diffuse out into the bulk. Thus, we come across all the five steps shown in Table 15.1 in the heterogeneous biocatalytic systems.

Interaction between Mass Transfer and Biochemical Reaction The mass transfer is always proportional to the concentration gradients; and the concentration gradients are established by the biochemical reaction taking place at the active sites. In other words, we can say that the more is the reaction rate, the more is the mass transfer rate. Thus, mass transfer rate and bio-chemical reaction rate are complementary to each other. More specifically, mass transfer rates will increase with the biochemical reaction rate, what we call traditionally in classical chemical reaction engineering terminology as "Mass transfer with chemical reaction". However, it may be noted that biochemical reaction rate *per se* is not improved by the mass transfer. It is dependent more on thermodynamic considerations, temperature (as given by Arrhenius equation), and reactor configuration. At the most, the biochemical reaction rate can be constrained if adequate mass is not being transferred. In such cases, we call the reaction is **mass transfer controlled or diffusion controlled**.

To understand the situation better, we introduce the following terms:

- mass transfer rate
- intrinsic biochemical reaction rate
- overall biochemical reaction rate.

Up to now we have dealt with the mass transfer rate. Let us see the other two terms.

Intrinsic biochemical reaction rate The rate of a reaction is dependent upon the concentration of the reactant (Eq. (7.2)). The concentration of the reactant (substrate) is reduced by the diffusional limitations, as we have seen in Fig. 15.2. Thus, in many cases, the real biochemical reaction rate is **clouded** by the mass transfer (or diffusional) considerations. The intrinsic chemical reaction rate is defined as the one which is free from mass transfer limitations. We also call it as the **true reaction rate**, and the reaction rate constant (k) obtained with it is called as the **true reaction rate constant.** Obtaining this reaction rate may be possible in laboratory-scale reactor, where all the mass transfer limitations can be overcome. The external mass transfer resistance and internal mass transfer resistance are avoided or reduced considerably by employing high fluid flow rates, and employing various particle sizes to make the biocatalyst pellets.

Overall biochemical reaction rate Since it is difficult to measure the concentration of the substrate at the active site where the biochemical reaction takes place, we use bulk concentration as the real concentration of the substrate. We measure the biochemical reaction rate, which we call as the observed reaction rate. The reaction rate constant obtained with C_{Af} and the observed reaction rate is known as the **overall reaction rate constant.** Obviously, the observed reaction rate might have been clouded or affected by the diffusional limitations.

The ratio between the observed reaction rate and the true reaction rate is known as **effectiveness factor**, which is always less than one. It will be discussed in more detail in the following sections.

15.2 INTRA-PARTICLE DIFFUSION AND REACTION RATE

We shall analyse the intra-particle diffusion and biochemical reaction simultaneously with the help of the mass balance equation (Eq. (8.2)). For this purpose, let us consider a biocatalyst pellet in spherical form (outer radius r_o), as shown in Fig. 15.4. Diffusion of the substrate takes place from outside to the centre of the particle. Let us consider an elemental sphere with a radius r and thickness Δr. Diffusion takes place through the elemental sphere. The mass balance equation for the elemental sphere in a unit time is as follows:

Mass inflow (1)	–	Mass outflow (2)	–	Mass consumed within (3)	+	Mass generated in the system (4)	=	Mass accumulated within the elemental sphere (5)

(8.2)

For applying the above equation, we presume the following assumptions to hold good:

- mass transfer takes place by diffusion alone
- diffusion process can be adequately explained by Fick's law
- diffusivity can be represented by a term known as **effective diffusivity** (D_{Ae}), as against the actual diffusivity which is difficult to measure
- the entire catalyst pellet is homogeneous
- the entire particle is in isothermal condition (in biochemical systems, the thermal effects are not very significant)

- the particle has attained the steady state, so that the mass accumulation term is zero
- substrate concentration varies linearly along the radial direction.

Now, with these assumptions armoured in our hands, we shall try to evaluate the various terms in the mass balance equation.

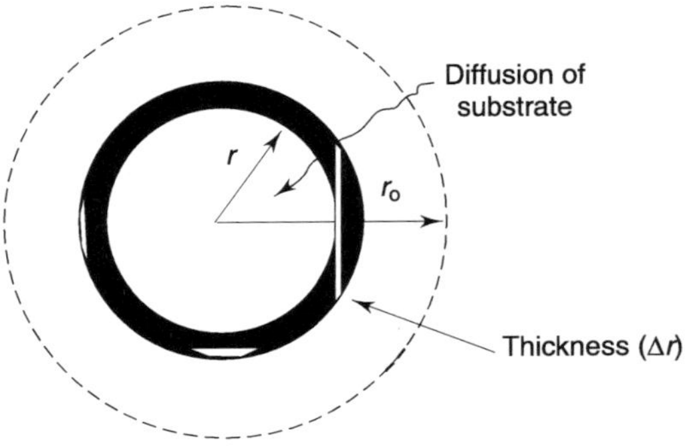

Fig. 15.4 *Mass balance on a spherical particle*

Mass inflow rate into the element at $r + \Delta r$

= (mass flux) × (cross-sectional area)

$$= D_{Ae} \left| \frac{dC_A}{dr} \times 4\pi r^2 \right|_{at\, r+\Delta r} \tag{15.1}$$

Similarly mass outflow rate from the element at r

$$= D_{Ae} \left| \frac{dC_A}{dr} \times 4\pi r^2 \right|_{at\, r} \tag{15.2}$$

Rate of mass consumption within the element

= rate of consumption × volume of the element

$$= r_A \times 4\pi r^2 (\Delta r) \tag{15.3}$$

Mass generated in the element = 0

Mass accumulated in the element = 0, because of steady state assumption.

Thus, the mass balance equation becomes

$$\left| D_{Ae} \frac{dC_A}{dr} \times 4\pi r^2 \right|_{at\, r+\Delta r} - \left| D_{Ae} \frac{dC_A}{dr} \times 4\pi r^2 \right|_{at\, r} - r_A\, 4\pi r^2 (\Delta r) = 0 \tag{15.4}$$

Dividing throughout by $4\pi\Delta r$ and taking the limit as $\Delta r \rightarrow 0$, Eq. (15.4) becomes

$$D_{Ae} \frac{d}{dr}\left(\frac{d\left(C_A \times r^2\right)}{dr} \right) - r_A r^2 = 0$$

or

$$D_{Ae} \left(\frac{(r^2)d^2 C_A}{dr^2} + (2r)\frac{dC_A}{dr} \right) - r_A r^2 = 0 \tag{15.5}$$

Equation (15.5), on integration, can yield a solution for the concentration profiles of A along the radius. Unfortunately, the integration is not straightforward as r_A depends on C_A in most cases. Sometimes only numerical solutions are possible.

In most of the cases, the reactions may be considered as first-order reaction, in which case (from Eq. (7.6))

$$r_A = k_1 C_A$$

or the biochemical reaction may follow Michaelis–Menten kinetics, in which case (from Eq. (4.14))

$$\upsilon = \left(\frac{\upsilon_{max}}{K_m + C_A} \right) C_A$$

15.2.1 Concentration Profiles for First-order Kinetics

Combining Eqs. (15.5) and (7.6), we get

$$D_{Ae}\left(\frac{d^2 C_A (r^2)}{dr^2} + 2r\frac{dC_A}{dr}\right) - k_1 C_A r^2 = 0 \tag{15.6}$$

in which k_1 is the first-order reaction rate constant and is independent of r.

The boundary conditions for Eq. (15.6) are:

when $r = r_0$, $C_A = C_{AS}$

and when $r = 0$, $dC_A/dr = 0$, which is evident from Fig. 15.2.

Integration of Eq. (15.6) gives (Van't Riet and Tramper, 1991):

$$C_A = C_{AS}\frac{r_0 \sinh\left(r\sqrt{k_1/D_{Ae}}\right)}{r \sinh\left(r_0\sqrt{k_1/D_{Ae}}\right)} \tag{15.7}$$

$$= C_{As}\frac{r_0 \sinh(M)}{r \sinh(M_0)} \tag{15.8}$$

Here, $$M = \left(r\sqrt{k_1/D_{Ae}}\right) \text{ and } M_0 = r_0\sqrt{k_1/D_{Ae}} \tag{15.9}$$

where M and M_0 are dimensionless numbers.

The hyperbolic sine,

$$\sinh(M) = \frac{e^M - e^{-M}}{2} \tag{15.10}$$

Example 15.1

An aerobic fermentation is being carried out using Aspergillus niger in the form of spherical pellets of 6.4-mm dia. The oxygen concentration in the bulk is 6.32 g/m³, and oxygen diffusivity is 1.8×10^{-9} m²s⁻¹. The oxygen consumption by the cells may be approximated to a first-order kinetics with specific reaction rate constant of 1.6×10^{-3}s⁻¹. Neglecting external mass transfer resistance, tabulate the steady-state concentration of oxygen in the pellet as a function of particle radius.

Solution For finding out the concentration profile along the radius, we use Eq. (15.8) for different values of r. First, we calculate M and M_O using Eq. 15.9.

Given:

Radius, $r_o = 6.4/2 \text{ mm} = 3.2 \times 10^{-3}$ m

$C_{AS} = 6.32 \times 10^{-3}$ kg/m³

$k_1 = 1.6 \times 10^{-3}$ s⁻¹

$D_{O2} = 1.8 \times 10^{-9}$ m²/s

$$M_O = \left(r_O\sqrt{k_1/D_{02}}\right) = 3.2 \times 10^{-3} \times ((1.6 \times 10^{-3} / 1.8 \times 10^{-9}))^{0.5}$$

$$= 3.2 \times 10^{-3} (0.9428 \times 10^3) = 3.017$$

$$M = r(k_1/D_{O2})^{1/2} = r(1.6 \times 10^{-3}/1.8 \times 10^{-9})^{1/2} = (942.8)\, r$$

$$\sinh(M_O) = \frac{(e^{M_O} - e^{-M_O})}{2} = 10.19$$

$$(r_O)\sinh(M_O) = 3.2 \times 10^{-3} \times 10.19 = 0.033$$

By Eq. (15.8)

$$C_A = C_{AS}\frac{(r_O)\sinh(M_O)}{(r)\sinh(M)} = \frac{6.32\times10^{-3}\times0.033}{(r)\sinh(M)}$$

$$= \frac{2.06\times10^{-4}}{(r)\sinh(M)} \tag{A}$$

For different values of r, sinh (M) is calculated using Eq. (15.9) and Eq. (15.10). Later by using (A), we calculate C_A for the corresponding r and are tabulated in Table E15.1

Table E15.1 *Data for E15.1*

r (m)	*M*	*sinh (M)*	*sinh (M) r*	C_A *kg/m*^{3}
0.5×10^{-3}	0.4714	0.489	978	1.94×10^{-3}
1.0×10^{-3}	0.9428	1.089	1089	2.14×10^{-3}
2.0×10^{-3}	1.8856	3.22	1609	3.195×10^{-3}
3.0×10^{-3}	2.8284	8.43	2810	5.58×10^{-3}
3.2×10^{-3}	3.017	10.19	3184	6.32×10^{-3}

15.2.2 Concentration Profiles for Michaelis–Menten Kinetics

Combining Eqs. (15.5) and (4.14), we get

$$D_{Ae}\left((r^2)\frac{d^2C_A}{dr^2} + 2r\frac{dC_A}{dr}\right) - \frac{v_{max}C_A r^2}{K_m + C_A} = 0 \tag{15.11}$$

It is not possible to find an analytical solution for the above because of the non-linearity of Michaelis–Menten expression. C_A as a function of r can be evaluated only numerically.

15.2.3 Concentration Profiles for Other Geometries

It is possible to find the concentration profiles for other geometries, viz. slab, cylindrical pore, etc. (Doran, 1995; Levenspiel, 1974). The details are not considered here.

15.3 Effectiveness Factor and Thiele Modulus

As has been mentioned earlier, the observed reaction rates are always less than the intrinsic reaction rates, based on concentration C_{AS} (or C_{Af}) being prevalent in the whole reaction zone. The observed reaction rate, $r_{A,\,obs}$, is based on the summation of all the reaction rates with concentrations varying from C_{Af} in the bulk to any other lowest concentration in the pores. It is equivalent to integrating the rate expression. Thus, for a 'first-order kinetics', we can write (Doran, 1995):

$$r_{A,\,obs} = \pi 4 r_0 D_{Ae} C_{AS}\,[M_0 \coth(M_0) - 1] \tag{15.12}$$

where hyperbolic cotangent, $\coth M = \dfrac{e^M + e^{-M}}{e^M - e^{-M}}$

Now we introduce the term **effectiveness factor** which is the ratio of the rate of substrate consumption with diffusional limitations to that without the diffusional limitations.

The effectiveness factor (η) is defined as

$$\eta = \frac{r_{\mathrm{A,obs}}}{r_{\mathrm{ideal}}} = \frac{\text{[actual observed reaction rate]}}{\text{[rate of reaction not suffering from any pore diffusional limitation, i.e. ideal reaction rate, if the concentration of A was } C_{\mathrm{Af}} \text{ throughout the catalyst pellet]}} \quad (15.13)$$

where r_{ideal} is defined as the reaction rate obtained if the substrate concentration were constant throughout the pellet and is equal to C_{Af}.

∴ Rate of consumption of the substrate under ideal conditions

$$= \text{volume of pellet} \times \text{rate of reaction}$$

$$= (4/3)\pi r_0^3 \times k_1 C_{\mathrm{Af}} \quad (15.14)$$

for first-order kinetics.

And

$$\eta_1 = (3/M_0^2)\,[M_0 \coth (M_0) - 1] \quad (15.15)$$

where η_1 stands for effectiveness factor for a first-order reaction.

Equation (15.15) shows that η_1 is dependent only on the dimensionless parameter M_0, which is a form of the Thiele modulus (Thiele, 1939), popularly abbreviated as ϕ.

The generalised expression for ϕ for any shape of particle and for any reaction kinetics is given by (Aris, 1965; Doran, 1995):

$$\phi = \frac{V_{\mathrm{p}}}{A_{\mathrm{S}}} \frac{(r_{\mathrm{A}})_{C_{\mathrm{A}} = C_{\mathrm{AS}}}}{\sqrt{2}} \left(\int_{C_{\mathrm{A}}}^{C_{\mathrm{AS}}} D_{\mathrm{Ae}} \times r_{\mathrm{A}}\, \mathrm{d}C_{\mathrm{A}} \right) \quad (15.16)$$

In the above equation, V_{p} is the volume of catalyst pellet of any shape (need not be sphere only) and A_{S} is the external surface area of the pellet; $(r_{\mathrm{A}})_{C_{\mathrm{A}} = C_{\mathrm{AS}}}$ represents the rate of reaction of A when the concentration of A is equal to C_{AS}, which is the concentration of A prevailing at the surface.

For a spherical geometry,

$$\frac{V_{\mathrm{p}}}{A_{\mathrm{S}}} = \frac{4\pi r_0^3}{3 \times 4\pi r_0^2} = \frac{r_0}{3}$$

Substituting this, we find Thiele modulus for spherical pellet:

$$M_0 = r_0 \sqrt{k_1 / D_{\mathrm{Ae}}} = 3\phi_1 \quad (15.17)$$

Thus, Eq. (15.15) becomes

$$\eta_1 = \frac{(3\phi_1 \coth 3\phi_1 - 1)}{3\phi_1^2} \quad (15.18)$$

Equation (15.18) is also written customarily as follows:

$$\eta_1 = \frac{1}{\phi_1}\left(\frac{1}{\tanh 3\phi_1} - \frac{1}{3\phi_1} \right) \quad (15.19)$$

in which $\tanh (x) = \dfrac{e^{x} - e^{-x}}{e^{x} + e^{-x}}$

Equations (15.18) and (15.19) show expressions for effectiveness factor for a spherical particle with first-order reaction kinetics, and Eq. (15.17) for Thiele modulus. Generalised expression for Thiele moduli for different shapes of pellets (viz. sphere and flat plate) and for different reaction kinetics (viz. zero-order, first-order and Michaelis–Menten reaction kinetics) are reproduced from Doran (1995) and presented in Table 15.2.

Table 15.2 *Generalized Thiele moduli*

First-order reaction: $r_A = k_1 C_A$

$$\phi_1 = \frac{V_p}{A_S}\sqrt{\frac{k_1}{D_{Ae}}}$$

Sphere $\phi_1 = \dfrac{r_0}{3}\sqrt{\dfrac{k_1}{D_{Ae}}}$

Flat plate $\phi_1 = b\sqrt{\dfrac{k_1}{D_{Ae}}}$

Zero-order reaction: $r_A = k_0$

$$\phi_0 = \frac{1}{\sqrt{2}}\frac{V_p}{A_S}\sqrt{\frac{k_0}{D_{Ae}\,C_{AS}}}$$

Sphere $\phi_0 = \dfrac{r_0}{3\sqrt{2}}\sqrt{\dfrac{k_0}{D_{Ae}\,C_{AS}}}$

Flat plate $\phi_0 = \dfrac{b}{\sqrt{2}}\sqrt{\dfrac{k_0}{D_{Ae}\,C_{AS}}}$

Michaelis-Menten reaction: $r_A = \dfrac{\upsilon_{max}\,C_A}{K_m + C_A}$

$$\phi_m = \frac{1}{\sqrt{2}}\frac{V_p}{A_S}\sqrt{\frac{\upsilon_{max}}{D_{Ae}\,C_{AS}}}\left(\frac{1}{1+\beta}\right)\left\{1+\beta\ln\left(\frac{\beta}{1+\beta}\right)\right\}^{-1/2}$$

$\beta = K_m/C_{As}$

Sphere $\phi_m = \dfrac{r_0}{3\sqrt{2}}\sqrt{\dfrac{\upsilon_{max}}{D_{Ae}\,C_{AS}}}\left(\dfrac{1}{1+\beta}\right)\left\{1+\beta\ln\left(\dfrac{\beta}{1+\beta}\right)\right\}^{-1/2}$

Flat plate $\phi_m = \dfrac{b}{\sqrt{2}}\sqrt{\dfrac{\upsilon_{max}}{D_{Ae}\,C_{AS}}}\left(\dfrac{1}{1+\beta}\right)\left\{1+\beta\ln\left(\dfrac{\beta}{1+\beta}\right)\right\}^{-1/2}$

(Reprinted from "Bioprocess Engineering Principles", Doran P.M., p. 310, Copyright © (1995), with permission from Elsevier.)

Figure 15.5 shows plots of effectiveness factor versus Thiele moduli for different shaped particles for first-order kinetics. Figure 15.6 shows a typical effectiveness factor for a flat biofilm as a function of dimensionless initial concentration of the substrate (β).

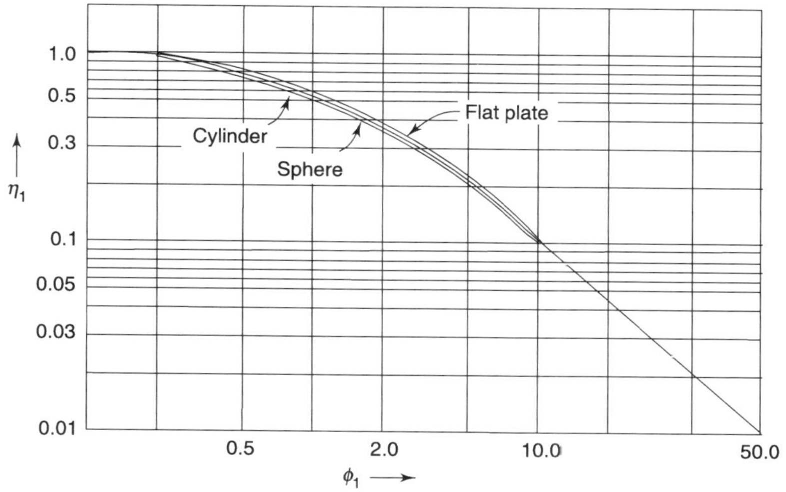

Fig. 15.5 *η versus ϕ for different shaped catalysts for first-order kinetics*
(Reprinted from "On shape factors for irregular particles. I. The steady state problem. Diffusion and reaction", Aris, R., Chem. Eng. Sci., Vol. 6, p. 265, Copyright © (1957), with permission from Elsevier)

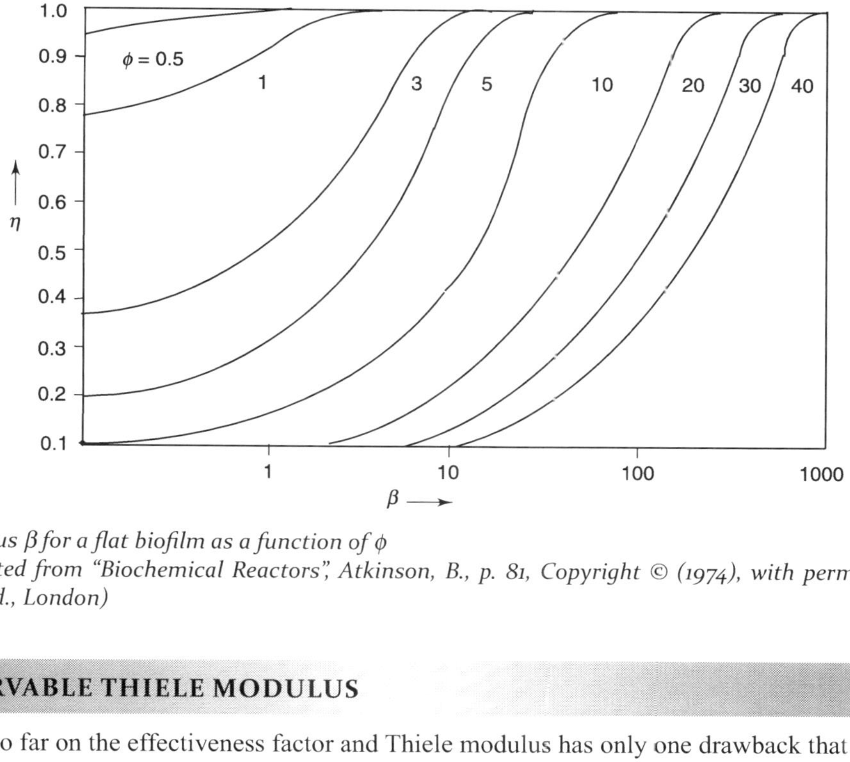

Fig. 15.6 *η versus β for a flat biofilm as a function of ϕ*
(Reprinted from "Biochemical Reactors", Atkinson, B., p. 81, Copyright © (1974), with permission from Pion Ltd., London)

15.4 OBSERVABLE THIELE MODULUS

The discussion so far on the effectiveness factor and Thiele modulus has only one drawback that the expressions for Thiele modulus employ the true reaction rate constants, which is difficult to evaluate as long as the true kinetics is masked by the diffusional limitations. To overcome this difficulty, we define another Thiele modulus known as **observable Thiele modulus** (Φ). It is also known as the Weisz's modulus (Weisz, 1973) and is defined as

$$\Phi = \frac{V_p^2}{A_S^2} \frac{r_{A,obs}}{D_{Ae} C_{AS}} \tag{15.20}$$

in which $r_{A,obs}$ is the observed reaction rate per unit volume of the catalyst pellet. The relationships between the two moduli (ϕ and Φ) for spherical particles are as follows (Doran, 1995):

For first-order reaction kinetics,

$$\Phi = \phi_1^2 \eta_1 \tag{15.21}$$

For Michaelis–Menten kinetics,

$$\Phi = 2\phi^2 \eta(1+\beta)\left(1+\beta \ln \frac{\beta}{(1+\beta)}\right) \tag{15.22}$$

where $$\beta = K_m/C_{AS}. \tag{15.23}$$

Criteria for Finding Mass Transfer Limitation The following criteria of the Weisz's modulus will help us identify whether the internal mass transfer limitations are significant.

Φ	η	*Internal mass transfer*
<0.3	≈1.0	Insignificant
>3.0	<1	Significant

For first-order kinetics,

if $$\Phi > 10, \eta_1 = 1/\Phi \tag{15.24}$$

Example 15.2

In a laboratory packed-bed fermentation process for the production of ethanol from glucose using S. cerevisiae in an immobilised form, the following data were obtained. As a first approximation, the reaction was assumed to follow first-order kinetics with the rate constant of 0.2h^{-1}. The average diameter of the particle is 2.5 mm. The diffusivity coefficient of the substrate is 5.2 × 10^{-10}m^{2}s^{-1}. Find the effectiveness factor, observable Thiele modulus (Weisz's modulus) and ascertain whether the reaction systems is suffering from any mass-transfer limitations.

Solution Given:

$$r_O = \frac{2.5 \times 10^{-3}}{2} = 1.25 \times 10^{-3} \text{ m}$$

$$D_{Ae} = 5.2 \times 10^{-10} \text{ m}^2/\text{s}$$

$$k_1 = 0.2 \text{ h}^{-1} = 5.5 \times 10^{-5} \text{ s}^{-1}$$

$$M_O = r_O (k_1/D_{Ae})^{1/2} = 1.23 \times 10^{-3} ((5.5 \times 10^{-5})/(5.2 \times 10^{-10}))^{1/2}$$

$$= 0.408$$

By Eq. (15.17), we have the Thiele Modulus

$$\phi_1 = M_O/3 = 0.136$$

$$\coth(M_O) = \frac{e^{M_O} + e^{-M_O}}{e^{M_O} - e^{-M_O}} = \frac{1.5046 + 0.6646}{1.5046 - 0.6646}$$

$$= 2.5822$$

By Eq. (15.18) and Eq. (15.17), we have the effectiveness factor.

$$\eta_1 = \frac{3}{M_O^2}\,[M_O \coth(M_O) - 1] = 0.965$$

Observable Thiele Modulus (Φ) is calculated by Eq. (15.21).

$$\Phi = \phi_1^2\,\eta_1 = (0.136)^2 \times 0.965 = 0.0178$$

Since $\eta_1 \approx 1.0$ and $\Phi < 0.3$, the internal mass-transfer limitations are negligible.

15.5 BIOREACTOR SELECTION CRITERIA

The reactor system should be akin to those used for fluid–solid catalyst reaction systems, since the immobilised cells or enzymes act as solid biocatalysts in the form of pellets. Generally, various types of base materials are used for the immobilisation of enzymes (Section 4.6). Most of them are fragile and lack mechanical strength. Thus, the basic consideration for the selection of the bioreactor for immobilised systems is low hydrodynamic shear. Hence, the recommended bioreactors are:

- packed bed bioreactors
- fluidised bed bioreactors
- air-lift bioreactors
- recycle bioreactors.

If a stirred tank fermenter were to be used due to any other operational advantages, care should be taken to see that the shearing effects are minimum. Alternatively, the immobilised solid matrix should be chosen, if possible, in such a way that it is strong and durable. The reactor can also be operated in a perfusion mode, in which a part of the solution or the nutrient medium may be passed through a recycle system as shown in Fig. 15.7.

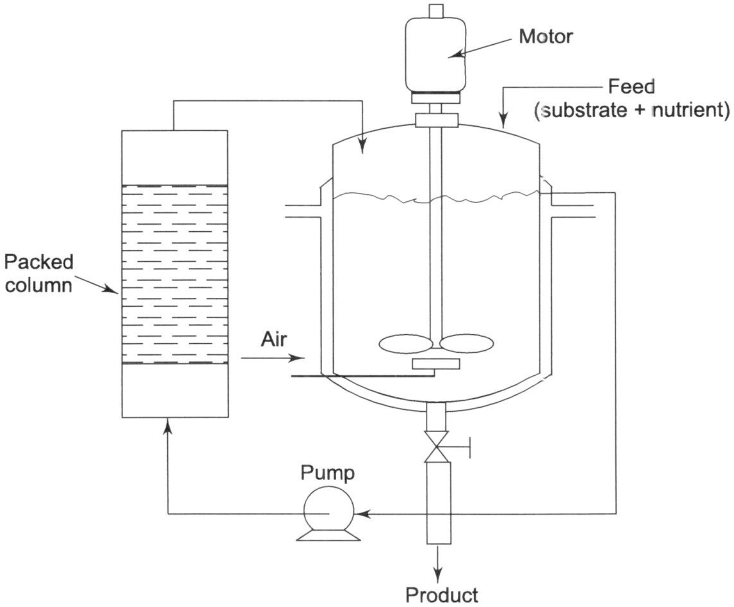

Fig. 15.7 *Recycle bioreactor system with a packed-bed containing the immobilized biocatalyst pellets*

Symbols

A_S	: surface area of catalyst pellet of any shape
b	: plate thickness (m)
C_A	: concentration of A (g mol/m^3 or kg/m^3)
C_A^*	: equilibrium concentration of substrate
C_{Af}	: concentration of A in the bulk fluid
C_{AS}	: concentration of A at the surface
D_{Ae}	: effective diffusivity of A (m^2/s)
k	: reaction rate constant
k_0	: reaction rate constant for zero-order kinetics
k_1	: reaction rate constant for first-order kinetics
K_m	: Michaelis–Menten constant
M	: dimensionless parameter [Eq. (15.9)]
M_0	: dimensionless parameter [Eq. (15.9)]
r	: radius of the elemental particle (m)
Δr	: thickness of the elemental space (m)
r_A	: rate of reaction of A
r_O	: outer radius of the sphere
r_{ideal}	: ideal reaction rate [Eq. (15.14)]
V_p	: volume of catalyst pellet of any shape

Subscripts

A	: component A
max	: maximum
obs	: observed
0	: zero-order reaction kinetics
1	: first-order reaction kinetics
m	: Michaelis–Menten kinetics

Greek Letters

β	: constant defined by Eq. (15.23)
υ	: rate of enzyme reaction
η	: effectiveness factor [Eq. (15.13)]
ϕ	: Thiele modulus
Φ	: observable Thiele modulus or Weisz's modulus

References

15.1 Aris, R. (1965), "A normalization for Thiele modulus", *Ind. Eng. Chem. Fund.*, 4, 227–229.
15.2 Atkinson, B. (1974), *Biochemical Reactors*, Pion Ltd., London, p. 81.
15.3 Doran, P. (1995), *Bioprocess Engineering Principles*, Academic Press, London, 297–332.
15.4 Levenspiel, O. (1974), *Chemical Reaction Engineering*, Wiley Eastern Pvt. Ltd., New Delhi, 460–491.

15.5 Thiele, E.W. (1939), "Relation between catalytic activity and size of particle", *Ind. Eng. Chem.*, 31, 916–920.
15.6 Van't Riet. K., and Tramper, J. (1991), *Basic Bioreactor Design,* Marcel Dekker, New York.
15.7 Weisz, P.B. (1973), "Diffusion and Chemical reaction: an interdisciplinary excursion", *Science,* 179, 433–440.

Review Questions

15.1 Describe the importance of heterogeneous biocatalytic reaction systems in bioprocessing.
15.2 What are the various steps that the substrate undergoes in a biochemical reaction being catalysed by an immobilised enzyme system?
15.3 What is meant by
- external mass transfer resistance?
- internal mass transfer resistance?

15.4 Describe the interaction between the mass transfer and biochemical reactions.
15.5 What is meant by
- intrinsic reaction rate constant?
- observed reaction rate constant?

15.6 Show diagrammatically the concentration profiles of the substrate in a spherical catalyst pellet.
15.7 What are the various assumptions you need to make for applying the material balance equation to heterogeneous catalytic systems?
15.8 Derive an expression for the concentration profiles along the radius in a spherical catalyst pellet.
15.9 What is meant by effectiveness factor and Thiele modulus? How are they related?
15.10 Show diagrammatically the relationship between effectiveness factor and Thiele modulus for different shapes of catalyst pellets and for first-order kinetics.
15.11 What is meant by Weisz's modulus? How is it related to Thiele modulus?
15.12 Describe the criteria for choice of a bioreactor for immobilised systems. What are the various types of bioreactors suitable?
15.13 Describe a recycle bioreactor.

chapter 16

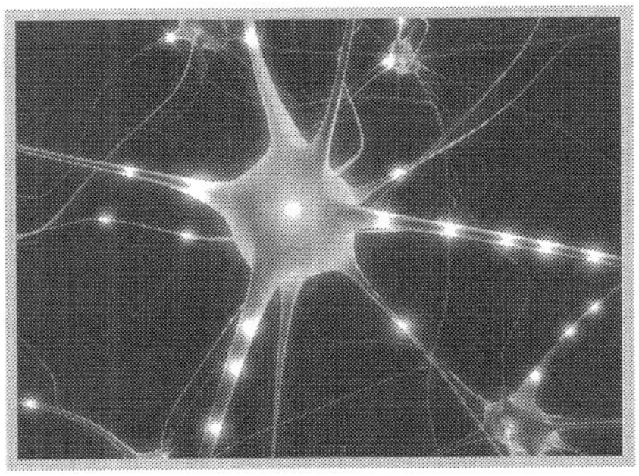

Bioreactors and Fermentation

The general concepts of the fermentation process have been described with some details in Chapter 5, along with various types of bioreactors used in fermentation processes. The design features of different bioreactors in terms of their ideal flow characteristics were illustrated in Chapter 8. In the present chapter, we shall discuss the classification of bioreactors and their applications, some of the process parameters which need to be controlled and monitored in bioreactors for effective running of fermentation process, and various accessories that are required to achieve the controlling of process parameters. In a nut shell, in this chapter we shall try to blend the bioreactor operations and fermentation processes.

16.1 BIOREACTORS

Atkinson's book (1973) on bioreactors is an excellent treatise on the subject. A bioreactor is a vessel in which the fermentation process is carried out with a substrate, in the presence of a nutrient media and carbon source, by the action of microorganisms, under certain ideal conditions. Thus, the fermentation process proceeds with the involvement of

- a microorganism
- a nutrient medium
- gas sparging (aeration for aerobic system)
- optimum process conditions.

Since many variables are involved in the fermentation process, the vessel which holds them will have to provide a congenial atmosphere, so that the living organisms (cells) can grow, multiply, proliferate and perform their job by consuming the substrate in the presence of nutrients under optimal process conditions. The vessel, which we call as **bioreactor** (or fermenter), is expected to cope with the requirements in terms of adequate

- mass transfer
- heat transfer
- fluid flow properties
- process conditions:

- pH
- temperature
- dissolved oxygen (gas) concentration
- agitation
- gas flow rate
- fluid–fluid mixing, etc.

This task cannot be performed by a simple vessel which is chosen arbitrarily, and this calls for a certain amount of systematic design with reference to different

- configurations of bioreactors
- flow characteristics
- mixing/blending methodologies for various phases
- techniques to maintain and control the process conditions
- methods to maintain asepsis.

16.1.1 Classification

The concepts of various types of bioreactors are borrowed from Chemical Reactor Theory (Denbigh and Turner, 1971). The classification is based on the ideal flow characteristics in the bioreactors (Fig. 5.6). They are:

- batch reactor
- plug flow reactor
- completely mixed flow reactor (CSTR)
- semi-batch reactors
- fluidised bed fermenters.

General description of the above types of reactors was already given in Section 5.7 and illustrations in Figs 5.6, 5.7 and 5.8.

The bioreactor is essentially expected to maintain the microbial culture in a stable and suspended form, which has made it necessary to develop various types of bioreactors. To keep the cells in suspension, we need to have a mixing system. The CSTR or CSTF (continuous stirred tank fermenter) is the obvious choice, but it has the drawback that the cells are washed away in the outlet continuously. The plug flow systems generally do not suit, as they do not allow the cells to be in a state of suspension. The batch operation and semi-batch operations have the disadvantage that they do not allow the product to be withdrawn, with the result that the cells and products will be accumulating in these unsteady systems. A system in which the microorganisms (or cells) are not washed away, but the substrate and nutrient media will be entering continuously, and the product is withdrawn continuously, is desirable. The answer to this expectation of ours could be a fluidised bed system in which only the cells or flocks of cells are kept in suspension (or fluidisation) under the flow of the entering fluid, and the products are withdrawn from the upper side of the fluidised bed reactor after the disengaging height (Fig. 5.8). Further details about the fluidised bed bioreactors will be given in Section 19.3.6. Some of the characteristic features of various bioreactors are given in Table 16.1.

16.1.2 Applications

Various bioreactors are used for the following range of fermentation processes:

(i) for production of microbial biomass
(ii) for production of microbial enzymes
(iii) for bioconversion of organic feed using microorganisms
(iv) for production of secondary metabolites.

Mostly, a semi-batch (fed-batch) or a batch fermenter is used for the production of a large number of fermentation products.

Table 16.1 *Characteristic features of various bioreactors*

Characteristic features	*Batch*	*PFR*	*CSTF*	*Fed- batch*	*Fluidised bed*
Flow conditions	Batch	Continuous	Continuous	Batch/Continuous	Continuous
Inflow of feed & cells	No	Yes	Yes	Yes	Yes
Outflow of products	No	Yes	Yes	Yes	Yes
Outflow of cells	No	Yes	Yes	Yes	No
Ideal/Non-ideal	Ideal	Ideal	Ideal	Non-ideal	Non-ideal
Steady/Unsteady state	Unsteady state	Steady state	Steady state	Unsteady state	Unsteady state
Maintenance of pH	Difficult	Difficult	Easy	Easy	Easy
Application	Very much used for small batches	Not used much, except as packed beds	Extensively used	Extensively used	Used with immobilised cells
Chief industrial application (Atkinson, 1974, p.42)	Most commercial fermentations	Waste-water treatment, Vinegar production	Waste-water treatment, Microbial protein production	Similar to batch fermenters	Beer and cider production

16.2 MONITORING AND CONTROL OF FERMENTATION PROCESSES

Controlling of process parameters in fermentation operation is one of the most crucial steps in successful operation of the fermenter. Since living organisms are at job in the fermenter, their ability in terms of their growth, multiplication, proliferation and performance to bring out a certain biochemical conversion is dependent upon the environment in the bioreactor. The environment in the bioreactor is dictated by physical, chemical and biochemical parameters (Doran,1995), and is detailed in Fig. 16.1.

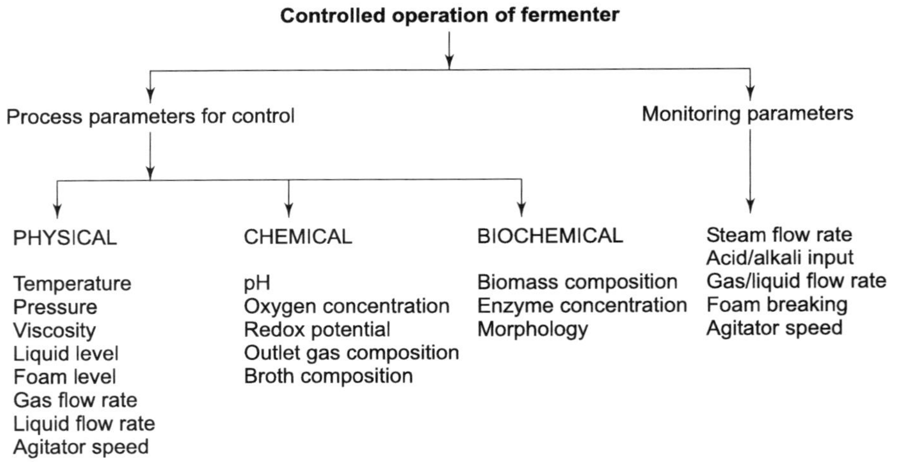

Fig. 16.1 *Monitoring and control of fermenter*

16.2.1 On-line and Off-line Control

The measurement and control of the parameters can be done on-line or off-line. On-line measurement is the method of measuring the process parameters within the proximity while the fermenter is working. For example, the temperature in a fermenter can be measured with a thermocouple on a digital indicator while the fermentation process is going on. The observed value will be compared with the set/desired value. Accordingly, the corrective action can be initiated. If the temperature is rising, the steam valve will be adjusted to decrease the flow rate, or electrical input will be reduced to control the heating. Similarly, the dissolved oxygen (DO) concentration can be measured with a DO analyzer and will be compared with the desired value. Based on the set point and comparison, the oxygen concentration will be monitored either by adjusting the oxygen flow rate or the agitator speed. Thus, the on-line measurement reduces the time-gap between the measurement and control, and hence is desired. Figure 16.2 shows a schematic diagram of a fermenter with various instrumentation for on-line controlling (Erickson and Stephanopoulos, 1987).

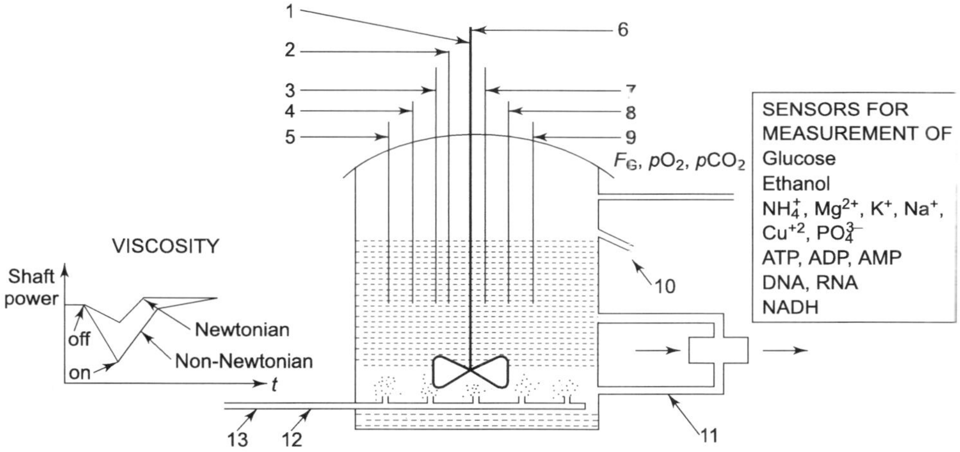

1. **SHAFT POWER**
   Torsion dynamometer (external)
   Strain gauge measure (internal)
2. **F_L: LIQUID FEED RATE**
   Drop count
   Addition vessel on load cell
   Electromagnetic flowmeter
3. **PRESSURE**
   Simple diaphragm gauge
4. **FOAM DETECTION**
   Capacitance
   Conductance
   CONTROL
   Mechanical system
   Silicone emulsion antifoam
   Oil based (sterility problem)
5. **TEMPERATURE**
   Thermometer bulb
   Thermocouple
   Thermistors
6. **IMPELLER SPEED**
   Generator-type tachometer
   RPM ↓ as V ↑ but impeller tip speed is the same ΠD (RPM)
7. **pH**
   Glass reference electrode
8. **REDOX**
   Combined P ↑ and reference electrode
   (measurement hard to interpret)
   Interaction with DO
9. **DISSOLVED OXYGEN**
   Polarographic sterilizable electrode
   $O_2 + 4H^+ + 4e^- \rightarrow 2H_2O$
10. **PRODUCT REMOVAL**
    Overflow, level control, load cell
11. **TURBIDITY**
    Spectrophotometer Biomass measurement
    (not for dense culture)
12. **pO_2**
    Paramagnetic O_2 anal. Electrochemical cell
    pCO_2 (IR analyzer)
13. **F_G GAS FLOW RATE**
    Rotameters, Mass flowmeters

Fig. 16.2 *Biochemical reactor instrumentation for online control*
(Reprinted from "Biological Reactors", Chemical Reaction and Reactor Engineering Erickson, L.E., and Stephanopoulos, G.; Carberry, J.J., and Varma, A. (Eds.), p. 799, 1987, by courtesy of Marcel Dekker, Inc., NY.)

However, all the process parameters cannot be measured on-line. Then they will be measured off-line, in which case, the sample is collected from the fermenter and analyzed separately, may be in a separate room. The results of analysis can be made available after 2–24 hours only. For this purpose, the samples will be collected once in every 4–8 hours and sent for analysis. Since the results of the analysis are available only after some time, the corrective measures cannot be implemented immediately. There is a time-lag between the measurement and control, i.e. the monitoring is delayed. Sometimes, by the time the corrective measure is implemented, the environment in the fermenter may also change and need a different type of control. Thus, the off-line measurement is not desired. For example, the pH of the broth can be measured on-line with a pH measuring electrode, and accordingly acid or alkali can be added to the broth to control the pH to the desired level by operating a solenoid valve connected to either the alkali tank or the acid tank. The same can be done off-line by collecting a sample from the fermenter, measuring the acidity or alkalinity by titration with a suitable titrant in the laboratory. Then the control measure of adding alkali or acid is done. Thus, there is a time-lag or delay in monitoring.

16.2.2 Controlling System

The measurement, monitoring and controlling steps involve the following four basic components, irrespective of whether it is on-line or off-line controlling operation:

- the process parameter
- the measuring element
- comparing with the set point
- corrective measure.

For example, the cell concentration is needed to be measured. If the cell concentration is correlated to the viscosity of the broth, the process parameter is viscosity. The viscosity of the broth will be measured off-line by usual techniques (by using a viscometer). The measured viscosity is compared to the set or desired value. If the value is higher, it is an indication that the cell concentration is on the higher side and vice versa. One method of controlling the cell population is by adjusting the oxygen flow rate to the broth. This can be done either by adjusting the air flow rate or by adjusting the agitator speed. Usually, the air flow rate is adjusted. This will control the cell population, which can be gauged by the changes in viscosity of the broth.

The controlling can be done in the following ways:

(i) manual control
(ii) automatic conrol
(iii) computer control.

(i) Manual Control In the case of manual control, the process parameter, after its measurement, is compared with the desired value (set point). Depending upon whether the measured value is higher or lower than the set point, the corrective action is initiated until the measured value matches with the set value. For example, in the case of measurement and control of temperature of the fermentation broth, initially the temperature of the broth is measured with a thermometer or a temperature indicator. The measured value is compared with the set value. If the temperature is higher, the steam valve supplying steam is adjusted to reduce/stop the steam supply manually, until the set point is reached (Fig. 16.3(a)). Thus, the manual control is subject to errors and delay in operation. It also depends upon the skill of the operator. The running costs are higher for it, but the initial capital investment is negligible.

(ii) Automatic Control The automatic controlling systems are better than the manual control systems, since they avoid human-biased error and skill. The controlling is done by an instrument. For this purpose, the measuring device should produce a signal or impulse. This will be compared with the set value in a loop. Accordingly an output signal is generated which implements the controlling action. Now consider-

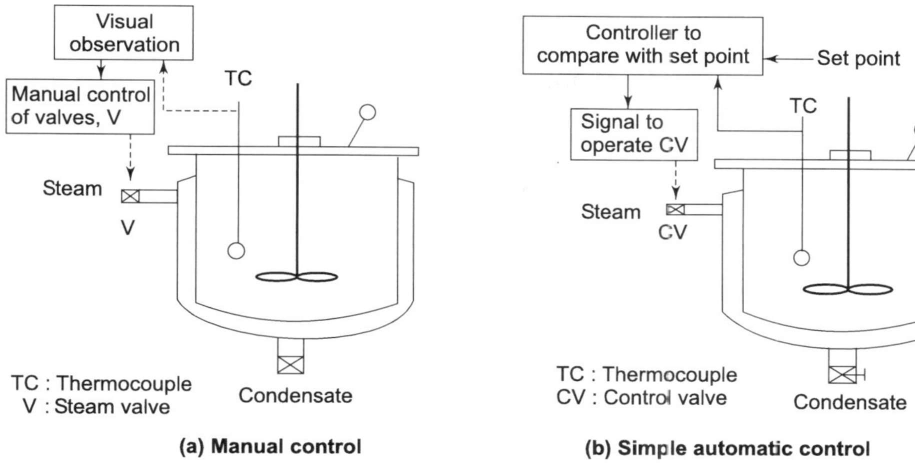

Fig. 16.3 *Control of temperature by adjusting steam valve*

ing the earlier example of temperature control, the temperature is measured with a thermocouple which gives impulse in the form of 'emf' (electro motive force). The thermocouple is connected to a temperature controller, which in turn generates some output which will operate the solenoid valve connected in the steam line to increase/reduce the temperature depending upon the set value. The automatic control operation is illustrated in Fig. 16.3(b).

The automatic control systems can be of the following four types, depending upon how the controlling is done:

- on/off controller
- proportional controller
- integral controller
- derivative controller.

A brief description on the above is available in Stanbury and Whitaker (1993(c)), but details are discussed in any standard textbook on instrumentation and process control (Coughanowr, 1991).

(iii) Computer Control of Fermentation Systems In the last two–three decades, computers have changed the methodology of controlling process parameters in fermenters. The speed of operation in the on-line control systems has considerably improved. The computer with its huge memory capacity (compared to pneumatic and electronic control systems) has become the nerve centre of control systems. A large number of process parameters can be controlled from a single control room which is remotely placed, and connected to the fermenter through softwiring. (Obviously, the fermenter has a large number of process parameters to be controlled.) Then the computer is interphased to the fermenter with an analogue-to-digital (A/D) converter.

All the physiological changes in the fermenter are manifested in the form of some physical or chemical environmental parameters in on-line experiments (Hampel, 1979). For example, the changes in cell growth will result in increase/decrease of dissolved oxygen concentration. The DO analyzer senses the oxygen concentration in the broth and gives the output in the form of millivolts, which in turn is fed to the computer. The input details are fed to computer. The input details, converted into millivolts, will be compared to the stimulus from the system by the CPU (central processing unit), and instructions will be issued to the actuator. In the present case, if the dissolved oxygen concentration is less, the air/oxygen inlet valve is activated to counter/correct the error, i.e. the solenoid valve admitting air/oxygen will be opened more to admit more gas. Thus, the computer interfacing can be classified to meet the following objectives:

Operator–computer	: instruction
Sensor–computer	: input
Computer–operator	: information
Computer–actuator	: manipulation
Computer–output devices	: communication.

The interfacing of a PC to a fermenter to control the temperature by operating the electrical heating system is shown in Fig. 16.4.

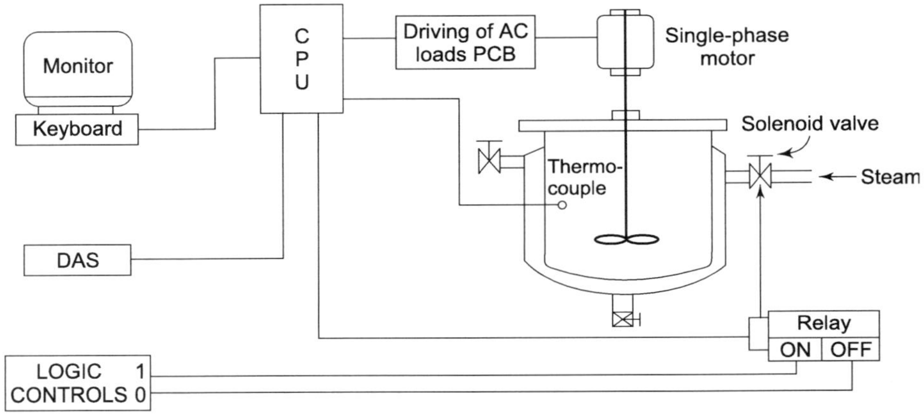

Fig. 16.4 *Block diagram for PC-controlled heating system for fermenter*

16.2.3 Biosensors for Fermentation Control

So far, various control techniques have been described for on-line control of the fermentation processes. Generally, spectrophotometric or chromatographic techniques are used for the on-line measurements of concentration of the components. These methods take relatively longer times for responding to the process change. To overcome these difficulties, **biosensors** are one useful alternative. Biosensors are developed as a result of the wedding of enzyme/microbial activity or reactivity with microelectronics.

Thus, a biosensor consists of:

- a biological sensing element which is chemically receptive
- a signal transducer which gives the output in the form of a measurable signal.

The biological sensing element is usually an enzyme or a microorganism or an antibody. To give physical strength and rigidity to the sensor, the biological molecule is immobilized on a probe. In view of its industrial importance, a good amount of studies was made on the biosensors. Even textbooks were published on the subject (Hall, 1991). However, we restrict our discussion for the present to describe only the basic principle of biosensors.

The tip of the biosensor is provided with a biologically sensitive material like an enzyme. If it comes across a compatible analyte, it would give out signal indicating positive response. The quantity of the signal also depends upon the number of the analytes, which in turn is proportional to the concentration of the analyte (Fig. 16.5(a)). An incompatible analyte is not detected or sensed by the tip, and hence does not give any response in the form of a signal output (Fig. 16.5(b)).

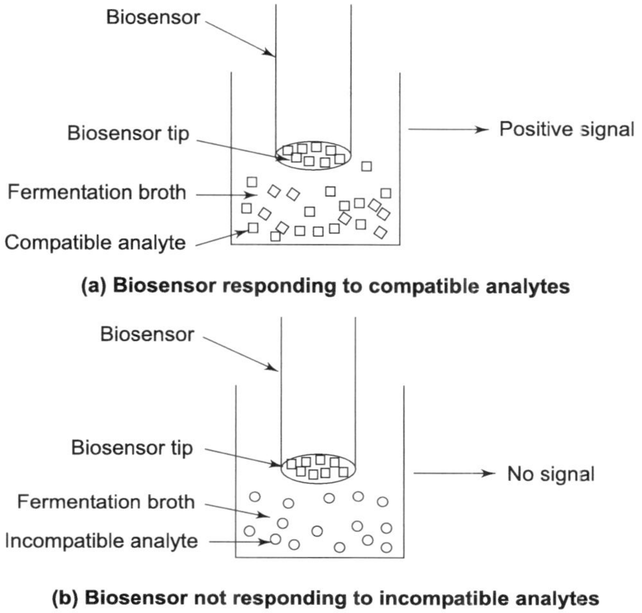

Fig. 16.5 *Schematic representation of the working of biosensor*

Since the enzymes are generally expensive and are also unstable to be used as sensing elements in bioprocess control, some microorganisms, which are known for their specificity to certain molecules can as well be employed for recognizing or sensing the biomolecules or analytes. Such sensors are termed as microbial sensors. For example, a cell membrane from the acetic bacterium **Gluconobacter suboxydans** can be used as an alcohol sensor. Karube and Sode (1989) have described the use of biosensors for sensing the concentration of:

- sugars (glucose, fructose and sucrose)
- alcohol
- glutamic acid
- carbon dioxide.

Schugerl (2000) observes that a good number of sensors was developed in this area in the last two decades, but their applications are restricted to mostly laboratory investigations. A short analysis time (response time) is a prerequisite for process control purposes; particularly the response times of the order of few minutes will suit the purpose of on-line monitoring.

Biosensors can be used for both qualitative and quantitative measurement. However, their use is more restricted to qualitative testing than quantitative estimation. Their prohibitive costs restrict their use in industry. They are being more used for medical or diagnostic purposes. A continuous research going on in this area for their industrial applications would make them highly selective, sensitive and reliable, in addition to making them cost-effective, which would make them suitable as disposable-type sensors.

16.2.4 Parameters for Control

Various parameters for monitoring and control have been described in the above Section. Since the fermentation process proceeds with the help of living cells (microorganisms) or enzymes, it is very important how closely the environment in the fermenter is controlled/monitored so that the living organisms find congenial atmosphere for bioconversion of the organic feed or biotransformation. So, it is imperative that the process parameters, like temperature, pH, flow rates, DO concentration, should be controlled as closely as possible. Deindoerfer (1960), almost four decades back, observed "*There is no reason whatsoever why batch fermentations should be run at constant temperature, constant pH, etc., when very likely a controlled variation of these influencing factors might have improved yields*". This observation puts us to rethink our concept of fermenter controlling/monitoring. This means, it is obvious that some process variables are critical and some are non-critical, which need to be controlled or otherwise.

16.2.5 Critical Parameters

The critical parameters are those which have a direct bearing on the progress of fermentation. It could be by the way of

- cell growth, or
- conversion of substrate.

Some process parameters which are important are:

- temperature
- pH
- DO concentration
- substrate flow rate
- air flow rate
- agitator speed
- redox potential.

Generally, the critical parameters are controlled online to avoid time delay by automatic methods or by computer-controlled techniques. The time delay in their control is critical and makes the process run away from control.

16.2.6 Non-critical Parameters

Non-critical parameters are those which need to be controlled, but the time delay in their controlling does not affect the progress of fermentation process drastically.

They are:

- pressure
- viscosity of the broth
- liquid level
- foam control
- outlet gas composition.

They are the process parameters which are not directly involved with the fermentation process. They are the outcome of some changes in the process. For example, the viscosity of the broth increases with the increase in mycelium production, or the liquid level in the fermenter alters due to some failure in the inlet feed systems. The pressure control does not play much role in the biochemical processes, as the pressure fluctuations do not normally occur. The pressure effects are generally felt in terms of oxygen (or gas) solubility.

16.3 VARIOUS ACCESSORIES

Various accessories mounted on the bioreactor are shown in Fig. 5.5. They are needed to measure and/or control the following process parameters:

- temperature
- pH
- dissolved oxygen
- pressure
- agitator speed
- liquid height
- feed flow rates
- foam control.

Table 16.2 shows the process parameters and the accessories required to measure and/or control them. The details of these accessories can be found in many textbooks (Stanbury and Whitaker, 1993; McCabe et al., 1993).

Table 16.2 *Accessories of fermenter to measure* and/or control process parameters

Parameter	*Accessory*
Temperature	Mercury thermometers (M) Bimetallic thermometers(M&C) Pressure bulb thermometers (M) Thermocouples (M&C) Thermistors (M&C)
pH	pH meter (M) pH electrode (M)
Dissolved oxygen	DO analyser (M)
Pressure	Bourdon pressure gauge (M) Pressure transducers (M&C)
Agitator speed	Torsion dynamometer (M) Tachometer (M)
Liquid height	Liquid level sensor (M) Solenoid valve (C)
Foam	Foam sensor (M) Adding antifoaming agents (C) Mechanical foam breakers (C)
Feed flow	Orifice meter (M) Rotameter (M) Venturi meter (M) Needle valve (C) Solenoid valve (C)
Flow accessories	Valves (C) Bends Elbows Pumps (C)
Weight	Load cells (M&C)
Outlet gas analysis	Titrometric methods (M) Gas chromatography (M)

M : *measuring*
C : *controlling*

16.4 CULTIVATION OF ORGANISMS

Cultivation of microorganisms is one of the most important activities of any fermentation process. Success of the fermentation process lies in our ability to maintain high cell density, particularly it is more so with growth-associated processes. As we have seen earlier, fermentations may be carried out in various types of reactors with different flow characteristics, such as:

- steady flow
- unsteady flow
- ideal flow with total mixing
- ideal flow with total non-mixing
- non-ideal flow, etc.

Depending upon the type of the fermenter, the growth characteristics also vary. Cultivation and growth of organisms were dealt with in Section 7.1. The growth curve was shown in Fig. 7.1. In this section we see the growth characteristics of microorganisms in the following three types of reactor systems:

- batch
- fed-batch
- continuous.

16.4.1 Batch Systems

A batch system is characterised by the flow pattern in which there will not be any flow rate of inlet or outlet. That is, all the contents are fed at once initially, into which a small quantity of the culture is inoculated, and the inlet and outlet valves are closed. Thus, the cells live in an environment initially which is rich in nutrients, medium and substrate. This gives a very congenial environment for the organisms to grow after initially passing through the acclimatisation phase, which we call as 'lag phase'. Subsequently, the cells start multiplying very fast exponentially which is given by Malthu's law:

$$\frac{dx}{dt} = \mu x$$

where dx/dt is the rate of change of cell mass, which we call as the **cell growth rate**, and μ is a coefficient, called as specific growth rate with the unit of time^{-1}.

As the cell growth rate increases, the number of cells will increase with time, by consuming the nutrients, medium and the substrate. In other words, with the progress of time, the concentration of nutrients and substrate will decrease, with the result that the cells are starved of the nutrients, and their growth rate will decelerate, which is shown as deceleration zone in Fig. 7.1. The deceleration will also occur if the substrate or one of the substrates is limited in supply. Such substrates are called limiting substrates and their concentrations are known as **limiting substrate concentrations.** As the limiting substrate concentration increases, the specific growth rate also increases, as shown in Fig. 7.5, up to a maximum value, and further increase in s does not show any increase in μ, which we call as μ_{max}. The relationship between the growth limiting substrate and the specific growth rate is well represented by Monod's equation (Eq. 7.16).

16.4.2 Continuous Systems

As can be seen in Fig. 7.1, after the log phase (exponential) growth of the cell culture, we notice that cell growth reaches a stationary phase as the medium concentration decreases. Suppose we devise a system in which the medium is added continuously, then obviously the cells never feel the dearth of medium/nutrients, and hence they grow continuously. However, there is a difficulty that the fermenter gets filled up with medium and cells after some time. We may contemplate a plan by which we remove the cells or products of equal volume as that of the medium entering, then the volume of the contents in the fermenter is constant. This is a situation which we call as 'continuous flow systems'. If the cells and contents in the fermenter are kept in a state of continuous suspension by agitation/stirring, we call the system as CSTR or CSTF. If the flow rates of the contents, both into and out of the fermenter, are adjusted, the cell growth also will be stabilised after a certain amount of time. Then we realise the advantage of a CSTF over a batch system where there are fluctuations in the growth of the cells, and hence fluctuations in the formation of the products, particularly those which are growth-associated. The through flows (capacities) also can be high, as the system runs continuously.

However, the operation of CSTF is not as rosy as it looks out to be. Sometimes, it takes even days, together with biological systems, for the stabilisation for CSTF to operate under stable conditions. Another serious disadvantage is that any amount of fluctuations in the flow or any type of contamination in the medium or the inlet flow (it is quite likely to happen in industrial processing, because the through-flows are high, and media etc. used may not be pure and sterile) will destabilise the growth process, in which case, it again takes days together to reach stability.

In spite of the above difficulties, continuous culture systems operate in a few bioprocessing industries like production of baker's yeast, brewing and waste-water treatments, etc. The volumetric feed flow rate (υ) into the fermenter is related to its volume (V) by a term known as **dilution rate**, D, where

$$D = \frac{\upsilon}{V} \tag{16.1}$$

This is also known as the **space-velocity** defined by Eq. (8.21). The *dilution rate is defined as the number of tank liquid volumes (reactor volumes) that pass through the vessel per unit time.*

The mass balance of the cells in a fermenter of volume V, is given by

$$V\frac{dx}{dt} = \text{cell growth} - \text{cells leaving the fermenter} \tag{16.2}$$

where dx/dt is the net rate of change of cell concentration,

i.e.
$$V\frac{dx}{dt} = V\mu x - \upsilon x$$

Dividing both sides by V, and replacing υ/V by D, we get

$$\frac{dx}{dt} = \mu x - (D)x$$
$$= x(\mu - D) \tag{16.3}$$

Under steady conditions, $dx/dt = 0$. Then Eq. (16.3) shows that

$$\mu = D \tag{16.4}$$

which means that the specific growth rate is more or less controlled by the dilution rate. Substituting for μ from the Monod equation into Eq. (16.3), we get

$$\frac{dx}{dt} = x\left(\frac{\mu_{max}\, s}{K_s + s} - D\right) \tag{16.5}$$

The steady state mass balance of the substrate in a CSTF can be arrived at by applying Eq. (8.2), in which the last term (accumulation term) is equal to zero.

i.e.
$$V(ds/dt) = \upsilon S_R - \upsilon s - \frac{V\mu x}{Y_{x/s}} \tag{16.6}$$

where $Y_{x/s}$ is the yield factor defined by Eq. (7.14), and S_R is the initial concentration of substrate.

The above equation can be written as

$$\frac{ds}{dt} = \frac{v}{V}S_R - \frac{v}{V}s - \frac{1}{Y_{x/s}}\mu x$$

$$= (D)S_R - (D)s - \frac{\mu x}{Y_{x/s}} \tag{16.7}$$

where μ can be replaced by Monod equation.

$$\frac{ds}{dt} = (D)S_R - (D)s - \mu_{max}\frac{x}{Y_{x/s}}\left(\frac{s}{K_s + s}\right) \tag{16.8}$$

The steady state cell concentration (x_{ss}) is given by Eq. (8.5):

$$x_{ss} = Y_{x/s}(S_R - s_{ss})$$

This can also be derived from Eq. (16.7), by replacing μ with D (Eq. (16.4)):

$$\frac{ds}{dt} = (D)S_R - (D)s - \frac{(D)x_{ss}}{Y_{x/s}} = 0$$

Hence,

$$S_R - s_{ss} = \frac{x_{ss}}{Y_{x/s}}$$

or

$$x_{ss} = Y_{x/s}\,(S_R - s_{ss}) \tag{16.9}$$

The steady state limiting substrate concentration, s_{ss}, can be derived from Eq. (16.8) by equating $ds/dt = 0$

$$(D)S_R - (D)s_{ss} - \mu_{max}\frac{x_{ss}}{Y_{x/s}}\left[\frac{s_{ss}}{K_s + s_{ss}}\right] = 0$$

$$(D)(S_R - s_{ss}) = \mu_{max}\frac{x_{ss}}{Y_{x/s}}\left[\frac{s_{ss}}{K_s + s_{ss}}\right]$$

By replacing $(S_R - s_{ss})$ by $(x_{ss}/Y_{x/s})$ from Eq. (16.9), we have

$$(D)\frac{x_{ss}}{Y_{x/s}} = \mu_{max}\frac{x_{ss}}{Y_{x/s}}\left[\frac{s_{ss}}{K_s + s_{ss}}\right]$$

$$(D)\,(K_s + s_{ss}) = \mu_{max}\, s_{ss}$$

$$K_s(D) = (\mu_{max} - D)s_{ss}$$

$$s_{ss} = \frac{K_s\,(D)}{\mu_{max} - D} \tag{16.10}$$

The above equation gives the relationship which shows how D controls s_{ss} . s_{ss} is the residual limiting substrate concentration which just supports the growth of the cells, so that it is stabilised.

16.4.3 Fed-batch Systems

Fed-batch systems are one of the most commonly used flow systems in bioprocessing. The semi-batch reactor systems, which we described earlier in Section 5.7.1 (Fig. 5.7), are also known as fed-batch systems. The word 'fed-batch' is derived from the fact that the culture materials are fed continuously to a batch system, and hence the name (Yoshida et al., 1973). Presently, we refer the fed-batch to mean semibatch operations in which

- one of the reactants is fed continuously (it could be culture inoculum, medium, nutrients or substrate)
- one of the products is removed continuously
- all the reactants are added continuously
- aeration alone is done continuously and unused air is removed continuously.

All the four types of flow systems were shown in Fig. 5.7.

The fed-batch system has some distinct advantages:

(1) Under substrate-inhibition conditions or catabolic repression conditions, fed-batch system is extremely useful. It does not allow the substrate concentration to build up by maintaining a limited supply of it.

(2) If one of the essential components of reaction has toxic effects on the reaction or on the microorganism (for example, sodium phenyl acetate, a precursor of the penicillin molecule is toxic to *Penicillin chrysogenum,* and hence is fed in small increments or continuously in a semibatch manner in penicillin production to keep its concentration below the inhibitory levels), fed-batch system is ideally suited.

In addition to the above significant advantages, the fed-batch system has the following general advantages also:

(1) The cells can be kept in a constant state of suspension by agitation, and thus we realise the advantages of a CSTF (or CSTR).

(2) Since one of the reactants (it could be cell cultures, media, nutrients or the substrate) is limited in supply, it is easy to engineer the reaction path and control the reaction to the desired products.

(3) If aeration is needed to be given continuously to the batch reaction system, the fed-batch systems are most ideally suited.

The mathematical description of the system is slightly difficult. We may have to use batch reactor models for components which are fed as batch, and continuous reactor models for those components which are added continuously, and interconnect the two with yield parameters.

In the case of a system where biomass generation is limited by the substrate supply, we can get an expression for the biomass concentration (Stanbury and Whitaker, 1993)

$$x = x_0 + Y_{x/s}(S_R - s) \tag{16.11}$$

where x_0 is the inoculum concentration, and is a very small quantity. Thus, maximum x can be obtained when s = 0

or
$$x_{max} = Y_{x/s}\, S_R \tag{16.12}$$

If at this stage, the medium feed is started, we can avail the full potential of the cells and all the substrate will be converted into the desired products. Then,

$$\upsilon_m S_R = \mu \frac{Vx}{Y_{x/s}} \tag{16.13}$$

where υ_m is the volumetric feed rate of the medium, and V is the volume of the reactor, so that Vx would give the total biomass in the fermenter.

16.5 MEDIA OPTIMISATION

The proper choice of a medium and its composition play a very vital role in the progress of the fermentation. It can even make or mar a fermentation process by pushing it to economic unviability. As has been mentioned in Section 5.4, the medium supplies

- nutrients for growth
- energy for the bioprocess
- building up of cell substances
- biosynthesis of fermentation products.

In turn, the fermentation medium should have

- carbon source
- nitrogen source
- minerals
- nutrients, like vitamins, etc.
- water
- energy source.

In view of the functions to be performed by the medium, and its composition, it is evident that an optimally balanced culture medium is mandatory. Sec. 5.4 highlights some of the naturally available sources for carbon and nitrogen.

Because of the cost considerations, most of the industrial media are crude in nature, and mostly they are the byproducts of some other industry, viz. molasses, soya meal, etc. Such types of crude media may also contain the inhibitors for microbial growth. Even though sterilisation of the medium is done to inactivate most of the inhibitory factors, it is not always effective.

A good medium with rich source of carbon and nitrogen is not always the solution for optimum production of best metabolic products. The reason is very interesting. Microorganisms, provided with rich and excessive nutrients under ideal growth conditions, would become greedy, and utilise more substrate than they normally do under constrained conditions (Casida Jr., 1996). The carbon substrate is partly used for cell growth, whereas the rest of the carbon source will be degraded by cell respiration to carbon dioxide and water.

If the stoichiometry of the metabolic product formation is known, it would be easy to design the medium based on the product requirement. Unfortunately, it is seldom possible in biological processes. Hence, the media optimisation procedure is difficult. A macroscopic look at the process indicates to us that the cell environment contains elements that should encourage the formation of cells; and the metabolic products should be such that their free energy of formation should be less than the free energy of the substrate. However, media composition can be optimised by trial-and-error experimentation in the laboratory. Keeping the yield of the metabolic product as the objective function, we shall try to optimise by varying different components of the medium. Various sources of carbon and nitrogen should be tried, and the product yield is to be evaluated in each case. It will afford us to look for an economical medium, because it was reported that, depending upon a particular process, 25–70% of the total fermentation cost may be due to the carbohydrate source alone (Crueger and Crueger, 1989). This exercise will give us leverage for alternate sources of carbohydrates.

The high yielding media are selected. They are further studied for their various components. Some of the components and nutrients in them are varied and the yield of the product is evaluated. Then the residual medium components are assayed to determine whether the critical carbon and nitrogen sources are too quickly exhausted, or they are remaining without being exhausted. This information will hint us to adjust these sources for optimal utilisation, and for best yields of the products. This information will also reveal to us whether diauxie phenomenon is occurring during fermentation process, in which one source of carbon is totally exhausted before a second source of carbon requiring enzymatic adaptation is utilised (Chapter 9).

It is also essential to assess the effect of foam formation or any inhibitory effect of the antifoaming agents on the medium. The effect should be assessed at laboratory-scale before larger-scale experiments. The small experiments would also help us assess the effect of aeration on the dispersed medium. Cells, present in the suspended form, will interfere with the break-up of air bubbles; proteins and other molecules will get absorbed at the gas–liquid interface and cause **interfacial blanketing**, which in turn would reduce the contact area between the gas and liquid phases.

Reactor configuration and flow pattern would also have a tremendous effect on the performance of the medium. As has been mentioned, if the availability of all the medium with rich sources of nutrients, etc. is detrimental for the 'active performance' of the cells to yield the desired products, it may be worth to try in a fed-batch reactor. The cell culture along with the substrate is taken initially as a batch in the fermenter, into which the medium is added slowly or in a programmed way, which would help its better utilisation.

In any commercial venture, ultimately the cost of the various components would dictate its economic viability, and hence its technical feasibility for a commercial reality. Thus, obviously the optimum medium should be also available economically and easily. We should also look into the difficulty as to whether the medium components would interfere with the product, making its recovery uneconomical.

Thus, the media optimisation is based on its

- availability
- affordability
- applicability
- adaptability.

Symbols

D	: dilution rate, υ/V (s^{-1})
K_S	: substrate utilisation constant
s	: substrate concentration
S_R	: original substrate concentration
t	: time (s or h)
V	: volume (m^3)
x	: cell concentration
$Y_{x/s}$	: yield factor of cells based on substrate (dimensionless)

Subscripts

0	: initial condition
m	: medium
max	: maximum
ss	: steady state

Greek Letters

μ	: specific growth rate (h^{-1})
υ	: volumetric flow rate (m^3/s)

References

16.1 Atkinson, B. (1974), *Biochemical Reactors*, Pion Ltd., London.

16.2 Coughanowr, D.R (1991), *Process Systems Analysis and Control*, McGraw-Hill, Inc., New York, pp. 111–163.

16.3 Casida Jr., L.E. (1996), *Industrial Microbiology*, New Age Int. Pub., New Delhi, p. 117, p. 134.

16.4 Crueger, W., and Crueger, A. (1989), *Biotechnology: A Textbook of Industrial Microbiology*, Brock, T.D. (Ed.), 2nd Edn, Sinauer Associates, Inc., Sunderland, USA, p. 59.

16.5 Deindoerfer, F.H. (1960), "Fermentation kinetics and model processes", *Advances in Applied Microbiology*, Umbreit, W.W. (Ed.), vol. 2, Academic Press, New York, pp. 321–334.

16.6 Denbigh, K.G., and Turner, J.C.R. (1971), *Chemical Reactor Theory*, 2nd Edn, Cambridge University Press, Cambridge.
16.7 Doran, P.M. (1995), Bioprocess Engineering Principles, Academic Press, London, p. 143.
16.8 Erickson, L.E., and Stephanopoulos, G. (1987), *Chemical Reaction and Reactor Engineering*, Carberry, J.J., and Varma, A.(Eds.), Marcel Dekker, Inc., New York, Ch. 12, p. 799.
16.9 Hall, E.A.H. (1990), *Biosensors*, Open University Press, Buckingham, pp. 2–29.
16.10 Hampel, W. A. (1979), "Applications of microcomputers in the study of microbial processes", Advances in Biochemical Engineering, Ghose, T.K., Fiechter, A, and Blakebrough, N. (Eds.), Springer Verlag, Berlin, pp. 1–33.
16.11 Karube, I., and Sode, K. (1989), "Biosensors for fermentation process control", *Bioprocess Engineering: The First Generation*, Ghose, T.K. (Ed.), Ellis Horwood, Chichester, pp. 207–219.
16.12 McCabe, W.L., Smith, J.C., and Harriott, P. (1993), *Unit Operation of Chemical Engineering*, 5th Edn, McGraw-Hill, Inc., New York, Ch. 8, pp. 181–231.
16.13 Schugerl, K. (2000), "Development of bioreaction engineering", *Advances in Biochemical Engineering and Biotechnology.*, Vol. 70, Fiechter, A. (Ed.), Springer-Verlag, Berlin, pp. 49–50.
16.14 Stanbury, P.F., and Whitaker, A. (1993), *Principles of Fermentation Technology*, Pergamon Press, Oxford, pp. 153–163, 206–208.
16.15 Yoshida, F., Yamana, T., and Nakamoto, K. (1973), "Fedbatch hydrocarbon fermentations with colloidal emulsion feed", *Biotech. Bioeng.*, 15, 257–270.

Review Questions

16.1 List the various process conditions which are important in the operation of a bioreactor.
16.2 Describe the various characteristic features of different types of bioreactors.
16.3 Describe a broad classification of a range of fermentation processes for which different types of flow bioreactors are used.
16.4 Distinguish between on-line and off-line control of fermenters.
16.5 What are the various parameters that you need to control for the successful operation of a fermenter?
16.6 What are the different methods of controlling fermenter process conditions? Describe them briefly.
16.7 Describe the various automatic controlling methods.
16.8 Describe how do you interphase a computer for controlling a fermenter.
16.9 List the critical and non-critical process parameters which need to be controlled in a bioreactor.
16.10 Define the term "dilution rate".
16.11 Derive an expression for steady state concentration of the limiting substrate in a continuous flow bioreactor.
16.12 What are the distinct advantages of the fed-batch systems in bioprocessing?
16.13 What are the essential requirements of a fermentation medium?
16.14 Describe how you optimise the medium.
16.15 Enumerate the various accessories used in fermenters to measure/control the process parameters.

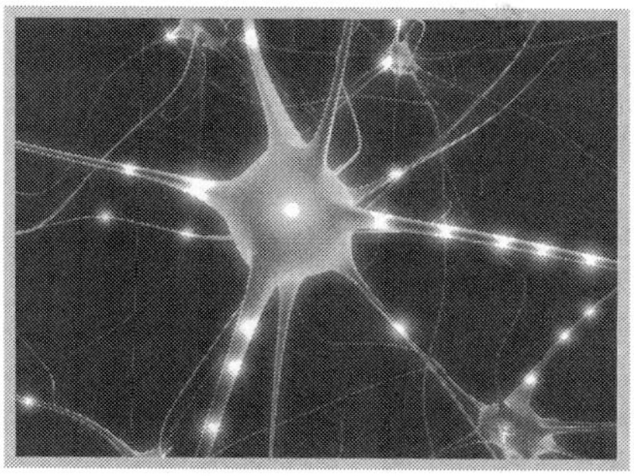

chapter 17

Product Recovery

Product recovery in bioprocessing plays a vital role. Sometimes, the product recovery costs may even decide the economic viability of the process. They vary anywhere between 20–60% of the product cost. In some extreme cases, the costs may go up to 90%, as in the case of recombinant DNA fermentation products.

The product recovery operations are also known as **downstream processing steps.** They include all the steps we take up after the fermentation step is completed in the bioreactor. They play a very crucial role because the products that come out of the bioreactor may consist of the following, in addition to the metabolic product:

- microorganisms
- whole cells
- cell debris/fragments
- soluble and insoluble medium products
- pellets of aggregated proteins
- undissolved nutrient components, etc.

In view of the large number of unnecessary products being present in the fermented broth, downstream processing steps are important. They may be broadly classified into

- initial isolation
- product recovery
- purification and concentration.

The type of the downstream processing step to be adopted for a particular fermentation process depends upon the product, its value and the use for which it is meant. A very costly product to be used for medicinal purposes in small quantities will be purified and concentrated using very sophisticated downstream processing steps; whereas those meant for industrial use in bulk quantities, may be recovered with some of the initial isolation steps. If the solids are present in suspension, they will be separated by simple sedimentation or filtration or centrifugation. It also depends upon the location of the product in the presence of cells, viz. extracellular or intracellular, i.e. whether the product is independently existing of the cells or the product is trapped inside the cells. In the latter case, we may have to go for some of the steps for disrupting the cells. The cell disruption techniques may be classified as:

- physical methods
- chemical methods.

Various unit operations involved in product isolation and purification are highlighted as Step 6 in Fig. 1.1. The specific choice of the recovery of the product may be summarised as follows (Stanbury and Whitaker, 1993):

- the location of the product, viz. intracellular or extracellular, and the heat labile nature of the product
- the quantity of the product in the fermentation broth, i.e. whether it is available in a concentrated form or in a very dilute form
- the cost of the product
- the end use of the product
- the minimal acceptable standards of the product
- the impurities present in the fermentation product, and their nature of interference with the product recovery steps.

Table 17.1 shows various unit operations in bioprocessing for separation of different phases based on the fact that in which phase the product is available and in which form (whether soluble or suspended). Table 17.2 classifies various unit operations on the basis of physical and chemical properties of the solids. The unit operations will be described in the following sections.

Table 17.1 *Various unit operations in bioprocessing for separation of different phases*

System	*Type*	*With phase change*	*Without phase change*
Solid–liquid	Soluble	Drying Evaporation Crystallization Adsorption	Ultra–filtration Reverse osmosis
	Insoluble	Drying	Filtration Sedimentation Centrifugation
Liquid–liquid	Soluble	Distillation Extraction	Chromatography
	Insoluble		Sedimentation Centrifugation
Solid–Solid	Miscible	Cell disruption (high pressure homogenisation, grinding with abrasives)	
	Immiscible	Selective solubilisation.	Air aspiration Sieving Screening Winnowing
Liquid–liquid–solid	Miscible	Adsorption	
	Immiscible		Centrifugation Sedimentation Decanting

17.1 REMOVAL OF SUSPENDED SOLIDS

The suspended solids could be microbial cells or some kind of undesirable solids. In certain cases, like in the case of production of baker's yeast, the biomass (the suspended solids) is the desired product and needs to be separated in a pure dry form. The following unit operations are used for the separation of solids:

Table 17.2 *Concentration process using different chemical and physical properties of particles and molecules*

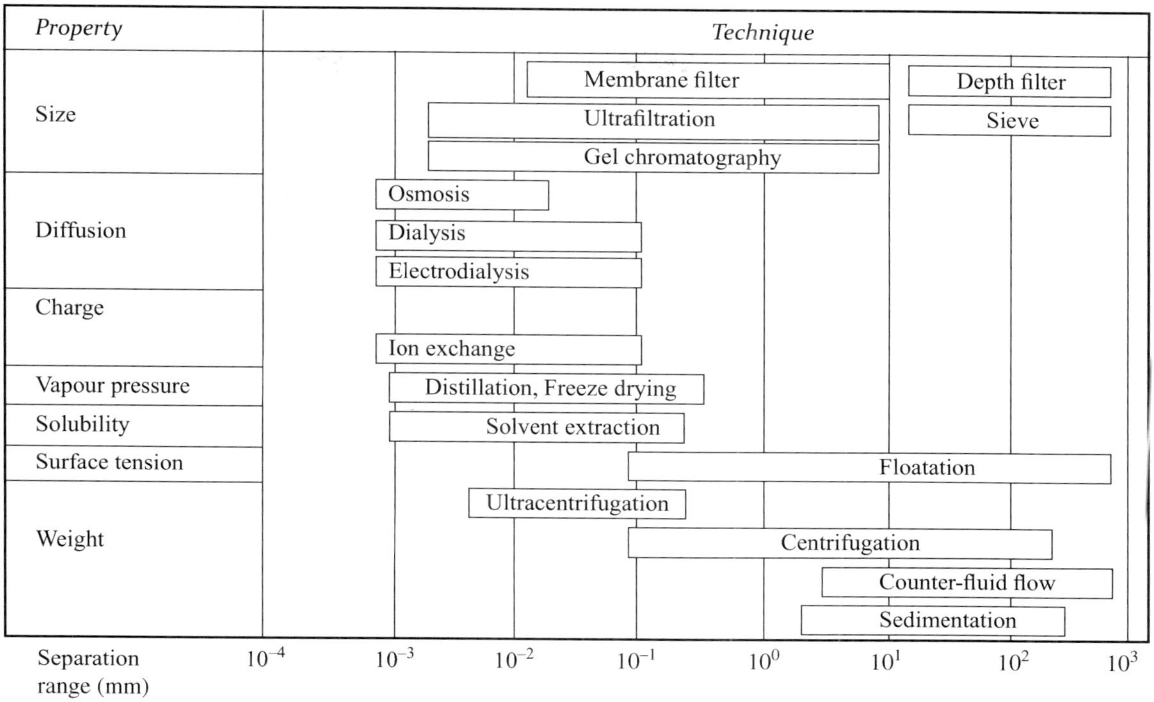

(Reprinted from "Biotechnology: A Textbook of Industrial Microbiology", Crueger, W., and Crueger, A., Brock, T.D. (Ed.), 2nd Edn (English), p. 113 Copyright © (1989), with permission from Oldenbourg, Industrieverlag, GmbH, Munchen.)

- filtration
- sedimentation
- centrifugation
- foam separation
- precipitation.

In this section, we describe foam separation and precipitation.

17.1.1 Foam Separation

Foaming is a surface phenomenon and is based on the differences in the surface activity. It is applicable to such products which have a tendency to form foams or to those which attach themselves with the foam. Air/gas is usually blown through the liquid medium containing some kind of foaming agents, viz. fatty acids or quaternary ammonium salts such as:

- lauric acid
- strearyl amine
- *t*-octyl amine
- ethyl hexadecyl dimethyl ammonium bromide.

The suspended solids adhere to the foam and raise to the upper layer along with the foam. Later the foam is separated and discarded. If the foam contains some active or useful solids, usually the foam is broken by applying some mechanical foam-breaking techniques. Generally, foaming is used for the removal of impurities.

17.1.2 Precipitation

It is also a very useful method for product recovery. By adding the precipitating agent, the compounds form the corresponding salts or the suspending solids which would settle down to the bottom. This method is extensively used for product recovery of protein isolates or protein concentrates from oil seeds. In one process of making the groundnut (peanut) protein isolate, the groundnut seeds are treated with an alkali to extract the protein. Later the extract is neutralised with an acid to precipitate the groundnut protein. It is later isolated, dried and packed for ready to use as a protein source in baby foods, etc.

17.2 FILTRATION

Filtration is a physical method of separation of suspended particles of any size from the liquid medium. Here, the only criterion is that the particle size should be larger than the molecular size of the liquid, which is always true. It is a very effective and convenient method of separating the suspended solids from the slurries. In this method of separation, either the solids or the liquid or both could be the desired products.

The liquid along with the suspended solids will be transferred on to a filter medium. The filter medium will retain all the particles which are bigger than the size of its aperture, and will pass on all the liquid medium along with some of the very fine suspended solid particles. All the material remaining on the filter medium is known as **residue** or **filter cake,** and all the material passing through (the filter medium) is known as filtrate (Fig. 17.1). The filter medium, usually known as filter cloth, is kept on a perforated screen to provide mechanical support to the filter aid. As the filtration process continues, more and more solids build up in the filter cake and offer more resistance for the flow of the fluids. But as the filter cake builds up, the filtration efficiency will also improve, that is to say that still finer particles can be retained in the residue and the filtrate can be more clear of the solids. Thus, filtration efficiency will improve, but the filtration rate will reduce.

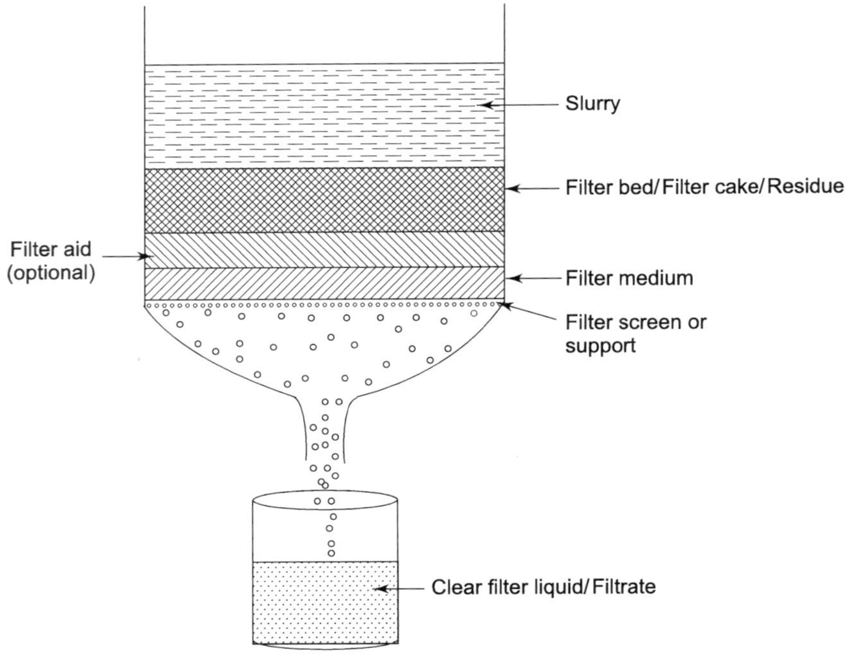

Fig. 17.1 *Filtration process*

Initially, when there is no thickness of the filter cake, the filtration efficiency is the least. To overcome this, we add initially some filter aid which does not interfere or react with the filtering slurry. It could be cellulose or diatomaceous earth or any inert chemical (such as **supercel**). However, it is to be noted that the addition of filter aid is purely optional. It is normally added when the residue is not the desired component. If the residue is the desired product and if we wish that the residue should not be contaminated with the filter aid, we may allow initially the filtration to proceed for sometime until a good amount of filter bed builds up, and the filtrate that has passed through during this period will be recycled. If the filtrate is the desired product, then the filter aid can be safely used.

Since a large number of industrial products is to be separated by filtration method, a good number of variety of industrial filters has been developed. In all these cases, the fluid passes through the filter bed and the filter medium by means of pressure differential across the medium. The pressure differential could be achieved by applying pressure on the upstream side of the filter medium or by applying vacuum on the downstream side. Based on the industrial need, they could be operated in continuous or semi-continuous manner. In a semi-batch (semi-continuous) manner, the slurry flows continuously, the filtrate comes out continuously, but the solids (residue) can be taken only intermittently. In a continuous operation, the slurry is fed continuously, both the residue and the filtrate are discharged continuously. The filter will be in operation continuously. It will be stopped only once in a while to clean the filter medium, when the filter medium pores are blocked.

Filters are divided into three main categories as follows:

- cake filters
- clarifying filters
- cross-flow filters.

Slurries with high solid concentration are operated in cake filters. The cake build-up is more, and offers the desired resistance for the flow, and thereby allows the solids to be retained on the filter medium as filter cake. Very dilute liquids are passed through the clarifying filters. They are more used for clarification purposes like clarified fruit juices, etc. The cross-flow filters are used for concentrating the solutions by using filtration technique. They are so named because the slurry flows on the filter medium in a cross-flow direction (Fig. 17.2). In this type of filtration, a clear demarcation is not made as residue and filtrate. The filtration process is essentially meant for concentrating a slurry, i.e. the feed contains dilute solids, and after filtration, the feed leaves the filtration unit in the form of concentrated slurry, as in the case of **ultrafiltration**.

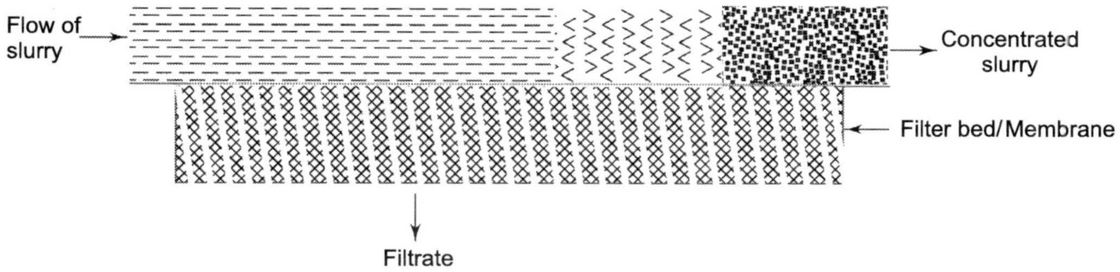

Fig. 17.2 *Cross-flow filter*

17.2.1 Theory of Filtration

Any rate process is directly proportional to the driving force and inversely proportional to the resistance. Thus,

$$\begin{aligned}\text{Rate of filtration} &\propto \text{Driving force}\\ &\propto \text{Cross-sectional area}\\ &\propto 1/(\text{Resistance to flow})\end{aligned} \tag{17.1}$$

In filtration process, the driving force is pressure drop (Δp). Resistance to flow mainly occurs from

- viscosity of the slurry
- resistance offered by the filter cake
- resistance offered by filter medium, filter aid, etc.

Thus,

$$R_f = \mu\alpha(L_c + L_m) \tag{17.2}$$

where R_f is the resistance to filtration
α is the specific cake resistance, which is defined as the resistance offered by the cake of unit thickness
L_m is the fictitious equivalent thickness of the filter cloth and filter aid, etc.
L_c is cake thickness (in m) and is given by

$$L_c = \frac{wV}{\rho A} \tag{17.3}$$

Equation (17.1) can now be written as (Earle, 1966):

$$\frac{dV}{dt} = \frac{A(\Delta p)}{\mu\alpha\left[\frac{w}{\rho}\frac{V}{A} + L_m\right]} \tag{17.4}$$

where dV/dt is the rate of filtration measured as the volumetric flow rate of the filtrate. Normally the filtration is carried out at constant pressure, and hence Eq. (17.4) can be rewritten for constant Δp as follows:

$$\mu\alpha\left[\frac{w}{\rho}\frac{V}{A} + L_m\right]dV = A\,(\Delta p)(dt) \tag{17.5}$$

On integration,

$$\frac{t}{(V/A)} = [\mu\alpha(w/\rho)/(2(\Delta p))]\,(V/A) + \mu\alpha L_m/(\Delta p) \tag{17.6}$$

In the above equation, V/A stands for the height or thickness of the filter cake. The solid concentration in the slurry can be experimentally measured by evaporating a small known quantity of the slurry for removal of all filtrate. μ is the viscosity of the filtrate, which is generally water. Equation (17.6) is the design equation for constant pressure filtration. By experimentally measuring the volumetric flow rate of the filtrate with time, we can calculate α (the specific cake resistance) and L_m (the equivalent cake thickness) as follows, by the graphical method. Once the α and L_m are known, the filtration process can be scaled up.

Graphical method for evaluating α and L_m From the experimental data, $t/(V/A)$ is plotted against (V/A) as shown in Fig. 17.3. The slope and intercept are measured.

$$\text{Slope} = \mu\alpha(w/\rho)/[2(\Delta p)] \tag{17.7}$$

Knowing μ, (w/ρ) and (Δp), α can be calculated in the units of $1/m^2$.

$$\text{Intercept} = \mu\alpha\, L_m/\Delta p \tag{17.8}$$

from which

$$L_m = \frac{\text{Intercept}\,(\Delta p)}{\mu\alpha} \tag{17.9}$$

Knowing α from Eq. (17.7), L_m can be calculated in units of m.

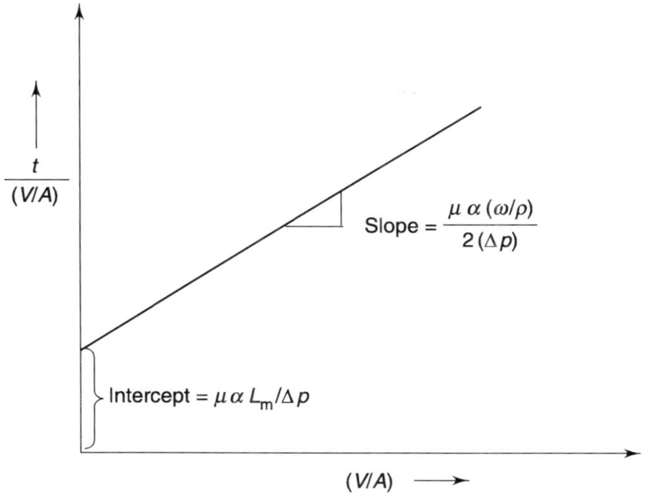

Fig. 17.3 *Procedure for graphical evaluation of α and L_m*

Example 17.1

In the downstream processing of a biochemical process for the recovery of the biomass, a constant pressure filtration technique is used. The fermentation broth contains 10 g of cells per litre of the slurry, and the cells have a density of 900 kg/m³. The viscosity of the filtrate may be taken as that of water. Data on volume of filtrate with time was collected at a gauge pressure of 0.1 MN/m² and presented in Table E17.1. Find the specific cake resistance and equivalent cake thickness, if the cross-sectional area of the filter press is 0.18 m².

Table E17.1

Time, min	8	27	54	90	140
Vol., lit	20	40	60	80	100

Solution It is a direct application of Eq. (17.6) to Eq. (17.9) and drawing the plot of Fig. 17.3. The data in Table E17.1 are used to calculate $t/(V/A)$ and (V/A), and are tabulated in Table E17.2.

Given:

$$w = 10 \text{ g/l} = 10 \text{ g/1000g} = 10 \text{ kg/m}^3 \text{ of slurry}$$
$$\rho = 900 \text{ kg/m}^3$$
$$\mu = 1 \times 10^{-3} \text{ Pa s}$$
$$\Delta p = 0.1 \text{ MN/m}^2 = 1 \times 10^5 \text{ N/m}^2 = 1 \times 10^5 \text{ Pa s}$$
$$A = 0.18 \text{ m}^2$$

Table E17.2 *Data of E17.1*

time(min)	*Vol.(l)*	*t(s)*	*V m³*	*V/A(m)*	*t/(V/A)(s/m)*
8	20	480	0.02	0.11	4324
27	40	1620	0.04	0.22	7363
54	60	3240	0.06	0.33	9728
90	80	5400	0.08	0.44	12273
140	100	8400	0.1	0.55	15273

The data in last two columns are plotted in Fig. E17.1

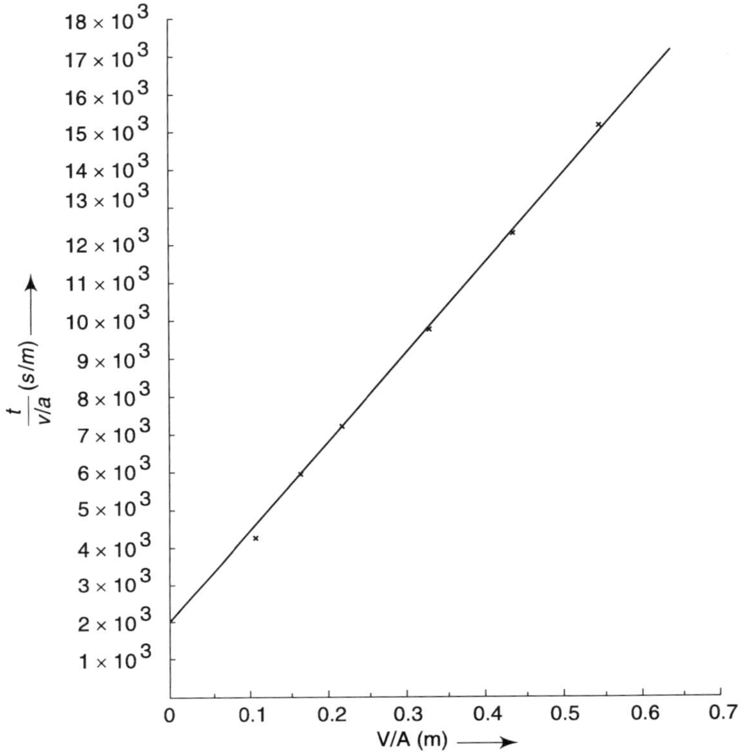

Fig. E17.1 *Data of E17.1*

$$\text{Slope} = \frac{(12-6)\times 10^3}{(0.42-0.165)} = 23529 \text{ s/m}^2$$

$$= \frac{\mu \alpha w}{2(\Delta p)\rho} = \frac{\left(1\times 10^{-3}\right)\times(\alpha)\times 10}{2\times\left(1\times 10^5\right)\times 900} = (5.55 \times 10^{-11})(\alpha)$$

$$\therefore \quad \alpha = \frac{23529}{5.5\times.10^{-11}} = 4.235 \times 10^{14}\ (1/\text{m}^2)^{\#}$$

$$\text{Intercept} = 2.1 \times 10^3 = \mu\,\alpha\,\frac{L_m}{\Delta p}$$

$$= \frac{1\times 10^{-3}\times 4.235\times 10^{14}\ (L_m\ )}{1\times 10^5} = 4.235 \times 10^6\ (L_m)$$

$$\therefore \quad L_m = \frac{2.1\times 10^3}{4.235\times 10^6} = 4.96 \times 10^{-4} \text{ m}$$

The units of α are dependent upon how Δp is defined. If it is taken in the units of kg/m^2 and if μ is defined in kg/m.s, the units of α will be in s^2/m^3 and its value will be less by 9.81 which is the value for g.

17.2.2 Filtration Equipment

After having understood briefly the theory and concept of constant pressure filtration, let us see some of the industrial filters.

Plate-and-frame filter press This is one of the most widely used filter presses in the industry in view of its versatility and ease of operation. It is normally operated under pressure. The operation of it is similar to the one described schematically in Fig. 17.1, where we have a filter medium on which the slurry with the suspended solids is fed to the filter press. The filtration rate is proportional to the filter area and is restricted. Instead of having only one filter medium, we can have more of them in a consecutive manner, so that the filtration area increases many times, and accordingly the filtration rate also increases. Such an arrangement is shown in Fig. 17.4.

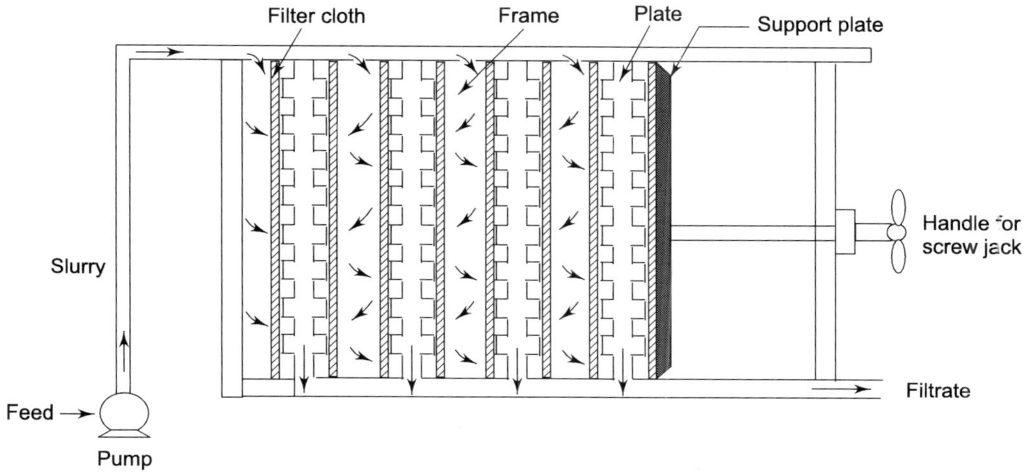

Fig. 17.4 *Plate-and-frame filter press*

The plate-and-frame filter press consists of a set of square plates separated by hollow frames to make compartments. The plates are 6–50 mm thick and the frames are 6–200 mm thick. The plates are covered with filter medium. The plates and frames are vertically stacked and tightly held in position by a screw or hydraulic ram. The plates will have a small projection all along the circumference, which sits into the depression provided in the frame. Thus, perfect compartments are made in the frame portions. The slurry is introduced into each plate and frame portion. The slurry channel is made in such a way that it enters every frame chamber, passes through the filter medium, and the clarified filtrate is passed through the grooves or cavities provided on the frame into the filtrate outlet channel.

After the assembly is made with the filter medium in position, and all the plates and frames are stacked tightly by tightening the hydraulic ram or the screw, the filter press is ready for operation. The assembled plate-and-frame filter press is shown in Fig. 17.5. The slurry is pumped through the press by using a pump at a pressure of 3–10 atm. As the filtration continues, the solids accumulate on the filter cloths. As time passes, they may even jam the filter cloths, with the result that the filtrate flow comes down drastically. It is the time for stopping the filtration process, dismantle the filter press, and scrape the solids from the filter cloth. Sometimes, before dismantling itself, the solids (residue) are washed with hot water or steam by injecting it in the other direction. The scraped solids are collected in bins and transported to the central collection zone. If the residue is the desired product, the solids are dried and packed. If the filtrate is the desired product and solids are not required, the residue is discarded.

Fig. 17.5 *Plate and Frame Filter Press used to separate calcium phytate cake from corn steep liquid (CSL) which is the starting raw material for manufacture of Inositol (a Vitamin B). The filter press size is 36″ × 36″ × 2″, No. of plates is 36, and capcity is 1000 kg of CSL cake (Courtesy: M/s Siddartha Enterprises, Nizamabad, AP)*

In industrial-scale, most of these operations are done automatically. The washing of the filter press thoroughly, including the filter cloth, is a very time-consuming process. So, if the solid content is high in the slurry, the press gets jammed frequently. Cleaning and reassembling will then be very tedious.

Pressure leaf filter press It is also similar to the plate-and-frame filter press, with an additional advantage of operating it under high pressures. It consists of a number of filter leaves stacked horizontally (or vertically) in a retractable rack. The filter leaves assembly is put in a horizontal cylindrical drum, into which the slurry is pumped under high pressure (Fig. 17.6). The slurry passes through the filter leaves and discharges out through a filtrate discharge manifold.

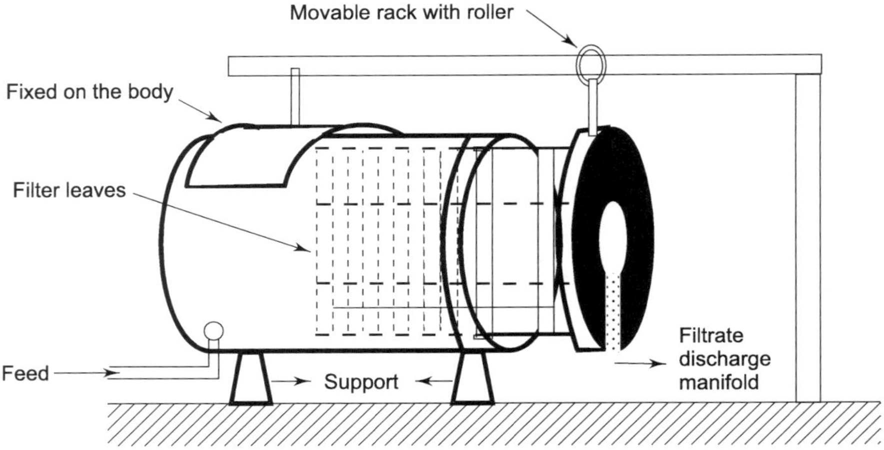

Fig. 17.6 *Pressure leaf filter press*

Continuous rotary vacuum filter press It is a continuous filtration unit in which both the residue and filtrate will be obtained continuously. Problems associated with the choking of the filter bed are not present in this filter. It operates under vacuum unlike the other filters. Obviously, the mechanical problems associated with maintaining vacuum cannot be avoided.

It consists of a large circular hollow drum of about 50–100 cm width and having a diameter of 2–3 m (Fig. 17.7). The slurry flows continuously into a trough, into which a segment of the filter press is dipped. The drum is provided with the filter medium all along the rim, and rotates continuously at 0.1–2 rpm speed with vacuum inside. When the drum dips into the slurry in the trough, the slurry is sucked because of the vacuum, and is filtered. The solids remain on the surface of the drum, while the filtrate is sucked into the drum which is separately discharged through a valve at the axis.

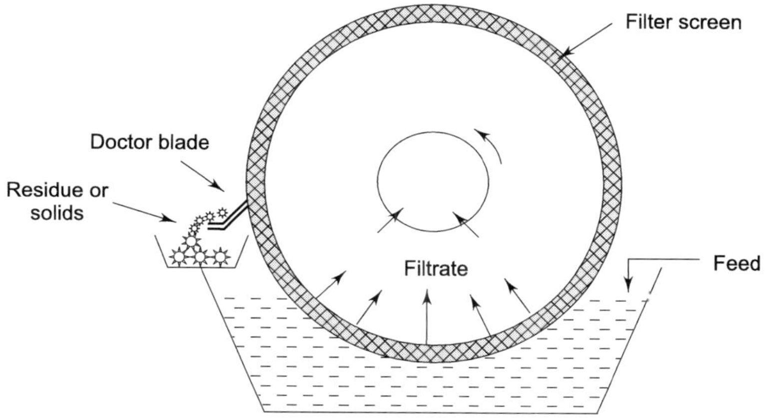

Fig. 17.7 *Continuous rotary vacuum filter press*

There can also be a provision for washing of the cake when it comes on to the top position. The solids are collected by scraping with a doctor blade into a tray from which the contents are discharged.

Some filters will have provision for consecutively applying vacuum to some portion of the drum, later it is blown with dry air from inside. This will help dry the solids. The screen slightly bloats, which helps crack the residue cake, and can be easily scraped for collecting the solids. During the process of pressurising and blowing from inside, there will not be any vacuum applied to the drum and hence there will not be any filtration during this period.

The drum is normally submerged into the trough to an extent of 30%, and in some cases, when washing of the cake is not desired, it may go up to 60–70% (McCabe et al., 1993 (a)). The cake thickness in the industrial filters varies from 3 to 40 mm, depending upon the solid content and nature of solids in the feed slurry.

17.3 SEDIMENTATION

Sedimentation is a method of separation of solids from liquids or liquid globules from immiscible liquids. The only force acting downwards on a solid particle suspended in a fluid is the gravitational force. This is opposite in direction to the buoyancy forces which act upwards. The settling velocity is dictated by the difference in the densities of the particle and fluid ($\rho_p - \rho_f$) and the viscosity of the fluid. In view of the small values of ($\rho_p - \rho_f$), the settling velocities are usually very small. Thus, sedimentation takes a lot of time for the suspensions to settle, probably of the order of a few days. Hence, normally, sedimentation is not used as an effective tool for product recovery. It is generally used in the effluent treatment for separation of the solid suspensions in the slurries, before the clear liquid is disposed off to the environment.

Theory of Sedimentation A particle falling from rest in a fluid, under gravity, will be having increasing velocities. A stage comes where the drag force on the particle acting in the upward direction will be equal to the gravitational pull. Afterwards, the particle attains a maximum constant velocity, which is known as **terminal velocity.** The terminal velocity depends upon the

- size and shape of the particle
- density of the particle
- density and viscosity of the fluid
- the gravitational pull which is usually constant.

The forces acting on a sedimenting particle are shown in Fig. 17.8.

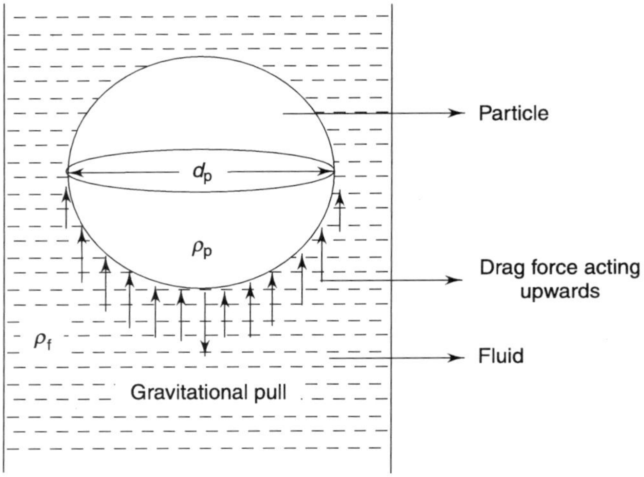

Fig. 17.8 *Various forces acting upon a sedimenting particle*

The net force acting on the particle is given by (Earle, 1966):

$$F_s = V_p \frac{g}{g_c} (\rho_p - \rho_f) \tag{17.10}$$

And the drag force F_d is given by

$$F_d = C_D \rho_f u^2 A_p / 2g_c \tag{17.11}$$

where C_D is the drag coefficient which is obtained by

Stoke's law in the streamline flow:

$$C_D = 24/N_{Re,p} \tag{17.12}$$

In the above equation

$$V_p = \frac{\pi}{6} d_p^3 \tag{17.13}$$

$$A_p = \frac{\pi}{4} d_p^2 \tag{17.14}$$

$$N_{Re,p} = \frac{\rho d_p u}{\mu} \tag{17.15}$$

When the particle velocity approaches the terminal velocity or settling velocity, the net gravitational force and drag force will be equal. Equating Eqs. (17.10) and (17.11) and substituting for various quantities, we get an expression for the terminal velocity of the particle.

$$u_t = d_p^2 g(\rho_p - \rho_f)/18\mu \tag{17.16}$$

The above equation is useful in process design to find out the minimum surface area of the thickener or the sedimentation tank.

If the slurry is held in a cylindrical tank, it exhibits various zones. Initially the suspension is uniform as shown in Fig. 17.9(a). As time progresses, the solid particles start settling down resulting in clear supernatant liquid on the top. In the intermediate stage (Fig. 17.2(b)), almost four zones will be existing (McCabe, et al., 1993 (b)). The topmost layer is the clear liquid. The bottommost layer is the thick solids. Over the thick solids, the slurry will have a composition which is richer in solids and poorer in liquid. Above this, there will be a layer which has the composition similar to that of the feed slurry. Slowly these zones vanish, and ultimately we find only two zones (Fig. 17.9(c)), the upper layer is a clear liquid and the bottom layer is thick solids which may be removed.

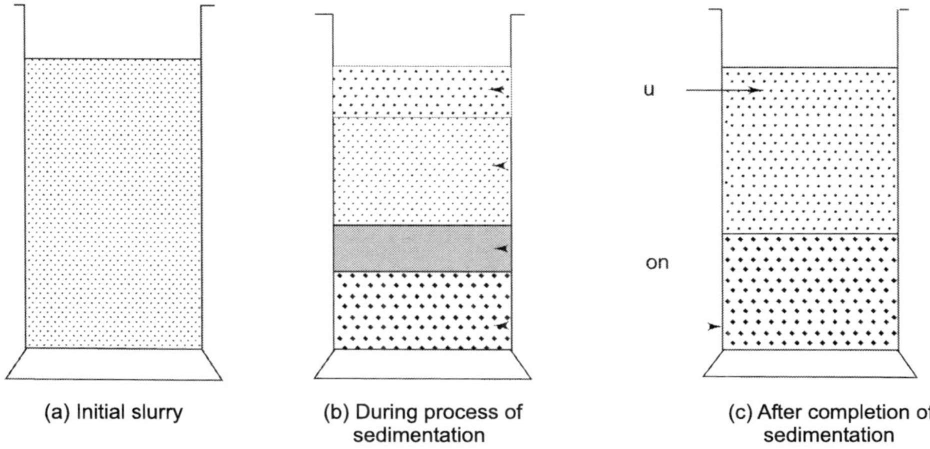

Fig. 17.9 *Settling of solids in a batch sedimentation process*

The design equations are arrived at to find the minimum area of a continuous thickener or sedimentation tank, so that the solids (or the sludge) are withdrawn from the bottom continuously and the clear liquid can be decanted from the top. For a continuous thickener, the rate of sedimentation can be equated to the counter-flow velocity of the rising fluid (Earle, 1966).

$$u_u = \frac{(F - L)r_s}{A\rho_f} \tag{17.17}$$

where u_u is the upward velocity of flow of liquid, which is also equal to the settling velocity of the particle (u_t); F is the ratio of liquid to the solids in the feed and L is that in the exit stream; and r_s is the mass feed rate of the slurry. Equation (17.17) is obtained by simple mass balance. It can be used to calculate the minimum area (A) of the sedimentation tank (or thickener).

$$A = \frac{(F - L)r_s}{u_t\,\rho_f} \tag{17.18}$$

The above equation can be used to evaluate the minimum cross-sectional area of the thickener, provided we know the average particle size d_p, which is the major drawback. The sedimentation slurry may not have all particles of same diameter, in which case, we may have to find out the mean particle size by averaging out various particle sizes and their mass fractions, which is also a difficult task. Mostly, the sedimentation operations are highly time-consuming. It takes a lot of time for the solids to settle.

Example 17.2 *A sedimentation tank is being used for settling of solid dirt particles in the downstream processing of effluent from a distillery after fermentation of molasses. The feed has a solid content of 1% and the outlet stream is free of solids. The feed rate is 10 tonnes/hour and the average diameter of the solid particle is 50 microns and its specific gravity is 1.1. Estimate the necessary area of the tank.*

Solution It is a direct application of Eq. (17.18)

Given: $d_p = 50 \text{ microns} = 5 \times 10^{-5}$ m

sp. gravity of particles = 1.1

$$\mu = 1 \times 10^{-3} \text{ Pa s}$$

$$\rho_p = 1.1 \times 10^3 \text{ kg/m}^3$$

$$\rho_f = 1000 \text{ kg/m}^3$$

Feed has a solid content of 1%

$$F = 99$$

$$L = 0$$

Feed rate is 10 tph.

$$r_s = 10 \times 10^3 \times 0.01 = 100 \text{ kg/h} = 0.028 \text{ kg/s}$$

By Eq. (17.16), $u_t = d_p^2\, g\, (\rho_p - \rho_f)/(18\, \mu)$

$$\therefore \quad u_t = \frac{\left(50 \times 10^{-6}\right)^2 (9.81)\left(1.1 \times 10^3 - 1000\right)}{18 \times 1 \times 10^{-3}}$$

$$= 1.363 \times 10^{-4} \text{ m/s}$$

By Eq. (17.18), we have

$$A = \frac{(F - L) r_s}{u_t\, \rho_f} = \frac{(99 - 0) \times 0.028}{1.363 \times 10^{-4} \times 1000} = 20.32 \text{ m}^2$$

17.4 CENTRIFUGATION

In case of sedimentation, the solids will be settling to the bottom only under the influence of gravitational force minus the buoyancy forces. The settling velocity, as given by Eq. (17.16), can be enhanced by increasing the value of g. This is where the centrifugal force will be of great help to improve u_t by increasing g. Thus, centrifugation is a very common operation in bioprocessing for the separation of the solids. It could be for separation of the cells from the fermentation broth or for separation of biomass which is the desired product, or for separation of some of the metabolic products.

As compared to filtration, this method is both faster and continuous. Centrifugation gives the product, free of the filter aid, etc. Unlike filtration, centrifugation can be used even for separation of two immiscible liquids also. Centrifugation can alone be used for separation of two immiscible liquids and solid phase continuously and quickly. For separation of the coagulated suspensions, hazy particles and filamentous materials, centrifugation is again the only industrial process which can be operated with reasonable ease and fastness.

17.4.1 Theory of Centrifugation

The movement of the particles in a fluid medium is given by Stoke's law (Eq. (17.16)) under gravitational force g:

$$u_t = d_p^2\, g\,(\rho_p - \rho_f)/18\mu$$

For settling of particles with an acceleration of a', the above equation is modified for particles settling with velocity u

$$u = d_p^2\, a'\,(\rho_p - \rho_f)/18\mu \quad (17.19)$$

The centrifugal force acting on a particle of weight m, at a distance of r from the centre of rotation, at an angular velocity of ω, is given by

$$F_c = mr\omega^2 \quad (17.20)$$

and

$$\omega = 2\pi N \quad (17.21)$$

where N is the revolutions per second.

From Eq. (17.20), we can write that the acceleration a' is

$$a' = F_c/m = r\omega^2 \quad (17.22)$$

By combining Eqs. (17.19) and (17.22), we get

$$u = d_p^2\, r(\omega^2)(\rho_p - \rho_f)/18\mu$$
$$= d_p^2\, r(2\pi N)^2\,(\rho_p - \rho_f)/18\mu \quad (17.23)$$

The **centrifuge effect** or **g-number** of a centrifuge, which is usually represented by Z, is the ratio of the centrifugal force provided by the centrifuge to the gravitation force. That is, it is a measure of how effective is the centrifuge in increasing the settling acceleration, which otherwise would have been only due to gravitational force. It is also known as **g-number** of the centrifuge, which indicates the number of times the acceleration is increased over g, and is represented by

$$Z = \frac{r\omega^2}{g} \quad (17.24)$$

Another important factor, usually associated with centrifuge, to compare the performance of various centrifuges and also to be used as a criterion for scale-up, is **Sigma factor,** and is represented by Σ:

$$\Sigma = \frac{\upsilon}{2u_t} \quad (17.25)$$

where υ represents the volumetric feed rate and u_t is the terminal velocity of the particles in gravitational field. In other words, Σ represents the cross-sectional area of the gravity settler with the same sedimentation characteristics (Doran, 1995).

Example 17.3

In a fermentation process producing baker's yeast, the biomass is separated using a continuous centrifuge, operating at 4000 rpm, with a feed rate of 80 l/min. The solid particles may be assumed to have an average diameter of 0.05 mm. The density of the biomass is 1010 kg/m³. The fluid may be assumed to have properties of water. Find the sigma factor for the centrifuge.

Solution The governing equations are

$$u_t = d_p^2\, g\,(\rho_p - \rho_f)/18\,\mu \quad (17.16)$$

and $\Sigma = \upsilon/2u_t$ (17.25)

$d_p = 0.05$ mm $= 5 \times 10^{-5}$ m

$\rho_f = 1000$ kg/m^3

$\mu = 1 \times 10^{-3}$ N s/m^2

$\rho_p = 1010$ kg/m^3

$g = 9.81$ m/s^2

$\upsilon = 80$ l/min $= (80 \times 10^{-3})/60 = 1.33 \times 10^{-3}$ m^3/s.

$$\therefore \quad u_t = \frac{\left(5\times10^{-5}\right)^2 \times 9.81 \times (1010 - 1000)}{18 \times 1 \times 10^{-3}} = 1.3625 \times 10^{-5} \text{ m/s.}$$

and
$$\Sigma = \frac{1.33 \times 10^{-3}}{2 \times 1.3625 \times 10^{-5}} = 48.8 \text{ m}^2$$

Example 17.4

Find the g – number of a centrifuge with an effective radius of 10 cm and rotating at a speed of 30 rps.

Solution It is a direct application of Eq. (17.24)

$r = 10$ cm $= 0.1$ m

$N = 30$ rps

$\omega = 2\Pi N = 2\Pi \times 30 = 188.5$ s^{-1}

$g = 9.81$ m/s^2

By applying Eq. (17.24), we get

$$g\text{-number} = Z = \frac{r\,\omega^2}{g} = \frac{0.1 \times (188.5)^2}{9.81}$$

$$= 362.2 \approx 362$$

17.4.2 Centrifugation Equipment

After having seen the theory of centrifugation, we will describe some of the common centrifuges used in bioprocessing.

Tubular bowl centrifuge Tubular bowl centrifuge, also popularly known as tubular centrifuge, is commonly used in biochemical operations for the separation of immiscible phases of

- heavy liquid–light liquid
- solids–liquid
- heavy liquid–light liquid–solids

According to the type of operations, some adjustments are made in the internal arrangement for the separation of two liquid phases or liquid–solid phases. Generally, it is used for the separation of two liquid phases which are immiscible and have relative differences in densities. The two liquid phases are known as light-liquid phase and heavy-liquid phase. This centrifuge can be used for slurries containing up to 10% solids of 0.1–200 microns size.

It consists of a tall and narrow bowl of 10–15 cm diameter, rotating at a speed of 15,000 rpm. The laboratory centrifuges (with a bowl capacity of 200 cm^3) can operate up to 50,000 rpm with an air turbine to produce 62,000 g. The feed enters through a nozzle at the bottom (Fig. 17.10) into the tubular bowl, which rotates at very high speed. The feed liquid mixture is subjected to centrifugal force. Accordingly, the two phases slowly separate out. Heavy liquid is thrown away from the centre and the lighter liquid concentrates at the centre of the rotating bowl. The two phases slowly rise up in the bowl. An adjustable ring at the top separates both the phases, which discharge from the outlets provided at the top. If the solids are also present, they will be settling at the bottom and slowly rise up away from the centre. The solids are removed intermittently by flushing water through a nozzle provided for the purpose (not shown in the figure).

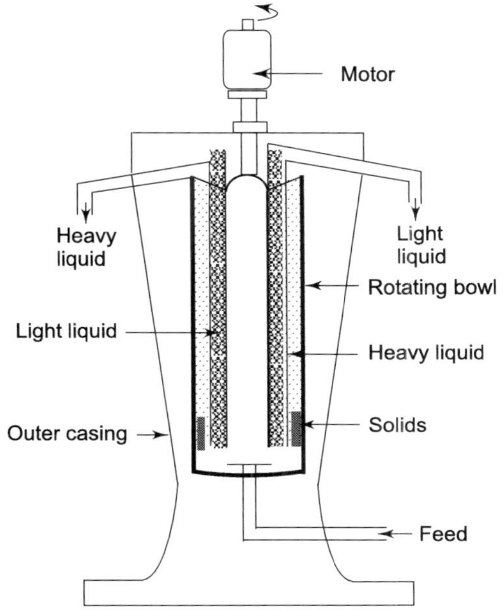

Fig. 17.10 *Tubular bowl centrifuge*

One of the major disadvantages with the tubular bowl centrifuge is that it is not efficient for solid separation or for heavy loads of solids. Particularly, it does not work effectively with three phases (solid–liquid–liquid).

Disc-bowl centrifuge The disc–bowl centrifuge (also known as disk centrifuge) consists of an outer bowl in which a large number of discs are stacked in series one over the other, with a small clearance of 3 mm. In fact, they are not discs; they are actually thin cones made up of a metallic sheet and having two holes on either side almost at the central portions (Fig. 17.11). The conical discs are so placed that the holes on both sides come in the same position. The bowl is usually of 20–50 cm in diameter and rotates at a high speed. The slurry is fed from the centre through a feed nozzle which is stationary. The feed comes and falls at the centre of the discs which are rotating at high speed along with the bowl. Due to centrifugal force, the heavier liquid separates from the lighter liquid and flows away to the periphery through the holes in the discs. The solids, if any, will be collected at the bottom along the periphery of the bowl. The lighter liquid reaches to the centre and rises up. This is discharged through the outlet. The heavy liquid also rises to the top along the periphery of the bowl and discharges through the outlet as shown in the Fig. 17.11.

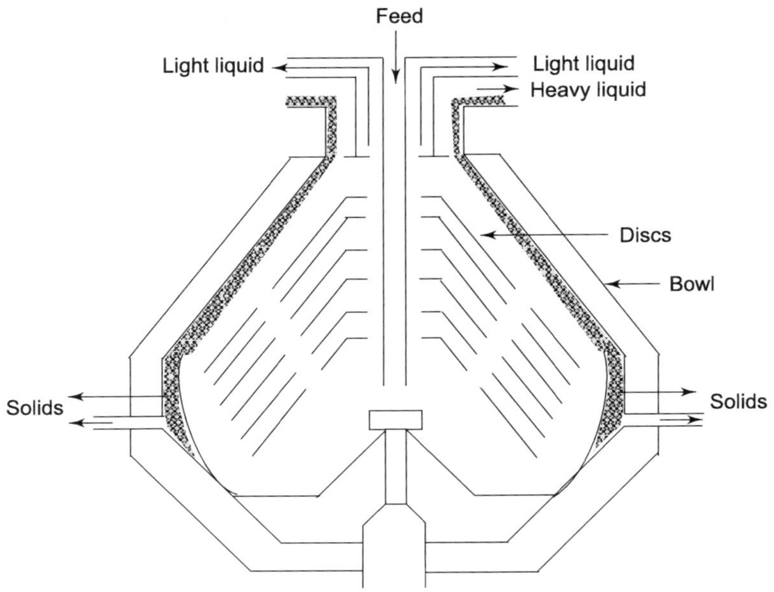

Fig. 17.11 *Disc–bowl centrifuge*

If the solids are little in quantity, they are discharged at the end of the centrifugation. In the nozzle discharge centrifuge, there is a provision for the removal of solids intermittently. Through the nozzle, a jet of liquid is introduced. The force of the jet is sufficient enough to lift the whole stack of discs. The stack of discs becomes very light during centrifugation. The liquid jet washes the solids and discharges through the bottom outlet.

This centrifuge is also used mainly for separating liquid phases or slurries with very low concentration of solids. It is mainly used for concentrating the liquid phases. One of the most classical applications of disc centrifuge is separation of cream from milk.

Basket centrifuge The basket centrifuge (also known as top suspended basket centrifuge) is generally used for separating solids of good concentration of liquids. The baskets are of different diameters (75–120 cm) and depths (45–75 cm). The basket has a large number of perforations. It rotates at speeds up to 4000 rpm. The perforated basket is held in position and rotated by a motor from top (in which case it is called as suspended basket centrifuge) or from the bottom. Figure 17.12 shows the latter case and the feed is fed from the top. The moment the feed falls at the centre of the centrifuge from the top into the rotating bowl, it gets centrifuged. The solids which are denser will move towards the walls of the centrifuge, the liquid collects at the centre at the bottom, from where it is discharged continuously. The solids collect at the wall of the perforated basket. If the solid particles are smaller in size, sometimes a filter medium is used. The filter medium could be simply a filter cloth.

An adjustable loader knife is used sometimes to scrape the solids from the filter medium. The scraped solids are washed with water or solvent through the wash inlet. The wash inlet sprinkles water on the cake, which serves two purposes: (i) to wash off the cake from the traces of the fermentation fluid, (ii) to discharge the cake from the centrifuge. The solid cake is discharged intermittently at the bottom along with the wash water. If the desired product is in the form of solid, as biomass in the case of production of baker's yeast, we may only wash the solids, but we do not scrape intermittently. We shall collect the washed cake only after the centrifuge is stopped after complete centrifugation.

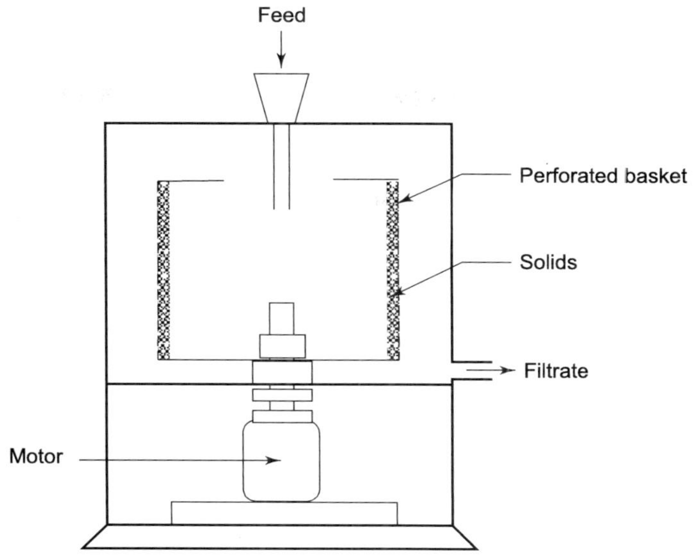

Fig. 17.12 *Basket centrifuge*

17.5 CELL DISRUPTION

After the fermentation process is successfully completed, it is necessary to separate the product from the solids (usually cells). This is an important step in the downstream processing side. We generally resort to classical solid–liquid separation techniques like filtration or centrifugation. This is possible only if the product is extracellular (i.e. the product is not held up by the cells), as in the case of production of alcohol or citric acid, etc. Thus, the solid–liquid separation techniques can be directly applied for the recovery of the product, if it is excreted from the cells.

Some of the enzymes and recombinant proteins are held by the microorganisms (which we call as intracellular products). Generally, the microorganisms are protected by tough cell walls, which need to be disrupted to isolate the metabolic product. In such cases, we need to resort to some of the cell disruption techniques described in detail in the literature (Hughes et al., 1971). There is a large number of such methods available, but only a very few can be practised at industrial scale. They are shown in Fig. 1.1 as follows:

Mechanical methods:
- high-speed agitation
- grinding with abrasives
- high-pressure pumping

Non-mechanical methods:
- osmotic shock
- treatment with solvents and detergents
- freezing and thawing
- enzymatic digestion of cell walls.

17.5.1 Mechanical Methods of Cell Disruption

High-pressure homogenisation is one of the most widely used techniques in mechanical methods at industrial scale in view of the availability of equipment. A schematic representation of the same is shown in Fig. 17.13.

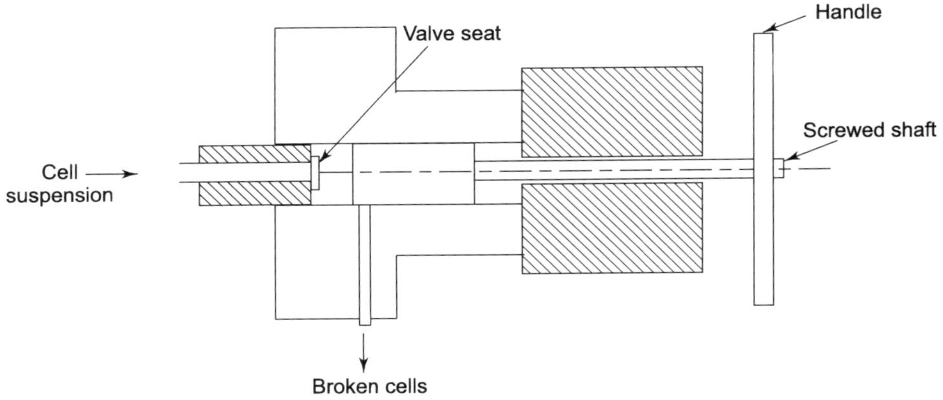

Fig. 17.13 *Cell disruption in high-pressure homogeniser*

The feed as a cell suspension enters from one side, and it is pressurised in the homogeniser with a pressure valve which is operated with a high-pressure pump. The valve forces the cell suspension through an orifice which develops very high pressure of the order of 550 atm. While passing through this high-pressure system, the cells get disrupted and release the product. The released products are isolated from cell debris by using usual solid–liquid separation techniques.

Cells can also be disrupted by mechanical means, either by subjecting the slurry to a high-speed agitation or by grinding the slurry with abrasive materials like glass beads. The slurry is agitated between high-speed discs with glass ballotini.

17.5.2 Non-mechanical Methods of Cell Disruption

The non-mechanical methods of cell disruption include chemical and biological methods. Osmotic shock treatment is caused by sudden change in salt concentration. This will cause the cell disruption. This technique was used for the extraction of luciferase from *Photobacterium fischeri*. Detergents are also used for disrupting the cells. The chemical detergents like quaternary ammonium compounds and sodium lauryl sulphate will damage the lipoproteins of the cell walls (membrane) and release the intracellular product. Sudden freezing and thawing will also cause cell disruption, because the freezing causes the ice crystals to form, which melt on thawing, and break the cell walls. Some enzymes are also used to hydrolyse specific bonds in cell walls. Lysozyme and enzyme extracts from leucocytes have such enzymatic activity.

17.6 EXTRACTION

Extraction is one of the most frequently used separation and isolation techniques of two components in chemical, biochemical and pharmaceutical industries. One of the most important applications in biotechnology is the extraction of penicillin from the fermentation broth using butyl acetate or amyl acetate solvent, after lowering the pH to get favourable partition coefficient. The active component (solute) present in one solvent medium/phase is extracted into another solvent/phase in which the solute has got preferential solubility. The process can be described in simple words as follows. The solvent–solute mixture in one phase is thoroughly mixed with another preferential solvent. The solute equilibrates itself between the two phases. The phases are kept in a settling tank where the two phases separate out as upper layer and bottom layer (or lower layer). It is schematically shown in Fig. 17.14. The most common procedure which we use in the

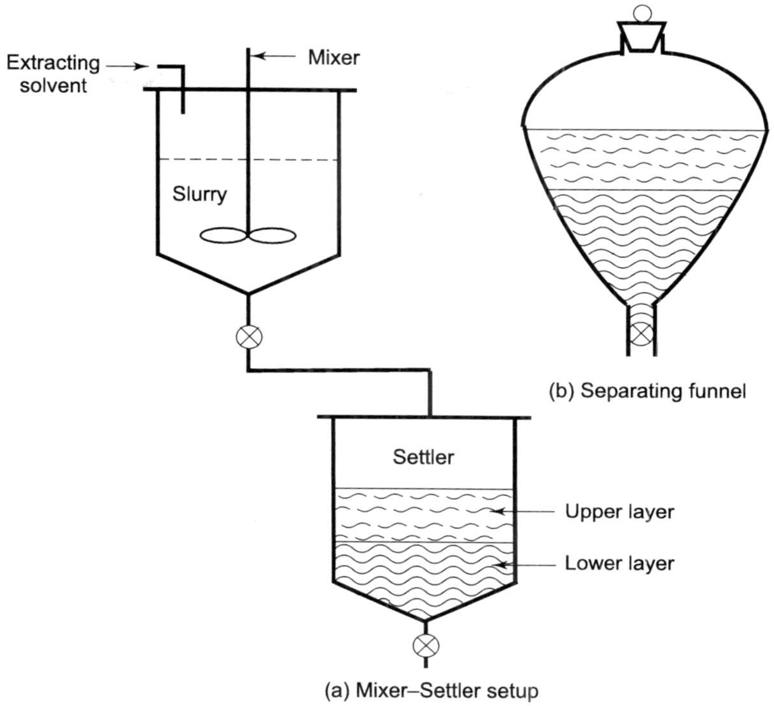

Fig. 17.14 *Simple extraction setup*

laboratory is a separating funnel (Fig. 17.14(b)) in which the solution and the extracting solvent are taken together, thoroughly shaken, and kept for settling to take place. Later, the lower layer is carefully withdrawn by opening the pinch-cock. When once the lower layer is discharged fully in one container, the upper layer will be discharged into another container.

Extraction process can be of two types:

(i) Liquid–liquid extraction
(ii) Solid–liquid extraction

The latter case is commonly referred to in chemical engineering literature as **leaching.** The most common industrial process is extraction of oil from oil seeds using a hydrocarbon solvent. In biochemical process, we generally come across liquid–liquid extraction process, and hence will be considered here. Some important applications of extraction process in biochemical operations are (Doran, 1995):

- isolation of erythromycin using pentyl or amyl acetate
- recovery of steroids
- purification of B_{12} from microbial sources
- isolation of alkaloids (morphin and codeine, etc.)

After the extraction process is over, the solvent is removed from the **raffinate** (the extracted solute + solvent) by distillation. Thus, extraction is always followed by distillation, without which it is not possible to recover the product (solute). That is why if a separation-cum-recovery can be achieved by distillation, we go for distillation. Only if the separation is not possible by distillation alone, we resort to solvent extraction procedure.

Principles of Extraction The extraction process proceeds with the preferential solubility of the solute in two solvent phases which are immiscible. This characteristic feature is termed as the **partition coefficient** (K) or **distribution coefficient**

$$K = C_{A1}/C_{A2} \tag{17.26}$$

Depending on the value of K, whether it is >1 or <1, we can say in which phase (phase 1 or phase 2) the solute has got preferential solubility.

The product recovery or yield parameter Y is defined as the amount of solute extracted into that phase over the total amount of the solute present in the solution initially, i.e.

$$Y_1 = \frac{\text{quantity of A in phase 1}}{\text{total quantity of A}} = \frac{V_1 \times C_{A1}}{V_0 \times C_{A0}} \tag{17.27}$$

$$= \frac{V_1 \times C_{A1}}{V_1\, C_{A1} + V_2 \times C_{A2}}$$

Similarly,

$$Y_2 = \frac{V_2 \times C_{A2}}{V_0 \times C_{A0}}$$

Y and K can be related as follows:

$$Y_1 = \frac{V_1}{V_1 + V_2 \dfrac{C_{A2}}{C_{A1}}} = \frac{V_1}{V_1 + \dfrac{V_2}{K}} \tag{17.28}$$

Rate of Extraction Rate of extraction may be viewed as a transfer process which is proportional to the driving force and the contact area. Thus,

$$\frac{dm}{dt} = K_E \times A \times (C_{Ai} - C_A) \tag{17.29}$$

where dm/dt is the rate of extraction

K_E is the extraction coefficient

$(C_{Ai} - C_A)$ is the driving force which is the difference between the concentration at the interface (C_{Ai}) and the concentration in the bulk (C_A). Normally, the concentration at the interface is equal to the equilibrium solubility denoted by $C_A^{\times}$.

$$dm = dC_A\,(V)$$

Hence, Eq. (17.29) can be written as

$$V\frac{dC_A}{dt} = K_E\,(A)(C_A^{\times} - C_A) \tag{17.30}$$

which on integration and applying limits

$$C_A = C_{A0}, \quad \text{when} \quad t = 0, \text{ give}$$

$$\text{In}\frac{\left(C_A^{\times} - C_{A0}\right)}{\left(C_A^{\times} - C_A\right)} = K_E\,(t)(A/V) \tag{17.31}$$

The above equation is more applicable for solid–liquid extraction and is complicated by the term A, which is difficult to measure. Equation (17.31) indicates that the approach to equilibrium is an exponential process in time, and hence takes a pretty long time to achieve the equilibrium concentration limits. One of the envisaged methodologies for the above is to use a counter-current extraction process.

Counter-Current Extraction Counter-current extraction is a very useful process for the extraction of solute from the solids (leaching) as in the extraction of oils from the oil seeds or extraction of natural colours from the seeds or pods. In a battery of extractors or contacting stages, as we move from the first extractor to the subsequent units, the concentration of the solute in the feed (solids) will be becoming lesser and lesser. So, it would be advantageous if we use the fresh solvent from the other direction, and the extract (solvent + solute) will be moving backwards. It will be getting richer in the solute. The counter-current battery is shown in Fig. 17.15. If the feed is a liquid mixture, the extract and the raffinate (extracted solution) will be moving (pumped) in the opposite directions, and the extraction batteries will be stationary. If the feed is a solid (as in the case of leaching), the spent solids are not disturbed. On the contrary, the extraction battery will be moving, and the extract will be added counter-currently, i.e. the fresh solids will receive the extract (rich in the solute) which will be withdrawn subsequently as the final extract. Similarly, fresh solvent will be added to the battery containing solids which will be withdrawn subsequently as the spent solids or spent cake.

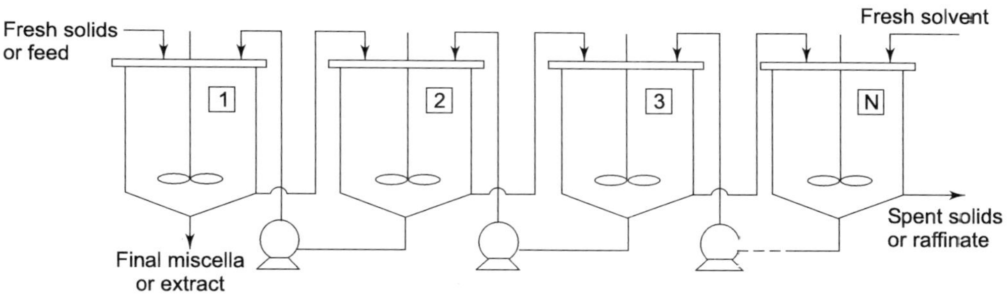

Fig. 17.15 *Counter-current extraction process*

17.7 MEMBRANE SEPARATION

Membrane separation process is almost akin to simple filtration process, except that the particle size is of micron level. The membrane acts as a filter medium, restricting all the particles having size bigger than the porosity of the membrane. Based on the range of particle sizes, the membrane processes are classified as follows (Crueger and Crueger, 1989):

Particle range (μm)	*Molecular weight range*	*Membrane process*
0.0001 – 0.001	<1000	Reverse osmosis
0.001 – 0.1	<1000	ultrafiltration
0.1 – 10		microfiltration

Some of the mycelia can be removed by means of vacuum drum filters, but single-celled bacteria need to be separated using membrane filtration alone.

Liquid mixtures also can be separated using membranes, in which case, the main criterion is the differences in the relative diffusivities of the solutes through the membranes. The classical example for it is **dialysis.** The membranes are made out of various dense polymeric materials such as:

- cellulose esters
- polyvinyl fluorides
- polycarbonates
- polysulphones
- cellulose, etc.

The membranes can be arranged in the filtration in a number of ways, depending upon the size of the membrane, and the use for which it is put to. The following are a few of such possibilities:

- as spiral wound modules
- in cassettes
- as cascades
- as bundles of tubes of 1–2 cm diameter
- as capillary bundles

The major basic parameters that need to be considered before selecting the membrane filtration process are highlighted in Table 17.3.

Table 17.3 *Parameters to be considered before choosing the membrane separation process*

S. No.	*Parameters*
1	Pore size of membrane
2	Particle selectivity to membrane material
3	Viscosity of the feed
4	Available membrane surface area
5	Ease of cleaning and regeneration of the filter membrane
6	Reusability of the membrane
7	Cost of the membrane vis-à-vis the cost of the product
8	Dead volume within the filter
9	Ease of sterilization of the system
10	Presence and nature of antifoam agents

The cross-flow filtration membranes are also used frequently, but mainly for concentrating the culture medium, rather than for separation or isolation as shown in Fig. 17.2.

Dialysis Dialysis is a process of selectively separating low molecular weight solutes from the solutions by using relative diffusivity coefficients. The driving force for separation is the concentration differences; generally there will not be any pressure differentials. The solute is diffused to a lower concentration zone by using a thin membrane. Relative diffusivities help in separating the solutes of low molecular weights which will have high diffusivities. Solutes of high molecular weights are retained in the feed solution.

The classical application of dialysis in human system is artificial kidneys. The kidney performs the function of dialysis and removes the waste products like urine, etc. from the blood, but retains salts, proteins, etc. For persons suffering from malfunctioning of kidneys, the separation does not take place in the natural course, and needs dialysis by using a hollow fibre cellulosic membrane. The blood is passed through the fibres, while saline water flows outside. Urea and waste materials, which are low molecular weight compounds, diffuse through the membrane to a lower concentration zone outside the blood; while high molecular weight compounds like proteins and cells are retained in the blood. Since the dialysing fluid contains added glucose and salts, they (glucose and salts) are prevented to diffuse from the blood.

17.8 CHROMATOGRAPHY

Chromatography is a method of separation of components based on their relative adsorption and desorption capacities on an adsorbent. Obviously, the system calls for a suitable

- adsorbent
- eluant

By and large, the absorption–desorption can be characterised by their molecular weights. The low molecular weight compounds elute faster than the high molecular weight compounds. The process is similar to separation using air aspiration technique. Suppose there is a mixture of materials of different densities. If they are all suspended freely and a breeze of air is blown over them, materials of low density will move to farther distances as compared to the materials of high density. This is how the solids can be separated. This can be schematically represented by Fig. 17.16.

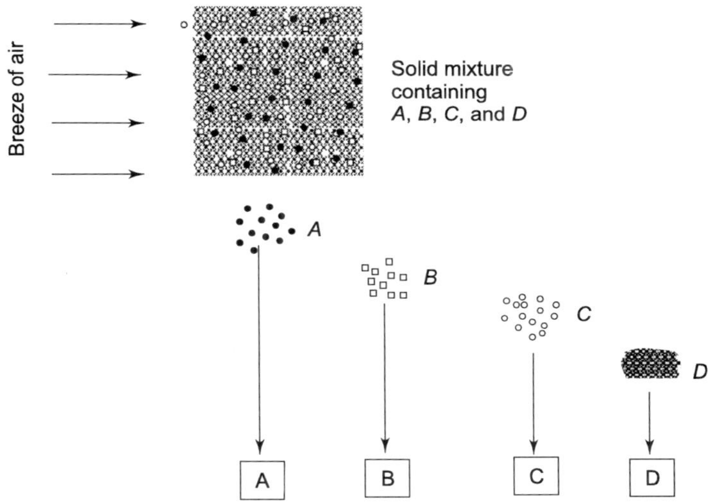

Fig. 17.16 *Schematic representation of air aspiration technique for separation of solids based on their densities*

In a chromatographic separation process, we use a polar adsorbent, viz.

- a silica gel
- alumina
- diatomaceous earth
- charcoal, etc.

The adsorbent is taken in a column and is suitably wetted with a solvent. Later the mixture to be separated is added at the top of the column, and is loosely plugged with cotton at both the top and bottom. The arrangement is shown in Fig. 17.17. A suitable solvent, known as the **eluant,** is added at the top, either continuously or until filling up to the top of the column. The eluant starts moving through the column. The mixture is adsorbed on the silica gel (Fig. 17.17(b)). Slowly the eluant starts desorbing the compounds from the adsorbent. The least desorbed or low molecular weight compound C will come down along with the solvent (eluant) and get collected in the bottom flask (Fig. 17.17(c)). Similarly, the compounds B and A are also desorbed and collected separately along with the eluant. Now the eluant is removed by distillation, and the pure compounds A, B and C can be separately collected. This method is popularly known as column chromatography.

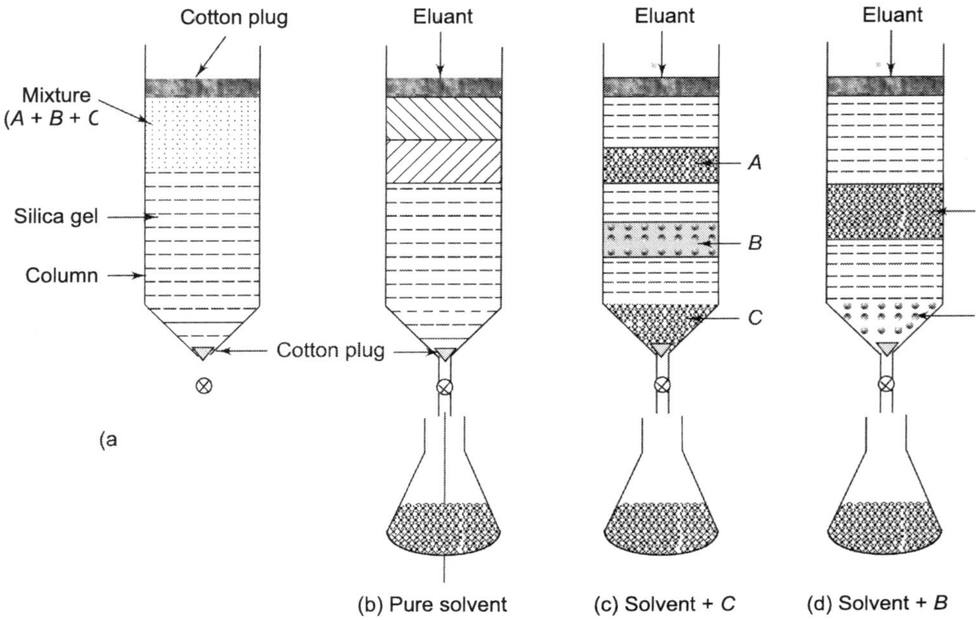

Fig. 17.17 *Schematic representation of column chromatography*

This chromatographic separation technique is highly involved and time-consuming. This is a high-resolution technique, and hence is selectively used (Doran, 1995) for:

- recovery of high-purity pharmaceuticals and therapeutics
- purification of proteins
- purification of peptides
- purification of amino acids
- purification of nucleic acids
- purification of alkaloids
- purification of vitamins
- purification of steroids, etc.

One of the classical applications of chromatography is gas chromatograph (GC), which is used as an analytical equipment for identifying the compounds rather than as a separation technique. In gas chromatograph, a suitable adsorbent is taken in a thin (capillary) long column and maintained at high temperature. The eluant is a neutral gas. The mixture to be identified is injected in microlitre quantities into the hot column. Separation takes place in the column, and the separated compounds come to the outlet, where a suitable detector is installed. The detector detects the compounds and gives separate peaks for different compounds. The output, in the form of peaks, is known as **chromatograms**.

For the separation of mixtures into their individual components, the following types of chromatographic techniques are available:

- adsorption chromatography
- ion-exchange chromatography
- gel filtration chromatography
- affinity chromatography.

They will be described in the following sections.

17.8.1 Adsorption Chromatography

In adsorption chromatography, we exploit the tendency of biological molecules to adsorb on to polar solid adsorbents by weak van der Waals' forces. The adsorbents are generally silica gel, aluminium hydroxide, aluminium oxide, etc. When the mixture is eluted with a suitable molecular polarity, the mixture separates into individual components based on their relative adsorptive–desorptive tendencies. Some small-scale antibiotics can be purified by this method. Dihydro streptomycin can be extracted from the filtrate on a charcoal column. Charcoal adsorbent is also extensively used for the removal of some of the pigments (Stanbury and Whitaker, 1993). Some of the resinous adsorbents like Amberlite can be used for the recovery of cephalosporin C, cefotiam, desferrioxamine B, etc. (Voser, 1982).

17.8.2 Ion-exchange Chromatography

Ion-exchange chromatography works on the principle of electrostatic attraction between the solute and the charged groupings on the adsorbent in the column. It is a reversible exchange of ions between the liquid phase and the solute. The adsorbent is an ion-exchange resin, like divinyl styrene copolymer. To this are attached the reactive groups, which have cationic or anionic exchange properties.

Anionic ion-exchange resins	*Cationic ion-exchange resins*
Secondary amines	Sulphonic acid group
Quaternary amines	Carboxylic acid group
Quaternary ammonium active groups	Phosphonic acid group

The resin properties are dependent upon:

- bead size
- pore size
- resin capacity
- reactive groups
- ion-exchange rate
- life of the resin.

This method can be used for the fractionation at commercial scale of antibiotics and proteins. The ion-exchange process can be schematically represented for the recovery of streptomycin as follows (Stanbury and Whitaker, 1993):

Ion-exchange reaction

$$\underset{\text{(resin)}}{\text{R-COO}'\,\text{Na}^{+}} + \underset{\text{(antibiotic)}}{\text{Streptomycin}} \longrightarrow \underset{\text{(resin-antibiotic complex)}}{\text{R–COO}'\,\text{Streptomycin}^{+}} + \text{NaOH} \tag{17.32}$$

Recovery of antibiotic

$$\text{Resin–antibiotic complex} \xrightarrow{\text{HCl washing}} \text{R-COOH} + \text{Streptomycin}^{+}\,\text{Cl}' \tag{17.33}$$

Regeneration of resin

$$\text{R-COOH} \xrightarrow[\text{through column}]{\text{running NaOH}} \underset{\text{(resin)}}{\text{R-COO}'\,\text{Na}^{+}} + H_2O \tag{17.34}$$

17.8.3 Gel Chromatography

Gel chromatography is also known as **gel filtration chromatography,** which separates molecules on the basis of their size. The gel material is packed into the column. The gels used in this technique are generally

- cross-linked dextrans
- agarose
- polyacrylamide.

The mixture to be separated is fed to the column. Molecules of smaller size diffuse through the gel matrix and penetrate more into it, whereas the larger molecules remain outside the gel. When the column is eluted with a solvent, the bigger molecules (which do not penetrate into the gel matrix) will elute first and get separated from the smaller molecules. They (the smaller molecules) elute out subsequently. That is how the separation is affected based on the molecular size. This technique can be effectively used for the separation of proteins and lipophilic compounds, purification of vaccines, etc.

17.8.4 Affinity Chromatography

It is a very potential technique for the separation and purification of most of the biological materials. This is based on the reversible binding specifity of biomolecules to the binding substance of the column matrix, known as **ligands.** The pairs of biological materials like

- enzyme–substrate
- enzyme–inhibitor
- antigen–antibody, etc.

can be separated by this method. Some of the popularly used polymers for binding matrices are:

- agarose
- cellulose
- dextrose
- polyacrylamide.

The matrix gels and liquids are coupled by using support materials like cyanogen bromide, etc. Agarose activated with cyanogen bromide is a very commonly used support for coupling the ligands.

Affinity chromatography methods are generally used for purification and isolation of proteins, for separation of enzyme–inhibitors, etc. Affinity chromatography technique is used for purifying nucleotide adenine dinucleotide (NAD) by allowing it to bind to a carrier containing dehydrogenase enzyme.

17.9 CRYSTALLISATION

Crystallisation is a process of separation of dissolved solids from the liquid solution based on the solubility. Hence, if the solubility is brought down either by changing the temperature or the concentration of the mother liquor, we find crystals to be separating out of the mother liquor. Crystallisation is an important industrial process. Most of the products are marketed in crystalline form. Crystallisation also yields the products in a pure form. The crystals that separate out of the mother liquor will be having the mother liquor on the surface even after the separation of the solids by filtration or centrifugation. They need to be washed with water or a solvent, and to be dried before packing in the powder (crystal) form.

In the production of citric acid, the filtered broth is treated with lime to precipitate out the citric acid as calcium citrate. This is later hydrolysed with sulphuric acid to regenerate the citric acid in the solution and the insoluble calcium sulphate is precipitated. The whole mixture is filtered to separate out a liquor which contains citric acid. The aqueous citric acid is concentrated when the citric acid crystals precipitate out (See Fig. 20.7). Some of the amino acids are also separated out by crystallisation.

17.9.1 Principles of Crystallisation

After the solids separate out as crystals, the mother liquor with crystals is withdrawn from the crystallizer. This is known as **magma.**

Solubility of the solids in the liquid plays a major role in crystallisation, which increases with temperature. In many cases, the solvent is water, and water is also held by the crystals in the form of water of crystallisation.

Crystallisation can take place only when the liquor is super saturated. This can be achieved by:

- concentrating the mother liquor
- cooling the mother liquor
- concentration and cooling.

Another important step in crystallisation is 'nucleation' which acts as a "seed" on which the crystals will start growing. Thus, in the formation of crystals, two steps are important, viz.

- the birth of a new particle (nucleation)
- its growth to macroscopic size.

The sequence of steps may be summarised as follows (McCabe et al., 1993(c)).

$$\text{cluster} \rightarrow \text{embryo} \rightarrow \text{nucleus} \rightarrow \text{crystal}$$

A small crystal that forms will be in equilibrium with the supersaturated solution. However, this equilibrium is unstable. Any large crystal that is present in the super saturated solution will further grow, making the smaller crystals to dissolve. This phenomenon is known as **Ostwald ripening.**

Kelvin equation This equation shows a relationship between solubility and particle size:

$$\ln \alpha' = \frac{4V_M \sigma}{\gamma RTL} \tag{17.35}$$

where L is the crystal size

V_M is the molar volume of crystal

α' is the ratio of concentrations of super saturated solution to saturated solution

γ is the number of ions per molecule of solute, which is normally 1.0 for molecular crystals.

σ is the interfacial tension between solid and liquid.

17.9.2 Crystallisation Equipment

Industrial crystallisation equipment may be broadly classified into two categories:

- those in which the mother liquor is cooled, so that the crystals start forming, which is possible in case of systems whose solubility is very much dependent on temperatures, and such equipments are called **crystallisers**;
- those in which the mother liquor is concentrated by evaporation, so that the super saturation may result in separating the crystals, and such equipments are known as **evaporators.**

In some intermediate cases, both evaporation and cooling are used for effective crystallisation of solids.

Crystallisers are operated both in a batch manner and in continuous manner depending upon the capacity; however, continuous crystallisation equipment is preferred in view of their ability to control crystal growth in addition to their capacity to process large throughputs. A **batch crystalliser** is an open tank in which the hot solution is kept to lose heat to surroundings. Sometimes they may be provided with a cooling coil or a cooling jacket and a slow agitation, or both cooling and agitation to assist in removing the heat and in formation of crystals. Since there is no control on nucleation, generally large crystals form. Thus, a batch crystalliser is simply a vessel provided with either an internal cooling coil or an external cooling jacket through which a coolant passes. The vessel is provided with a slowly moving internal scraper to scrape the crystals forming on the walls of the vessel and also to assist in heat transfer. A schematic arrangement is shown in Fig. 17.18, which in similar to Fig. 13.13.

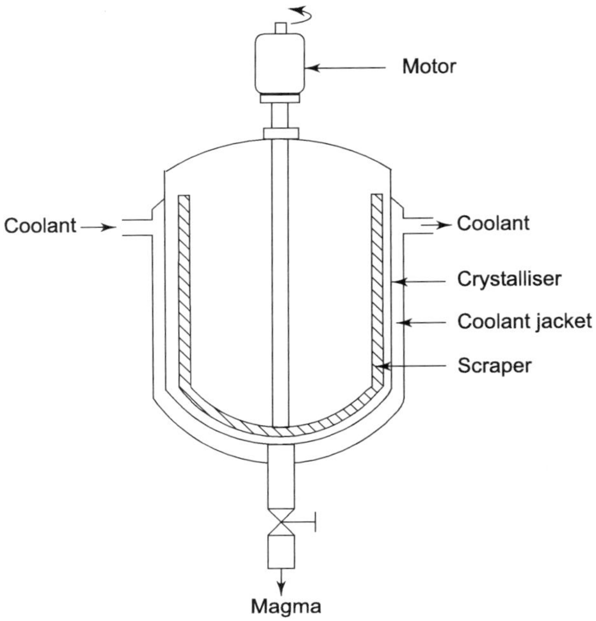

Fig. 17.18 *Batch crystalliser*

Draft tube baffle crystalliser A draft tube baffle crystalliser is a continuous crystalliser and is shown in Fig. 17.19 (Randolph, 1970). The feed is fed continuously through a circulating pump which pumps a part of the recirculating fluid (from the crystalliser) through a heat exchanger. The crystalliser is connected to a barometric condenser and a vacuum system to concentrate the feed solution. A propeller-type agitator is used to agitate the contents and also to push the solution through a draft tube. The draft tube helps in reduction of energy requirements for mixing, and at the same time, maintaining good circulation of slurry. The draft tubes generally have diameters equal to 0.7 times of the tank diameter. The minimum velocity of the slurry should be of the order of 2.1 m/s in the transfer pipes to reduce settling of the solids (Genck, 2003). This circulation will help the separation of larger crystals which slip into the elutriation leg, while the fines through the baffle zone go into the recirculation pipe. The slurry with fines is mixed with the feed, and on heating it becomes a homogeneous solution. It is then subjected to vacuum in the crystalliser and the process continues. The product, in the form of larger crystals, is withdrawn through the product discharge line. By recirculating a large fraction of the mother liquor from the crystalliser, the magma densities are increased sharply. About 30–50% rise in magma densities are achieved (McCabe et al., 1993 (c)).

17.10 DRYING

Drying is a general term referred to mean the removal of last traces of solvent to recover the product in the form of dry solids. It generally includes only drying of solids, even though it is possible to dry the solvents, etc. also. **Dehydration** is a specific term used to mean the removal of moisture from the solids, i.e. if the solvent is water, we use the term dehydration. Since in most of the cases, the solvent happens to be moisture, the terms drying and dehydration are used interchangeably.

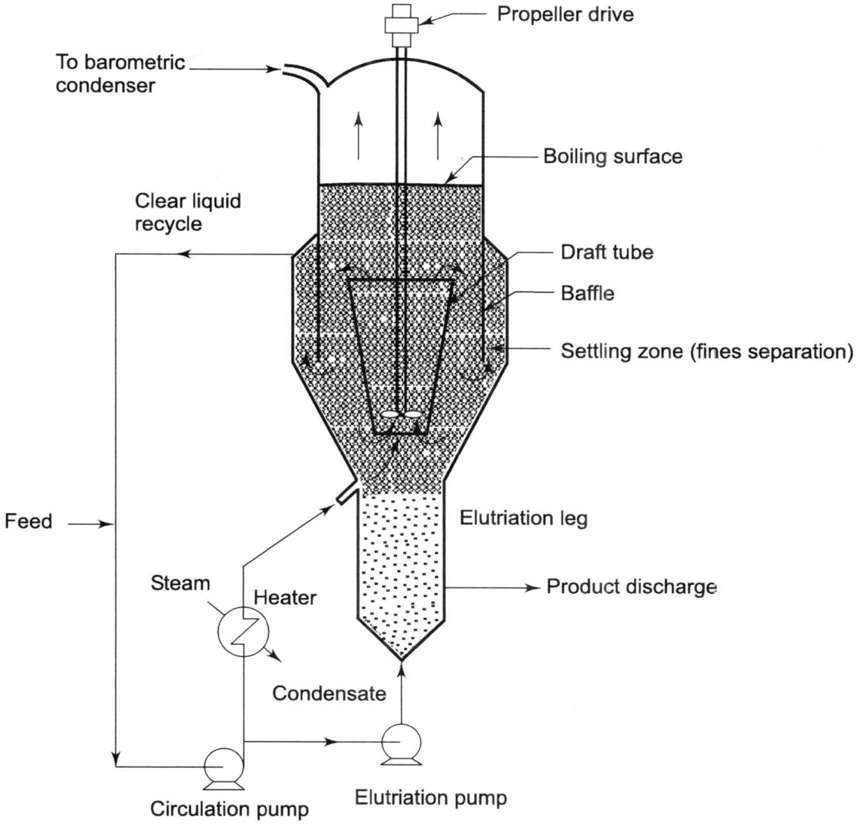

Fig. 17.19 *Draft tube baffle crystalliser*
(Excerpted from "Crystallization", Randolph, A.D., Chemical Engineering, Vol. 77(10) (May), p. 86, Copyright © (1970), with special permission from Chemical Week Associates, New York, NY 10038)

Drying is generally used for the removal of the last traces of adhering moisture from the solids, so that the product is recovered in the form of dry and free-flowing solids. If all the moisture is removed, the product is said to be **bonedry.** Drying has certain specific advantages which make it invariably an inevitable last unit operation in the product recovery. The main advantages are:

(i) The water activity, which is a measure of the moisture content, when reduced below certain levels, will make the biological materials free from the survival and growth of microorganisms, and hence will help in increasing the shelf-life of the material.
(ii) Drying will reduce both the weight and volume of the product, which will facilitate easy packaging and transportation.
(iii) The shelf-life of dry products will increase considerably.

Such an important activity of drying involves simultaneous heat and mass transfer operations and can take place by means of the following five major processes:

(i) **Convective** and contact drying takes place by means of the direct contact of the wet product with drying air, which takes up moisture from wet material by convection; or the heat may be transferred from hot metallic surface by **conduction**.

(ii) **Vacuum drying** takes place by subjecting the wet solids to vacuum, where the advantage is derived for the removal of moisture by vaporisation under reduced pressures.

(iii) **Freeze drying** is a process in which the moisture in the product is frozen by reducing temperature and pressure below the triple-point of water, and subsequently the moisture is removed by raising the temperature and keeping the vacuum constant, so that the moisture in the form of ice sublimes to vapour without going through the liquid phase.

(iv) **Radiation heating** removes moisture by supplying radiant heat from the heat sources in which the heat transfer is neither by conduction, nor by convection, but by radiation.

(v) **Microwave heating** is altogether a different heat transfer method, in which heat is generated in the wet solids by molecular friction by continuously changing the polarity of the water molecules by applying electromagnetic waves.

17.10.1 Principles of Drying

Drying of wet solids is achieved by exposing the solids to a wave of hot air which removes the moisture by convection from an initial moisture content to a desired final moisture content. In the process, both the heat transfer and mass transfer take place. Heat is transferred from hot air to the solids, and in turn, the moisture diffuses out of the solids and gets transferred into the stream of relatively dry air. The following are certain terms which are frequently referred to in the drying literature, and are described with the equilibrium moisture content (EMC)–relative humidity (RH) curve shown in Fig. 17.20.

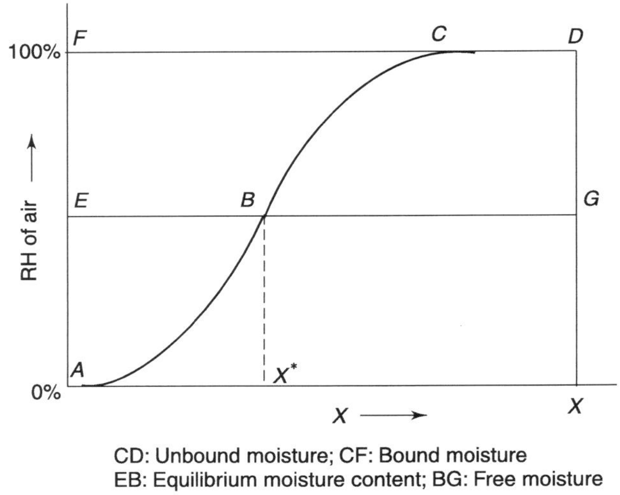

Fig. 17.20 *EMC–RH curve*

Moisture content The moisture content of a solid is the ratio of moisture present initially in the solid to the weight of the wet solid, i.e.

$$X_w = \frac{\text{kg of moisture}}{\text{kg of wet solid}} \tag{17.36}$$

$$= \frac{\text{kg of moisture}}{\text{kg of moisture} + \text{kg of the dry solids}}$$

This moisture content is defined on wet basis.

Moisture content on dry weight basis (X)

We generally prefer to define the moisture content on dry weight basis because of its ease in expression. The denominator of it does not change with the drying process.

$$X = \frac{\text{kg of moisture}}{\text{kg of dry weight of solids}} = \frac{X_w}{1 - X_w} \tag{17.37}$$

Conversely,

$$X_w = \frac{X}{X + 1} \tag{17.38}$$

Equilibrium moisture content (EMC) It is defined as the moisture content present in the solid which is in equilibrium with the partial pressure of the surrounding vapour, and is generally represented by X^*. Thus, in Fig. 17.20, EB represents the EMC. This is also the moisture content which cannot be removed by the surrounding vapour or gas, unless its relative humidity is changed.

Free moisture It is defined as the moisture contained in excess of EMC, i.e. it is equal to $X - X^*$. Obviously, we can remove only the free moisture content out of the total moisture content. Thus, BG represents the free moisture.

Unbound moisture The unbound moisture can be easily removed from the wet solids. The equilibrium vapour pressure exerted by the moisture in the solid is equal to the saturated vapour pressure of the pure liquid. CD represents the unbound moisture.

Bound moisture The moisture present in excess of the unbound moisture is all the bound moisture. It is represented by CF in Fig. 17.20. Only a part of the bound moisture can be removed. All of it cannot be removed.

Drying curve When a wet solid is exposed to a stream of dry air, initially the moisture on the surface of the solids goes into the air. During this period, the solid behaves as if it is surrounded by a pool of liquid on its surface, and the moisture goes into the air at a constant rate. If we draw a plot between the rate of drying versus the moisture content present (Fig. 17.21), the curve is known as 'drying curve'. The rate of drying is constant initially, and we get a horizontal line BC which is known as the **constant rate period.** The initial portion AB of the drying curve is during the period the material gets acclimatised to the environment.

When once all the moisture present on the surface and in the outermost portion of the pores is exhausted, the rate of drying will no more be constant and it will be falling during CD. The period is known as **falling rate period.** During this period, the moisture present in the pores will be transferred into the drying air by capillary forces. Hence, adequate amount of moisture will not be available, with the result that the rate of drying will be falling as the moisture content is reduced. The change in drying rates, from constant drying rate to falling drying rate, occurs at point C, and the moisture content present at point C is known as the **critical moisture content** (X_c).

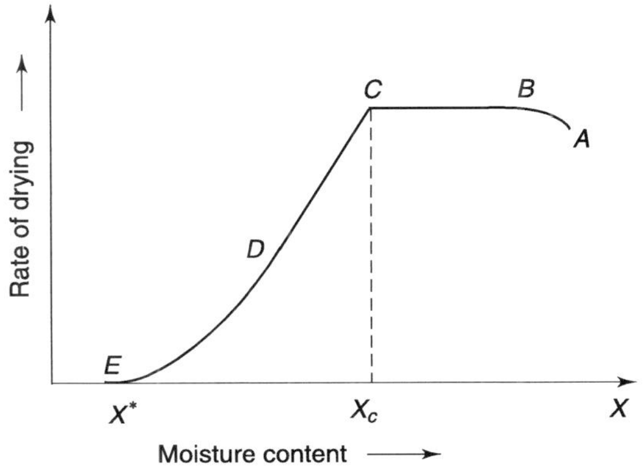

Fig. 17.21 *Drying curve*

After point D, for some solids, it may so happen that still some amount of liquid would be trapped in the pores, and this will be covered with a layer of drying air. In such cases, the moisture has to **diffuse** out of the air film, and this results in a different falling rate period up to point E. The period DE is curved and is known as **second falling rate period.** At point E, the moisture content reaches the **equilibrium moisture content** (X^*), and there will not be any more loss in moisture content. The drying rate falls to zero when the moisture content reaches the equilibrium moisture content.

17.10.2 Drying Equipment

Various types of dryers are available commercially. Many times, many types of dryers will be suitable for a particular type of operation. The selection is based on the cost of equipment and the cost of the product. The physico-chemical properties of the product also determine the type of the dryer. The choice of a particular type of dryer is a very critical parameter, but is unfortunately done on trial-and-error basis. Some tips for the selection of a suitable dryer were given by Forrest (1968).

A large number of dryers are available industrially. Table 17.4 highlights some of them. However, we describe here only a few of them for general comprehension.

Table 17.4 *Various types of industrial dryers.*

1	Fixed tray dryers	(i) (ii) (iii)	Cabinet tray dryer Tunnel dryer Bin dryer
2	Vacuum dryers	(i) (ii) (iii)	Vacuum shelf dryer agitated pan dryer continuous belt dryer
3	Moving bed dryers	(i) (ii) (iii)	Pneumatic conveyor dryer screw conveyor dryer belt conveyor dryer
4	Product suspended in hot air	(i) (ii) (iii) (iv) (v) (vi)	Spray dryer pneumatic dryer ring dryer fluidised bed dryer moving bed fluidised dryer Falling particle dryer
5	Roller dryers	(i) (ii)	Drum dryer rotary dryer
6	Freeze dryers	(i) (ii)	Batch freeze dryer continuous freeze dryer

Cabinet tray dryer This is one of the most common and most versatile dryers with a very little investment. The operating costs are also reasonable. The product quality may not be good, and hence is generally used for products which are not of high value. The time taken for drying is also more, usually of the order of few hours. It is a batch dryer.

It consists of a heating chamber into which a trolley carrying trays is introduced (Fig. 17.22). The trays are generally of the size 40 cm × 100 cm × 2.5 cm. The dryer consists of trays in multiples of 12. The trays may or may not have perforations. The material is loaded into the trays outside and then pushed into the trolley. When all the trays are loaded, the trolley is pushed into the heating chamber (cabinet). The chamber is generally heated electrically or by means of steam coils. It has provision for thermostatic control. Air is sucked with an electric blower, and is heated by contact with the heating electric coils or steam coils. There is a provision for certain amount of recirculation of air in the dryer.

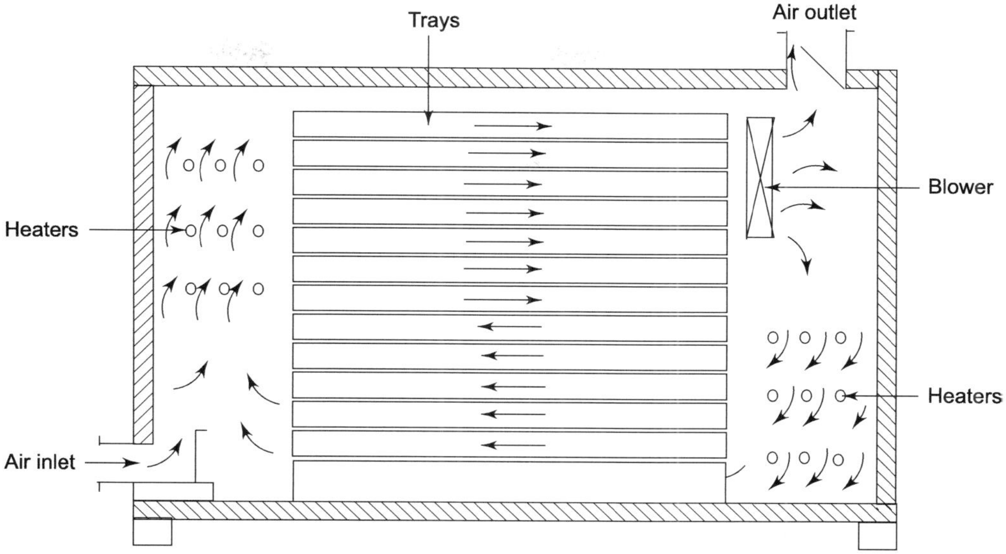

Fig. 17.22 *Cabinet tray dryer*

The material is dried in the dryer under controlled temperature conditions. Occasionally the dryer is stopped, the trays are interchanged, and the material is mixed to reshuffle. Even-though care is taken to see that the air distribution is uniform, it is seldom possible. Hence, uniform heating is never achieved, with the result that the product quality is not always good. It is generally used for products of low cost, and also as a first step in the drying operation.

Vacuum shelf dryer Operation-wise the vacuum shelf dryer is similar to a cabinet tray dryer. Here, the advantage is taken of the low-temperature evaporation of the moisture (or solvent) by applying vacuum. It does not contain any trolley. It has flat solid metallic plates which are heated from inside by steam, or radiation heating is used by providing heating electrical coils. Generally, the direct heating by naked electric coils which radiate heat is discouraged to prevent any ignition of the organic solvents in the product. The material is loaded into the trays which are kept on the solid plates. Conductive heat from the plates is transferred to the material. There will not be convective mass transfer as air is not blown on to the materials. Moisture is removed by evaporation under vacuum.

It is an excellent drying method, particularly for heat-sensitive products. The product quality is also good, as contraction of the tissues does not take place. Except for the energy consumption for creating vacuum, and associated problems of preventing leaks, etc., vacuum drying is an excellent method.

Spray dryer Spray dryer is a continuous dryer used for removing moisture from slurries to get the product in the form of fine powder with very low bulk density. Spray dried milk powder and some of the fruit juice powders are very common examples of the spray drying processes.

In a spray dryer, the slurry (containing solids approximately 20–30%) is sprayed from the top in the form of fine droplets (Fig. 17.23). The feed is atomised into fine droplets using pressure nozzles or dropping the feed on a rotating disc. The former is generally used. Design of spray nozzles is a subject matter for discussion in its own way. Since the feed is atomised into fine droplets, the surface area is considerably increased. These droplets while travelling in the downward direction in the spray drying chamber will come in contact counter-currently with hot air which is introduced from the bottom. The drying air will be at temperatures

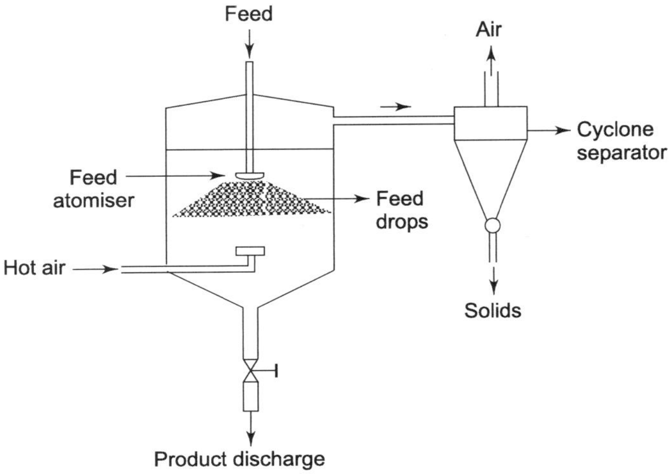

Fig. 17.23 *Spray dryer*

as high as 220–250 °C, but the contact time is only a few seconds. As such, the temperature of the product does not increase appreciably and will always be less than the boiling temperature of water. By the time the droplets reach the bottom surface, they are dried, and fine powder forms in the dry form which is highly hygroscopic. The material is collected from the bottom. The trajectile of the falling droplets is so controlled that they do not touch the spray dryer surface which is at a considerably high temperature, lest the product chars. The powder is directly collected at the bottom continuously. The outgoing air from the top passes through a cyclone separator where any carried materials are separated and collected at the bottom as dust.

Spray dryers are excellent for slurries which are not very viscous and which do not contain high amounts of sugars. The initial investment is high, but the operating costs are low, and the product quality is excellent. Drying rates are very high, and hence the drying times are very short. Thus, the throughputs can be high. The bulk density of the product is very low, and hence its miscibility on reconstitution is excellent.

Freeze dryer Freeze drying provides maximum protection to the biological material being dried, because drying takes place at low temperatures and at very low pressures. The moisture is removed by sublimation, i.e. initially the water is frozen to ice in the material and later it is sublimed to vapour without going through the liquid phase. Because of sublimation, contraction of the tissues of the biological materials does not take place, with the result that the shape and structure of the product are retained.

The phase diagram for water is shown in Fig. 17.24. Moisture exists in the liquid phase in the zone BOC and as ice in the zone AOB, whereas it exists as vapour in the zone beyond AOC. The point O is known as *triple-point* where the moisture exists in all the three phases. If the moisture is brought to condition Q and if the temperature is increased along QQ′, the moisture passes through all the three phases. On the contrary, if the moisture is brought to the point P, and if the temperature is increased along PP′, keeping the pressure constant, the moisture directly sublimes from solid (ice) phase to vapour phase, without passing through the liquid phase. Freeze dryers are operated in this region. Initially, the pressure and temperature are brought down so that the material goes below the triple-point zone (AO region). The moisture in the biological material solidifies to ice. Later, keeping the vacuum constant, the temperature is increased when the moisture leaves in the form of vapours. This is how freeze drying takes place.

Schematic representation of freeze dryer is shown in Fig. 17.25. The biological material to be dried is kept in trays which are placed on the shelves. High vacuum (100–1000 μm of mercury) is applied to the drying

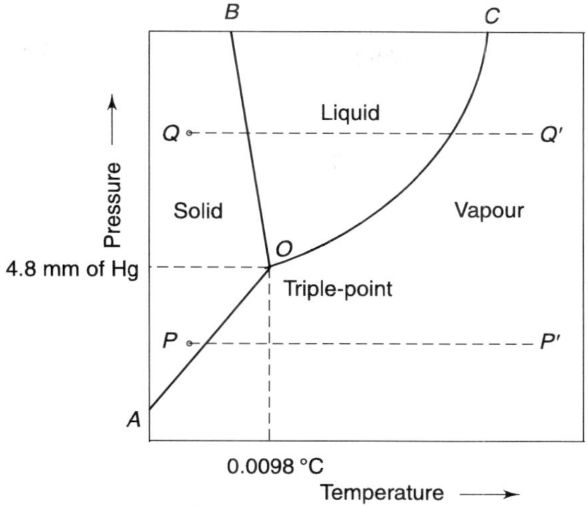

Fig. 17.24 *Phase diagram of water*

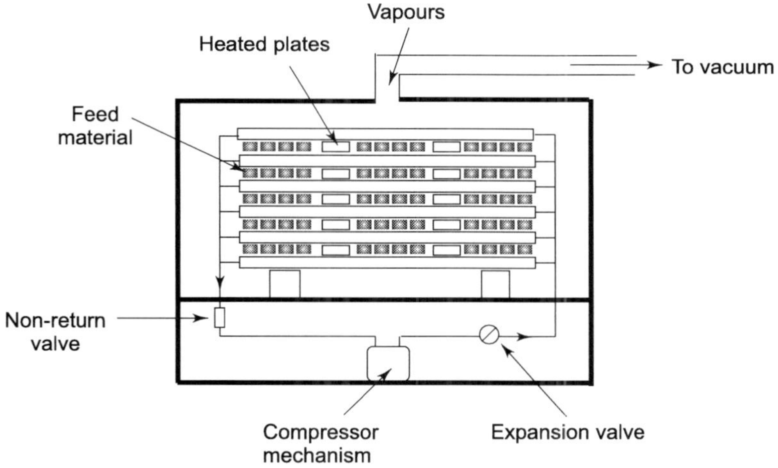

Fig. 17.25 *Freeze dryer*

chamber. The plates (shelves) are cooled with a refrigerant flowing through them. The material is slowly brought to the conditions similar to P in Fig. 17.24. Once these conditions are reached, slowly the temperature of the shelves is raised so that the material passes through line PP′, and dries. Later the vacuum and temperature are released and the product is discharged.

Except for the high initial cost and high operating costs, freeze drying is an excellent method for drying, particularly heat-sensitive high-valued products. Freeze dryers are used for drying of blood or vaccines, etc. **Lyophilization** technique, which is frequently used in biotechnological processes is, in fact, freeze drying. Most of the enzymes and microorganisms are preserved by lyophilization for using subsequently whenever they are required.

Symbols

a'	: acceleration (m/s^2)
A	: cross-sectional area (m^2)
A_P	: projected area of the particle
C_A	: concentration of component A or solute A (kg/m^3 or kg mol/m^3)
C_A^*	: equilibrium solubility
C_{Ai}	: concentration of A at the interface (kg/m^3 or kg mol/m^3)
C_D	: drag coefficient
d_p	: diameter of particle
F	: ratio of liquid to solids in the feed [Eq. (17.17)]
F_c	: centrifugal force
F_d	: drag force
F_s	: settling force
g_c	: gravitational constant
g	: acceleration due to gravity (m/s^2)
K	: partition coefficient [Eq. (17.26)]
K_E	: extraction coefficient
L	: ratio of liquid to solids in the exit [Eq. (17.17)]; crystal size [Eq. (17.35)]
L_c	: cake thickness (m)
L_m	: fictitious thickness of filter medium, filter aid, etc. [Eq. (17.2)]
m	: mass, usually mass of solute (kg)
N	: rotations per second (s^{-1})
$N_{Re,p}$	: particle Reynolds number
Δ_p	: pressure-drop across the filter press
r	: distance of the particle from the centre of the rotation in a centrifuge (m)
r_s	: slurry feed rate (kg/s or kg/h)
R	: universal gas constant
R_f	: resistance for filtration [Eq. (17.2)]
t	: time (s)
T	: temperature (K)
u	: velocity of the particle (m/s)
u_t	: terminal velocity of the particle (m/s)
u_u	: upward velocity of flow of liquid
V	: volume (m^3)
V_M	: molar volume of crystal (m^3/(kg mol))
w	: weight of the solids present per unit volume of slurry (kg/m^3)
X	: moisture content on dry basis
X^*	: equilibrium moisture content
X_c	: critical moisture content
X_w	: moisture content on wet basis
Y	: yield parameter
Z	: a parameter in centrifugation [Eq. (17.24)]

Subscripts

0	: initial
1	: phase 1
2	: phase 2
f	: fluid
p	: particle

Greek Letters

μ	: viscocity of slurry
α	: specific cake resistance ($1/m^2$)
α'	: ratio of concentration of super saturated solution to saturated solutions [Eq. (17.35)]
Σ	: sigma factor of centrifuge (m^2)
ω	: angular velocity (s^{-1})
ρ	: density of the solids (kg/m^3)
γ	: number of ions per molecule of solute [Eq. (17.35)]
σ	: interfacial tension between solid and liquid [Eq. (17.35)]
υ	: volumetric feed rate (m^3/s).

References

17.1 Crueger, W., and Crueger, A. (1989), *Biotechnology: A Textbook of Industrial Microbiology,* Brock, T.D. (Ed.), 2nd Edn, Sinauer Associates, Inc., Sunderland, MA, pp. 114–115.

17.2 Doran, P. (1995), *Bioprocess Engineering Principles,* Academic Press, London, pp. 219–252.

17.3 Earle, R.L. (1966), *Unit Operations in Food Processing,* Pergamon Press, Oxford, pp. 212–282.

17.4 Forrest, J.C. (1968), "Drying Processes", *Biochemical and Biological Engineering Science,* Blakebrough, N. (Ed.), Vol. 2, Academic Press, London, pp. 97–135.

17.5 Genck, W.J. (2003), "Optimizing crystallizer scale-up", *Chem. Eng. Prog.*, 99(6), 36–44.

17.6 Hughes, D.E., Wimpenny, J.W.T., and Lloyd, D. (1971), "The disintegration of microorganisms", *Methods in Microbiology,* Norris, J.R., and Ribbons, D.W. (Eds.), Vol. 5-B, Academic Press, London, pp. 1–54.

17.7 McCabe, W.L., Smith, J.C., and Harriott, P. (1993), *Unit Operations of Chemical Engineering,* 5th Edn, McGraw-Hill, Inc., New York, (a) pp. 1002–1010, (b) pp. 1052–1055, (c) pp. 882–918.

17.8 Randolph, A.D. (1970), "Crystallization", *Chem. Eng.*, May, vol 77(10), 80–96.

17.9 Stanbury, P.F., and Whitaker, A. (1993), *Principles of Fermentation Technology,* Pergamon Press, Oxford, pp. 193–219.

17.10 Voser, W. (1982), "Isolation of hydrophilic fermentation products by adsorption chromatography", *J. Chem. Tech. Biotechnol.*, 32, 109–118.

Review Questions

17.1 Describe the improtance of downstream process steps in bioprocessing.

17.2 What are the various unit operations you come across in downstream processing?

17.3 What are the criteria for choosing a particular product recovery step?

17.4 What are the various unit operations for the separation of suspended solids?
17.5 Describe foam separation process as a tool for downstream processing.
17.6 Distinguish between precipitation and sedimentation.
17.7 Compare and contrast sedimentation and centrifugation processes.
17.8 Describe a constant-pressure filtration process.
17.9 Write short notes on:
  (i) specific cake resistance
  (ii) equivalent cake thickness
  (iii) filter aid
  (iv) factors that affect the filtration rate.
17.10 Derive an expression for specific cake resistance.
17.11 How do you find out the specific cake resistance and equivalent cake thickness in a constant-pressure filtration?
17.12 Describe the plate and frame filter press with a neat diagram.
17.13 Describe a continuous pressure filter.
17.14 Describe the working of a leaf filter.
17.15 Define terminal velocity in freely settling bodies. What are the various factors that affect the terminal velocity?
17.16 Sketch the various forces acting on a sedimenting solid.
17.17 Derive an expression for settling velocity.
17.18 Compare and contrast filtration and centrifugation.
17.19 Define g-number or centrifuge effect.
17.20 Define Σ factor of a centrifuge.
17.21 Describe the working of a tubular centrifuge with a neat diagram.
17.22 Describe basket centrifuge to be used in downstream processing.
17.23 Describe the working of a disk-bowl centrifuge for the separation of a solid–liquid–liquid system.
17.24 Describe the various cell disruption techniques.
17.25 Define a partition coefficient or distribution coefficient.
17.26 Describe the liquid–liquid extraction process in general, as a tool for downstream processing.
17.27 Describe a counter-current extraction process with a neat schematic representation.
17.28 Describe the membrane separation process.
17.29 What are the various polymeric materials that can be used as membranes in the membrane separation process?
17.30 What are the various parameters you think as important for choosing the membrane separation process?
17.31 Describe the principles of chromatographic separation process.
17.32 What are the various applications of chromatographic separation processes?
17.33 Describe the various chromatographic separation processes.
17.34 Explain the working of a gas chromatograph.
17.35 Describe the process of crystallisation.
17.36 Write short notes on:
  (i) magma
  (ii) Ostwald ripening
  (iii) Kelvin equation.
17.37 Describe the working of a batch crystalliser.
17.38 Describe the working of a continuous crystalliser (draft tube baffle crystalliser).

17.39 Describe the process of drying as a tool for downstream processing steps.

17.40 Write short notes on the following:

(i) initial moisture content
(ii) critical moisture content
(iii) free moisture content
(iv) EMC–RH data
(v) constant rate period
(vi) falling rate period.

17.41 Draw a typical drying curve indicating constant rate period and falling rate period.

17.42 Draw the pressure–temperature diagram for water, indicating the triple-point. Explain the principle of freeze dryer.

17.43 Describe a freeze dryer with a neat diagram.

17.44 Describe a spray dryer. What are its relative advantages?

17.45 List out various industrial dryers available for downstream processing.

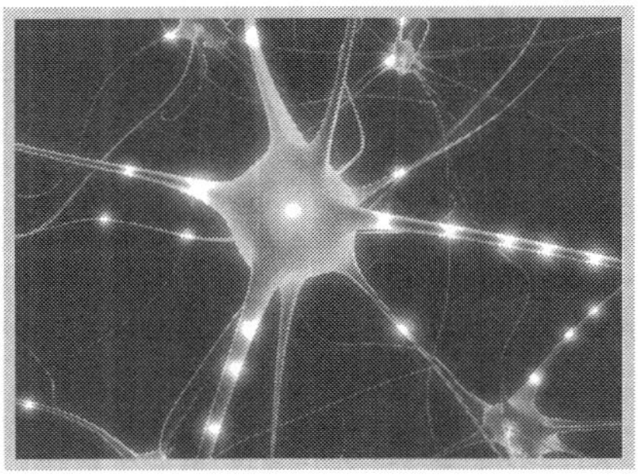

chapter 18

Effluent Treatment

18.1 NEED FOR EFFLUENT TREATMENT

Virtually any process industry releases effluents, but they are more predominant in chemical and biochemical industries. In a fermentation process, a large number of raw materials is consumed to produce a variety of products. In the process, they also release a lot of wastes, effluents and waters, etc. Depending upon the nature of the industry, the nature of wastes also varies. These wastes typically include:

- microbial cells
- unconverted raw materials
- organic media components
- suspended solids
- filter aids
- wash waters
- cooling water
- water containing trace solvents, acids, etc.

In the good old days, these effluents were being dumped in nearby rivers, which resulted in heavy water pollution. It is now no more possible to dump these effluents in the flowing water sources without giving adequate pretreatment. It has become virtually mandatory now for any processing plant to have an effluent treatment plant. Of late, some of the local governments came up with the concept of centralised effluent treatment plants (CETP), where a cluster of industries groups together and constructs a centralised effluent treatment plant. Figure 18.1 shows one such plant.

The organic matter that is left in the effluent, mixes with outside water and consumes the oxygen in the water, which is typically measured as biological oxygen demand (BOD). It is a measure of the quantity of oxygen required for the oxidation of organic matter in the water by the microorganisms present, in a given time period, at a given temperature. If the BOD is very high, the water is depleted of oxygen content, which affects the survival of most of the aquatic species. Erichsen Jones (1964) reported that the dissolved oxygen concentration should not fall below 4 mg/l for the healthy survival of fish.

Fig. 18.1 *Dissolved Air Floatation unit with outlet chamber of Central Effluent Treatment Plant (Courtesy: M/s. Pattancheru Enviro-Tech. Ltd., Pattancheru, AP)*

The BOD is met by continuously aerating the effluent in the effluent-treatment plants, which take about 4–6 hours for normal effluents and 24–28 hours for higher strength industrial waste waters. Eckenfelder et al. (1972) reported that a systematically designed plant, utilising aeration and sludge, would be able to reduce the BOD from 1000 mg/l to 25 mg/l in a few hours (0.36 days). The CETP shown in Fig. 18.1 reduces the BOD to 90 ppm level by various treatment methods. Stanbury and Whitaker (1993) presented the data in a tabular form for a variety of effluents (Table 18.1).

Table 18.1 *Data of BOD strengths of various effluents, (mg/dm^3)*

Effluent		*BOD*	*Reference*
Domestic sewage		350	Boruff (1953)
Sulphite liquor from paper mill		20,000 – 45,000	
Beer:	(a) spent-grain press	15,000	Abson and Todhunter (1967)
	(b) hop-press liquor	7430	
	(c) mash filter cloth wash wastes	4930	
	(d) yeast wash water	7400	
	(e) spoilt water	Up to 1,00,000	
	(f) Bottle washing	550	
Maltings:	(a) suspended solids	1240	Koziorowski and Kucharski (1972)
	(b) wastes	20 – 204	
	(c) grain washing	1500	
Industrial alcohol stillage (molasses)		10,000 – 25,000	Blaine (1965)
Distillery stillage		10,000 – 25,000	Jackson (1960)
Yeast production		3000 – 14,000	Boruff (1953)
Antibiotic wastes		5000 – 30,000	Jackson (1960)
Penicillin:	(a) wet mycelium from filter	40,000 – 70,000	Boruff (1953)
	(b) filtrate	2150 – 10,000	
	(c) wash water	210 – 13,800	
Streptomycin spent liquor		2450 – 5900	Koziorowski and Kucharski (1972)
Aureomycin spent liquor		4000 – 7000	
Solvents		Up to 20,00,000	

Treatment Methodology Industrial fermentation processes release large amounts of effluents. They contain varying amounts of organic matter, salts, etc., which are highly hazardous to the environment. They need to be processed/treated before being released into the outside world. In some occasions, even the effluents and their hazardous nature speak on the technical and economic viability of the product. The various effluent treatment techniques can be classified as:

(i) physical
(ii) chemical
(iii) biological.

The final choice depends upon the individual cases and local circumstances (Stanbury and Whitaker, 1993). Sometimes, the strategic approach may be a combination of them. The physical methods are used initially for filtering the suspended solids, etc., before going for biological treatments. The chemical methods are based on flocculation, and coagulation of the solids. Sometimes, the acidic/alkaline effluents are neutralised before being released into the environment, to reduce the salinity of the land. The biological treatment methods are based on aerobic or anaerobic processes. The BOD of the effluents is considerably reduced by various contacting methods before they are let out.

18.2 PHYSICAL METHODS

The physical methods are normally used for the removal of suspended solids. The most commonly used approach is **sedimentation** (Section 17.3). The suspended solids are allowed to settle to the bottom by gravitational force alone. It is a very economical and convenient method, but it takes a long time for sedimentation to take place. It does not work if the solid suspensions are very small in size. Laboratory techniques are available to evaluate the sedimentation velocity, and at what height the sedimented solids will attain limiting compressible strengths.

18.2.1 Open Ponds/Lagoons

Another slightly modified approach to sedimentation is to go for holding ponds, known as **lagoons.** If the suspended solids are not hazardous, the effluents are let into open ponds, where the solids settle to the bottom, and the upper water layer slowly evaporates. Even though it is a very economical treatment technique, the application of this method is subjected to the following:

- the suspended solids are not hazardous
- the space available is in abundance
- the local factory regulations do not prohibit this treatment.

18.2.2 Filtration Through Loose Beds

If the effluent contains very small quantities of fats/oils, as in the case of effluents coming from meat processing industries, they can be clarified by passing the effluent through a bed of loosely packed bale of hay and held in position in a flexible frame. The loose packing does not offer much resistance for fluid flow. When the filtering medium gets saturated/chocked, it is removed and incinerated.

18.3 CHEMICAL METHODS

These methods are also akin to physical methods, except that a chemical agent is used to affect the treatment of the effluent. Some of the methods include:

- flocculation (coagulation)
- adjustment of pH

- chlorination
- activated sludge treatment
- oxidation in trickle filters.

18.3.1 Flocculation/Coagulation

If the suspended solids are very fine and take a very long time to settle by sedimentation, we use a flocculating agent like alum (aluminium sulphate) or chitosan [(1,4)-2-amino-2-deoxy-β-D-glucan]. The flocculating agent reorients the electrostatic forces and flocculates the suspensions. They form bigger particles and hence will settle down easily. Coagulation is also similar to flocculation, except that this is an instantaneous process of settling, and does not require any agitation.

18.3.2 Chlorination

Effluents with low BOD are sometimes treated by passing chlorine gas through them or by treatment with calcium hypochlorite. This method is very effective if the treated water is to be reused (Webb, 1964). The chlorine levels are restricted to 1–10 ppm. If the dosages are high, it becomes costly, and the treated water will have an obnoxious smell of chlorine, which is normally discouraged.

18.3.3 Adjustment of pH

If the effluent waters go into nearby fields or into any rivers/sea, the pH of the effluent is to be adjusted to make it neutral, in the absence of which it may increase the alkalinity or acidity of the soil. Depending on the fact whether the effluent is acidic or basic in pH, it is neutralised to make neutral pH. Even though it is a costly process, it needs to be done to meet the statutory requirements.

18.3.4 Activated Sludge Treatment

This method is very common and very useful for treating industrial effluents and sewage. The effluent is vigorously aerated in an aeration tank (described in Section 18.4) with a sludge. The sludge is usually an inorganic salt like calcium hydroxide (lime) or ferrous/ferric sulphate, or zinc chloride, etc. This will enable the undesirable components in the effluent (pollutants) to precipitate as flocs. A part of the sludge is recirculated for effective utilisation.

18.3.5 Oxidation in Trickle Filters

The trickle filter is similar to a trickle-bed reactor, and owes its origin to trickle beds. A trickle bed operation involves the contacting of a gas and liquid in the presence of solids. The trickle bed filters also operate in the similar manner for the treatment of effluents of low BOD. The effluent is passed from the top through a fixed bed or loosely packed bed containing stones, or pebbles or some polypropylene or polyethylene grids of irregular shapes. Air is blown from the bottom. It is essentially a method of gas–liquid contacting with inert solid surfaces to increase the contact area. In addition to acting only as a facilitator for providing better gas–liquid contacting, the solid surfaces assist to form a **zoogloeal** film containing a wide range of oxidative microorganisms which help in reducing the BOD load considerably. It is a continuous process in which the effluent with low BOD load enters from the top of the trickle filter and the treated effluent is discharged from the bottom. It is a convenient and economical process (with less energy consumption) for treating the effluents with low BOD loads.

18.4 BIOLOGICAL METHODS

Most of the organic effluents are treated biologically using microorganisms to utilise the organic matter in the effluent. In the process, the microorganisms produce some useful products/gases which can be used as fuels. The basic purpose of it is not to make really useful products, but to degrade the organic matter in the effluents, so that it reduces the BOD load. The biological treatment methods are broadly classified as:

- anaerobic processes
- aerobic processes.

The type of fermentation depends upon the fermentable solute load (Webb, 1964). If the fermentable solute load is between 1 and 3%, anaerobic fermentation is economical and is frequently used. If the solute load is below 1%, one of the aerobic fermentation methods can be applied; and if the load is above 3%, it may be better to go for natural slow drying of the solids in yards/farms, and dried solids can be used as cattle-feed/poultry-feed. Alternatively, the effluent waste solid material can be digested using a blue-green algae for the production of biomass, which in turn can be used as fish-feed. This is an indirect way of converting the wastes from some of the fruit-processing industries into useful proteinaceous sources like fish for human consumption (Sunita and Rao, 2003). These methods will be discussed briefly in the following sections.

18.4.1 Anaerobic Fermentation

The anaerobic fermentation method is one of the classical and economical methods to digest the process effluents to produce methane gas using *Methano bacterium*. This is popularly known as **biogas** generation unit or **biogas plants.** The biogas is a mixture of CH_4 and CO_2 in the proportions of 1:3 to 1:1, with a little proportion of hydrogen gas. Biogas is produced under strict anaerobic conditions by many organisms, including the most popular *E. coli.* The biogas generation from the food industry wastes/effluents is described as follows.

18.4.2 Biogas Generation

Food-processing industries, particularly fruits and vegetable-processing industries generate a solid waste and some effluents. The solid wastes are almost 50% of the raw material processed. Their disposal poses a heavy burden on the processing units. Biogas generation is a prospective alternative which helps in mitigating the fuel crisis at domestic level as well as at industrial scale. This also helps in reducing the environmental pollution.

The fruit and vegetable-processing wastes contain (Krishna Nand, 1999):

Cellulose	:	30–40%
Hemicellulose	:	20–40%
Lignin	:	20–30%.

Thus, it is a good source of carbon and can be used to produce biogas by methanogenisis. Some of the important strains of methanogens (Krishna Nand, 1999) are:

Methanobacterium ruminantium
Methanothermo autotrophicum
Methanospirillium hungatti
Methanobacterium formicicum
Methanococcus vannieli
Methanosapcina barkeri
Methylomonas albus

Methylobacter capsulatus
Methylosinus trichosporium, etc.

The digester, maintained at 35–40 °C, produces the biogas after about 20 days retention time. The optimum hydraulic retention time is reported to be 25–30 days. The main substrates for methane generation are acetate, carbon dioxide, hydrogen, methanol, carbon monoxide, etc. But essentially for all of them, the precursor is acetic acid.

$$\underset{\text{(Acetic acid)}}{CH_3COOH} \longrightarrow \underset{\text{(Methane)}}{CH_4 + CO_2} \quad (18.1)$$

The biogas yield with most of the fruit and vegetable processing wastes is reported to be of the order of 0.6 m^3 of solids digested. Figure 18.2 shows bioconversion of organic residues into methane and carbon dioxide. Figure 18.3 shows an industrial biogas digester of a distillery with installed capacity of 4600 m^3 to produce about 10,000 m^3 of biogas per day from 300 m^3 of residual liquor of the industry.

The biogas generation units vary in size, depending upon the level of operation. At domestic scale it could be 4 m^3, and at industrial scale it could be anywhere between 100 and 500 m^3. The handling of wastes is in the range of 64 kg to 8 tonnes per day, yielding a gas production of 8–1000 m^3. The techno-economics of biogas production was given by Krishna Nand (1999), according to which, the cost of production of 1 m^3 of biogas varies between Rs. 1.5 and 1.8.

18.4.3 Bioconversion Using Microalgae

Wastes from some of the processing industries, particularly fruit-processing industries, contain a good amount of proteins and organic matter which can be a potential source for the growth of some blue-green algae. The biomass, so generated, in turn can be a good aqua-feed for growth of fish. Thus, it is a very potential indirect method of converting food processing industries wastes into useful proteinaceous source for human consumption in the form of fish. Sunita and Rao (2003) reported a method for bioconversion of mango-processing waste to fish-feed by microalgae isolated from the soil to which the fruit-processing industrial effluents were being discharged. The trials were restricted to laboratory-scale studies. The wastes from mango-processing industries were suitably diluted with water to make the overall solid contents to 20%. They were treated with blue-green algae (mostly filamentous form), viz.

- *Chlorella*
- *Gleotrichia*
- *Anabaena*
- *Oscillatoria*
- *Nostoc.*

The trials were carried out for 15–20 days to produce enough biomass in the laboratory in sterilised carboys of 20 l capacity in the presence of light (approx. 4000 lux). This resulted in reducing the COD to 6000 mg/l from an initial 96,000 mg/l. The harvested biomass (blue-green algae) was fed to a model fish *Tilapia mossambica.* The fish relished the biomass. The resulting increase in weight was considerable, and it was almost 2.5–3.0 times as compared to those fed on control diet.

Thus, blue-green algae, which have the potential to convert the wastes from the fruit-processing industries into biomass, are extremely useful for converting the wastes into useful products, and at the same time helping to reduce industrial pollution. This is an indirect route of converting the wastes into a protein source through aqua-feed for human consumption.

18.4.4 Suspended Solids Separation Techniques

The suspended solids in raw water and industrial waste-waters as well as sewage are removed to purify the above waters and subject them for further treatment. In water treatment, the suspended solids are objection-

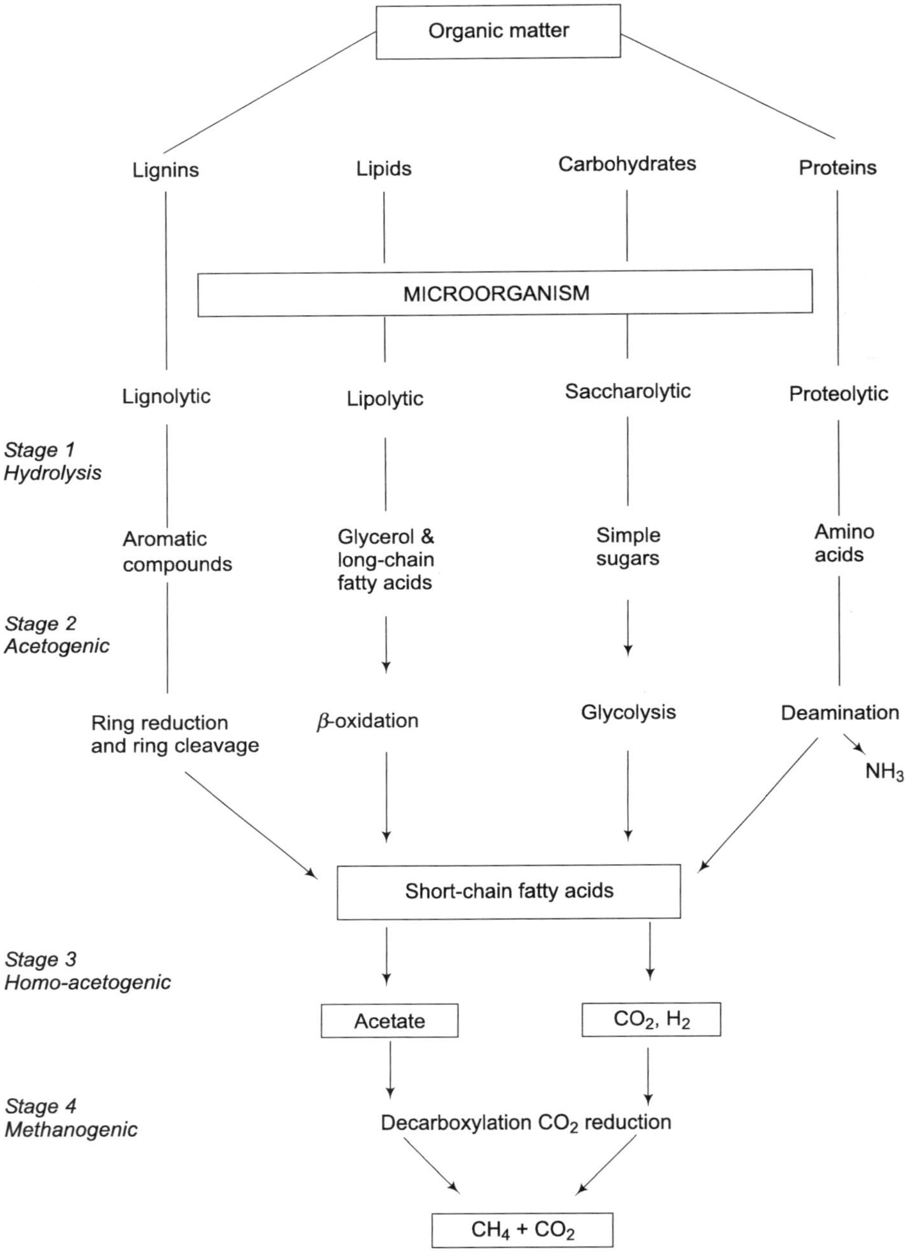

Fig. 18.2 *Bioconversion of organic residues into CH_4 and CO_2*
(Reprinted from "Biotechnology: Food Fermentation", Joshi , V.K., and Pandey, A., p. 1353, Copyright © (1999), with permission from Asiatech Publishers, New Delhi)

Fig. 18.3 *Industrial biogas digester of a distillery, with installed capacity of 4600 m³ to produce about 10,000 m³ of biogas per day*
(Courtesy: M/s. Nava Bharat Ferro Alloys Ltd., Deccan Sugars Division, Samalkot, AP)

able in drinking water supply. Hence, the following operations and mechanisms are used for simple gravity separation technique.

Settling tank These are used especially when the suspended solid concentration is more than 5000 mg/l during rainy seasons in the natural river waters. Here, the settling tanks are used with surface loading of 50–80 m^3/m^2 per day with detention time of 1 h. As the concentration of suspended solids is generally high, 90% of the suspended solids are removed in this process.

Clarifier The clarifier, as per nomenclature, clarifies the water from suspended solids in primary treatment of water and waste-waters from approx. 1000 mg/l. The efficiency of this process also depends on the addition of coagulant and pH adjustment with preferably 5% dilute lime solution. The surface loading for sizing of the unit in water treatment is in the range 24–40 m^3/m^2 per day. However, when the same unit is used as clarifier in sewage treatment or in waste-water treatment, it is called as primary clarifier. In case of primary clarifier in sewage and waste-water treatment plants, the surface loading will be between 20 and 30 m^3/m^2 per day, depending on the source industry and its characteristics. The primary clarifiers are also called as primary settling tanks. The detention time varies from 2 to 3 h.

The suspended solids are removed after the biological treatment, either by activated sludge process or by trickling filters. The surface loading will be lower as the biological solids to be removed will be lighter in density when compared to primary solids. Hence, it varies from 15 to 25 m^3/m^2 per day.

Tilted Plate Separator and Tubular Settler techniques were subsequently developed. They have the following advantages:

(i) less area requirements
(ii) minimum power/no power requirement
(iii) more efficient
(iv) minimum maintenance and operational costs.

Dissolved air floatation During the 1970's, dissolved air floatation process was used for the removal of suspended solids. It is similar to the air floatation technique used in mineral processing for extraction of minerals. The advantages are:

- less area requirement compared to clarifier
- settling is more efficient.

However, it has the disadvantage of high power requirements.

A common effluent treatment plant (CETP) concept Every processing industry needs to have an effluent-treatment plant to treat its effluents before letting into the environment. But, economically it may not be affordable for small processing units to have a separate effluent-treatment plant. Hence, the concept of common effluent-treatment plant (CETP) was mooted by the small-scale processing units. Thus, M/s. Pattancheru Enviro-Tech. Ltd., Hyderabad, a CETP for industrial wastes, mainly from pharmaceutical industries, has come into existence, and is operating with dissolved air floatation system which was supplied by Krofta Engineering, for the removal of suspended solids from the effluents received from a number of industries (Fig. 18.1). Here, a coagulant (alum) is used for helping in the removal, along with small dosages of polyelectrolyte. The efficiency of the treatment with respect to suspended solids is 60–80%, and the removal of BOD & COD is 5–10% (Akella, 2003). Detention time is only 2–3 min as against 2–3 h in clarifiers. The depth in the tank will be 0.6–0.7 m as against 2–4 m in clarifiers.

Description of components of the dissolved air floatation unit
(Ref.: Operation manual of Krofta Supercell)[1]

Inlet flow Air is dissolved into the water by means of Krofta air dissolving tube. Unclarified water is released through a valve. The water flows in at the exact centre through a rotary joint and into the distribution duct. Excess air is released through a vent pipe in the duct. The flow is regulated to eliminate turbulence.

Clarified water outlet The clarified water, near the bottom, is removed by slotted extraction pipes. The pipes are attached to the wall of the moving centre part. The water flows into the central section, which is isolated by a simple rubber seal at the bottom. The clean water then flows by gravity to the overflow weir located at the edge of the unit.

Removal of floated material The floated material/suspended solids are recovered from the top surface by means of the patented Krofta spiral scoop. The scoop is designed to remove the floated material at the highest possible consistency, with minimum surface disturbance. The scoop itself is moving on a welded flange, driven by a variable-speed drive motor and a drive wheel with sprocket and chain, while on the opposite side, the scoop is supported on wheels (Fig. 18.4).

The scoop discharges into a stationary central section, and the floated material flows from the centre, by gravity, to a sludge collection sump or storage tank. Units with more than 3.6 m^3 /MT capacity are equipped with two gear motors. In case of separate motors, the scoop is supported on the drive side as well, while the second drive goes directly on the drive wheel.

The level of water in the unit installation is controlled and adjusted by a telescopic tube. The level adjustment determines the spiral scoop dip and thus determines the quantity of sludge to be removed. There is a provision for a further adjustment of the sludge discharge. The adjustment is done according to stock quantity and desired stock consistency. It is important that floated stock/sludge is scooped and removed from the system before excess stock is built up on the surface.

---
[1] Based on the personal correspondence with Dr. V.R. Akella of CETP, Pattancheru, Hyderabad.

Fig. 18.4 *Sludge Decanter to scoop floated material from the Dissolved Air Floatation unit of Central Effluent Treatment Plant*
(Courtesy: M/s. Pattancheru Enviro-Tech. Ltd., Pattancheru, AP)

Removal of settled material Every floatation process produces a small amount of settled material which must be removed from the bottom to prevent clogging and improve water clarity. This is especially important when any sand, grit or scale is present. There is also a tendency for scum to build up on the sides of the tank.

The unit has rubber wipe blades built-in to remove any scum that accumulates. The settled material drops into the sump, specially provided and built in the tank. The solids, which collect in the sump, are purged through the bottom opening by means of an automatic pneumatic self-cleaning unit during continuous operation. The pneumatic valve is controlled by a three-way solenoid valve and connected with a timer to maintain the frequency and duration of discharge.

Installation support The unit is delivered with support which is welded to the tub and coated with paint. The support serves the installation as a stiffening base during the transport, and is welded to the segments of the unit in order to be flanged together at the customer's site. The steel girder frame, with the support and the unit itself, can be placed on pillars, and can be above floor construction. Maximum ground load is approx. 0.7 t/m^3, and there is practically no loss of operation space for stock and water recovery.

References

18.1 Abson, J.W., and Todhunter, K.H. (1967), "Effluent treatment", *Biochemical and Biological Engineering Science,* Blakebough, N. (Ed.), Vol. 1, Academic Press, London, pp. 309–341.

18.2 Akella, V.R. (2003), Personal Correspondence.

18.3 Blaine, R.K. (1965), "Fermentation products", *Industrial Waste Water Control,* Gurnham, C.F. (Ed.), Academic Press, New York, pp. 147–166.

18.4 Boruff, C.S. (1953), "Fermentation industries", *Industrial Wastes—Their disposal and treatment*, Rudolfs, W. (Ed.), Reinhold, New York, pp. 99–131.

18.5 Eckenfelder Jr., W.W., Goodman, B.L., and Englande, A.J. (1972), "Scale up of biological waste water treatment reactors", *Advances in Biochemical Engineering,* Ghose, T.K., Fiechter, A., and Blakebrough, N. (Eds.), Vol. 2, Springer-Verlag, Berlin, pp. 146–180.

18.6 Erichsen Jones, J.R. (1964), *Fish and River Pollution,* Butterworths, London.

18.7 Jackson, J. (1960), "The treatment of distillery and antibiotics wastes", *Waste Treatment,* Isaac, P.C.G. (Ed.), Pergamon Press, Oxford, pp. 226–239.

18.8 Koziorowski, B., and Kucharski, J. (1972), *Industrial Waste Disposal,* Pergamon Press, Oxford.

18.9 Krishna Nand (1999), "Biomethanation from agro-industrial and food processing wastes", *Biotechnology: Food Fermentation,* Joshi, V.K., and Pandey, A. (Eds.), Vol. 2, Educational Publishers and Distributors, New Delhi, pp. 1349–1372.

18.10 Stanbury, P.F., and Whitaker, A. (1993), *Principles of Fermentation Technology,* Pergamon Press, Oxford, pp. 220–225.

18.11 Sunita, M., and Rao, D.G. (2003), "Bioconversion of mango processing waste to fish-feed by micro-algae isolated from fruit processing industrial effluents", *J. Sci. Ind. Res.*, 62, 344–347.

18.12 Webb, F.C. (1964), *Biochemical Engineering,* D. Van Nostrand Co. Ltd., London, pp. 704–732.

Review Questions

18.1 Describe the importance of effluent treatment in bioprocessing.

18.2 What are the various effluent-treatment methods? Describe them briefly.

18.3 Describe the various physical methods of effluent treatment.

18.4 Describe the various chemical methods of effluent treatment.

18.5 Describe briefly the various biological methods of effluent treatment.

18.6 Explain anaerobic fermentation process.

18.7 Describe a biogas generation unit.

18.8 Describe how do you convert the wastes from some food-processing industries to a useful proteinaceous source.

18.9 Briefly explain a CETP concept.

18.10 Describe a suspended solids separation technique for processing industrial effluents.

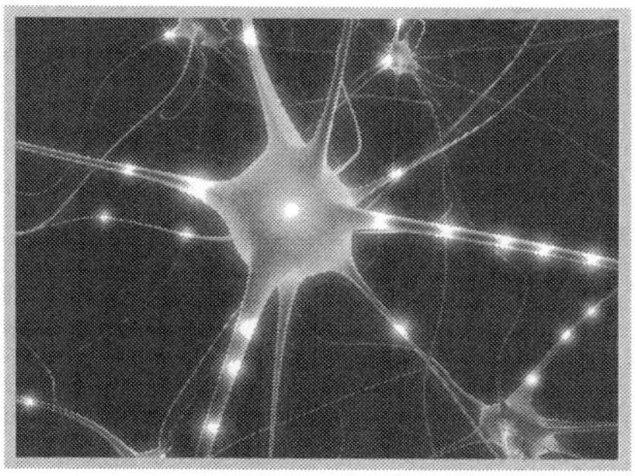

chapter 19

Design and Analysis of Bioreactors

The home for bioreactor design is the laboratory fermenter in which the kinetic data are collected. In general, the laboratory fermenter can be operated as close to ideal flow systems as possible under isothermal conditions, and the mass transfer interactions are deciphered. This calls for a situation where fluid dynamics are set aside for the time being, and reaction kinetics alone are considered. The laboratory data are analysed and processed to find out the kinetic model as described in Chapter 7.

Fogler (1986) gives an excellent treatise on the collection and analysis of reactor data, and suggests that one of the most important strategies to be adopted while collecting kinetic data in a laboratory fermenter is to employ the extreme (maximum and minimum setting) range of controlled variables. Frequently what is done (as against what should have been done) is that the choice of laboratory fermenter is not done on rational basis. Submerged laboratory fermentations are studied in a variety of equipment such as shake flasks, aerated jars and mechanically agitated fermenters, etc. Most of the laboratory fermenters are unsatisfactory because the operating conditions deviate drastically from commercial fermenters. What is required is that the laboratory fermenter should reasonably be a prototype of the commercial fermenter. Bartholomew et al. (1950) described the design and operational features of a laboratory fermenter (Fig. 5.4), and could achieve 6% more yield of streptomycin as compared to that in control shake flasks.

Subsequently, the kinetic model is to be combined with a transport model for the overall 'design' (Nauman, 1987). The primary objective of a fermenter design is that the fermenter should be designed in such a way that it should operate reliably, and be controlled so that optimum conditions prevail in spite of small fluctuations in operating parameters.

The systematically generated data of the laboratory fermenter is modelled by applying classical principles of chemical reaction engineering to account for multiphase growth. Some such parameters are:

- residence time
- segregation
- surface adsorption
- transport limitation
- reaction kinetics.

Attention needs to be focussed on certain characteristic features of biological systems and processes, viz.

- energitics

- stoichiometry
- asepsis
- kinetics
  - of cell growth
  - of substrate consumption
- morphology of cells.

Most of the above are not typical of chemical systems.

19.1 STABILITY AND ANALYSIS OF BIOREACTORS

The stability of a bioreactor is of vital importance in designing its control systems. A rigorous design procedure may lead us to an optimal design of a bioreactor, but if it is too sensitive for any of the process parameters, it gets destabilised frequently for any small change in parameters and becomes a nightmare for operation. Stable bioreactor operation is dictated by:

- physical parameters, viz. temperature, pH, DO concentration, etc.
- biochemical parameters, viz. cell growth kinetics and cell morphology.

19.1.1 Fermenter Dynamics

Let us first consider the dynamic characteristics of the bioreactors. We consider a CSTF (CSTR) for studying the dynamics. The procedure is applicable for any other type of flow of the fermenters also. A suitable dynamic model is proposed on the basis of unsteady state mass balance (Bailey and Ollis, 1986(a)):

$$\frac{d(VC_i)}{dt} = \upsilon(C_{i_0} - C_i) + Vr_{fi} \tag{19.1}$$

where V is the volume of the reactor
υ is the volumetric feed rate
r_{fi} is the rate of formation of component i in the reactor
C_i is the local concentration of component i in the reactor
C_{i0} is the concentration of component i in the feed.

The LHS term in Eq. (19.1) indicates the rate of accumulation of component i in the reactor. The first term in the RHS gives the difference in inlet and outlet flow of component i, whereas the second term in the RHS stands for the rate of formation of component i. Thus, Eq. (19.1) is the material balance equation (Eq. (8.2)). Dividing throughout by V and equating $\upsilon/V = D$ (the dilution rate), we get

$$\frac{dC_i}{dt} = D\,(C_{i0} - C_i) + r_{fi} \tag{19.2}$$

The above equation holds good for either cell growth or substrate conversion or any metabolic product formation. r_{fi}, which stands for rate of formation of component i, will have to be suitably modified for its sign (+ve or –ve), if it is rate of conversion. Thus, the unsteady state material equation forms the starting point for the fermenter dynamics.

The dynamic characteristics of the fermenter is a

- function of concentration changes ($f(C_i)$)
- function of parameter changes.

We shall examine, as an example, how the single limiting substrate concentration limits the growth of cells and the performance of the fermenter. Using Monod's expression for specific growth rate, and applying mass balance for cell growth and substrate consumption, we get

$$\frac{dx}{dt} = D(x_0 - x) + \frac{\mu_{max}\, sx}{K_s + s} \tag{19.3}$$

$$\frac{ds}{dt} = D(s_0 - s) - \frac{1}{Y_{x/s}} \frac{\mu_{max}\, sx}{K_s + s} \qquad \text{[Eq. (16.8)]}$$

In Eq. (19.3), if $x_0 = 0$, it indicates that the feed is sterile. This may give rise to two cases, what we call as

(i) non-trivial steady state solutions, given earlier by Eqs. (16.9) and (16.10)

or

(ii) a "wash-out" steady state situation in which $x = 0$ and $s = s_0$, i.e. there is no build-up of cells in the fermenter, and hence $x = 0$. Then, all the substrate with concentration s_0 entering the fermenter will be leaving as it is, so that $s = s_0$.

This calls for two situations, in which

(i) $$D > \frac{\mu_{max}\, s_0}{K_s + s_0}$$

or

(ii) $$D < \frac{\mu_{max}\, s_0}{K_s + s_0}$$

The steady state criteria for a Monod chemostat model under steady state with total wash-out conditions are (Bailey and Ollies, 1986 (a))

(i) if $D > \dfrac{\mu_{max}\, s_0}{K_s + s_0}$,

the chemostat performance is stable. Otherwise,

(ii) the chemostat will be unstable when $D < \dfrac{\mu_{max}\, s_0}{K_s + s_0}$

19.1.2 Biomass Production and Dilution Rate

In Eq. (19.3), at steady state conditions, $dx/dt = 0$ and hence,

$$\left[\frac{\mu_{max}\, s}{K_s + s} - D\right] x + D x_0 = 0 \tag{19.4}$$

For a sterile feed, $x_0 = 0$. Hence, Eqs. (19.4) and (16.8) can be solved to yield Eqs. (16.9) and (16.10)

$$x_{ss} = Y_{X/S}\,(S_R - s_{ss}) \qquad \text{[Eq. (16.9)]}$$

$$s_{ss} = \frac{K_S\, D}{\mu_{max} - D} \qquad \text{[Eq. (16.10)]}$$

The steady state conditions are dependent on the dilution rate D, and hence on the flow rate. If $D \to 0$, s_{ss} tends to zero. Then, the cell mass concentration x_{ss} is equal to $Y_{x/s}\,(S_R)$.

Let us see what happens if D increases. As D increases, s increases proportionately in the initial stages. After a certain stage (at point P in Fig. 19.1), it increases sharply because of the combined influence of increasing numerator value and decreasing denominator value in Eq. (16.10) It is shown in Fig. 19.1. Similarly, the cell mass concentration will also be decreasing linearly in the initial stages. After a certain stage (point P), since s_{ss} increases sharply, x_{ss} decreases sharply. As D approaches μ_{max}, s_{ss} becomes very high, and

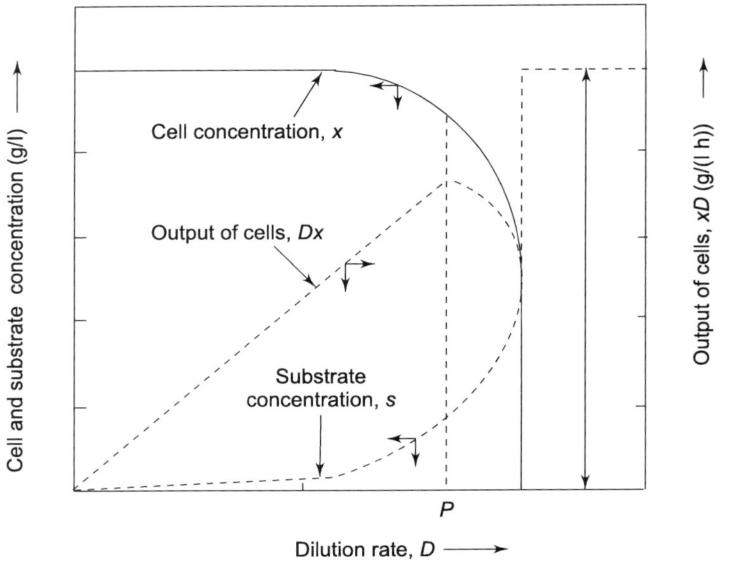

Fig. 19.1 *Effect of D on x, s and output of cells*
(Reprinted from "Biochemical Engineering Fundamentals", Bailey, J.E., and Ollis, D.F., p. 387, Copyright © (1986), with permission from McGraw-Hill, Inc., New York)

hence x_{ss} becomes almost zero. This indicates that all the cells are lost at steady state; this situation is known as 'wash-out'. Here, D_{max} will be

$$D_{max} = \frac{\mu_{max}\, S_R}{K_S + S_R} \tag{19.5}$$

Similarly, the output of the cells, which is a product of D and x, will increase sharply up to point P, and then falls steeply. Hence, Bailey and Ollis (1986(b)) suggest that we should forgo the maximum value of cell production to avoid the region of large sensitivity. Any increase in the value of D at this stage results in sharp fall of cell production. The value of D at point P may be considered as **critical dilution rate.**

19.1.3 Thermal Stability

Thermal stability of CSTR was described to some extent in Section 9.4.2. Figure 9.11 shows a heat generation curve in a sigmoidal form for a CSTR (or CSTF) given by Eq. (9.41), and Fig. 9.12 shows the heat removal lines given by Eq. (9.43). In Fig. 9.12, for example, point f indicates very low heat generation rate. Denbigh and Turner (1971 (a)) analysed this type of state which is favoured by

- low t_{Av}
- low reaction rate constants
- low heat of reaction (ΔH_r)
- low operating temperatures
- higher heat removal rates by means of high values of A or U.

Similar type of observations hold good even for point g (Fig. 9.12) in the opposite direction.

If we analyse line 2 with multiple points a, b and c, point a is similar to point f, and point c is similar to point g. But the analysis of point b is very pertinent. No doubt that the situation is stable here. But any slight increase in temperature at point b would make the heat generation rate to be more than the heat removal rate, and push the stability to point c. Similarly, any small fall in temperature at point b would decrease the temperature, and bring the stability level to point a. Thus, point b exhibits very unusual features in terms of thermal stability.

Mostly we do not come across the thermal stability problems in fermenters, since the biochemical reactions are not highly exothermic. The type of thermal stability analysis presented here is more relevant for highly exothermic reactions like oxidation of hydrocarbons, manufacture of phthalic anhydride, etc.

19.2 APPLICATION OF DESIGN TO CONTINUOUS STERILIZERS

Continuous sterilization process was discussed in Section 6.5, including its advantages and disadvantages, and various types of equipment used. In this section, we shall consider some of the design features. In the continuous sterilisation process, the medium or substrate (in this section, we mostly consider crude medium only for sterilisation) is passed through a tubular section in which the contents are sterilised by heating (Fig. 6.9). It is equivalent to flow through pipes like a PFR. As can be expected, when the medium is passing through a tubular section, different elements of the medium will have different ages of their stay in the tube, and hence will be exposed to heat to different time periods. The situation is almost similar to non-ideal flow systems in which different elements or packets of fluids will have different residence times (Aiba et al., 1965(a)).

19.2.1 RTD Concept

In other words, we come across distribution of residence times, which we call as residence time distribution (RTD) for different elements of fluids. The details of RTD studies were discussed in Chapter 10.

The velocity distribution for fluids flowing through pipes is a known concept. For viscous fluids, the mean velocity (average velocity) is half of the maximum velocity which is at the centre of the pipe. In turbulent flow conditions, the average velocity is 82% of the maximum velocity (Coulson et al., 1993). The velocity profiles for flow through pipes are similar to those shown in Fig. 10.5 for the dispersion model. For this type of flow situation, we expect that the dispersion model (Section 10.4.1) for non-ideal flow would be adequate to describe the flow pattern and the RTD. The dispersion model for the tracer concentration is given by Eq. (10.16):

$$\frac{\partial C}{\partial \theta} = \left[\frac{D_e}{uL}\right]\frac{\partial^2 C}{\partial z^2} - \frac{\partial C}{\partial z}$$

in which the tracer concentration is simulated to the microbial concentration. The kinetics of sterilisation is described by Eq. (6.1) and the solution is given by Eq. (6.2):

$$\frac{N_t}{N_0} = e^{-kt}$$

which is equivalent to first-order kinetics for which we write

$$\frac{C_A}{C_{A0}} = e^{-kt}$$

where C_{A0} and C_A are, respectively, initial concentration and concentration of reactants present at any time. They are equivalent to N_0 and N_t in Eq. (6.2) which are, respectively, the initial number of cells present per unit volume and the number of cells present at time t per unit volume. Hence, the concentration profile based on the dispersion model equation (Eq. (10.16)) for first-order kinetics is given by Eq. (10.25). The fractional concentration (N_t/N_0) versus the reduced time θ ($= t/t_{Av}$) can be drawn in the form of curves which form a range between the PFR (when $N_{Pe} = \infty$) and CSTR (when $N_{Pe} = 0$) (Aiba et al., 1965(a)).

$$\frac{N_t}{N_0} = \frac{4q \exp(N_{pe}/2)}{(1+q)^2 \exp(\delta) - (1-q)^2 \exp(-\delta)} \tag{19.6}$$

where $q = (1 + 4(N_{Da}/N_{Pe}))^{1/2}$

and $\delta = N_{Pe}\, q/2$

The degree of sterility, as calculated by the above equation, is plotted against N_{Da} for various Pecklet numbers, and is shown in Fig. 19.2 (Aiba et al., 1965(a)).

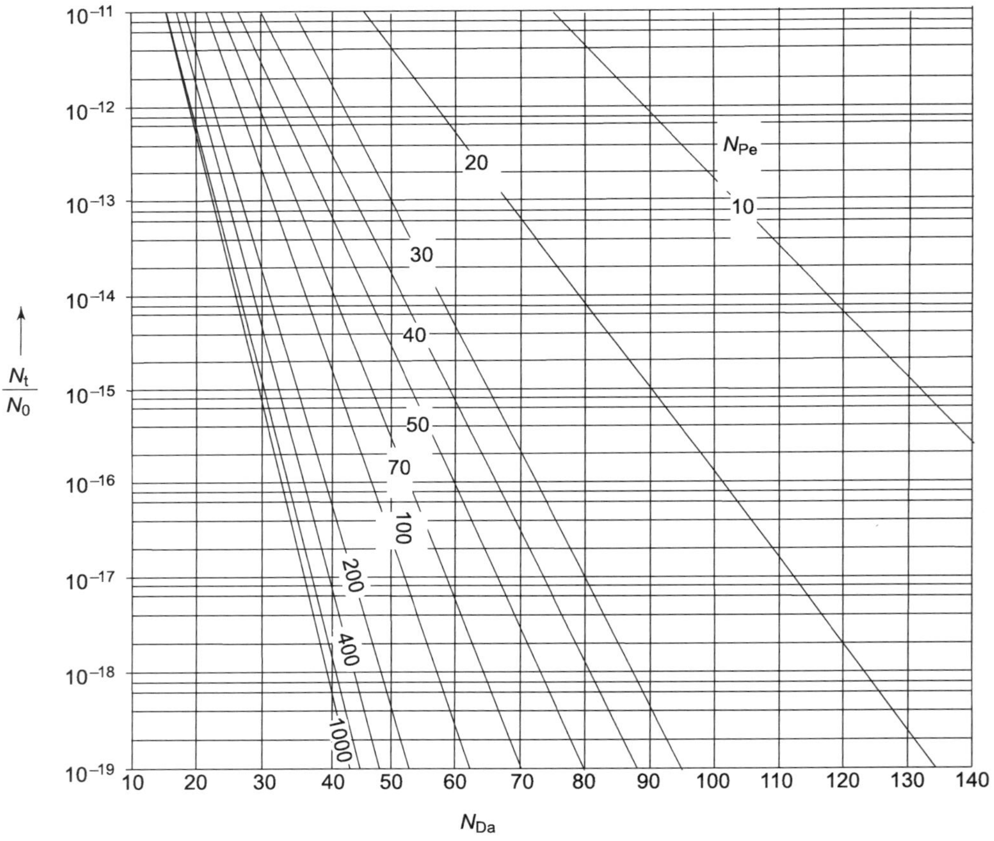

Fig. 19.2 *N_t/N_0 versus N_{Da} data for different N_{Pe}*
(Reprinted from "Biochemical Engineering", Aiba, S., Humphrey, A.E., and Millis, N.F., p. 212, Copyright © (1965), with permission from Elsevier)

19.2.2 Applications of Design Principles

The application of design principles was well-described by Aiba et al. (1965(a)), with an example for the evaluation of heat requirement (or steam requirement) for achieving a given level of sterility.

From the volumetric flow rate (or mean flow rate) of the medium, we can calculate u_{Av} by dividing the flow rate with the cross-sectional area of the pipe (A).

Then, $$u_{Av} = \frac{v}{A}$$

From the physico-chemical parameters of the medium, the Reynold's number (N_{Re}) is calculated as

$$N_{Re} = \frac{\rho d u}{\mu_f} \tag{19.7}$$

where ρ is the density, μ_f the viscosity, u the velocity of the medium, and d is the diameter of the pipe. From the published data in Fig. 19.3 on N_{Re} versus D_e/ud, by Levenspiel (1958), we can find the value of D_e/ud. With a knowledge of the diameter of the pipe and length of the pipe, D_e/uL can be found by

$$\frac{D_e}{uL} = \frac{D_e}{ud} \times \frac{d}{L}$$

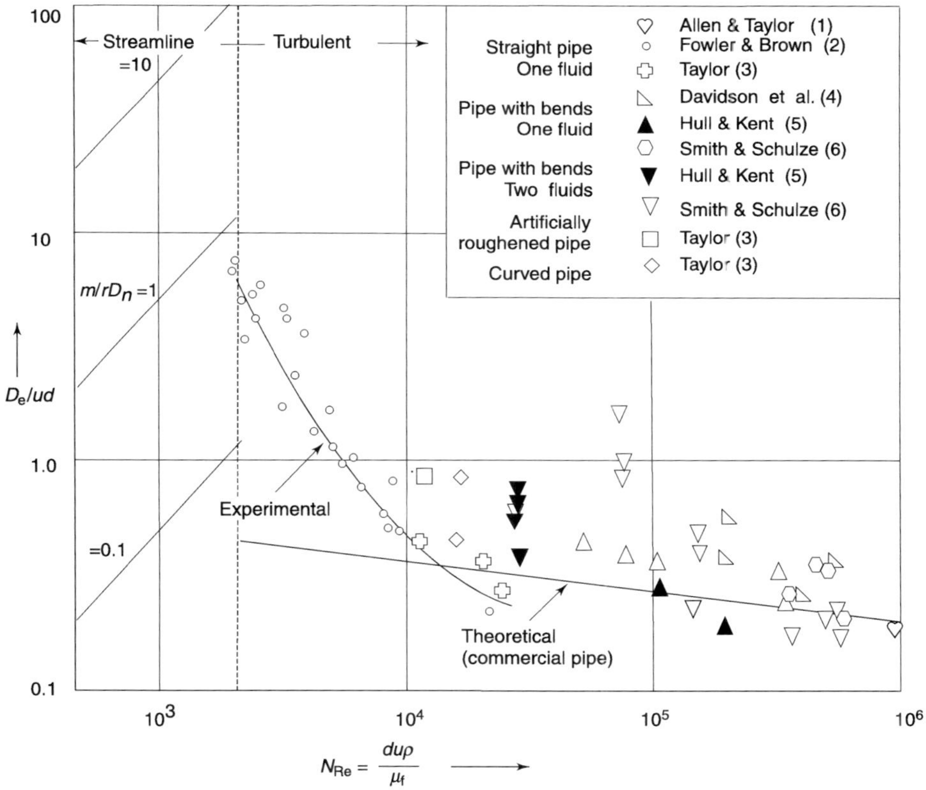

(1) Allen, C.M.,and Taylor, E.A. (1923) *Trans. Am. Soc. Mech. Engrs.*, 45, 285;
(2) Flower, F.C., and Brown G.G. (1943), *Trans. Am. Inst. Chem., Engrs.* 39, 491.
(3) Taylor, G.I. (1954), *Proc. Roy. Soc*, 223 A, 446.
(4) Davidson, J.F., Farquharson, D.C., Picken, J.Q., and Taylor, D.C. (1955), *Chem. Eng. Sci.*, 4, 201;
(5) Hull, D.E., and Kent, J.W. (1952), *I & EC*, 44, 2745.
(6) Smith, S.S., and Schulze, R.K. (1948), *Petroleum Engr.* 19, 94.

Fig. 19.3 *Correlation for longitudinal dispersion number as a function of N_{Re} (Reprinted from "Longitudinal mixing of fluids flowing in circular pipes", Levenspiel, O., I & EC, Vol. 50, p. 345, Copyright © (1958), with permission from American Chemical Society)*

From the sterility level required (it is normally 10^{-3}, which means a contamination of 1 in 1000) and the contamination level present, N/N_0 can be calculated. Using N/N_0 values and the Pecklet number, we can find out N_{Da} from Fig. 19.2.

$$N_{Da} = \frac{D_e}{ud}$$

from which k (with the units of min^{-1}) can be calculated. From the value of k, we can calculate the temperature by Arrhenius equation (Section 6.4.2)

$$k = k_0 e^{-E/RT}$$

If the feed is an unsterile medium containing predominantly *Bacillus stearothermophilus* spores, the following values for E and k_0 can be used (Deindoerfer and Humphrey, 1959):

$$E = 283 \text{ kJ/(g mol)} = 2.83 \times 10^5 \text{ J/(g mol)}$$

and

$$k_0 = 1 \times 10^{36.2} \text{ s}^{-1} = 9.51 \times 10^{37} \text{ min}^{-1}$$

$$R = 8.3144 \text{ J/(g mol K)}$$

Having known the temperature of sterilisation (T_s) and the initial feed (medium) temperature (T_0), we can calculate the total heat requirement (q_s, in kJ) for sterilisation of the medium by the expression

$$q_s = mC_p(T_s - T_0) \tag{19.8}$$

where m is the quantity of medium in kg
C_p is the specific heat of the medium in kJ/(kg °C).

From the value of q_s, we can find out the steam requirement for sterilisation based on the latent heat of vapourisation of steam at its available pressure, which is normally of the order of 0.5–0.9 MN/m^2,

$$q_s = m_s \times \lambda_s \tag{19.9}$$

where m_s is the quantity of steam requirement (in kg) and λ_s is the latent heat of vapourisation of steam (in kJ/kg).

Example 19.1

A particular fermentation medium is sterilised in a continuous sterilization system by passing the medium at the rate of 80 l/min through a hot pipe of 5 cm diameter. The medium has the following physico-chemical properties:

density of the fluid : 1015 kg/m³

viscosity of the fluid : 0.01 Pas

Find the dispersion coefficient for the flow using Fig. 19.3.

Solution Given:

$$\rho = 1015 \text{ kg/m}^3$$

$$\mu = 0.01 \text{ Pa s}$$

$$d = 5 \text{ cm} = 0.05 \text{ m}$$

$$v = 80 \text{ lpm} = 4.8 \text{ m}^3\text{/h}$$

$$A = \frac{\Pi}{4} d^2 = \frac{\Pi}{4}(0.05)^2 = 1.963 \times 10^{-3} \text{ m}^2$$

$$u = v/A = 4.8/(2 \times 10^{-3}) = 2.39 \times 10^3 \text{ m/h} = 0.66 \text{ m/s}$$

$$N_{Re} = \frac{\rho d u}{\mu} = \frac{1015 \times 0.05 \times 0.66}{0.01} = 3349.5 \approx 3350$$

$\frac{D_e}{ud}$ is noted from Fig. 19.3 for N_{Re} corresponding to 3350 *ud*

$$\frac{D_e}{ud} = 1.1$$

$$\therefore \quad D_e = 1.1 \times 0.66 \times 0.05 = 0.0363 \text{ m}^2/\text{s}$$
$$= 3.63 \times 10^{-2} \text{ m}^2/\text{s}$$

Example 19.2 *The above steriliser, with a pipe length of 50 m, raises the temperature of the feed instantaneously to sterilisation temperature. The feed predominantly contains Bacillus stearothermophilus at a level of 1.72 × 10⁵ number of cells/ml and needs to be sterilised to a tolerable contamination level (1 in 1000). Find the sterilisation temperature.*

Solution Given:

$$N_o = 1.72 \times 10^5 \text{ cells/ml} = 1.72 \times 10^{11} \text{ cells/m}^3$$
$$N_t = 1 \times 10^{-3}$$
$$L = 50 \text{ m}$$
$$u = 0.66 \text{ m/s} \quad \text{(from E19.1)}$$
$$d = 0.05 \text{ m} \quad \text{(from E19.1)}$$
$$\frac{D_e}{ud} = 1.1 \quad \text{(from E19.1)}$$
$$= \frac{D_e}{u \times L} \times \frac{L}{D} = \frac{D_e}{uL} \times \frac{50}{0.05}$$

$$\therefore \quad \frac{D_e}{uL} = \frac{1.1 \times 0.05}{50} = \frac{1}{1100}$$

$$\text{Peklet No.} = N_{Pe} = \frac{uL}{D_e} = 1100$$

$$\frac{N_t}{N_o} = \frac{1 \times 10^{-3}}{1.72 \times 10^{11}} = 5.81 \times 10^{-15}$$

From Fig. 19.2, we note the value for Damkohler number (N_{Da}) corresponding to $N_t/N_o = 5.81 \times 10^{-15}$ and for Pecklet No. 1100

$$N_{Da} = 32$$
$$= k\,L/u = k \times 50/0.66$$

$$\therefore \quad k = \frac{32 \times 0.66}{50} = 0.422 \text{ s}^{-1} = 25.3 \text{ min}^{-1}$$

k, in turn, is equal to $k_o\, e^{-E/RT}$ by Arrhenius equation; where

$$k_o = 9.51 \times 10^{37} \text{ min}^{-1}$$
$$E = 2.83 \times 10^5 \text{ J/(g mol)}$$
$$R = 8.3144 \text{ J/(g mol K)}$$

$$\therefore \quad k = 25.3 = 9.51 \times 10^{37} \times \exp(-2.83 \times 10^5/(8.3144T))$$

$$-\frac{34037}{T} = -84.2$$

$$\therefore \quad T = 404 \text{ K} = 131°\text{C}$$

Example 19.3 *For the above steriliser, find the quantity of heat required for sterilisation if the initial feed temperature is 28°C and has an average specific heat of 4.5 kJ/kg°C.*

Solution Given:

$C_P = 4.5$ kJ/kg°C

Feed Temp. = 28°C

Vol. flow rate of medium = 4.8 m^3/h

Density of medium = 1015 kg/m^3

Mass flow rate, $m = 4.8 \times 1015 = 4872$ kg/h

Sterilization temp. = 131°C

$$q = \dot{m}\ C_P\,(\Delta T)$$

$$= 4872 \times 4.5 \times (131 - 28) = 2.26 \times 10^6 \text{ kJ/h}$$

19.3 DESIGN AND OPERATION OF BIOREACTORS

Bioreactor is the heart of any biochemical process in which the microbial, enzymatic, fungal, mammalian or plant cell systems are grown and engaged for the production of various useful and economic products, by exploiting the nature of living cells and constraining them by the environmental conditions to perform and produce useful metabolic products from the substrate under mild (mostly) conditions. Biochemical reactors or bioreactors can be broadly classified into the following two main categories:

(i) microbial fermenters
(ii) enzyme (cell-free) reactors.

Both the types have some distinct features and some similar characteristics. In case of microbial processes, the cell concentration will also be increasing at the expense of the substrate as the fermentation continues, and hence a mixed flow reactor is more suitable. In case of enzymatic reactions, the enzyme concentration does not change with the progress of time, which is an indication of non-autocatalytic type, and hence a plug flow reactor is more suitable. The time versus concentration profiles are shown in Fig. 19.4.

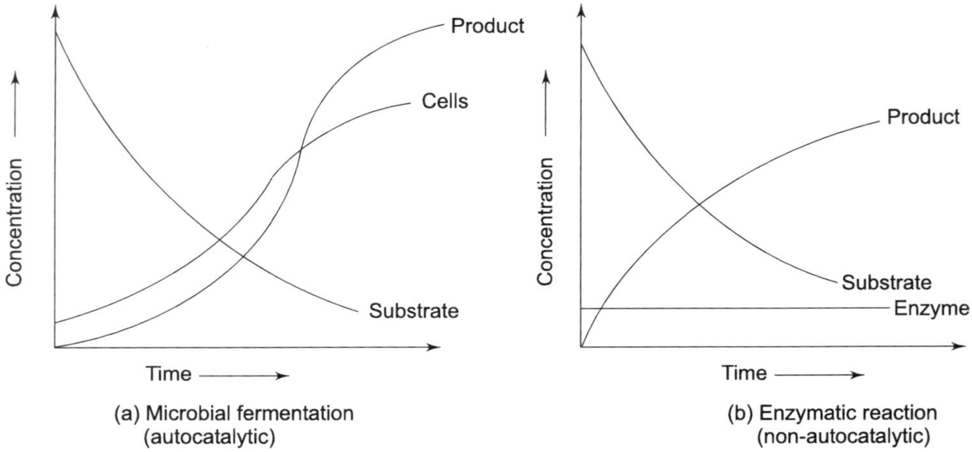

Fig. 19.4 *Concentration profile in a batch fermenter for (a) microbial fermentations, and (b) enzymatic reaction process*

Depending upon the process requirements they may be classified as

- aerobic
- anaerobic
- solid state
- immobilised.

The various steps involved in the biotechnological processes can be schematically represented by Fig. 19.5, which is similar to Fig. 1.1. Since various factors are involved in the performance of a bioreactor, its operation is complicated and depends upon the following functions (Cliffe, 1988; Moo-Young and Blanch, 1981):

(i) cell growth rate and concentration, which need to be high in case of growth-associated fermentation processes
(ii) maintenance of sterile conditions
(iii) effective mixing by agitating systems, so that the substrate and microorganisms are kept in a state of suspension
(iv) effective heating/heat removal systems
(v) maintenance of optimal shear rates (higher shear rates are detrimental for the cells, whereas lower shear rates will result in cells flocculating together, with the result that their entire surface area is not effectively utilised)
(vi) adequate control of various process parameters for effective function
(vii) the bulk densities of the suspended cells and the substrate particles are comparable to that of the liquid environment, and hence relative flow between the dispersed phase and continuous phase becomes difficult
(viii) the microbial cells are very small in size (few microns), and hence attaining high particle Reynolds numbers is not possible, which in turn makes it difficult to maintain turbulent conditions
(ix) many multicellular microorganisms (especially fungi) form large cell lumps such as mycelia, and they complicate the intra-particle mass transfer
(x) the concentrations of the reactants/products are low, which pose problems in mass transfer as the concentration driving force is very small
(xi) unlike chemical reaction rates, the microbial growth rates are very low, and hence they require large reactor volumes and residence times.

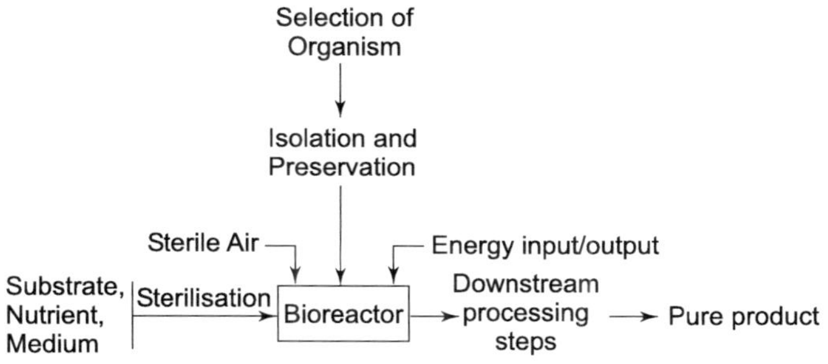

Fig. 19.5 *Schematic representation of bioprocess with the bioreactor occupying the central place*

The mass transfer rates are low, and diffusional limitations invariably complicate the mass transfer process. Particularly in case of aerobic processes, the mass transfer rate, measured by k_La, varies between 0.02 and 0.25 s^{-1} only, and the bubble size should be between 1 and 3 mm.

The design features of bioreactors may be categorised as:

- mass transfer aspects
- bioreactor type, choice and design.

19.3.1 Mass Transfer Aspects

The basic design equation for mass transfer in the case of a gas–liquid system is

$$N_A = k_L a\,(C_{AL}^* - C_{AL}) \tag{19.10}$$

where N_A = mass transfer rate (kg or mol/m^3 s)

k_L = liquid side mass transfer coefficient (m/s)

a = interfacial area (m^2/m^3)

C_{AL}^* = equilibrium concentration of the gaseous component A in the liquid phase (kg or kg mol/m^3)

C_{AL} = concentration of gaseous component A in the liquid phase (kg or kg mol/m^3)

The $k_L a$ values are dependent on various factors, viz.

- agitation rate
- gas velocity
- bubble size
- gas sparging system
- temperature
- viscosity of the broth
- cell type
- cell morphology.

The mass transfer aspects in the design of bioreactors were dealt in detail by Moo-Young and Blanch (1981) and Erickson and Stephanopoulos (1987). Some details were given in Ch. 14.

19.3.2 Bioreactor Type and Design

The bioreactors are classified according to the type of flow conditions as follows, which is akin to chemical reactor theory:

- batch fermenter (BF)
- continuous stirred tank fermenter (CSTF)
- tubular reactor (fermenter (TF))
- fed-batch fermenter
- fluidised bed fermenter (FBF).

A brief introduction about the various types of bioreactors was given in Section 5.7. They were discussed in detail, along with the design equations, in Chapter 8. Multiple reactor systems (combination of reactors) were also highlighted. The choice of type of reactor for different types of reaction schemes was indicated in Chapter 9.

The basis of design is to find out the size of the fermenter and time of fermentation, which are in turn obtained from the material balance and energy balance equations (Eqs. (8.2) and (8.3)).

We shall now briefly discuss here some bioreactor types.

19.3.3 Continuous Stirred Tank Bioreactor

A brief description about the continuous stirred tank reactor (CSTR) was given in Section 5.7.1 and a schematic representation of it was given in Fig. 5.6(c). Subsequently, the design equations based on the material balance were described in Section 8.2.2. The volume of the CSTF is given by Eq. (8.28) and the residence time of the reactants in the fermenter by Eq. (8.27).

As has been mentioned earlier, the basic assumption in the case of an ideal CSTF is that, the moment the contents (it could be the sterile media, nutrients or the substrate) enter the CSTF, they are thoroughly mixed, and the concentration of the reactants fall suddenly to the concentration level existing in the fermenter, which also happens to be the lowest, and outlet concentration of the reactants/components. This assumption leads to the conclusion that the concentration of the reactants are lowest in the reactor, and hence the rates of reaction will be lowest in the CSTF as compared to the batch or plug flow reactor (PFR). In fact, for certain types of parallel reactions (Section 9.1.2), in which the reaction order of the desired reaction is less than that of the undesired reactions ($a_1 < a_2$), CSTF is a better choice than other types, since it maintains lowest concentrations. The best part of CSTF is that, since the contents are thoroughly mixed, the cells are kept in a very good state of suspension, and hence can be effectively utilised. Another point which weighs in favour of CSTF is its accessibility to clean inner parts (unlike PFR), so that high degree of asepsis and cleanliness can be maintained.

Bischoff (1966) described the procedure for an optimal design of a continuous reactor, based on the assumption that microbial processes are akin to autocatalytic reactions of chemical reaction engineering. The residence time in the reactor is calculated based on the material balance equation

$$t_m = \frac{x - x_0}{r_X(x, s)} = \frac{s - S_R}{r_s(x, s)} \tag{19.11}$$

where x is the concentration of the cells in the CSTF
x_0 is the initial concentration of the cells in the CSTF
s is the substrate concentration
S_R is the initial substrate concentration
$r_x(x, s)$ is the rate of cell multiplication
$r_s(x, s)$ is the rate of conversion of substrate.

Bischoff (1966) suggests that it is reasonable to assume that

$$r_x(x, s) = r_x = \text{rate of microorganisms formation} \tag{19.12}$$

Then, the equation for a CSTF is

$$t_{m,Av} = \frac{x - x_0}{r_x} \tag{19.13}$$

Usually, a stoichiometric relationship between the microorganisms and substrate concentration can be represented as follows (Herbert, 1961):

$$\frac{dx}{ds} = -\beta(x, s) \tag{19.14}$$

where β is a stoichiometric ratio. Equation (4.14), on integration, gives a relationship for the substrate concentration as a function of the concentration of microorganism

$$s = s(x) \tag{19.15}$$

The Michaelis–Menten form of rate equation (Eq. (4.14)) can be used for autocatalytic biological reactions, and by incorporating this into Eq. (19.13), we get (Bischoff, 1966):

$$t_{m,\,Av} = \frac{x - x_0}{v_{max}\, x} \times \frac{\beta(K_m + \sigma) - x}{\beta\sigma - x} \tag{19.16}$$

where σ is the constant of integration obtained by integrating Eq. (19.14). For a case of single substrate and addition of microorganism at the initial concentration, we have

$$\sigma = S_R + x_0/\beta \qquad (19.17)$$

Thus, Eq. (19.16) may be considered as the design equation for the average residence time after incorporating the Michaelis–Menten kinetic equation into the usual material balance equation for a CSTF.

Bischoff (1966) explained this with a numerical example, for a typical case of lactic acid fermentation, using *Lactobacillus delbrueckii*, for a single CSTF and a combination of reactors. Value of K_m/σ was taken as 0.5, and conversion of input substrate to organisms was 95% with 3% initial concentration of microorganisms. He reported the total holding time (residence time) for different combinations of reactor systems as given in Table 19.1.

Table 19.1

Reactor combinations	*Total time in reactors* = $v_{max}\, t_{Av}$ *(dimensionless)*
One CSTF	11
Two equal CSTFs	5.48
Two optimum CSTFs	5.23
CSTF + PFR	3.97

It can be noted from the above table that, two CSTFs are better than single back mix reactor. However, holding time is not much different for two equal-sized CSTFs and two optimum-sized CSTFs. A combination of back mix reactor and plug flow reactor (CSTF + PFR) is the best combination with least residence time, and hence obviously the optimum-sized reactor configurations.

Example 19.4

A continuous stirred tank (fully mixed) reactor is used for an enzymatic process in which the conversion of the substrate follows the Michael–Menten kinetic path with

$K_m = 1.44$ *kg/m*^{3}

maximum specific growth rate (v_{max}) = 0.801*g/m*^{3}*.s*

$x_o = 0.6$ *kg/m*^{3}

$x = 16.5$ *kg/m*^{3}

The original substrate concentration (S_R) = 12.6 *kg/m*^{3}

The stoichiometric ratio factor (β) = 0.45

Find the average residence time in the CSTR.

Solution It is a direct application of Eq. (19.17) and Eq. (19.16)

Given:

$$K_m = 1.44 \text{ kg/m}^3$$
$$v_{max} = 0.801 \text{ g/(m}^3\text{ s)} = 0.801 \times 10^{-3} \text{ kg/(m}^3\text{ s)}$$
$$x = 16.5 \text{ kg/m}^3$$
$$x_o = 0.6 \text{ kg/m}^3$$
$$S_R = 12.6 \text{ kg/m}^3$$

Stoichiometric ratio factor, $\beta = 0.45$

By Eq. (19.17), we have

$$\sigma = S_R + (x_o/\beta)$$
$$= 12.6 + (0.6 / 0.45) = 13.93 \text{ kg/m}^3$$

By Eq. (19.16), we have

$$t_{m,AV} = \frac{x - x_o}{v_{max} \cdot x} \times \frac{\beta(K_m + \sigma) - x}{\beta\sigma - x}$$

$$= \frac{16.5 - 0.6}{16.5 \times 0.801 \times 10^{-3}} \times \frac{[0.45(1.44 + 13.93)] - 16.5}{(0.45 \times 1393) - 16.5}$$

$$= 1126.8 \text{ s} = 0.312 \text{ h}$$

19.3.4 Fed-batch Bioreactor

Fed-batch or semi-batch bioreactor was described earlier in Section 5.7.1 and various feeding conditions were shown in Fig. 5.7. They have some of the obvious advantages of a CSTF, while eliminating some of its disadvantages. Since all the reactants are not fed at one time, fed-batch operation results in higher conversion to the desired intermediate product as in series reactions/series–parallel reactions (Section 9.3). Hence, it will not be discussed here, except mentioning the material balance on component i (Bailey and Ollis, 1986(a)):

$$\frac{dC_i}{dt} = \frac{\upsilon(t)}{V_R}(C_{i0} - C_i) + r_{fi} \tag{19.18}$$

where $\upsilon(t)$ indicates the volumetric feed rate at time t, V_R is the culture volume and is a function of time, and r_{fi} is the rate of reaction of component i. If a suitable analytical expression is available for rfi, Eq. (19.18) can be integrated.

19.3.5 Air-lift Bioreactor

Air-lift bioreactor or fermenter is yet another class of reactor in which air is essentially used as the dispersion phase and is used to circulate the contents (it could be medium or cells) in the reactor. Thus, air-lift bioreactors are used only in aerobic processes. Some description about air-lift reactors was given in Section 5.7.2. As is shown in Fig. 5.9, they contain typically a draught tube through which air enters at the bottom and passes at a high velocity pushing through the fermentation broth. The buoyancy force exerted by the rising bubbles causes a drag on the surrounding fluid and drives the broth up the tube. This causes adequate mixing. When the air reaches to the top, much of it gets disengaged from the bulk liquid and escapes through the outlet. After the air leaves the bulk liquid, the density of the liquid increases, and hence it comes down along with a minimum amount of dissolved air through other channels, causing adequate recirculation.

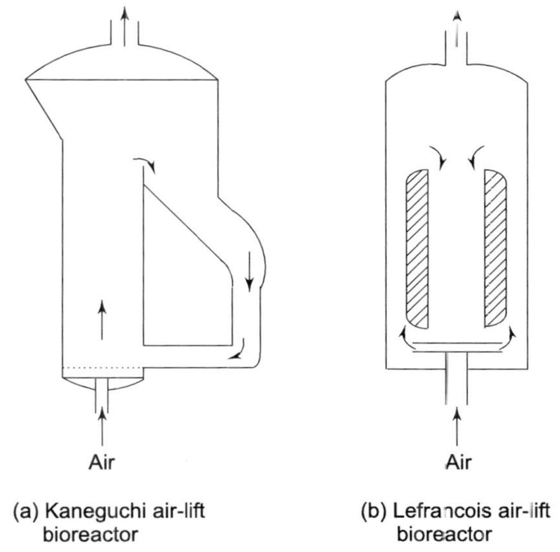

Fig. 19.6 *Various types of air-lift bioreactors*

Two designs of air-lift reactors are shown in Fig. 19.6 (Cliffe, 1988). These are in addition to those already shown in Fig. 5.9. The basic feature for the maintenance of the air-lift conditions is that the flow due to forced convection should be more than the internal back mixing. In the absence of it, the reactor operates like a bubble column reactor. Sittig and Faust

(1982) mentioned the mixing data for air-lift loops of different volumes. It can be seen from Table 19.2 that the mixing time is almost six times the circulation time, and the circulation time does not increase proportionately with the reactor volume.

Table 19.2

Reactor volume (m^3)	*Circulation time (s)*	*Mixing time (s)*
0.4	10	63
2.0	14	80
27.0	17	101

It indicates that at least six passages through the bioreactor are required to attain homogeneity.

ICI Company, UK, has used air-lift bioreactor systems for the production of 60,000 tpa single-cell protein (SCP) from methanol as substrate using *Methylophilus methylotrophus* (Cliffe, 1988).

19.3.6 Fluidised Bed Bioreactor

Fluidised bed reactors occupy an important place in chemical engineering literature. There are textbooks written on fluidisation and fluidisation engineering. A detailed discussion on fluidised beds is beyond the scope of this book. General concepts of fluidised bed bioreactor were described in Section 5.7.2 with a schematic representation in Fig. 5.8. Fluidised beds are used where intimate contact between the solid particles and fluid is required. The solid particles could be flocculated organisms, or pellets of immobilised enzymes or cells. The fluid could be the fermentation broth liquid. Pressure drop versus the fluid velocity was shown graphically, and the concept of minimum fluidisation velocity, u_m, at the onset of fluidisation, was explained. If the velocities are higher than u_m, three types of behaviour can be envisaged (Denbigh and Turner, 1971(b)):

(i) When the velocity of the fluid is equal to u_m, the fluidisation is called **incipient fluidisation.**

(ii) When $u > u_m$, and if the fluid is liquid (which is mostly the case in biotechnological processes), the bed expands further because of the particles moving apart uniformly along with the liquid, and the fluidisation is called **particulate fluidisation.**

(iii) When $u > u_m$, and if the fluid is a gas, the gas moves in the form of bubbles, aggregating and bursting depending upon the local conditions. The beds look like a boiling liquid, and the fluidisation is called **aggregative** or **bubbling fluidisation.**

The fluidised beds are made in such a way (Fig. 5.8(c)) that the diameter of the tower is higher at the top, so that the linear fluid velocity (u) will come down and the upward buoyancy forces on the particles will fall down, with the result that the particles will slip into the fluidising zone. Thus, the concentration of the particles (mostly the immobilised enzymes) is kept constant. If the cells are flocculating and participate in the fluidisation, cells will be growing (produced) in the bed, and some of them will be washed out in the outlet along with the fluid. However, to maintain a constant microbial hold-up, it is essential that the rate of loss of microorganisms from the fermenter should be equal to their growth rate (Atkinson, 1974).

Since the bed contains a solid phase and a fluid phase, and both are moving, it is possible to have the following combinations:

(i) fully mixed conditions in both solid phase and fluid phase

(ii) plug flow conditions in both solid phase and fluid phase

(iii) fully mixed conditions in solid phase and plug flow in fluid phase

(iv) plug flow conditions in solid phase and mixed flow conditions in fluid phase.

Based on the above, fluid bed models have been proposed, considering the bed to consist of only two phases:

(i) bubble phase

(ii) emulsion phase.

Kunni–Levenspiel (1991) model introduced a third phase known as **Cloud and Wake** phase.

We shall restrict our discussion to the three-phase model of Kunni–Levenspiel. The three phases are shown in Fig. 19.7. They are:

(i) bubble phase
(ii) cloud and wake phase
(iii) emulsion phase.

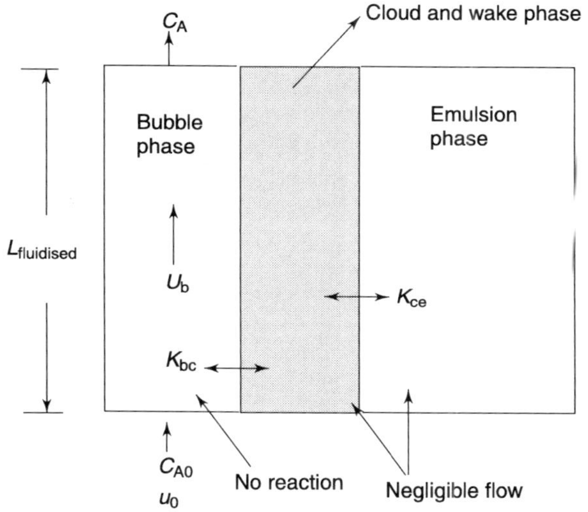

Fig. 19.7 *Fluidised bed model of Kunni–Levenspiel used for reaction applications (Reprinted from "Chemical Reaction Engineering", Levenspiel, O., 2nd Edn., p. 517, Copyright © (1974), with permission from John Wiley and Sons, Inc.)*

The reaction scheme is as follows:

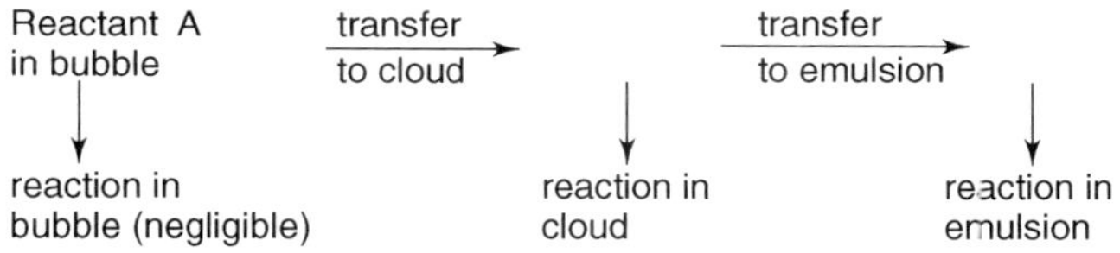

(19.19)

The reaction taking place in the bubble is assumed to be negligible. Mass transfer takes place from bubble to emulsion through the cloud, i.e. bubble to cloud and cloud to emulsion. Mostly the reaction takes place in the emulsion phase. Plug flow conditions are assumed for the fluid phase in the bed, and the particles are mixed in the bed.

Based on these assumptions, equations were developed for fractional conversions. Further details on fluidised bed reactor modelling are not described here, as these models are more applicable for gas fluidised beds which we do not normally come across in biotechnological processes.

However, for liquid fluidised beds which operate under particulate fluidisation, the mass transfer rates can be related to those in packed beds as follows (Denbigh and Turner, 1971(b)):

$$N_{Sh} = \frac{k_L d_p}{D} = \frac{0.81}{\varepsilon} N_{Re,p}^{1/2} N_{Sc}^{1/3} \tag{19.20}$$

where N_{Sh} is Sherwood number
$N_{Re,p}$ is particle Reynolds number = $\rho d_p u/\mu_f$
N_{Sc} is Schmidt number = $\mu_f/\rho D$
D is the diffusivity coefficient
d_p is average particle diameter

$$\varepsilon = \text{void fraction} = 1 - \frac{\text{volume of solids}}{\text{total volume of bed}} \tag{19.21}$$

Fluidised bed bioreactors find some industrial applications, particularly in the brewing industry using flocculating yeasts or in the waste-water treatment using activated sludge, etc. (Cliffe, 1988).

19.4 BIOREACTOR FOR PLANT AND ANIMAL CELLS

Plants have been a nature's gift to mankind to meet most of their requirements. In fact, man depended on the plant kingdom from time immemorial for his survival, may it be for the production of food in the ancient days, which has been later named as agriculture. Thus, plants have been a source of food. Subsequently, plants have become the starting materials for:

- food colours
- industrial colours (particularly azodyes for textiles)
- food flavours
- food fragrances
- pharmaceuticals (the whole science of Ayurveda and Homoeopathy)

The plant cells have peculiar behaviours which are not common to microbes. The structure of the plant cell was described in Chapter 2. The plant cell has a tendency to grow and multiply, even if it is differentiated from its parent culture, and generates the whole plant under appropriate physico-chemical/agroclimatic environment. The plant cell's ability to regenerate the whole parent plant is known as **totipotency** and is considered as an essential trait of plant cells.

The animal cells are eucaryotic cells with the structure and composition described earlier (Chapter 2). They play an important role in biotechnology, particularly for the production of vaccines and for the application of recombinant DNA techniques for the production of significant amounts of native cells. They are of various sizes (approximately 10–30 μm) and shapes (spherical, rods, ellipsoidal). The endoplasmic reticulum in them is critical for protein synthesis. It plays a vital role for post-transitional processing in the initial stages. The requirement of nutrients and metabolism of animal cells are very peculiar. Their growth medium contains glucose, glutamine, some essential amino acids, mineral salts, and a serum (horse or calf). The serum is a cell-free liquid recovered from blood. The use of serum in the medium poses a host of problems; the primary one being cost, and the next most important is the contamination that is associated with serum. Sometimes, the cost of serum constitutes almost 80% of the cost of production, even when it is used in 10% levels (Bailey and Ollis, 1986(c)).

We restrict our interest in the present section only to envisage some of the bioreactor aspects.

19.4.1 Bioreactors for Plant-cell Cultures

The plant cells are larger in size. Hence, high shear rates (high rpm of agitator) may damage the cells as they may turn and twist around the impeller blades; but lower shear rates affect the transport properties, and particularly the transport of nutrients to the cells. Fortunately their requirement of oxygen is also high, which incidentally makes an air-lift bioreactor a suitable candidate for low or moderate cell densities (below 20 g/l on dry basis). For high cell densities, when the viscosity also becomes prohibitively high, we may have to go for a paddle type or helical ribbon impellers (Shuler and Kargi, 2003).

Another difficulty with most of the plant cells is their low growth rates, which make the reactor volumes larger and reaction times higher. This is further complicated by the genetic instability of most cells. It is not very uncommon that a particular plant cell line produces a useful product at a considerable rate for some time (a few months) and subsequently loses activity.

Mixing also plays an important role in the performance of plant cells. Sometimes even a multiple reactor system may be a profitable strategy. The first reactor with a particular nutrient and medium composition may be used initially for the production and multiplication of plant cells. Subsequently, these cells are taken in a second reactor with a different medium and nutrient composition, so that the plant cells can be used to produce useful products.

19.4.2 Bioreactors for Animal Cells

Most of the animal cells grow when they are anchored to a surface, usually a biopolymer, by immobilisation. Such animal cells are known as **anchorage-dependent.** Others, which are not dependent on the anchorage and do not need a support for growth, are known as **non anchorage-dependent** cells. Their oxygen requirements are low, typically of the order of 5 times less than that of the plant cells. For example, Katinger and Scheirer (1982) reported the specific oxygen uptake rate (μ mol of oxygen/10^6 cells per hour) as: B H K 21c 13 baby hamster kidney cells require 85–110 × 10^{-3} of oxygen, and mouse L929 cells require 20–65 × 10^{-3} of oxygen, whereas human foreskin F5-4 requires 50–85 × 10^{-3} of oxygen.

The cells are not of long fibre type and hence they do not twist to the impellers. The viscosities are also very moderate. However, the anchorage-dependent cells require bioreactors with high surface area to volume ratio. Since the cells are immobilised, the configuration is similar to fluid–solid systems. Hence, the following type of traditional reactors would be suitable:

- packed beds
- fluidised beds
- CSTF

Some of the other types of high surface area to volume (S/V) ratio reactor configurations are shown in Fig. 19.8 (Bailey and Ollis (1986(c)).

Development of toxic metabolic products hinders the growth of animal cells. In such cases, a fed-batch operation is recommended. Reuveny et al. (1986) reported the results of a batch and fed-batch process for the production of monoclonal antibodies. Cells were grown in 100 ml flasks. Instead of adding all the growth medium initially, it was added in batches of 5–10% for 10^6 cells for every 24 hours in a fed-batch operation with a periodic harvesting of 30% of the grown cells (Fig. 19.9). Thus, antibody production could go up to approximately 275 μg/ml in 15 days as contrast to approximately 170 μg/ml in a batch operation.

19.5 SCALE-UP OF BIOREACTORS

Scale-up of operations is an engineering problem encountered in translating a successful laboratory-scale process into a commercial venture. Various steps involved in bioprocessing operations were highlighted in Fig. 1.1. The first four steps, viz.

(i) identification and isolation of the strain of the microorganism
(ii) preservation of strain
(iii) culturing or growing of inoculums
(iv) pre-fermentation culturing

are typical of laboratory-stage development, and do not require any efforts for scale-up. The remaining three steps, viz.

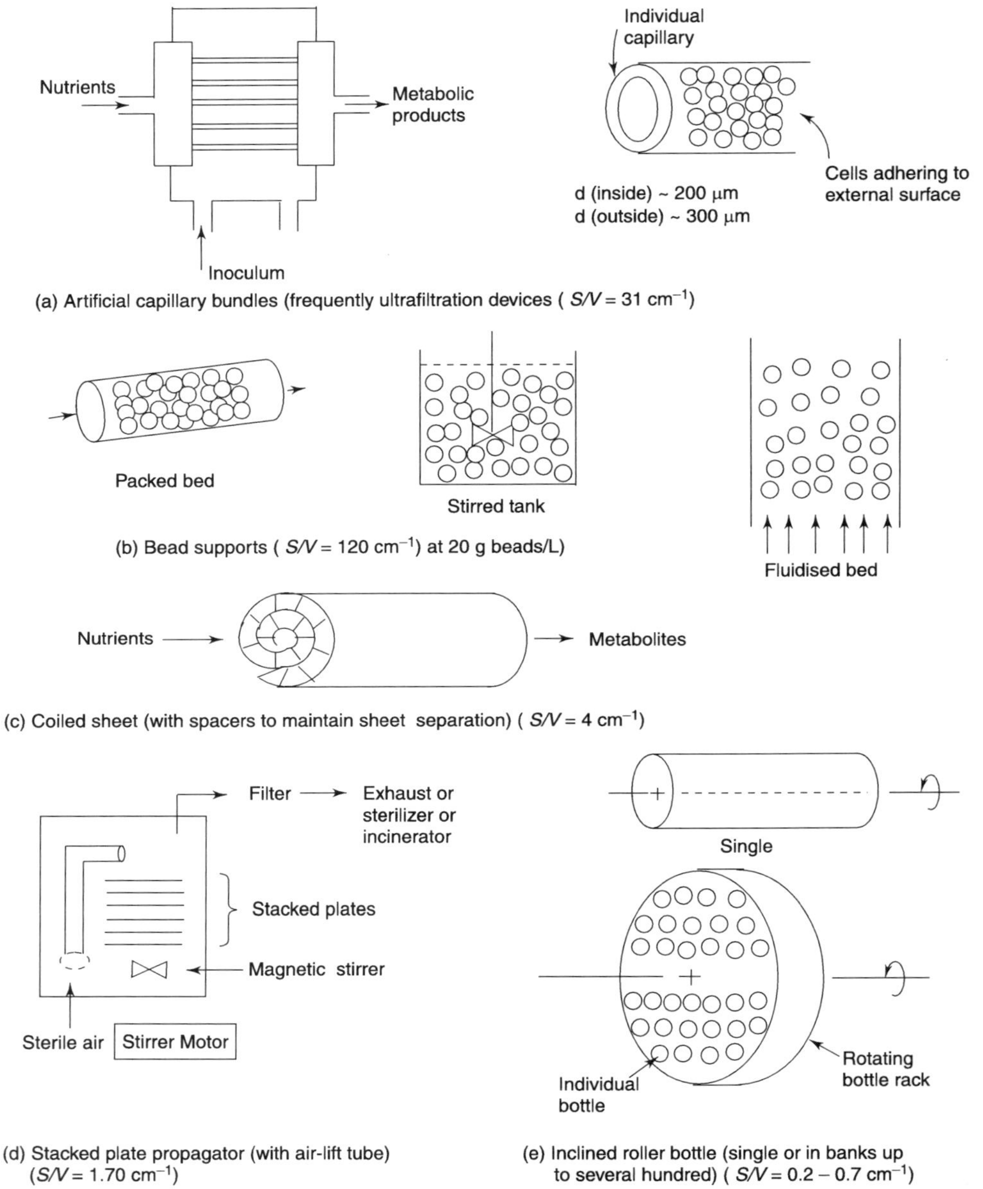

Fig. 19.8 *High surface area configurations for tissue culture*
(Reprinted from "Biochemical Engineering Fundamentals", Bailey, J.E. and Ollis, D.F., 2nd Edn., p. 637, Copyright © (1986), McGraw-Hill, Inc., New York.)

(v) fermentation process
(vi) recovery and purification of the product
(vii) treatment of effluents

all need high level of engineering inputs for scale-up. In fact, most of the scale-up operations are a part of chemical engineering studies, and are dealt at length in chemical engineering literature (Bisio and Kabel,

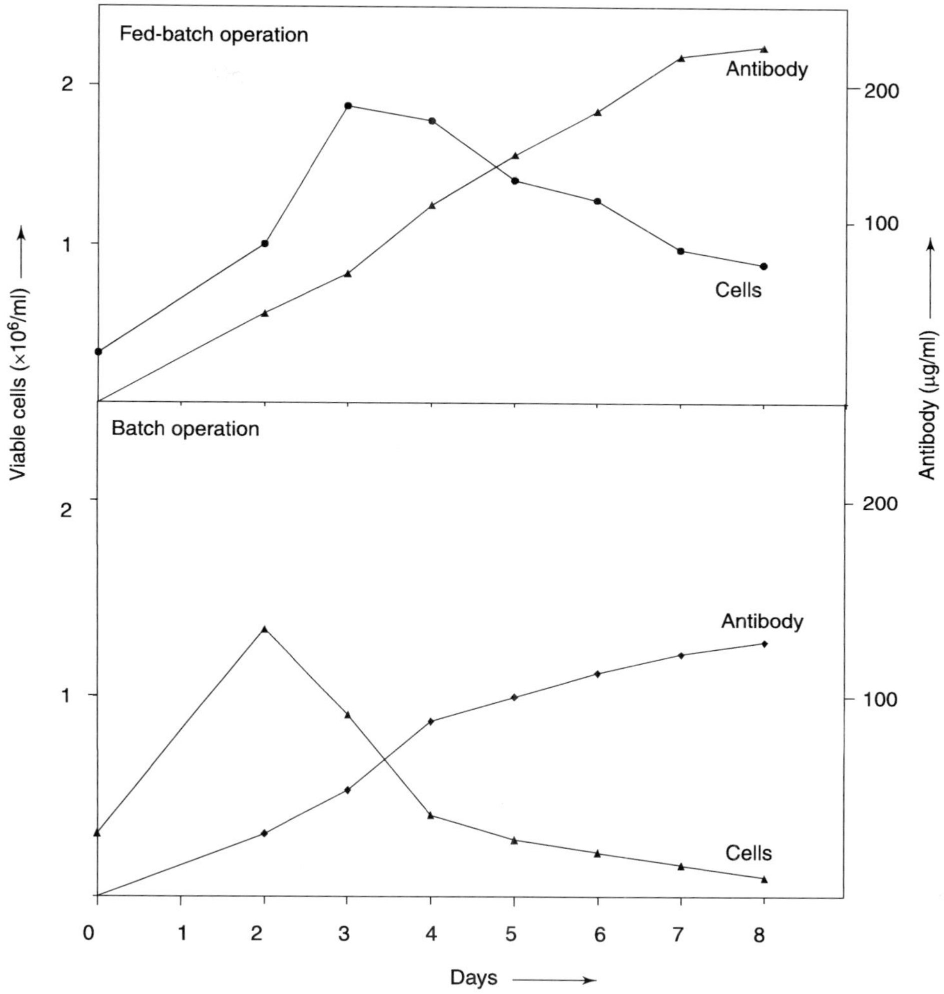

Fig. 19.9 *Comparison of cell growth and monoclonal antibody production in batch and fed-batch cultures (Reprinted from "Comparison of cell propagation methods for their effect on monoclonal antibody yield in fermenters", J. Immunol. Methods, Reuveny, S., Velez, D., Miller, L. and Macmillan, J.D., p.63, Copyright © (1986), with permission from Elsevier)*

1985). A fermentation process adds another dimension to the scale-up problem, i.e. the proliferation and performance of microorganisms by competitively consuming the substrate. Maintaining aseptic conditions is a newer dimension added to fermentation processes, which is not typical of chemical processes, and hence is not very familiar to chemical engineers. However, we attempt the problem in the present section as purely a case of interphase mass transfer and mixing, even though interphase heat transfer is also very important. Incidentally, the heat generated in a fermentation process (which needs to be removed), or heat requirement in a fermentation process (which needs to be supplied) are marginal. Thus, maintaining adequate heat transfer rates is not a problem for biochemical engineers. Hence, problems associated with heat transfer in biochemical processes are not discussed here.

We may establish the best aerating conditions in a particular fermentation process in the laboratory scale. To establish proper aeration rates and agitation in larger tanks envisage certain problems. For example, as the volume increases, the diameter and height of the bioreactor increases. Proportionately, the impeller diameter also increases. What is more important is the increase in the height of the reactor which matters as far as the trajectory of a bubble is concerned. Based on the physico-chemical properties of the fluid, the size of the bubble and its raising velocity (or terminal velocity) are the same in small and large bioreactors. Hence, the time spent by the bubble in the large reactor is more, which results in higher gas hold-up without adequately increasing the oxygen mass transfer rate. If adequate amount of oxygen is not transferred in the larger reactors, the cell growth will be reduced and accordingly the product yield will also reduce during scale-up. This means that oxygen mass transfer plays a crucial role in the scale-up, which we describe as aeration and agitation. The scale-up methods for aeration and agitation may be explained on the basis of the following three criteria:

(i) constant $k_L a$
(ii) constant ratio of power input to agitation per unit volume of fluid (P/V)
(iii) constant tip speed of the impeller.

There is no simple rule to say which of the above criteria is valid. It depends on the conditions (Karanth, 1998). For shear-sensitive fermentation broths, constant tip speed is the best approach. Methods based on constant $k_L a$ has limited applicability, as the prediction of $k_L a$ in production fermentations is difficult. Hence, the constant P/V concept has a wider applicability in scale-up. The pioneering work of Oldshue (1966) on mixing in fermentation equipment and scale-up is often quoted in biochemical engineering literature. The fermentation process is subdivided into a number of interrelated individual steps. Mixing in gas–liquid equipment, coupled with the effect of shearing and heat transfer, is very important. Table 19.3 highlights the variation of the various process parameters of fermenters on scale-up from 20 to 2500 gallons (Oldshue, 1966). The increase in volume is 125 times. Hence, the diameter of the impeller is increased by 5 times. The impeller speed, N_i, comes down by 3 times, but the d_i is increased by 5 times; hence the $\pi N_i d_i$, which is the tip speed of the impeller, will increase on scale-up. Typically, if the circulation time is kept constant, v/V is constant, as shown in line 6 in Table 19.3; it results in an abnormal increase in power input, which is about 3125 times, and it is impossible to achieve.

Table 19.3 *Properties of fluid mixer on scale-up*

Property	*Pilot scale, 20 gal*	*Plant scale, 2500 gal*			
P	1.0	125	3125	25	0.2
P/V	1.0	[1.0]	25	0.2	0.0016
N	1.0	0.34	1.0	0.2	0.04
d_i	1.0	5.0	5.0	5.0	5.0
υ	1.0	42.5	125	25	5.0
υ/V	1.0	0.34	[1.0]	0.2	0.04
$N_i d_i$	1.0	1.7	5.0	[1.0]	0.2
$N_{Re,i}$	1.0	8.5	25.0	5.0	[1.0]

(Adapted from "Biotechnology Bioengineering", Oldshue, J.Y. Vol. 8(1), p. 4, Copyright © (1966), Reprinted with permission of Wiley-Liss, Inc., a subsidiary of John Wiley & Sons, Inc.)

Now let us look in detail at the performance on scale-up for constant P/V and constant $k_L a$.

19.5.1 Constant P/V

The power input to the impeller, P, is given by

$$P \propto N_i^2 d_i^3 \quad \text{in laminar zone} \tag{19.22}$$
$$\propto N_i^3 d_i^5 \quad \text{in turbulent zone}$$

The volume of the reactor is proportional to the cube power of tank diameter, and hence cube power of impeller diameter, i.e.

$$V \propto d_i^3 \tag{19.23}$$

$$P/V = N_i^2 \quad \text{in laminar zone} \tag{19.24}$$
$$= N_i^3 d_i^2 \quad \text{in turbulent zone}$$

For two reactors of different volumes, say V_1 and V_2, with power inputs P_1 and P_2, we have

$$\frac{P_1/V_1}{P_2/V_2} = \frac{(N_{i1})^3 (d_{i1})^2}{(N_{i2})^3 (d_{i2})^2} \tag{19.25}$$

The above equation may be written as

$$\frac{P_1/V_1}{P_2/V_2} = \frac{(N_{i1} d_{i1})^3}{(N_{i2} d_{i2})^3} \left[\frac{d_{i2}}{d_{i1}}\right] \tag{19.26}$$

$N_i d_i$ is proportional to the tip speed of the stirrer. Thus, if we take the criterion of equal power input to unit volume, the LHS is equal to 1.0, and hence the RHS is also equal to 1.0.

Thus, $$\left[\frac{N_{i1} d_{i1}}{N_{i2} d_{i2}}\right]^3 \left[\frac{d_{i2}}{d_{i1}}\right] = 1.0$$

or $$\frac{(N_{i1} d_{i1})}{(N_{i2} d_{i2})} = \left[\frac{d_{i1}}{d_{i2}}\right]^{1/3} \tag{19.27}$$

$$\frac{(\pi N_{i1} d_{i1})}{(\pi N_{i2} d_{i2})} = \frac{\text{tip speed of impeller 1}}{\text{tip speed of impeller 2}} = \left[\frac{d_{i1}}{d_{i2}}\right]^{1/3} \tag{19.28}$$

Thus, equal P/V ratio leads us to employ higher speeds of the impeller tip in the scale-up of bioreactors. The higher tip speed may cause higher shearing effect, which in turn may be detrimental for the performance of microorganisms in the fermentations. Thus, even though constant P/V ratio is frequently referred to as a criterion for scale-up, experimental data in support of it is scarce (Aiba et al., 1965(b)). However, the data presented by Gaden (1961), in Fig. 19.10, on the yield of penicillin versus the power input per unit volume, are a good criterion for fermenters of sizes 5 lit., 200 gal and 10,000 gal. When $P/V = 1.0$ (hp/100 gal), the yield of penicillin in terms of units per volume is almost constant.

19.5.2 Constant $k_L a$

$k_L a$ stands for oxygen mass transfer rate. Thus, constant $k_L a$ implies maintaining equal mass transfer rates both in the small-scale and in the large-scale fermenters. Bartholomew (1960) described the yield of vitamin B_{12} versus mass transfer group, which he described as $k_L a\, p$, where p is total pressure. The pressure term is added here to take care of the oxygen solubility with increase in total pressure. The data are shown in Fig. 19.11. The figure shows that constant $k_L a$ is not very valid in scale-up; particularly at higher $k_L a$ values, it is interesting to note that the yield of vitamin B_{12} falls, indicating possible adverse effect of higher shear

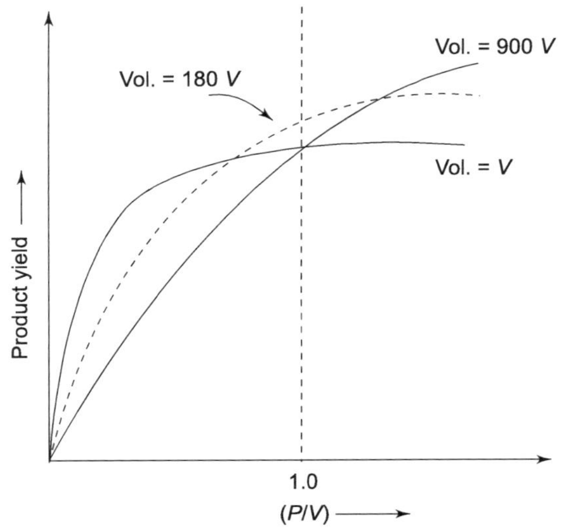

Fig. 19.10 *Product yield versus power input in various sized vessels (Gaden, 1961).*

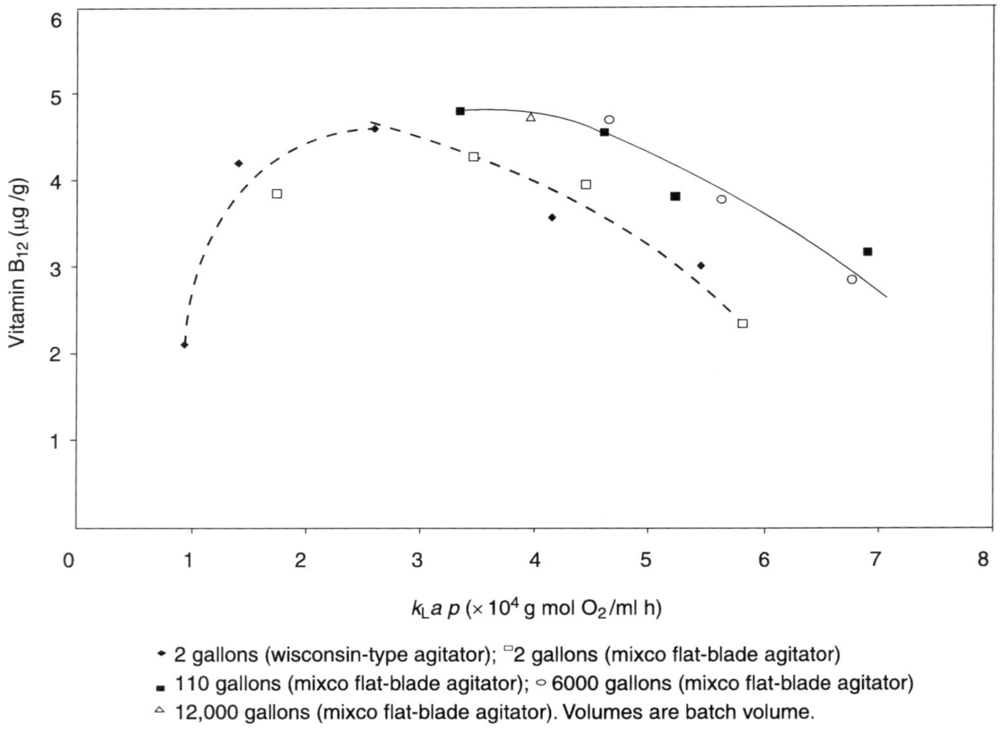

Fig. 19.11 *Vitamin B_{12} yield (μg/g) versus mass transfer group (k_L ap) Bertholomew (1960)*

rates on the survival of microorganisms. However, the data were found to correlate well for 420 l pilot plant vessel and 22.7 m^3 commercial fermenter at $k_L a\, p = 4.5 \times 10^{-4}$ or 7.0×10^{-4} or even at $5.2\text{–}5.5 \times 10^{-4}$ g mol of O_2 /(ml h).

Table 19.3 shows how the process parameters vary for different situations of equivalence. Thus, the scale-up of bioreactors is more of an art. The choice of transport property for equivalence should be the one which is more critical for the mixing process in gas–liquid systems. Oldshue (1966) described that the mixing can be broken down into the following interrelated areas:

(i) Process design
- fluid mechanics of impeller
- fluid regime for process scale-up
- hydraulic similarities

(ii) Impeller power characteristics
- diameter
- power requirement
- rpm

(iii) Mechanical design (problem associated with)
- impellers
- shafts
- motor hp and drive assembly

Scale-up techniques are applied keeping in view the above and their relative importance in scale-up after making the small and big fermenters geometrically similar. The yield of the product and its rate of production are the ultimate objectives which need to be consistent in both the fermenters.

19.6 SOME CRITERIA FOR SELECTION OF BIOREACTORS

The choice of a bioreactor for a particular biotechnological process is a very difficult task, and is dependent more on the experience rather than on any clear-cut rules. The proper choice is one of the most difficult tasks. Generally what is done is a routine practice of choosing a stirred tank operated in fed-batch conditions. The most commonly used stirred tank reactor is shown in Fig. 5.5. When a process is being developed at bench-scale level, the biochemical engineers are not involved. It is generally done by basic scientists, may be a biologist or a microbiologist. Engineers are involved only at the time of scale-up. Hence, the engineers face the lack of data with reference to utilities and their requirements, transport processes including interphase mass transfer, which are normally overlooked at bench-scale developmental levels by biologists. Hence, generally the choice/selection of a bioreactor is done based on the physico-chemical properties of the fermentation broth, interphase mass transfer/oxygen requirements in case of multiphase systems, stirring and agitation requirements depending upon the nature and morphology of cells, and the type of controlling devices to be incorporated for process monitoring.

Some of the tips for selection of a bioreactor are indicated in Table 19.4. The information in the tabular form is not exhaustive, it is only indicative. The choice is based on:

- cost
- convenience
- compatibility

of the bioreactors in case of multiple choice with all identical features.

Table 19.4 *Some tips for selection of bioreactors*

Physico-chemical property/cellular type	*Reactor configuration*	*Remarks*
Low-viscosity broths	Stirred tank, air-lift, bubble column	The selection does not matter much
Large fermenters (50 –500 m^3) with low-viscosity broths	Bubble columns	
Volume of the fermenter >500 m^3		Impossible to operate stirred tanks
High-viscosity broths	Stirred tanks	CSTR alone is suitable
Fermentation broths with particle attrition	Packed beds	Stirred tanks are not suitable
High heat of reaction systems (such as SCP from methanol)	Fluidised beds, bubble columns	Stirred tanks are not suitable
Fermentations with product inhibition/substrate inhibition	Fed-batch reactors	
Immobilized cells/enzymatic reactions	Packed towers, fluidised beds, CSTR	The system is akin to fluid–solid systems
Immobilized systems	Packed towers, fluidised beds, air-lift	Low hydrodynamic shear to be ensured
Liquid recycling systems	Packed beds, fluidised beds, Trickle beds	
Plant cells of low cell density	Air-lift reactors	
Plant cells of high cell density	Stirred tank with paddle/ helical ribbon impeller	If the plant cells twist around impeller, proper choice of impeller is very important
Anchorage-dependent animal cells	Fluidized beds, packed beds, CSTF	Similar to immobilised systems, high *S*/*V* ratios are desired
Non anchorage-dependent animal cells	Stirred tank, air-lift, bubble column	Viscosities are moderate
Animal cells used for production of monoclonal antibodies	Fed-batch reactors	
Dense cultures/filamentous organisms	Bioreactors with anchor-type stirrer/helical screw stirrer	The fermentation broth is non-Newtonian in nature
Organisms producing extracellular products	Bioreactors with anchor-type stirrer, helical screw stirrer	The fermentation broth is non-Newtonian in nature
Pelletised filamentous organisms	Packed beds, Fluidised beds	
Shear-sensitive mammalian/ insect/ plant cells	Stirred tanks are not suitable	Reactors with low shear but high mass transfer rates are preferred

Symbols

a	: interfacial area (m^{-1})
A	: cross-sectional area (m^2)
C	: concentration (kg/m^3 or kg mol/m^3)

C_{AL} : concentration of air in liquid phase (kg/m^3)
C_{AL}^* : concentration of air in liquid phase at equilibrium (kg/m^3)
Cp : specific heat (kJ/(kg K))
d : diameter of pipe (m)
d_i : impeller diameter (m)
d_p : diameter of particle
D : dilution rate (s^{-1} or h^{-1})
D_e : longitudinal dispersion coefficient (m^2/s)
D_v : molecular diffusion coefficient (m^2/s)
E : activation energy (J/mol)
ΔH_r : heat of reaction
k : first-order reaction rate constant (s^{-1})
k_0 : Arrhenius constant (s^{-1} or min^{-1})
k_L : liquid side mass transfer coefficient (m/s)
k_{ce} : mass transfer coefficient between cloud and emulsion phases (Fig. 19.7)
k_{bc} : mass transfer coefficient between bubble and cloud phases (Fig. 19.7)
K_m : substrate utilisation constant/Michaelis–Menten constant
K_s : Monod constant
L : length of the reactor (m)
m : quantity of medium (kg)
m_s : quantity of steam (kg)
$\dot{m}$: mass flow rate (kg/s or kg/h).
N : number of cells present per unit volume
N_A : mass transfer rate ($kg/(m^3\ s)$)
N_{Da} : Damkohler number
N_i : impeller speed (s^{-1})
N_{Pe} : Pecklet number
N_{Re} : Reynolds number
$N_{Re,i}$: impeller Reynolds number
$N_{Re,p}$: particle Reynolds number
N_{Sc} : Schmidt number
N_{Sh} : Sherwood number
N_t : number of cells at any time t in unit volume
p : total pressure
P : power input (W)
q : parameter [Eq. (19.6)], heat transfer rate (KW)
q_s : heat required for sterilisation (kJ)
r_f : rate of fermentation/rate of formation
r_s : rate of reaction of substrate ($kg/(m^3\ s)$)
r_x : rate of formation of cells ($kg/(m^3\ s)$)
R : universal gas constant (J/(mol K))
s : limiting substrate concentration
S : surface area (m^2).
S_R : original (initial) substrate concentration
t : time (s or h)

t_m	: residence time in mixed reactor
T	: temperature (K)
T_0	: initial temperature of medium (K)
T_s	: sterilisation temperature (K)
u	: linear velocity (m/s)
u_m	: minimum fluidisation velocity (m/s)
U	: heat transfer coefficient ($W/(m^2\ K)$)
V	: reactor volume (m^3)
V_R	: culture volume (m^3)
x	: concentration of cells
X	: distance in X-direction (m)
$Y_{x/s}$	: dimensionless yield factor of x based on s
z	: reduced length = X/L

Subscripts

0	: initial
1, 2	: vessel numbers
A	: component A
Av	: average
i	: component i
m	: mixed
max	: maximum
ss	: steady state
t	: at time t

Greek Letters

μ	: specific growth rate (h^{-1})
μ_f	: viscosity of fluid
θ	: reduced time = t/t_{Av}
δ	: parameter [Eq. (19.6)]
ρ	: density (kg/m^3)
λ_s	: latent heat of vapourisation of steam (kJ/kg)
β	: stoichiometric ratio [Eq. (19.14)]
σ	: constant [Eq. (19.17)]
D	: diffusivity coefficient (m^2/s)
ε	: void fraction (dimensionless)
V_{max}	: maximum specific growth rate in Michaelis–Menten rate equation
υ	: volumetric flow rate (m^3/s)
$\upsilon(t)$	: volumetric feed rate at time t

References

19.1 Aiba, S., Humphrey, A.E., and Millis, N.F. (1965), *Biochemical Engineering,* Academic Press, New York, (a) pp. 206–214, (b) pp. 163–177.

19.2 Atkinson, B. (1974), *Biochemical Reactors,* Pion Ltd., London, pp. 143–146.

19.3 Bailey, J.E., and Ollis, D.F. (1986), *Biochemical Engineering Fundamentals,* 2nd Edn, McGraw-Hill, Inc., New York, (a) pp. 536–549, (b) pp. 383–387, (c) pp. 630–641.

19.4 Bartholomew, W.H., Karow, E.O., and Sfat, M.R. (1950), "Design and operation of a laboratory fermenter", *I&EC*, 42, 1827–1830.

19.5 Bartholomew, W.H. (1960), "Scale up of submerged fermentations", *Advances in Applied Microbiology*, Umbreit, W.W. (Ed.), Vol. 2, Academic Press, New York, pp. 289–300.

19.6 Bischoff, K.B. (1966), "Optimal continuous fermentation reactor design", *Can. J. Chem. Eng.*, 44, 281–284.

19.7 Bisio, A., and Kabel, R. L. (1985), *Scale up of chemical processes*, John Wiley & Sons, Inc., New York.

19.8 Cliffe, K. (1988), "Bioreactors", *Biotechnology for Engineers*, Scragg, A.H. (Ed.), Ellis Horwood Ltd., Chichester, pp. 277–301.

19.9 Coulson, J.M., Richardson, J.F., Backhurst, J.R., and Harker, J.H. (1993), *Coulson & Richardson's Chemical Engineering*, 4th Edn, Vol. 1 Pergamon Press, Oxford, p. 67.

19.10 Deindoerfer, F.W., and Humphrey, A.E. (1959), "Analytical methods for calculating heat sterilization times", *Appl. Microbiol.*, 7, 256–264.

19.11 Denbigh, K.G., and Turner, J.C.R. (1971), *Chemical Reactor Theory*, 2nd Edn, Cambridge University Press, Cambridge, (a) pp. 166–169, pp. 197–198, (b) pp. 155–158.

19.12 Erickson, L.E., and Stephanopoulos, G. (1987), "Biological reactors", *Chemical Reaction and Reactor Engineering*, Carberry, J.J., and Varma, A. (Eds.), Marcel Dekker, Inc., New York, pp. 779–838.

19.13 Fogler, H.S. (1986), *Elements of Chemical Reaction Engineering*, Prentice-Hall, Englewood Cliff, New Jersey, pp. 179–217.

19.14 Gaden Jr., E.L. (1961), "Aeration and agitation in fermentation", *Sci. Rep. Instituto Superiore di Sanita*, 1, 161.

19.15 Herbert, D. (1961), "Continuous culture of microorganisms", *Soc. Chem. Ind., Monograph No. 12,* London, p. 21.

19.16 Karanth, N.G. (1998), "Process engineering considerations in design and scale up of bioreactors", *Biotechnology: Industrial perspectives, Proceedings of the Symposium on Biotechnology Industry*, Chand, S. and Jain, S.C. (Eds.), All India Biotech Association, New Delhi, pp. 265–273.

19.17 Katinger, H.W.D., and Scheirer, W. (1982), "Status and development of animal cell technology using suspension culture techniques", *Acta Biotechnologica*, 2, 3.

19.18 Kunni, D., and Levenspiel, O. (1991), *Fluidization Engineering*, 2nd Edn., Butterworth-Heinemann, Boston, MA.

19.19 Levenspiel, O. (1958), "Longitudinal mixing of fluids flowing in circular pipes", *I&EC*, 50, 343–346.

19.20 Levenspiel, O. (1974), *Chemical Reaction Engineering*, Wiley Eastern Pvt. Ltd., New Delhi, pp. 308–314, 516–522.

19.21 Moo-Young, M., and Blanch, H.W. (1981), "Design of biochemical reactors: Mass transfer criteria for simple and complex systems", *Advances in Biochemical Engineering*, Fiechter, A. (Ed.), Vol. 19 Springer-Verlag, Berlin, pp. 1–69.

19.22 Nauman, E.B. (1987), *Chemical Reactor Design*, John Wiley & Sons, New York, p. 255.

19.23 Oldshue, J.Y. (1966), "Fermentation mixing scale up techniques", *Biotech. Bioeng.*, 8, 3–24

19.24 Reuveny, S., Valez, D., Miller, L. and Macmillan, J. (1986), "Comparison of cell propogation methods for their effect on monoclonal antibody yield in formenters". *J Immunological Methods*, 86, 61–69.

19.25 Shuler, M.L., and Kargi, F. (2003), *Bioprocess Engineering:* Basic Concepts, 2nd Edn., Prentice-Hall India, New Delhi, p. 411.
19.26 Sittig, W., and Faust, U. (1982), "Biochemical applications of the air-lift loop reactor", *The Chemical Engineer,* 381, 230–232.

Review Questions

19.1 Describe some characteristic features of a fermenter which are needed to be considered while designing a fermenter.
19.2 Write a note on fermenter dynamics, and indicate the conditions for stability.
19.3 Write a note on "thermal stability of a fermenter".
19.4 Describe the procedure for designing a continuous steriliser.
19.5 Describe the various functions and characteristic features to be considered in the design of a fermenter.
19.6 Describe the mass transfer aspects you need to consider in the design of bioreactors.
19.7 Derive an expression for the average residence time for a CSTF incorporating Michaelis–Menten kinetics.
19.8 Write the design features of a CSTF.
19.9 Write a note on air-lift bioreactor with a neat diagram.
19.10 Explain the various types of fluidisation.
19.11 Describe the "cloud and wake" phase of Kunni–Levenspiel model of fluidised bed reactor.
19.12 Describe the Kunni–Levenspiel model of a fluidised bed reactor.
19.13 Describe a bioreactor for growth of plant cells.
19.14 Describe a bioreactor configuration for animal cells.
19.15 Distinguish between the bioreactor configuration for plant and animal cells.
19.16 Describe the various criteria for the scale-up of bioreactors.
19.17 Justify whether 'constant k_La' concept is a better approach for scale-up.
19.18 What are the various criteria for the selection of bioreactors?
19.19 Tabulate some tips for the selection of bioreactors.

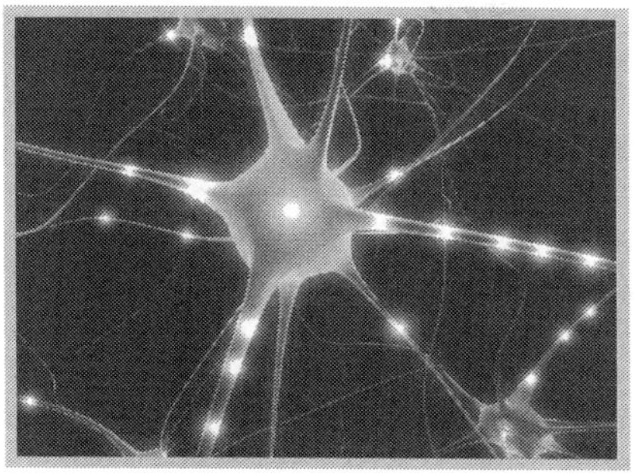

chapter 20

Bioprocess Economics

In this chapter, we look into the interplay of economics in the viability of a bioprocessing venture and its successful sustenance. The success of any process does not lie primarily in its technical viability alone, but also on its economic feasibility. The ultimate success of any R & D output to be translated into a commercial venture is gauged from the point of view of the profits it would spin. A technology or an R & D output which is not likely to spin out profits will invariably be shelved. At the very outset, when any project proposal comes for consideration, the management would like to have some preliminary estimates for the project. If the preliminary estimates are satisfactory, the management goes for a market survey for the product. The market survey indicates the profitability of the project. Then the management goes for a final detailed economic analysis and launching of the project. These management decisions are summarised in Fig. 20.1 in the form of a management decision tree. It is obvious that at every stage, if the response is negative, the project would be shelved. Hence, for a product to be pushed through from the management side, it needs to have a strong marketability. Thus, the second step is very important and crucial. Any faulty or hasty decision taken at this stage may jeopardise the financial interests of the company. There is no theoretical approach to make a decision of investment actually based on the management tree. It is mostly based on experience. However, the experience gained in this area should have a rational basis and a theoretical understanding of the marketing principles. Still, the methods based on experience do need to be revalidated in the light of the changing socioeconomic scenario.

The whole strategy is based on the fundamental concept of **economic balance,** which is based on realising maximum financial benefits from the manufacturing process. The basic criterion for financial benefits is profits. The approach for making a profit is based on various parameters which are a part of the strategy for the development / implementation of a new project or taking up of a successful process route. The basic objectives are summarised in Table 20.1 which are applicable to any chemical or food or bioproduct.

From the above discussions, it is evident that an appropriate strategy would come out of an amalgam of economics and engineering. The management planning is well projected by Rudd (1975) who proposed a suitable industrial development model shown in Fig. 20.2. However, the model is relatively more applicable for an intermediate chemical industry. Since the issues involved are akin, the above model could adequately represent a typical biotechnological industry as well.

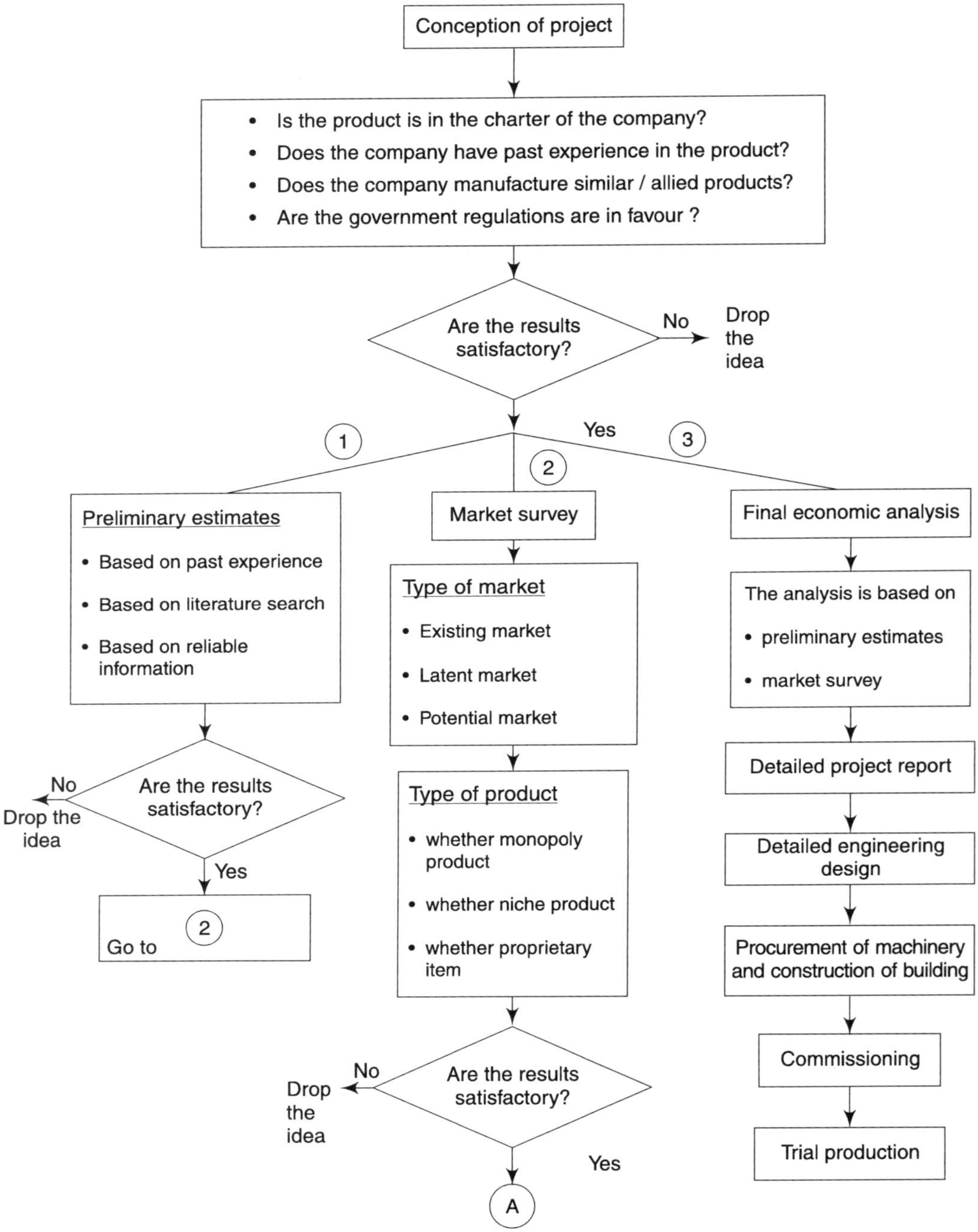
Conception of project
• Is the product is in the charter of the company?
• Does the company have past experience in the product?
• Does the company manufacture similar / allied products?
• Are the government regulations are in favour ?
Are the results satisfactory?
No
Drop the idea
Yes
1
2
3
Preliminary estimates
• Based on past experience
• Based on literature search
• Based on reliable information
Are the results satisfactory?
No
Drop the idea
Yes
Go to 2
Market survey
Type of market
• Existing market
• Latent market
• Potential market
Type of product
• whether monopoly product
• whether niche product
• whether proprietary item
Are the results satisfactory?
No
Drop the idea
Yes
A
Final economic analysis
The analysis is based on
• preliminary estimates
• market survey
Detailed project report
Detailed engineering design
Procurement of machinery and construction of building
Commissioning
Trial production

(Contd.)

Fig. 20.1 *Management – decision tree for implementing a project*

Table 20.1 *Strategy for implementation of a new project*

(i)	The product should have good market with good economic returns.
(ii)	The new project should result in a niche product with good profitability.
(iii)	The processing steps should be novel and may be economical.
(iv)	The process should result in as little amount of effluents as possible / effluents should be easily disposable.
(v)	The downstream processing steps should be minimum so that the product recovery is easy and economical.
(vi)	The new process route should result in saving of some labour / the process line should be amenable to automation which in turn reduces the labour costs.
(vii)	The bioproduct should satisfy the statutory regulations of the government without which the product cannot be marketed.
(viii)	The new product should not rise any ethical or moral issues which may result in banning of the product.
(ix)	It is advisable to see whether any special government concessions / incentives are applicable to the product so that the production costs are reduced or the marketability of the product could be easier.

The information on bioprocess economics is relatively restricted. Since most of the bioprocesses are akin to chemical processes (except for an additional step of *asepsis*), we draw equivalence from chemical engineering process economics to systematically explain the bioprocess economics. A good amount of literature is available in this area for chemical engineering process economics (Happel and Jordan(1975), Peters and Timmerhaus (1980) and Ulrich (1984)). However, the coverage of subject matter in these publications with reference of biological products is not straightforward, and hence is inadequate. They are more biased towards chemical products. But the basic process of economics principles can be of some use in bioprocess economics.

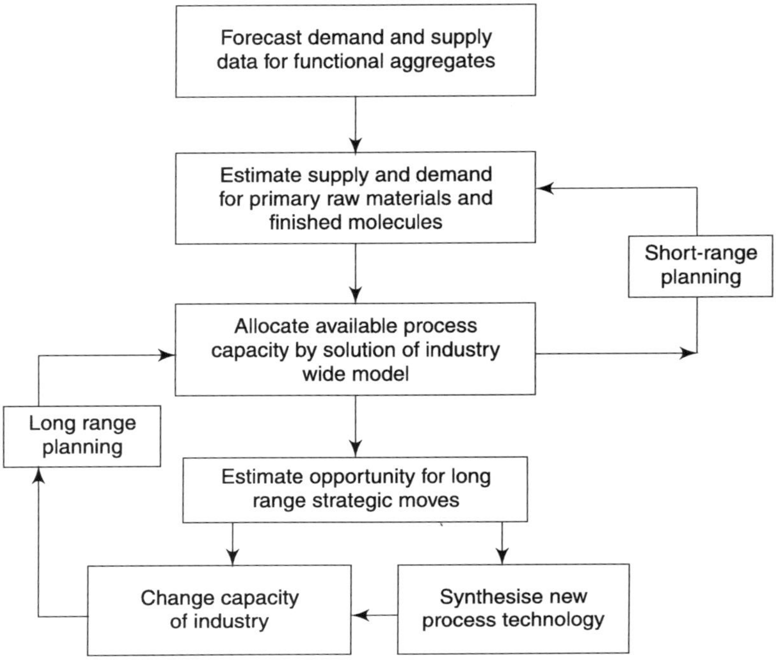

Fig. 20.2 *Model for strategic planning in industrial development (Reprinted from "Modelling the Development of Intermediate Chemical Industry", The Chem. Eng. Journal, vol. 9, p. 7, Rudd, D.F., Copyright @ 1975 with permission from Elsevier)*

20.1 PLANT DESIGN AND ECONOMICS

This stage of the project implantation step constitutes the step (3) in Fig. 20.1. The project idea for manufacturing a particular product may come from the sales team who during their direct interaction with the consumers may indicate the market potentiality for the product. Based on this experience, the management will take the first step in launching the project. Initially they go for making preliminary estimates. If the results are satisfactory, the management goes for a final economic analysis after making the following queries (Bailey and Ollis, 1986):

- Whether the product falls into the charter of the company or whether the company has any past or similar experience in the product line?
- Whether the company has adequate manpower and technology for the product, and if not, how to procure them?
- Whether the product line falls into the niche category and what could be the market potentiality and market threats?
- Whether any existing machinery or equipment could be used in the production activity of the new product?
- Whether the machinery and equipment in the proposed project will be flexible enough to be used effectively in another project or elsewhere?
- How reliable are the preliminary estimates made?

If the answers for the queries are satisfactory, a "go ahead" will be given for the project. A project engineering firm will be engaged / hired for preparing the final detailed engineering design. The various stages in the plant design project are shown in Table 20.2 (Bailey and Ollis).

Table 20.2 *Stages in plant design project*

(i)	Inception
(ii)	Preliminary evaluation of economics and market
(iii)	Development of data necessary for final design
(iv)	Final economic analysis
(v)	Detailed engineering design
(vi)	Procurement
(vii)	Construction
(viii)	Start – up and trial runs
(ix)	Production

(Reprinted from Bailey, JE and Ollis, DF (1986), *Fundamentals of Biochemical Engineering*, 2nd Ed., p.799, copyright © 1986, the material is reproduced with permission of The Mc Graw-Hill Companies, New York).

20.1.1 Financial Analysis

For deciding to invest money in any industrial activity, financial analysis is a useful tool to assess how effectively the finance could be used in the project to generate revenues. Investment is always made with the sole aim of getting some profits from the investment. Obviously, the profits should be much more than the returns which one would get in any safe investment like a fixed deposit in a bank. In other words we make an economic balance on the investment, something which calculates the options so as to determine the opportunities for maximising the financial returns from manufacturing operations. The returns are gauged in terms of

- average annual returns
- payout period
- net present value (NPV) of investment
- returns on investment (ROI)

For understanding the above, we need to define certain terms (Happel, 1958).

Net profit The net profit is the gross profit minus the taxes and depreciation♣.

If P is the net profit, R is the gross profit, I is the investment in facilities, we can write

$$P = R - Rt - dI \quad (20.1)$$

where t is the tax rate and d is the depreciation rate.

Turnover It is the ratio of sales realisation to the investment made.

Thus,

$$\text{turnover} = \frac{\text{Sales}}{\text{Investment}} \quad (20.2)$$

$$\text{Margin} = \frac{\text{Earnings}}{\text{Sales}} \quad (20.3)$$

♣ Depreciation is the term associated with the fixed assets like building, machinery and equipment which due to wear and tear during usage need replacement after certain periods. To account for it, a certain amount of money is set aside from accounting. In fact, the money is not taken away from the investor.

ROI The rate of return on the investment is the ratio of the earnings to the total investment.

$$\text{ROI} = \frac{\text{Earnings}}{\text{Total investment}} \tag{20.4}$$

$$= (\text{margin} \times \text{turnover})$$

ROI is one of the oldest and still frequently used concepts for deciding how attractive an investment is. If the ROI is high, the investment is good. We normally expect at least 10% ROI. In most of the processing ventures, an ROI up to 25 – 30% is considered to be satisfactory. Attempts have been made in chemical industries to find an ROI, if an additional equipment is added to the existing line, which in turn would save some money every year. Chou (1957) explained this with a chart which shows the inclusion of a turbogenerator expected to have a life of 30 years in an existing plant and at an assumed tax rate of 50%.

Pay-out time It is defined as the reciprocal of the gross yearly returns on investment. It is sometimes modified to take care of taxes also. Hence,

$$T = 1/P \tag{20.5}$$

where T = pay-out time

NPV The net present value (NPV) indicates what should be the investment to be made to earn a fixed earning over a period of years at the expected rate of return (ROI). This is explained with an example. Let us say that we expect to get Rs 100 every year for a period of 3 years at an expected ROI of 10%.

At the end of the first year, to get a return of Rs 100, we need to invest Rs 90.9 now.

or $$90.9 \times 1.1 = \text{Rs } 100$$

Similarly, Rs 82.6 will have to be invested at the beginning of the second year to get Rs 90.9 at the end of the second year.

or $$82.6 \times 1.1 = \text{Rs } 90.9$$

And Rs 75.1 is to be invested at the beginning of the third year to get Rs 82.6 at the end of the third year.

or $$75.1 \times 1.1 = \text{Rs } 82.6$$

Thus we need to invest today an amount of Rs 90.9 + 82.6 +75.1 = Rs 249 to get consistently an earning of Rs 100 at the end of every year for 3 years at the rate of 10% ROI. This is called ROI on a discounting basis and can be written in the form of a mathematical expression as follows (Valentas et al, 1991):

$$P_n = \frac{S_n}{(1+i)^n} \tag{20.6}$$

in which P_n = present value of investment in the nth year

S_n = expected cash flow at the end of every year

i = ROI

n = number of years

Thus, the present value = ΣP_n

The denominator in Eq. (20.6) is known as the present value factor. They are worked out for different discount rates and are shown in Table 20.3.

Table 20.3 *Present value factors $[1/(1+i)^n]$ at different discount rates*

n	2%	4%	6%	8%	10%	12%	14%	16%	18%	20%	22%	24%	26%	28%	39%
1	0.98	0.962	0.943	0.926	0.909	0.893	0.877	0.862	0.847	0.833	0.82	0.806	0.794	0.781	0.769
2	0.961	0.925	0.89	0.857	0.826	0.797	0.769	0.743	0.718	0.694	0.672	0.65	0.63	0.61	0.592
3	0.942	0.889	0.84	0.794	0.751	0.712	0.675	0.641	0.609	0.579	0.551	0.524	0.5	0.477	0.455
4	0.924	0.855	0.792	0.735	0.683	0.636	0.592	0.552	0.516	0.482	0.451	0.423	0.397	0.373	0.35
5	0.906	0.822	0.747	0.681	0.621	0.567	0.519	0.476	0.437	0.402	0.37	0.341	0.315	0.291	0.269
6	0.888	0.79	0.705	0.63	0.564	0.507	0.456	0.41	0.37	0.335	0.303	0.0275	0.25	0.227	0.207
7	0.871	0.76	0.665	0.583	0.513	0.452	0.4	0.354	0.314	0.279	0.249	0.222	0.198	0.178	0.159
8	0.853	0.731	0.627	0.54	0.467	0.404	0.351	0.305	0.266	0.233	0.204	0.179	0.157	0.139	0.123
9	0.837	0.703	0.592	0.5	0.424	0.361	0.308	0.263	0.225	0.194	0.167	0.144	0.125	0.108	0.094
10	0.82	0.676	0.558	0.463	0.386	0.322	0.27	0.227	0.191	0.162	0.137	0.116	0.099	0.085	0.073
11	0.804	0.65	0.527	0.429	0.35	0.287	0.237	0.195	0.162	0.135	0.112	0.094	0.079	0.066	0.056
12	0.788	0.625	0.497	0.397	0.319	0.257	0.208	0.168	0.137	0.112	0.092	0.076	0.062	0.052	0.043
13	0.773	0.601	0.469	0.368	0.29	0.229	0.182	0.145	0.116	0.093	0.075	0.061	0.05	0.04	0.033
14	0.758	0.577	0.442	0.34	0.263	0.205	0.16	0.125	0.099	0.078	0.062	0.049	0.039	0.032	0.025
15	0.743	0.555	0.417	0.315	0.239	0.183	0.14	0.108	0.084	0.065	0.051	0.04	0.031	0.025	0.02
16	0.728	0.534	0.394	0.292	0.218	0.163	0.123	0.093	0.071	0.054	0.042	0.032	0.025	0.019	0.015
17	0.714	0.513	0.371	0.27	0.198	0.146	0.108	0.08	0.06	0.045	0.034	0.026	0.02	0.015	0.012
18	0.7	0.494	0.35	0.25	0.18	0.13	0.095	0.069	0.051	0.038	0.028	0.021	0.016	0.012	0.009
19	0.686	0.475	0.331	0.232	0.164	0.116	0.083	0.06	0.043	0.031	0.023	0.017	0.012	0.009	0.007
20	0.673	0.456	0.312	0.215	0.149	0.104	0.073	0.051	0.037	0.026	0.019	0.014	0.01	0.007	0.005

20.2 COST OF PRODUCTION

So far we have seen how the decision for an investment is to be taken, and the basis for such decisions based on the returns on the investment. After having taken the decision to invest and start a processing unit, let us see what components constitute costs for production, based on which we can make out what are the crucial items to meddle with to reduce the cost of production or maximise the sales revenue so that the profits are pushed up.

For producing any item in an industry, we encounter two types of costs, the summation of which constitutes the total cost of production. They are

- fixed costs
- variable costs

Fixed costs are those costs associated with production which do not vary with the magnitude of production, i.e., even if there is no production, these costs will be incurred. They are the basis for production, viz., the costs involved for land, building, machinery, equipment or manpower. Similarly, variable costs are those costs which are almost directly proportional to the magnitude of production, i.e., if the production capacity is doubled, the variable costs would also double. They constitute such items as costs of raw materials, utilities and labour. Various items constituting the fixed costs and variable costs are shown in Fig. 20.3. An exhaustive list of various costs involved, classified as

- manufacturing costs, and
- general expenses

is shown in Fig. 20.4 (Peters and Timmerhaus, 1980).

The manufacturing costs are a summation of direct production costs + fixed charges + plant overhead costs. The percentage contribution of various components are shown in Table 20.4 for

- direct production charges
- fixed charges
- plant overheads
- administrative costs
- distribution and selling costs
- R & D costs
- interest component
- gross engineering costs

The percentage contributions in many cases are given as a wide range. They depend on many factors such as

- plant location
- type of process
- downstream processing steps involved
- company policies

However, in case of bioprocesses, the cost of raw materials is usually low; but the downstream processing costs could be high depending upon the end use of the product.

20.3 BREAK-EVEN ANALYSIS

Break-even analysis is an important concept in any project evaluation to find out what could be the viable capacity of production so that the project results in economic viability. As mentioned earlier, the cost of production is a combination of fixed costs and variable costs.

$$\text{Cost of production} = \text{Fixed cost} + \text{Variable cost} \tag{20.7}$$

Total cost

Fixed costs

- Interest on fixed assets
  - land
  - land development
  - building
  - machinery
  - equipment
  - auxiliary facilities
- Payback amount on fixed assets (usually distributed over 5–6 years)
- Depreciation on fixed assets (usually for accounting purpose only and is of the order of 10–15%)
- Salaries for production staff
- Administrative expenses
- Factory overheads
- Taxes
- R & D expenses
- Marketing expenses (long-range strategy)

Variable costs

- Cost of raw materials
- Chemicals
- Any additives
- Packaging materials
- Utilities
  - water
  - electricity
  - furnace oil
  - coal
- Wages to workers
- Sales expenses (short range strategy)
- Refrigeration charges
- Transportation
- Miscellaneous expenses
- Sundry expenses
- Interest on working capital (taken for meeting the above costs)

Fig. 20.3 *Various components of total cost*

The fixed costs do not change with the capacity as they are the minimum necessary costs the project incurs for its sustenance. The contribution of fixed-costs component per piece of production or per kg of the product can be reduced by increasing the production capacity. As the production capacity increases, the revenue due to sales will also increase. But the production capacity is constrained by the market forces. If the production is too high, the company may not be able to sell all the products, with the result that the inventory in the product stores will increase.

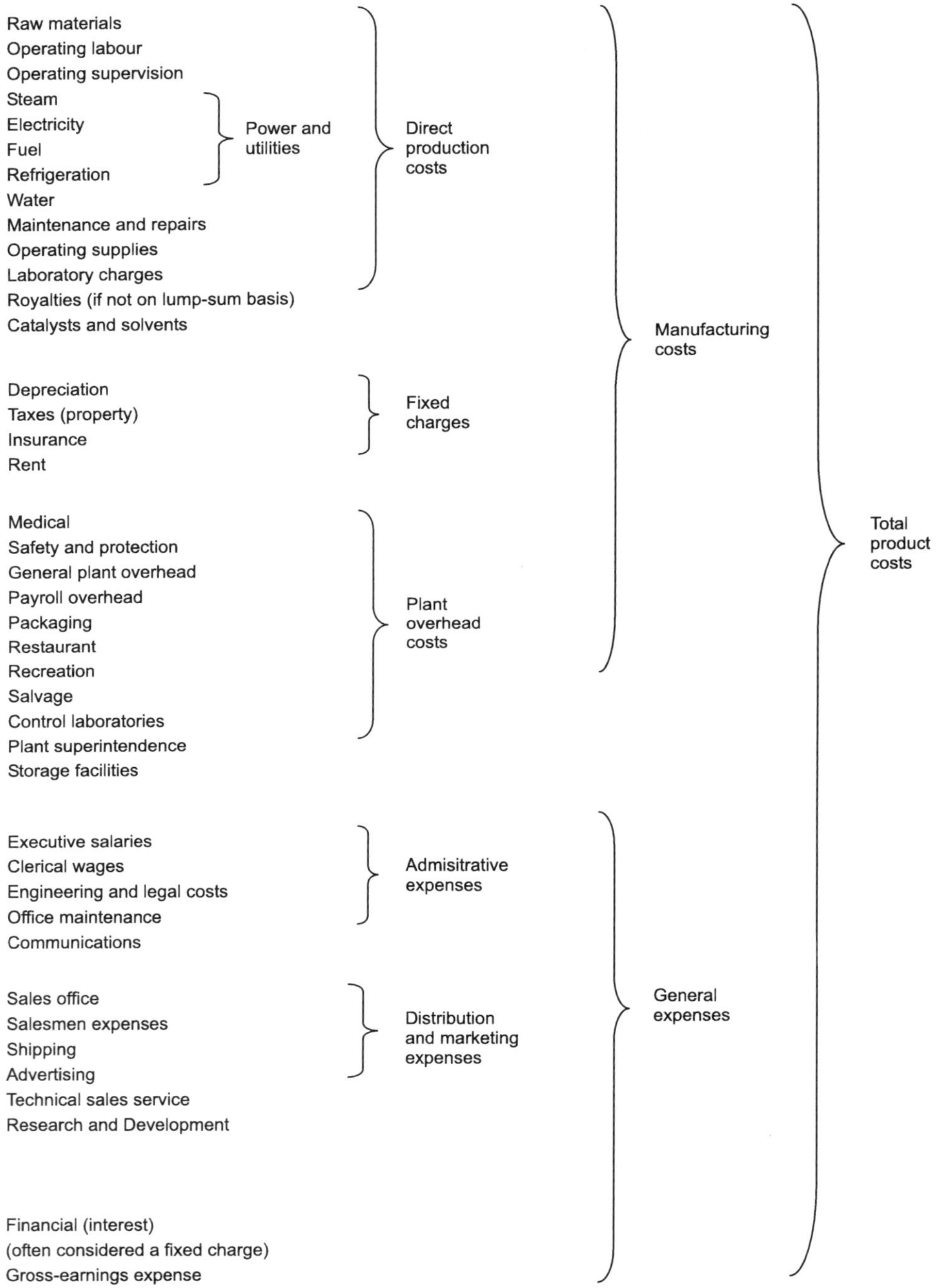

Fig. 20.4 *Costs involved in total product cost for a typical chemical process plant (Reprinted from Peters, M.S. and Timmerhaus, K.D.(1980), "Plant Design and Economics for Chemical Engineers", 3rd edition, p. 192, Copyright © (1980), the material is reproduced with permission of The Mc. Graw-Hill Companies, New York)*

Table 20.4 *Estimation of total product cost (showing individual components)*

I Manufacturing cost = direct production costs + fixed charges + plant overhead costs

A. Direct Production costs (about 60% of total product cost)
1. Raw materials (10–15% of total product cost)
2. Operating labour (10–20% of total product cost)
3. Direct supervisory and clerical labour (10–25% of operating labour)
4. Utilities (10–20% of total product cost)
5. Maintenance and repairs (2–10% of fixed-capital investment)
6. Operating supplies (10–20% of cost for maintenance and repairs, or 0.5–1% of fixed-capital investment)
7. Laboratory charges (10–20% of operating labour)
8. Patents and royalties (0–6% of total product cost)

B. Fixed charges (10–20% of total product cost)
1. Depreciation (depends on life period, salvage value, and method of calculation—about 10% of fixed-capital investment for machinery and equipment and 2 – 3% of building value for buildings)
2. Local taxes (1 – 4% of fixed-capital investment)
3. Insurance (0.4 – 1% of fixed-capital investment)
4. Rent (8 – 12% of value of rented land and buildings)

C. Plant overhead costs (50 – 70% of cost for operating labour, supervision and maintenance, or 5 – 15% of total product cost); include costs for the following: general plant upkeep and overhead, payroll overhead, packaging, medical services, safety and protection, restaurants, recreation, salvage, laboratories, and storage facilities.

II General expenses = administrative costs + distribution and selling costs + research and development costs

A. Administrative costs (about 15% of costs for operating labour, supervision, and maintenance, or 2 – 6% of total product cost); includes costs for executive salaries, clerical wages, legal fees, office supplies, and communications

B. Distribution and selling costs (2 – 20% of total product cost); includes costs for sales offices, salesman, shipping, and advertising

C. Research and development costs (2 – 5% of every sales dollar or about 5% of total product cost)

D. Financing (interest)♣ (0 – 10% of total capital investment)

III Total product cost♠ = manufacturing cost + general expenses

IV Gross-earnings cost (gross earnings = total income – total product cost; amount of gross-earnings cost depends on amount of gross earnings for entire company and income-tax regulations; a general range for gross-earnings cost is 30–60% of gross-earnings)

♣ interest on borrowed money is often considered as a fixed charge.

♠ if desired, a contingency factor can be included by increasing the total product cost by 1 – 5%.

(Reprinted from Peters, M.S. and Timmerhaus, K.D.(1980), "Plant Design and Economics for Chemical Engineers", 3rd edition, pp. 207–208, Copyright© (1980), the material is reproduced with permission of The McGraw-Hill Companies, New York)

The variable costs are almost proportional to the production capacity. Hence more the production, the more will be the variable cost. And the more will be the sales realisation by the products.

$$\text{Profits} = \text{total sales realisation} - \text{total cost of production} \tag{20.8}$$

Now we define the breakeven point as that point of production capacity where there is neither profit nor loss, i.e.,

$$\text{Total cost of production} = \text{total sales realisation} \tag{20.9}$$

The above concept is explained graphically by drawing a plot between the quantity of production on the x-axis and the cost of production/sales realisation on the y-axis as shown in Fig. 20.5. The fixed costs are shown by the line AB. Variable costs are shown by line OC. Line AD shows the total costs, whereas OE shows revenue by sales realisation. Lines AD and OE intersect at the point P which is known as the break-even point, and the production represented by the point P' indicates the Break-even production capacity. Any volume of production beyond the break-even point will result in profits. The more we go away from P, the more profits we get. Conversely, any volume of production below the Break-even point will result in losses. The more we drift away from P towards O, the more will be the losses.

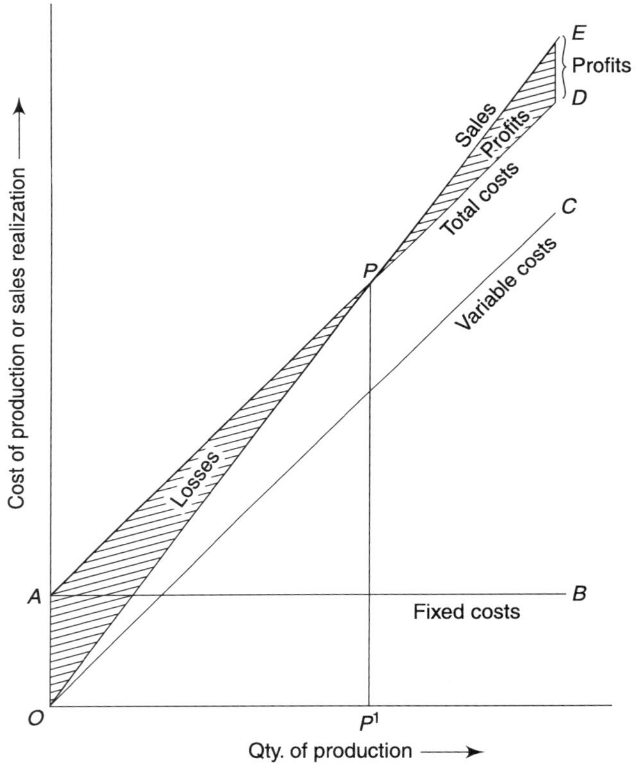

Fig. 20.5 *Break-even analysis*

Figure 20.5 is drawn based on the conditions prevailing with respect to production as well as sales. Any changes in these conditions would change the break-even point/break-even production. For profits to be more, the line DE should increase. It is possible either by decreasing the slope of AD or by increasing the slope of OE, i.e., either the production costs are to be reduced or the sale prices are to be increased or both. Increasing the sale price of the product is not always possible, unless it is a monopoly product. Decreasing the cost of production is possible by adopting some of the following steps:

- By using alternate raw materials which are of low cost
- By improving the efficiency of production

- By using superior technologies which reduce the
  - wastage
  - downstream process steps
  - utilities
  - processing time
- By increasing the efficiency of workers either by
  - motivating the workers
  - hiring skilled manpower

20.4 PROJECT ECONOMICS

Project economics is an important component of any commercial activity, more so if it is an industrial venture involving financial matters. It is to identify what should be the total investment for the industrial venture once a final decision has been taken to select the project line. Predesign fixed-capital investments involve the following steps (Ulrich, 1984):

(i) Determination of the purchase price of various equipments involved in the processing activity; it could be by way of collecting budgetary estimates or by direct discussions with the suppliers
(ii) Making proper provision for the escalation of prices by the time actual purchase orders are placed; the escalation could be 10–20% depending upon the time period
(iii) Making a provision for installation of machinery, electrical installations, piping and instrumentation (Table 20.5)
(iv) Total up the above to obtain the total cost; multiply this sum with a factor of 1.18 to account for contingencies

Table 20.5 *Pre design cost estimation**

Cost Item	*Average Multiplier*	*Range of values*
TOTAL PLANT DIRECT COST (TPDC)		
1. Equipment Purchase Cost (PC)		
2. Installation	0.50 × PC	0.2 – 1.5
3. Process Piping	0.40 × PC	0.3 – 0.6
4. Instrumentation	0.35 × PC	0.2 – 0.6
5. Insulation	0.03 × PC	0.01 – 0.05
6. Electrical	0.15 × PC	0.1 – 0.2
7. Buildings	0.45 × PC	0.1 – 2.0
8. Yard Improvement	0.15 × PC	0.05 – 0.2
9. Auxiliary Facilities	0.50 × PC	0.2 – 1.0
TOTAL PLANT INDIRECT COST (TPIC)		
10. Engineering	0.25 × TPDC	0.2 – 0.3
11. Construction	0.35 × TPDC	0.3 – 0.4
TOTAL PLANT COST (TPC)	TPDC + TPIC	
12. Contractor's fee	0.05 × TPC	0.03 – 0.08
13. Contingency	0.10 × TPC	0.07 – 0.15
DIRECT FIXED CAPITAL (DFC)	TPC + 12 + 13	

* Reprinted from Petrides (2003), Copyright © 2000, by Demitri Petrides

After having made the predesign estimates, the project estimates are prepared. For starting any industrial venture, the first and foremost important thing is to go for selection of a site, and development of the site. We may even need to go for creation of various infrastructure facilities♠ like

- roads and transportation links
- water lines or
- electrical lines.

They are all considered as a part of fixed capital for the project. The project cost is usually classified into the following♦ :

- Fixed capital
- Working capital

Now we discuss them briefly here.

20.4.1 Fixed Capital

The fixed capital is the investment required for creating capital equipment which are a part of the project. These facilities are fixed in nature and do not vary much with the capacity of production. They include mostly

- land and land development♣
- building
- machinery and equipment
- auxiliary equipment
- accessories for air conditioners, computers, office furniture
- preoperative expenses

The above fixed assets are a part of the fixed capital, which would be sanctioned by the financial institutions. They are available for resale even if the project is closed or withdrawn after some time. The fixed capital amount is fixed. It cannot be diverted for any other purpose. The amount is repaid to the financing institutions within a period of 5–6 years. The interest rates are different, and are usually low. The quicker repayment would reduce the financial burden for payment towards interest.

When we are considering the machinery and equipment, we also need to consider various associated costs in addition to their bare costs. The associated items to be considered along with the equipment were indicated by Guthrie (1969) and are shown in a tabular form (Table 20.6). Percentage contribution of some of them were shown already in Table 20.5.

♠ In developed areas these facilities would already exist; but it may be difficult to get land in such developed industrial areas, or even if the land is available it would be very costly.

♦ It is usually customary to make a provision for growth capital also in the project cost, which will take care of any future developments.

♣ In the industrial areas / parks, the charges for land development are the least, and need to be paid as a lump sum to the authorities.

Table 20.6 *Costs associated with the equipment*

PRIMARY COST ELEMENTS	
(i)	Equipment fob cost
(ii)	Direct material
(iii)	Direct field labor
(iv)	Direct M & L cost
(v)	Indirect costs
(vi)	Bare module cost
SECONDARY COST ELEMENTS	
(i)	Piping
(ii)	Concrete
(iii)	Steel
(iv)	Instruments
(v)	Electrical
(vi)	Insulation
(vii)	Paint
INSTALLATION COSTS	
(i)	Material erection
(ii)	Equipment setting
OTHER INDIRECT COSTS	
(i)	Freight insurance
(ii)	Construction overhead
(iii)	Engineering
(iv)	Contingency
(v)	Contractor fee

(*Capital Cost Estimating*, Guthrie, KM, (1969). Reprinted by special permission from CHEMICAL ENGINEERING, p.116, Vol 76 (March 24, 1969) Copyright © (1969), by Access Intelligence, New York, NY 10038)

In addition to the above, we also require at least 25% of the fixed capital investment to maintain the process plant (Ulrich, 1984). Various items in this are as follows:

S. No	*Cost item*	*% of fixed capital*
(i)	Maintenance and repairs	6
(ii)	Operating supplies	1
(iii)	Overheads	4
(iv)	Taxes and insurance	2
(v)	Depreciation	10
(vi)	General	1
	Total	**24**

Adopted from Ulrich, 1984.

20.4.2 Working Capital

As the name indicates, it is the capital required for day to day working of a factory. In other words, it is the fund required to meet the costs for operating the enterprise. It is also known as **circulating capital**. Various items covered under this are shown in Table 20.7. The requirement varies from day to day. Hence the interest rates are high for this capital, but the interest is calculated on a day-to-day basis. The capital is never returned to the financial institutions. The capital will circulate as long as the factory runs. The expenses are usually high during the first week of the month, and hence capital requirements are higher. Thus, the working capital is not a fixed amount; it varies based on the needs. So the financial institutions sanction the working capital not as a fixed amount but as a limit. Usually, it is of the order of 10–20 % of the total project cost; but in some cases it could be as high as 50% or more *for the companies producing products of seasonal demand because of large inventories* (Peters and Timmerhaus, 1980).

Table 20.7 *Items covered under working capital*

(i)	Basic raw materials
(ii)	Auxiliary raw materials like solvents, chemicals, enzymes etc
(iii)	Packaging materials
(iv)	Handling and transportation charges
(v)	Wages for the workers
(vi)	Salaries for the regular staff
(vii)	Marketing and sales expenses
(viii)	Taxes, insurance and security arrangements
(ix)	Credit extended to customers
(x)	Expenses to collect the receivables
(xi)	Cash to meet the sundry expenses
(xii)	Payment of telephone bills, power bills, water bills
(xiii)	Cash to purchase the utilities
(xiv)	Readily available cash to meet any emergencies and exigencies

Most of the items covered under the variable costs in Fig. 20.3 will constitute a part of the working capital. Management of working capital is a tricky art. Success in its management leads ultimately to the success of the industry. The following components constitute a working cycle:

- Raw materials
- Materials in process
- Finished goods
- Sales
- Debtors
- Cash

The working capital cycle is shown in Fig. 20.6. By reducing the working capital cycle, the same amount can be recirculated a greater number of times in a year. According to financial experts, it is good to *expedite the receipts and delay the payments as much as possible without spoiling the reputation of the company. This will help reduce the working capital cycle.*

20.5 DEPRECIATION

Loss in value of the equipment, machinery and building over a period of time is considered as depreciation. Land is not considered to depreciate because land appreciates over a period of time, and is perenially available. Thus, depreciation means only a loss in value; and hence a provision needs to be made in the project finance plan to replace the equipment once it totally wears off. Of course, a well maintained equipment may not need any replacement, but still a provision is made in the process economics. It is a means of building up of the funds needed to refinance the replacement of the worn-out machinery.

The definition of depreciation from a chemical engineer's point of view was given by Holland et al(1973) which is worth noting.

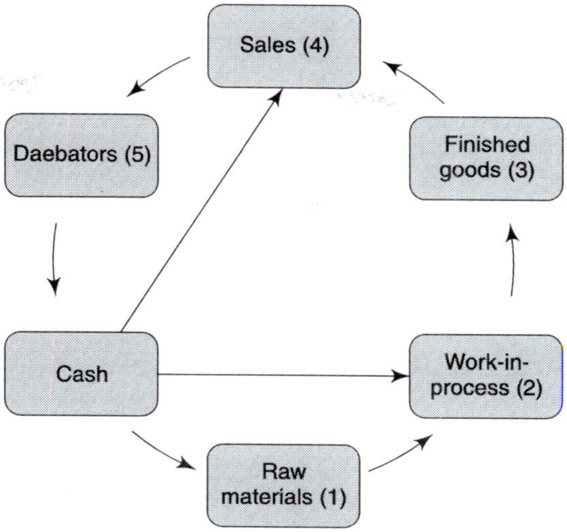

Fig. 20.6 *Working Capital cycle*

- Depreciation is a means of providing a tax allowance.
- Depreciation is a means of providing a cost for operation.
- Depreciation is a means of providing finance for replacement of plant.
- Depreciation is a means of providing an allowance for a falling value of the asset with time.

In reality, the amount is not taken out from the gross income. It is usually shown on paper as the amount deducted from the gross earnings. Later on, the taxes are paid to the local government as per statutory requirements. Then the net earnings are calculated. The depreciation amount also goes to the company along with the net earnings, but it is shown in books as transaction. Considering the fact that the machinery or equipment will be worn off within a period of 6–10 years, the depreciation is fixed at 10–15 % of the capital equipment cost. Since the annual depreciation costs are deducted from the gross earnings and shown as 'book entries', the percentage values are to be agreed upon by the local governmental agencies.

There are different methods of calculating depreciation charges. The prominent among them are

- Flat rate
- Diminishing rates

Calculation of the flat-rate method is simple and easy, and hence is mostly followed. According to this method, if the depreciation rate is 10%, the cost of the asset will decrease by 10% every year flatly, and hence the value will become zero at the end of 10 years. If the cost of the asset is Rs 10 lakhs, it will depreciate at the rate of Rs 1.0 lakh annually, and will become zero at the end of 10 years.

In the diminishing rate method, the value will depreciate at the fixed rate on the value of the asset present at that particular time. If 10% is rate of depreciation, a Rs 10.0 lakh worth of equipment will depreciate by one lakh at the end of the first year. The cost of the asset becomes Rs 9.0 lakhs after depreciation, and hence it will depreciate by 0.9 lakhs at the end of the second year. The cost of the equipment would be Rs 8.1 lakhs at the beginning of the third year which will depreciate by 10% of 8.1 lakhs. Thus, the value of the asset will never be zero. It will be Rs 3,48,680 as a contrast to zero at the end of ten years as per the previous method. The depreciation value and cost of the equipment by the two different methods for the same percentage of depreciation (10%) are shown in Table 20.8.

Table 20.8 *The depreciation and cost of the asset evaluated by two methods[#].*

Period at the end of (years)	*Flat rate of depreciation @ 10%*		*Diminishing rate of depreciation @ 10%*	
	Cost of the asset (Rs)	*Depreciation amount (Rs)*	*Cost of the asset (Rs)*	*Depreciation amount (Rs)*
0	10,00,000	0	1,00,0000	0
1	9,00,000	1,00,000	9,00,000	1,00,000
2	8,00,000	-do-	8,10,000	90,000
3	7,00,000	-do-	7,29,000	81,000
4	6,00,000	-do-	6,56,100	72,900
5	5,00,000	-do-	5,90,490	65,610
6	4,00,000	-do-	5,31,440	59,050
7	3,00,000	-do-	4,78,300	53,140
8	2,00,000	-do-	4,30,470	47,830
9	1,00,000	-do-	3,87,420	43,050
10	0	-do-	3,48,680	38,740

20.6 PROJECT ECONOMICS FOR CITRIC ACID MANUFACTURE[♣]

Process description Citric acid is a 6-carbon tricarboxylic acid which is extensively used in food processing, beverage making, pharmaceutical industries, manufacture of cosmetics and chemical industries. It was first prepared in the year 1794 in crystal form from lemon juice. Later, the developments in fermentation processes have replaced the lemon-juice process. The world production of citric acid is approximately 17 lakh tonns out of which approximately 99% is produced by submerged fermentation of molasses using the microorganism Aspargillus niger, and about 1% is produced from citrous juice. The flow sheet for the production process is shown in Fig. 20.7 which takes approximately 7 days for completion. In fact, the manufacturing process is taken up in seven batches in a staggered manner, so that there is production of citric acid everyday. Efforts are in progress for making a continuous production system. They are yet to reach commercial stature.

The inoculum is first cultured in a small-seed fermenter using an inoculum level of 2%. Later it is transferred to the second-seed fermenter which is about 10 times the volume of the first fermenter. The contents are then shifted into the main fermenter along with the substrate and nutrient media. The nutrients play a vital role in citric-acid production. They are dissolved in water and then added. Air is supplied to the fermenter at the rate of 0.15 to 1.0 VVM (volume of air per unit volume of liquid per minute). Cooling water is recirculated outside the fermenter to remove the exothermic heat (2990 kcal of heat/kg of citric acid) and maintain the temperature at 28°C. Later the contents are discharged into a holding tank from where the contents go for downstream processing to recover citric acid in the form ofcrystals.

The biomass is separated in a rotary vacuum filter. The clarified fermentation liquor is mixed with lime at the rate of one part of $Ca(OH)_2$ for every two parts of liquor and stirred slowly to precipitate calcium citrate. The precipitated calcium citrate is separated in the rotary vacuum filter 2. The filtrate is then discarded. The calcium citrate is taken in the stirred vessel 6 in which it is acidified with dilute sulphuric acid under controlled conditions of temperature and pH. The calcium is precipitated out as calcium sulphate (gypsum) and the citric acid is left in the dissolved form in solution. The contents are filtered in rotary vacuum filter 3, where the filter cake in the form of gypsum is discarded. The clarified solution of citric acid

[#] The values in the last two columns are adjusted to nearest ten rupees

[♣] The data presented here is only for the purpose of explaining the procedure; it is not authenticated. The actual costs have to be ascertained based on the market prices.

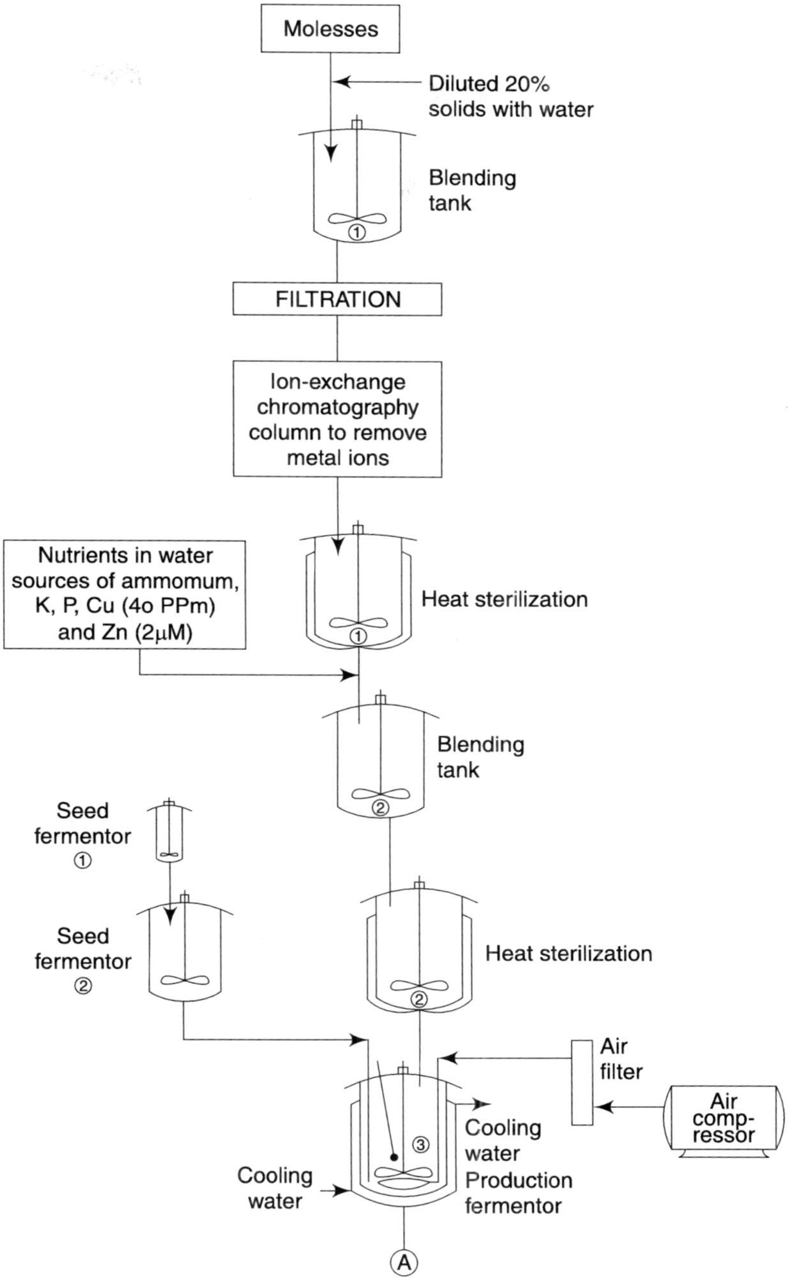
Molesses
Diluted 20%
solids with water
Blending
tank
①
FILTRATION
Ion-exchange
chromatography
column to remove
metal ions
Nutrients in water
sources of ammomum,
K, P, Cu (4o PPm)
and Zn (2μM)
Heat sterilization
①
Blending
tank
②
Seed
fermentor
①
Seed
fermentor
②
Heat sterilization
②
Air
filter
Air
comp-
ressor
Cooling
water
③
Cooling
water
Production
fermentor
A

(Contd.)

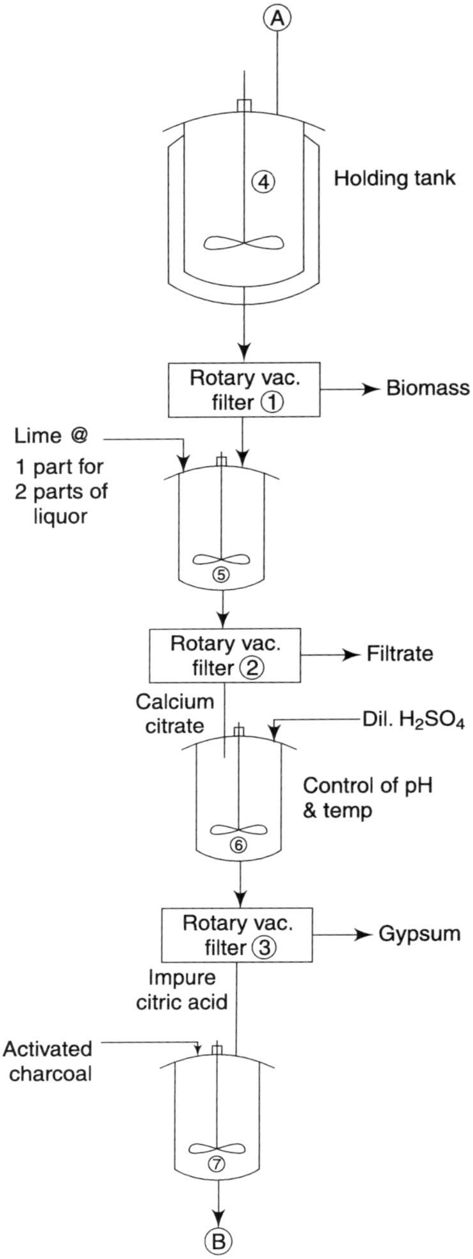
A
Holding tank
4
Rotary vac. filter 1
Biomass
Lime @
1 part for
2 parts of
liquor
5
Rotary vac. filter 2
Filtrate
Calcium
citrate
Dil. H_2SO_4
Control of pH
& temp
6
Rotary vac. filter 3
Gypsum
Impure
citric acid
Activated
charcoal
7
B

(Contd.)

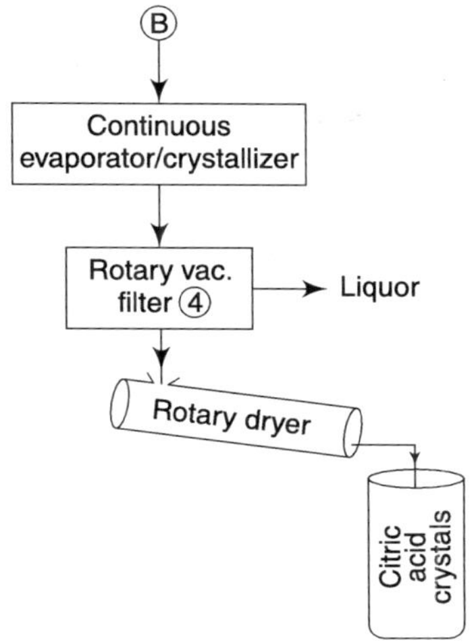

Fig. 20.7 *Flow-sheet for production of citric acid*

is bleached on charcoal to remove any undesirable colour. Later the solution is concentrated in a continuous evaporator to crystallise citric acid in the form of fine white crystals. The citric acid crystals are separated in the rotary vacuum filter 4, and the liquor is discarded. The crystals are dried in a rotary dryer to get dry crystals.

A detailed citric acid production flow sheet is available in Petrides (2003) and is shown as Fig. 20.8.

20.6.1 Project Economics

From process flow sheet given in Fig. 20.7 we can list out various equipment used, and this is tabulated in Table 20.9 with their costs. The cost of equipment and machinery is Rs 65.5 lakhs. The requirements of various materials for production of one tonne of citric acid are shown in Table 20.10.

Table 20.9 *Capital Equipment for citric acid production*#

	No.	*Cap*	*Cost (lakhs)*	*Total*
Seed fermenter (1)	1	10 lit	0.5	0.5
Seed fermenter (2)	1	100 lit	1.0	1.0
Blending tanks	7	1000 lit	2.5	17.5
Filtration units	4		5.0	20.0
Heat sterilizers	2		4.5	9.0
Air compressor	1		1.0	1.0
Air filter	1		3.0	3.0
Vac. Evaporators	2		5.0	10.0
Rotary dryer	1		3.5	3.5
				65.5

Generator set of 200 KVA capacity costing approximately Rs 15.0 lakhs may be included as an additional facility for stand – by power supply.

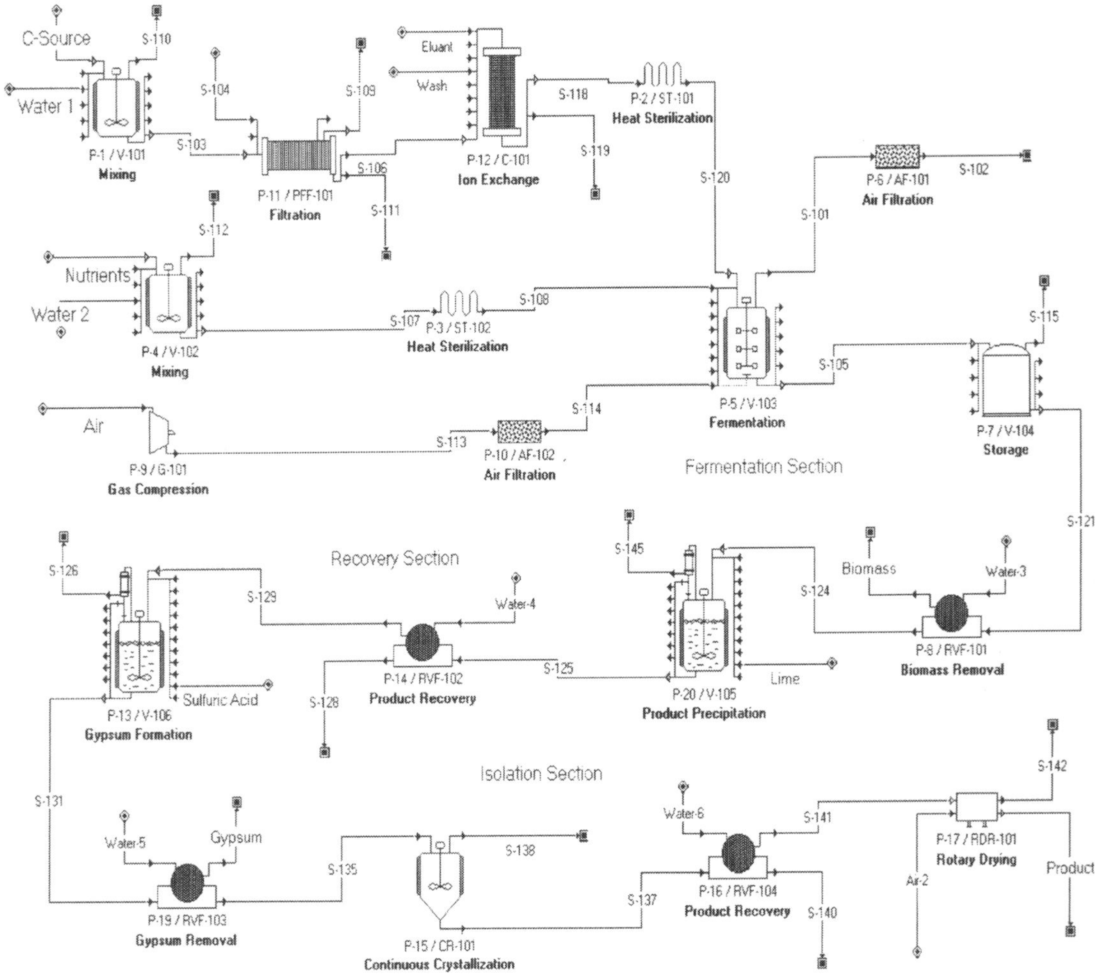

Fig. 20.8 *Flowsheet for Citric acid (Copyright © 2000, by Demitri Petrides)*

Table 20.10 *Requirement of various materials with their costs for production of one tonne of citric acid*

Material	*Quantity*	*Cost (Rs)*	*Total cost (Rs)*
Molasses	3000 kg	2500/*t*	7500
Sulphuric acid	800 kg	18/kg	14,400
Activated carbon	25 kg	20/kg	500
Nutrients	5 – 15 kg	20/kg	200
Lime	450 kg	3/kg	1350
Water	15 tonnes	40/*t*	600
Steam	15 tonnes	500/tonne	7500
Electricity	3000 kWh	2/kWh	6000
			38,050

Estimation of production costs Production costs are evaluated by the following two methods:

(i) Preliminary estimates

(ii) Firm estimates

Preliminary estimates, also known as predesign estimates, are based on the available information or experience. Hence the data is not very reliable, but is sufficient to go ahead with the decision on whether to start a processing industry or not. The firm estimates need a lot of data and hence the process is time consuming. Here, we follow the former method. The data is ephemeral and should not be taken as authenticated. It is only for the purpose of explaining the procedures of cost estimations and evaluating the break-even point.

Cost estimation *for 2 tpd citric acid production*

Sale price of citric acid ex-factory is taken as Rs 70 / kg.

Fixed capital (Rs in lakhs)

(i) Land (1000 m^2)	:	5.0	
(ii) Building (600 m^2)	:	16.0	
(iii) Equipment (Table 20.9)	:	65.5	
(iv) Miscellaneous Exp (20% of equipment cost)	:	13.1	
(v) Preoperative Exp (LS)	:	10.4	
Total		110.0	(20.10)

Operating costs per day (Rs)

(working capital requirement)

(i) Raw materials and utilities (Table 20.10)	:	76,100	(A)	
(ii) Packaging materials (LS)	:	5000	(B)	
(iii) Salaries (Table 20.11)	:	2240	(C)	
(iv) Wages (Table 20.11)	:	3680	(D)	
(v) Marketing and selling expenses (10% total product sale price)	:	14,000	(E)	
(vi) Interest on fixed assets (@ 12% pa) $(1{,}11{,}00{,}000 \times 0.12) / 300$	:	4,400	(F)	
(vii) Depreciation on building and Equipment @ 10% pa $(94{,}60{,}000 \times 0.1) / 300$	:	3,155	(G)	
(viii) Tax on gross earnings (sales revenue – depreciation) (@ 12% on gross earning) $(1{,}40{,}000 - 3155) \times 0.12$	:	16,420	(H)	
(ix) Factory overheads (LS)	:	1000	(I)	
(x) Interest on working capital a@ 16% (with four recycles per annum) $(A + B + C + D + E + F + H + I) \times 0.16/4$	:	4915	(J)	
(xi) Payback amount of fixed assets (in 10 years)	:	3,670	(K)	
Total	:	134580		(20.11)

Table 20.11 *Man power requirements*

	Particulars	*Nos*	*Monthly salary (Rs)*	*Total/month (Rs)*	*Category (salary/wage)*
(i)	Technologist	1	30,000	30,000	Salary
(ii)	Chemists	2	10,000	20,000	Salary
(iii)	Administrative staff	1	6000	6,000	Salary
(iv)	Mechanical operators	3	6000	18,000	Wage
(v)	Skilled workers	6	5000	30,000	Wage
(vi)	Semi-skilled workers	6	4000	24,000	Wage
(vii)	Helpers	10	2000	20,000	Wage
	Total exp.			1,48,000	
	Exp. per day @ 25 days a month			5920	

20.6.2 Profits

From the above for a 2 tpd plant of citric acid, we have

cost of production = Rs 1,34,580

The sale price of citric acid ex-factory is Rs 70 / kg.

$$\text{Sales realisation} = 70 \times 2 \times 1000 = 1{,}40{,}000 \tag{20.12}$$

Profits/day after paying taxes = sales realisation – cost of production

= 140,000 – 1,34,580 = Rs 5420

$$\text{Hence annual profits} = 5420 \times 300 = \text{Rs } 16{,}26{,}000 \tag{20.13}$$

20.6.3 Analysis of Costs

We try to break the cost of production into
- fixed costs
- variable costs

Fixed costs Fixed costs are shown in Fig. 20.3.

$$\text{Fixed costs} = C + F + G + I + K$$
$$= \text{Rs } 14{,}465$$

Variable costs They include all other items in the cost of production ($A + B + D + E + H + J$)
Variable costs for production of 2 tpd of citric acid = 1,34,580 – 14465 = Rs 1,20,115

∴ the variable costs for production of every tonne of citric acid per day

$$= 1{,}20{,}115/2 = \text{Rs } 60{,}058 \tag{20.14}$$

From the above data, we can make a breakeven analysis to find out beyond what production capacity we need to operate the plant.

20.6.4 Break-even Analysis

The principles of break-even analysis were discussed earlier. We apply them here for finding out the quantity of production / day at which the project breaks even.

The fixed costs per day= Rs 14,465

The variable costs for every ton of production per day is Rs 60,058 /-

We can draw a plot of the total cost of production and total sales realization per day in Fig. 20.9, from which we note that the break-even production is approximately 1.5 tpd.

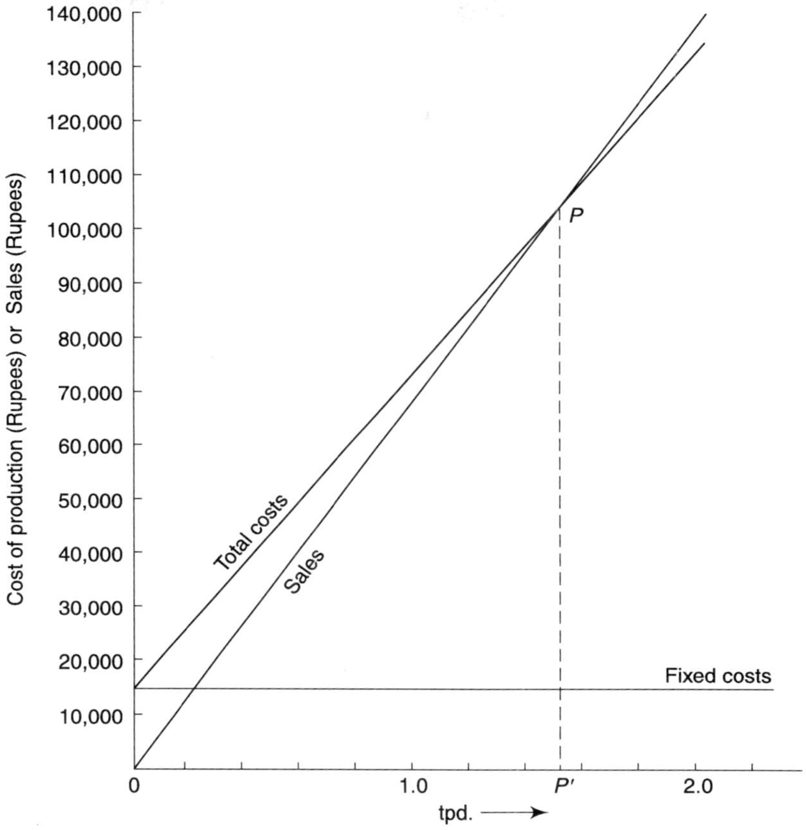

Fig. 20.9 *Break-even analysis for citric acid production*

We can also calculate the same from the data as follows. If q tonnes per day is the break-even production capacity,

variable costs of production = 60,058 (q)

fixed costs of production = 14,465

∴ total cost of production for q tpd plant = 14,465 + 60,058 q

Sales realization for q tpd plant = 70,000 q

At break-even point, 70,000 q = 14,465 + 60,058 q (20.15)

(70,000 – 60,058) q = 14,465

$$\therefore \quad q = \frac{14{,}465}{9942} = 1.46 \text{ tpd} \qquad (20.16)$$

Break-even analysis is sometimes considered in terms of percentage production. In the initial stages of functioning of the factory, it may not be possible to realise a 100% production capacity. For the first six months or one year or so, only 60% production is realised. In the subsequent period a 70% production is realised, and so on. Under such circumstances, it would be of interest to note at what production capacity the plant breaks even. It can be done as follows:

The Break-even plant capacity = 1.46 tpd
(from Eq. (20.16))

Production capacity = 2 tpd

∴ break-even occurs at (1.46/2) × 100 = 73%

20.6.5 Analysis of the Project

The project is analysed for its economic viability.

The working capital requirement/day is given by $(A + B + C + D + E + F + H + I)$
= Rs 1,22,840

We presume that the working capital will have a cycle of 75 days; i.e., it will have four recyclings per year.

∴	working capital = 1,22,840 × 75 =	92,13,000
	Fixed capital requirement (Eq. 20.10) =	1,10,00,000
∴	total project cost	Rs 2,02,13,000

Usually, the equity share is 1/3 of the project cost.

$$\therefore \quad \text{investor's share} = \frac{2{,}02{,}13{,}000}{3} = 67{,}40{,}000 \qquad (20.17)$$

Lending from financial institutions = 2,02,13,000 – 67,40,000 = 1,34,73,000

Annual profits are Rs 16,26,000 (after paying taxes). It is considered to be a rewarding♠ return on the investment of Rs 67,40,000 which is of the order of 24%. Hence, the project seems to be viable.

Some of the **attractive features** of the project are that the annual profits are calculated after taking into consideration a payback period of 10 years; i.e., one tenth of the fixed capital is paid every year. Thus the interest component on the fixed assets (given by item F in Sec. 20.6.1) will be reducing gradually and will become zero at the end of ten years. Later the profits would substantially increase.

The citric-acid requirement is likely to grow in the years to come in view of its extensive usage in food and beverage industries (70%). The estimated production in the year 2007 was approximately 17 lakh tonnes. The present plant is conceived for only 600 tpa. Hence there is a good scope to increase the production capacity which is likely to result in higher profits.

Thus the project seems to be viable and attractive. The bioprocess economics for any product can be done in a similar way to arrive at the profits and also to make break-even analysis.

Symbols

d	: depreciation rate
i	: returns on investment
I	: investment (Rs).
n	: number of years.
P	: net profit (Rs)

♠ The returns on the investment should be always more than the interest paid by the banks for fixed deposits; and is usually of the order of 25 – 40% for processing industries.

P_n	: present value of investment n[th] in the year (Rs)
q	: break-even production (kg or tons.)
n	: gross profit (Rs)
S_n	: expected cash flow rate
t	: tax rate (%)
T	: payout time (years).

References

20.1 Bailey, JE and Ollis, DF (1986), *Biochemical Engineering Fundamentals*, 2nd Ed., Mc Graw-Hill Inc., New York, Ch. 12, pp 798–853.

20.2 Cho (1957), *Chart for Net Return of Investment After Taxes*, Chem. Eng., Vol. 64 (1), p.258

20.3 Guthrie, K M (1969), *Capital Cost Estimating* Chemical Eng. (March 24, 1969), 114–142

20.4 Happel, J and Jordan, DG (1975), *Chemical Process Economics*, Marcel Dekkar, New York

20.5 Holland, FA, Watson, FA and Wilkinson, JK (1973), *Capital Costs and Depreciation*, Chem. Eng. (July 23), 118 – 121.

20.6 Peters, MS and Timmerhaus, KD (1980), *Plant Design and Economics for Chemical Engineers*, 3rd Ed., Mc Graw-Hill Int Book Co., Ch-5, pp. 147-208

20.7 Petrides, D(2003), *Bioprocess Design and Economics* in Bioseperations Science and Engineering, Ed. by RG Harrison, PW Todd, SR Rudge and D Petrides, Ch. 11, Oxford University Press, 2003

20.8 Rudd, DF (1975), *Modeling the Development of the Intermediate Chemical Industry*, The Chem. Eng. J, 9, 1–20

20.9 Ulrich, GD (1984), *A Guide to Chemical Engineering Process Design and Economics*, John Wiley & Sons, New York. Pp. 259–401

20.10 Valentas, KJ, Levine, L and Clark, JP (1991), *Food Processing Operations and Scale up*, Marcel Dekker Inc., New York, pp 30–49.

Review Questions

20.1 Describe the importance of bioprocess economics.
20.2 What should be the strategy for implementation of a new project?
20.3 What are the various stages in plant design?
20.4 How do you analyse a biotechnological project in terms of economic returns?
20.5 Describe a management decision tree while starting any new project venture.
20.6 What queries do you raise before deciding whether a project is economically viable or not?
20.7 Define and explain the following terms:
   (i) Net profit
   (ii) Pay-out period
   (iii) Turnover of a company
   (iv) Returns on the investment (ROI)
   (v) Pay-out time
   (vi) Net present value
20.8 How does financial analysis help to narrow down a project selection?
20.9 What is meant by fixed costs? What are the various components in it?
20.10 What is meant by variable costs? What are the various components in it?

20.11 What are the components of a product cost?
20.12 How do you make the predesign cost estimation?
20.13 What are the various costs associated with the equipment?
20.14 What is meant by fixed capital? What are the various components in it?
20.15 What is meant by working capital? What are the various components in it?
20.16 Draw a working capital cycle.
20.17 What is break-even analysis? What is its importance in project economics?
20.18 Describe diagrammatically how BEP is arrived at.
20.19 What is meant by depreciation? What is its role in project economics?
20.20 What are the different methods of calculation of depreciation? Describe them briefly.
20.21 What is meant by flat rate of depreciation?
20.22 What is meant by diminishing rate of depreciation?

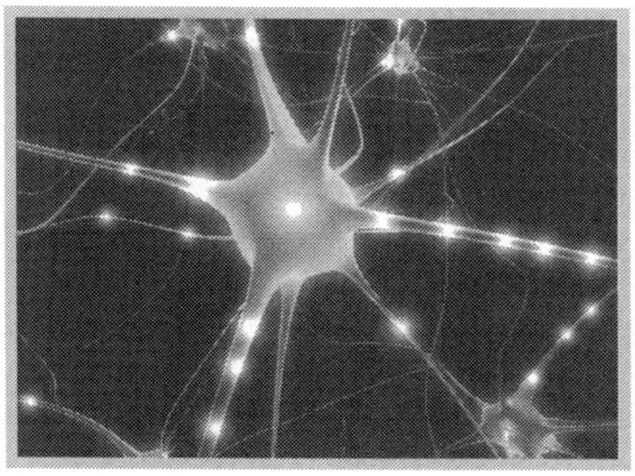

Index

Made in United States
Orlando, FL
04 November 2026

90201135R00275